Information Technology Project Management
Fourth Edition

Kathy Schwalbe, Ph.D., PMP

Augsburg College

THOMSON

COURSE TECHNOLOGY

Australia • Canada • Mexico • Singapore • Spain • United Kingdom • United States

THOMSON
COURSE TECHNOLOGY

Information Technology Project Management, Fourth Edition,
is published by Thomson Course Technology.

Executive Editor
Mac Mendelsohn

Associate Product Manager
Mirella Misiaszek

Copy Editor
Harold Johnson

Senior Acquisitions Editor
Maureen Martin

Editorial Assistant
Jennifer Smith

Proofreader
John Bosco

Senior Product Manager
Tricia Boyle

Senior Marketing Manager
Karen Seitz

Indexer
Michael Brackney

Developmental Editor
Karen Lyons

Text Designer
GEX Publishing Services

Cover Designer
Laura Rickenbach

Production Editor
Danielle Slade

For Dan, Anne, Bobby, and Scott

Preface

The future of many organizations depends on their ability to harness the power of information technology, and good project managers continue to be in high demand. Colleges have responded to this need by establishing courses in project management and making them part of the information technology, management, or engineering curriculum. Corporations are investing in continuing education to help develop effective project managers and project teams. This text provides a much-needed framework for teaching courses in project management, especially those that emphasize managing information technology projects. The first three editions of this text were extremely well received by people in academia and the workplace. The fourth edition builds on the strengths of the previous editions and adds new, important information and features.

It's impossible to read a newspaper, magazine, or Web page without hearing about the impact of information technology on our society. Information is traveling faster and being shared by more individuals than ever before. You can research and purchase many products online, receive e-mail with a mobile phone, or use a wireless Internet connection at your local coffee shop. Companies have linked their many systems together to help them fill orders on time and better serve their customers. Software companies are continually developing new products to help streamline our work and get better results. When technology works well, it is almost invisible. But perhaps it has occurred to you to ask, "Who makes these complex technologies and systems happen?"

Because you're reading this text, you must have an interest in the "behind-the-scenes" aspects of technology. If I've done my job correctly, as you read you'll begin to see the many innovations society is currently experiencing as the result of thousands of successful information technology projects. In this text, you'll read about projects around the world that were a success, like Motorola's Six Sigma project in the U.S., Canada's Dynamic Mutual Funds customer relationship management project, England's Boots Company's systems infrastructure project, and Kuala Lumpur's state-of-the-art Integrated Transport Information System (ITIS) project. Of course, not all projects are successful; factors such as time, money, and unrealistic expectations, among many others, can sabotage a promising effort if it is not properly managed. In this text, you'll also learn from the mistakes people made on many projects which were not successful. I have written this book in an effort to educate you, tomorrow's project managers, about what will help make a project succeed—and what can cause it to fail. Many readers tell me how much they enjoy reading these real-world examples in the "What Went Right?" and "What Went Wrong?" features.

The "Media Snapshot" is a new feature to this edition that will help you recognize projects in all aspects of your life, particularity in the media. People often don't realize that the world is full of projects; a careful observer will see examples of projects whenever they read the newspaper, watch a television show, or see a movie. Relating project management concepts to all types of projects will help you understand and see the importance of this growing field.

Although project management has been an established field for many years, managing information technology projects requires ideas and information that go beyond standard project management. For example, many information technology projects fail because of a lack of user input, incomplete and changing requirements, and an absence of executive support. This book includes suggestions on dealing with these issues. New technologies can also aid in managing information technology projects, and examples of using software to assist in project management are included throughout the book.

Information Technology Project Management, Fourth Edition, is still the only textbook to apply all nine project management knowledge areas—project integration, scope, time, cost, quality, human resource, communications, risk, and procurement management—and all five process groups—initiating, planning, executing, controlling, and closing—to information technology projects. This text builds on the *PMBOK® Guide 2004*, an American National Standard, to provide a solid framework and context for managing information technology projects. It also includes an appendix, *Guide to Using Microsoft Project 2003*, and a 120-day trial version of Microsoft Project 2003 software. A second appendix provides advice on earning and maintaining Project Management Professional (PMP) certification from the Project Management Institute (PMI) as well as information on other certification programs, such as CompTIA's Project+ certification. The fourth edition features three additional appendices to help you practice and apply the skills learned in this text by providing running cases, templates, and a popular project management simulation tool provided by Fissure.

Information Technology Project Management, Fourth Edition, provides practical lessons in project management for students and practitioners alike. By weaving together theory and practice, this text presents an understandable, integrated view of the many concepts, skills, tools, and techniques involved in information technology project management. The comprehensive design of the text provides a strong foundation for both students and practitioners of project management.

NEW TO THE FOURTH EDITION

Building on the success of the previous editions, *Information Technology Project Management, Fourth Edition,* introduces a uniquely effective combination of features. The main changes made to the fourth edition include the following:

- The Project Management Institute (PMI) released a new version of the *PMBOK® Guide* in 2004, and this text incorporates many of those changes. For example, there are now 44 processes in the nine project management knowledge groups instead of 39. This text describes each of these processes and puts them into the context of managing information technology projects. It includes hundreds of examples of real projects, real project documents, and suggested references to help you understand how to practice good project management.

- Appendix E is new and provides information on using a project management simulation tool developed by Fissure, a PMI Registered Education provider specializing in project management, software management, leadership, and organizational change training. Fissure has used an expanded version of this simulation tool to help thousands of people learn how to apply various project management concepts.

- Fissure software, to accompany Appendix E as described above, is included with the text.

- Appendix A has been updated for Microsoft Project 2003 Standard—the most widely used project management software tool today. This comprehensive appendix teaches students the fundamentals of Project 2003 in the context of the scope, time, cost, human resource, and communications management. Updated exercises at the end of the appendix will help test students' learning.

- The companion Web site is password protected. If you purchased a new copy of this book, you received a pass code. It is inserted into the CD-ROM envelope at the front of this book, with the Fissure simulation software.

- A new companion Web site for the fourth edition (*www.course.com/mis/schwalbe4e*) provides you with access to informative links from the Suggested Readings and footnotes, lecture notes, study aids, project management templates, the entire ResNet case (a real case study from the first and second editions of this text), additional running cases, and many other items to enhance your learning.

- The minicases at the end of most chapters have been replaced by a running case. The same project is used in the running case, with new scenario information and tasks you can perform based on information learned in each chapter. Appendix C and the companion Web site provide additional running cases in a similar format. Revised templates are also available in Appendix D and on the companion Web site.

- The "Media Snapshot" is a new feature to this edition that will help you recognize projects in all aspects of your life, particularity in the media. Relating project management concepts to all types of projects will help you understand and see the importance of this growing field.

- Updated examples are provided throughout the text. You'll notice several new examples in the fourth edition to reflect recent events in managing real information technology projects. Several of the What Went Right and What Went Wrong examples have been updated, and many new Suggested Readings have been added to keep you up-to-date. Results of new studies are also included throughout the text.

- User feedback is incorporated. Based on feedback from students, instructors, practitioners, and translators (this book has been translated into Chinese and Japanese), you'll see several additional changes, mostly additions, to help clarify information. For example, Chapter 7, Project Cost Management, includes a new cost estimate and a more practical example of determining earned value based on feedback from several practitioners and experts in the field.

APPROACH

Many people have been practicing some form of project management with little or no formal study in this area. New books and articles are being written each year as we discover more about the field of project management and as project management software continues to advance. Because the project management field and the technology industry change rapidly, you cannot assume that what worked even five years ago is still the best approach today. This text provides up-to-date information on how good project management and effective use of software can help you manage projects, especially information technology projects. Five distinct features of this text include its relationship to the Project Management Body of Knowledge, its bundling with Microsoft Project 2003, its value in preparing for Project Management Professional and other certification exams, its inclusion of running case studies, templates, and simulation software, and its companion Web site.

Based on the *PMBOK® Guide 2004*

The Project Management Institute (PMI) created the Guide to the Project Management Body of Knowledge (the *PMBOK® Guide 2004*) as a framework and starting point for understanding project management. It includes an introduction to project management, brief descriptions of all nine project management knowledge areas, and a glossary of terms. The *PMBOK® Guide 2004* is, however, just that—a guide. This text uses the *PMBOK® Guide 2004* as a foundation, but goes beyond it by providing more details, highlighting additional topics, and providing a real-world context for project management. *Information Technology Project Management, Fourth Edition*, explains project management specifically as it applies to managing information technology projects in the 21st century. It includes several unique features to bring you the excitement of this dynamic field (for more information on features, see the section entitled "Pedagogical Features").

Contains a Microsoft Project 2003 Trial and a Guide on How to Use It

Software has advanced tremendously in recent years, and it is important for project managers and their teams to use software to help manage information technology projects. Each copy of *Information Technology Project Management, Fourth Edition*, includes a 120-day trial version of the leading project management software on the market—Microsoft Project 2003. Examples using Project 2003 and other software tools are integrated throughout the text, not as an afterthought. Appendix A, *Guide to Using Microsoft Project 2003*, teaches you in a systematic way to use this powerful software to help in project scope, time, cost, human resource, and communications management. You do not need to buy a separate book to learn how to use Project 2003 effectively, but suggested resources are provided if you want to learn even more about Project 2003.

Resource for PMP and Other Certification Exams

Professional certification is an important factor in recognizing and ensuring quality in a profession. PMI provides certification as a Project Management Professional (PMP), and this text is an excellent resource for studying for the certification exam. This text will also help you pass other certification exams, such as CompTIA's Project+ exam. Keep in mind that having experience working on projects does not mean you can easily pass the PMP or other certification exams.

I like to tell my students a story about taking a driver's license test after moving to Minnesota. I had been driving very safely and without accidents for over sixteen years, so I thought I could just walk in and take the test. I was impressed by the sophisticated computer system used to administer the test. The questions were displayed on a large touch-screen monitor, often along with an image or video to illustrate different traffic signs or driving situations. I became concerned when I found I had no idea how to answer several questions, and I was perplexed when the test seemed to stop and a message was displayed saying, "Please see the person at the service counter." This was a polite way of saying I had failed the test! After controlling my embarrassment, I picked up one of the Minnesota driving test brochures, studied it for an hour or two that night, and successfully passed the test the next day.

This story illustrates how important it is to study information from the organization that creates the test and not be over-confident that your experience is enough. Because this text is based on PMI's *PMBOK® Guide 2004*, it provides a valuable reference for studying for PMP certification. It is also an excellent reference for CompTIA's Project+ exam. I have earned both of those certifications and kept them in mind while writing this text.

Provides Running Cases, Templates, Examples, and Simulation Software

Based on feedback from readers, the fourth edition has modified the running cases to more closely map to the materials in each chapter. There are additional templates, examples of real project documents, and simulation software you can use to actively practice your skills in managing a project. All of these features help the subject matter come alive and have more meaning.

Includes a Companion Web Site

A companion Web site provides you with a one-stop location to access informative links and tools to enhance your learning. Similar to other companion Web sites provided by Thomson Course Technology, this site will be a valuable resource as you view lecture notes, chapter study aids, templates, student files, important articles, and references. The companion Web site is password protected. If you purchased a new copy of this book, you received a pass code. It is inserted into the CD-ROM envelope at the front of this book, with the Fissure simulation software.

ORGANIZATION AND CONTENT

Information Technology Project Management, Fourth Edition, is organized into three main sections to provide a framework for project management, a detailed description of each project management knowledge area, and five appendices to provide practical information for applying project management. The first three chapters form the first section that introduces the project management framework and sets the stage for the remaining chapters.

Chapters 4 through 12 form the second section of the text that describes each of the project management knowledge areas—project integration, scope, time, cost, quality, human resource, communications, risk, and procurement manage-ment—in the context of information technology projects. An entire chapter is dedicated to each knowledge area. Each knowledge area chapter includes sec-tions that map to their major processes as described in the *PMBOK® Guide 2004*. For example, the chapter on project quality management includes sections on quality management, quality assurance, and quality control. Additional sections highlight other important concepts related to each knowledge area, such as Six Sigma, testing, maturity models, and using software to assist in project quality management. Each chapter also includes detailed examples of key project man-agement tools and techniques as applied to information technology projects. For example, the chapter on project integration management includes samples of various project-selection documents, such as net present value analyses, ROI cal-culations, payback analyses, and weighted scoring models. The project scope management chapter includes a sample project charter, part of a preliminary and detailed project scope statement, and several work breakdown structures for information technology projects.

Appendices A through E form the third section of the text, which provides practical information to help you apply project management skills on real or practice projects. You will learn how to use Project 2003 by following the detailed, step-by-step guide in Appendix A, which includes more than 70 screen shots to help you check your work. Appendix B summarizes what you need to know to earn PMP or other certifications related to project manage-ment. Appendix C provides additional running cases to help you practice your new skills, Appendix D provides useful templates you can start using immedi-ately, and Appendix E provides instructions on using Fissure's project manage-ment simulation tool.

PEDAGOGICAL FEATURES

Several pedagogical features are included in this text to enhance presentation of the materials so that you can more easily understand the concepts and apply them. Throughout the text, emphasis is placed on applying concepts to up-to-date, real-world information technology project management.

Learning Objectives, Chapter Summaries, Discussion Questions, Exercises, Running Cases, and Study Aids

Learning Objectives, Chapter Summaries, Discussion Questions, Exercises, Running Cases, and Study Aids are designed to function as integrated study tools. Learning Objectives reflect what you should be able to accomplish after completing each chapter. Chapter Summaries highlight key concepts you should master. The Discussion Questions help guide critical thinking about those key concepts. Exercises provide opportunities to practice important techniques, as do the Running Cases. The companion Web site provides several study aids, such as chapter review sheets.

Opening Case and Case Wrap-Up

To set the stage, each chapter begins with an opening case related to the materials in that chapter. These scenarios (most based on the author's experiences) spark student interest and introduce important concepts in a real-world context. As project management concepts and techniques are discussed, they are applied to the opening case and other similar scenarios. Each chapter then closes with a Case Wrap-Up—with some cases ending successfully and some, realistically, failing—to further illustrate the real world of project management.

What Went Right? and What Went Wrong?

Failures, as much as successes, can be valuable learning experiences. Each chapter of the text includes one or more examples of real information technology projects that went right as well as examples of projects that went wrong. These examples further illustrate the importance of mastering key concepts in each chapter.

Media Snapshots

The world is full of projects. Several televisions shows, movies, newspapers, Web sites, and other media highlight project results, good and bad. Relating project management concepts to all types of projects highlighted in the media will help you understand and see the importance of this growing field. Why not get people excited about studying project management by showing them how to recognize project management concepts in popular television shows, movies, or other media?

Suggested Readings

Every chapter ends with a list of Suggested Readings with annotations that offer opportunities for further research. Several exercises require you to review these readings or research related topics. The Web is widely referenced throughout the text as a research tool, and many of the Suggested Readings can be found on the Web.

Key Terms

The fields of information technology and project management both include many unique terms that are vital to creating a workable language when the two fields are combined. Key terms are displayed in bold face and are defined the first time they appear. Definitions of key terms are provided in alphabetical order at the end of each chapter and in a glossary at the end of the text.

Application Software

Learning becomes much more dynamic with hands-on practice using the top project management software tool in the industry, Microsoft Project 2003, as well as other tools, such as spreadsheet software and Internet browsers. Each chapter offers you many opportunities to get hands-on and build new software skills. This text is written from the point of view that reading about something only gets you so far; in order to understand project management, you have to do it for yourself. In addition to the exercises and running cases found at the end of chapters, several challenging projects are provided at the end of Appendix A, *Guide to Using Microsoft Project 2003*.

SUPPLEMENTS

The following supplemental materials are available when this text is used in a classroom setting. All of the teaching tools available with this text are provided to the instructor on a single CD-ROM.

- **Electronic Instructor's Manual** The Instructor's Manual that accompanies this textbook includes additional instructional material to assist in class preparation, including suggestions for lecture topics and additional discussion questions.
- **ExamView®** This textbook is accompanied by ExamView, a powerful testing software package that allows instructors to create and administer printed, computer (LAN-based), and Internet exams. ExamView includes hundreds of questions that correspond to the topics covered in this text, enabling students to generate detailed study guides that include page references for further review. The computer-based and Internet testing components allow students to take exams at their computers, and also save the instructor time by grading each exam automatically.
- **PowerPoint Presentations** This text comes with Microsoft PowerPoint slides for each chapter. These are included as a teaching aid for classroom presentation, to make available to students for chapter review, or to be printed for classroom distribution. Instructors can add their own slides for additional topics they introduce to the class.

- **Solution Files** Solutions to end-of-chapter questions can be found on the Instructor Resource CD-ROM and may also be found on the Thomson Course Technology Web site at *www.course.com*. The solutions are password protected.

- **Distance Learning** Thomson Course Technology is proud to present online courses in WebCT and Blackboard, to provide the most complete and dynamic learning experience possible. When you add online content to one of your courses, you're adding a lot: self tests, links, glossaries, and, most of all, a gateway to the 21st century's most important information resource. We hope you will make the most of your course, both online and offline. For more information on how to bring distance learning to your course, contact your Thomson Course Technology sales representative.

- **Student Online Companion** Visit the companion Web site for this text at *www.course.com/mis/schwalbe4e* for links to some of the Suggested Readings and footnotes, lecture notes, project management templates, study aids, Project 2003 student files used in Appendix A, the entire ResNet case, additional running cases, and other helpful resources.

ACKNOWLEDGEMENTS

I never would have taken on this project—writing this book, the first, second, third, or fourth edition—without the help of many people. I would like to thank the staff at Thomson Course Technology for their dedication and hard work in helping me produce this book and in doing such an excellent job of marketing it. Maureen Martin, Tricia Boyle, Karen Lyons, Danielle Slade, Karen Seitz and many more people did a great job in planning and executing all of the work involved in producing this book. It's amazing how many little details are involved in writing a good book, and Thomson Course Technology did a great job in producing a high quality text.

I thank my many colleagues and experts in the field who contributed information to this book. David Jones, Rachel Hollstadt, Cliff Sprague, Michael Branch, Barb Most, Jodi Curtis, Rita Mulcahy, Karen Boucher, Bill Munroe, Tess Galati, Joan Knutson, Neal Whitten, Brenda Taylor, Quentin Fleming, Nick Erndt, Dragan Milosevic, Bob Borlink, Arvid Lee, Kathy Christenson, Peeter Kivestu, and many other people who provided excellent materials included in the fourth edition of this book. Special thanks go to Jesse Freese and his colleagues at Fissure for providing their simulation software with this text. I really enjoy the network of project managers, authors, and consultants in this field who are very passionate about improving the theory and practice of project management.

I also want to thank my students and colleagues at Augsburg College and the University of Minnesota for providing feedback on the earlier editions of this book. I received many valuable comments from them on ways to improve the text and structure of my courses. I am also grateful for the articles and examples students provided and for the questions they asked in classes. I learn something new about project management and teaching all the time by interacting with students, faculty, and staff.

Three faculty reviewers provided excellent feedback for me in writing this fourth edition. Aileen Cater-Steel, University of Southern Queensland, Australia; Raffael Guidone, New York City College of Technology; and John McDowell, TAFE (Technical and Further Education) Department of Employment and Training, Queensland, Australia provided outstanding suggestions for improving the text in this fourth edition. I also wish to thank the many reviewers of the earlier editions of this text. I also thank the many other instructors and readers who have contacted me with praise as well as suggestions for improving this text. I really appreciate the feedback and do my best to incorporate as much as I can.

Most of all, I am grateful to my family. Without their support, I never could have written this book. My wonderful husband, Dan, was very patient and supportive of me working like crazy to write the first edition of this book. The later editions weren't quite as hectic, but I still needed support from Dan. Dan has written textbooks himself, so he provided some great advice. He even referred to parts of my book to help him in his job as a lead architect for software development with Com Squared Systems, Inc. and helped me in updating some of the sections on new software development programs and techniques. Our three children, Anne, Bobby, and Scott, continue to be very supportive of their mom's work. Our two older children, Anne and Bobby, are both in college now, and they actually think it's cool that their mom wrote a textbook. Our children all understand the main reason why I write—I have a passion for educating future leaders of the world, including them.

As always, I am eager to receive your feedback on this book. Please send all feedback to Thomson Course Technology at mis@course.com. My editors will make sure it gets to me!

Kathy Schwalbe, Ph.D., PMP
Augsburg College

ABOUT THE AUTHOR

Kathy Schwalbe is an Associate Professor in the Department of Business Administration at Augsburg College in Minneapolis, where she teaches courses in project management, problem solving for business, systems analysis and design, information systems projects, and electronic commerce. Kathy is also an adjunct faculty member at the University of Minnesota, where she teaches a graduate-level course in project management in the engineering department. She also provides training and consulting services to several organizations and speaks at several conferences. Kathy worked for ten years in the industry before entering academia in 1991. She was an Air Force officer, systems analyst, project manager, senior engineer, and information technology consultant. Kathy is an active member of PMI, having served as the Student Chapter Liaison for the Minnesota chapter of PMI, VP of Education for the Minnesota chapter, Editor of the ISSIG Review, and member of PMI's test-writing team. Kathy earned her Ph.D. in Higher Education at the University of Minnesota, her MBA at Northeastern University's High Technology MBA program, and her B.S. in mathematics at the University of Notre Dame.

Brief Contents

Contents

1

Introduction to Project Management

Objectives

After reading this chapter, you will be able to:

1. Understand the growing need for better project management, especially for information technology projects
2. Explain what a project is, provide examples of information technology projects, list various attributes of projects, and describe the triple constraint of projects
3. Describe project management and discuss key elements of the project management framework, including project stakeholders, the project management knowledge areas, common tools and techniques, and project success factors
4. Understand the role of the project manager by describing what project managers do, what skills they need, and what the career field is like for information technology project managers
5. Describe the project management profession, including its history, the role of professional organizations like the Project Management Institute, the importance of certification and ethics, and the growth of project management software

OPENING CASE

*A*nne Roberts, the new Director of the Project Management Office for a large retail chain, stood in front of five hundred people in the large corporate auditorium to explain the company's new strategies. She was also broadcasting to thousands of other employees, suppliers, and stockholders throughout the world via the Internet. The company had come a long way in implementing new information systems to improve inventory control, sell products using the Web, and streamline the sales and distribution processes.

However, the stock price was down, the nation's economy was weak, and people were anxious to hear about the company's new strategies.

Anne began to address the audience, "Good morning. As many of you know, our CEO promoted me to a new position as Director of the Project Management Office. Most of what we do in this department involves projects, and my role in this new position is to turn the company around by helping us effectively select and manage those projects. Our challenge is to develop a culture in which we all work together to provide high-quality goods and services to our consumers while earning a profit in this difficult market. To meet this challenge, we must collaborate to focus on finding solutions to complex problems. We must decide what projects will most benefit the company, how we can continue to leverage the power of information technology to support our business, and how we can exploit our human capital to successfully plan and execute those projects. If we succeed, we'll become a world-class corporation."

"And if we fail?" someone asked from the audience.

"Let's just say that failure is not an option," Anne replied.

INTRODUCTION

Many people and organizations today have a new or renewed interest in project management. Until the 1980s, project management primarily focused on providing schedule and resource data to top management in the military and construction industries. Today's project management involves much more, and people in every industry and every country manage projects. New technologies have become a significant factor in many businesses. Computer hardware, software, networks, and the use of interdisciplinary and global work teams have radically changed the work environment. The statistics below demonstrate the significance of project management in today's society, especially for projects involving information technology (IT):

- A 2001 report showed that the U.S. spends $2.3 trillion on projects every year, an amount equal to one-quarter of the nation's gross domestic product. The world as a whole spends nearly $10 trillion of its $40.7 trillion gross product on projects of all kinds. More than sixteen million people regard project management as their profession.[1]

[1] Project Management Institute (PMI), The PMI Project Management Fact Book, Second Edition, 2001.

- Worldwide IT spending is expected to grow by between 4 percent and 6 percent in the next few years. Forrester Research estimated that technology spending in the United States would total $752 billion in 2004, an increase of 4.4 percent over 2003. Forrester also predicts that U.S. IT spending will grow by another 5.7 percent in 2005, to reach $795 billion.[2]

- In 2003, the average senior project manager in the U.S. earned almost $90,000 per year. The average salary of a program manager was $103,464, just slightly less than the average Chief Information Officer (CIO) salary of $103,925. The average salary for a Project Management Office (PMO) Director was $118,633.[3]

- In the U.S., the number-one reality television show in 2004, *The Apprentice*, portrayed the important role project managers play in business. Each week of the show, teams would select a project manager to lead them in accomplishing that week's project. The project manager would be held partly responsible for the team's success or failure. Whether you're trying to make money by selling lemonade, running a golf tournament, or developing a new information system, project managers play a vital role in business success.

Today's companies, governments, and non-profit organizations are recognizing that to be successful, they need to be conversant with and use modern project management techniques. Individuals are realizing that to remain competitive, they must develop skills to become good project team members and project managers. They also realize that many of the concepts of project management will help them in their everyday lives as they work with people and technology on a day-to-day basis.

⊗ What Went Wrong?

In 1995, the Standish Group published an often-quoted study entitled "CHAOS". This prestigious consulting firm surveyed 365 information technology executive managers in the United States who managed more than 8,380 information technology application projects. As the title of the study suggests, the projects were in a state of chaos. United States companies spent more than $250 billion each year in the early 1990s on approximately 175,000 information technology application development projects. Examples of these projects included creating a new database for a state department of motor vehicles, developing a new system for car rental and hotel reservations, and implementing a client-server architecture for the banking industry. The survey found that the average cost of an information technology application development project for a large company was more than $2.3 million; for a medium company, it was more than $1.3 million; and for a small company, it was more than $434,000. Their study reported that the overall success rate of information technology projects was *only* 16.2 percent. The surveyors defined success as meeting project goals on time and on budget.

[2] Butler, Steve, *IT Spending*, Analyst Views, February 2004.

[3] Project Management Institute (PMI), *Project Management Salary Survey, Third Edition*, 2003.

The study also found that more than 31 percent of information technology projects were canceled before completion, costing U.S. companies and government agencies more than $81 billion. The authors of this study were adamant about the need for better project management in the information technology industry. They explained, "Software development projects are in chaos, and we can no longer imitate the three monkeys—hear no failures, see no failures, speak no failures."[4]

Many organizations claim that using project management provides advantages, such as:

RANK ?

- Better control of financial, physical, and human resources
- Improved customer relations
- Shorter development times
- Lower costs
- Higher quality and increased reliability
- Higher profit margins
- Improved productivity
- Better internal coordination
- Higher worker morale

This chapter introduces projects and project management, discusses the role of the project manager, and provides important background information on this growing profession. Although project management applies to many different industries and types of projects, this textbook focuses on applying project management to information technology projects.

WHAT IS A PROJECT?

To discuss project management, it is important to understand the concept of a project. A **project** is "a temporary endeavor undertaken to create a unique product, service, or result."[5] Operations, on the other hand, is work done in organizations to sustain the business. Projects are different from operations in that they end when their objectives have been reached or the project has been terminated.

Examples of Information Technology Projects

Projects can be large or small and involve one person or thousands of people. They can be done in one day or take years to complete. Information technology

[4] The Standish Group, "The CHAOS Report" (*www.standishgroup.com*) (1995). Another reference is Johnson, Jim, "CHAOS: The Dollar Drain of IT Project Failures," *Application Development Trends* (January 1995).

[5] Project Management Institute, Inc., *A Guide to the Project Management Body of Knowledge (PMBOK® Guide)* (2004), p. 5. Note: Page numbers are based on the early release of the third edition released on August 31, 2004.

projects involve using hardware, software, and/or networks to create a product, service, or result. Examples of information technology projects include the following:

- A help desk or technical worker replaces laptops for a small department
- A small software development team adds a new feature to an internal software application
- A college campus upgrades its technology infrastructure to provide wireless Internet access
- A cross-functional task force in a company decides what software to purchase and how it will be implemented
- A company develops a new system to increase sales force productivity
- A television network develops a system to allow viewers to vote for contestants and provide other feedback on programs
- The automobile industry develops a Web site to streamline procurement
- A government group develops a system to track child immunizations
- A large group of volunteers from organizations throughout the world develops standards for a new communications technology

Project Attributes

As you can see, projects come in all shapes and sizes. The following attributes help to define a project further:

- *A project has a unique purpose.* Every project should have a well-defined objective. For example, Anne Roberts, the Director of the Project Management Office in the opening case, might sponsor an information technology collaboration project to develop a list and initial analysis of potential information technology projects that might improve operations for the company. The unique purpose of this project would be to create a collaborative report with ideas from people throughout the company. The results would provide the basis for further discussions and projects. As in this example, projects result in a unique product, service, or result.
- *A project is temporary.* A project has a definite beginning and a definite end. In the information technology collaboration project, Anne might form a team of people to work immediately on the project, and then expect a report and an executive presentation of the results in one month.
- *A project is developed using progressive elaboration.* Projects are often defined broadly when they begin, and as time passes, the specific details of the project become more clear. Therefore, projects should be developed in increments. A project team should develop initial plans and then update them with more detail based on new information. For example, suppose a few people submitted ideas for the information technology collaboration project, but they did not clearly address how the ideas would support the

business strategy of improving operations. The project team might decide to prepare a questionnaire for people to fill in as they submit their ideas to improve the quality of the inputs.

■ *A project requires resources, often from various areas.* Resources include people, hardware, software, or other assets. Many projects cross departmental or other boundaries to achieve their unique purposes. For the information technology collaboration project, people from information technology, marketing, sales, distribution, and other areas of the company would need to work together to develop ideas. The company might also hire outside consultants to provide input. Once the project team has selected key projects for implementation, they will probably require additional hardware, software, and network resources. People from other companies—product suppliers and consulting companies—will become resources for meeting new project objectives. Resources, however, are limited. They must be used effectively to meet project and other corporate goals.

■ *A project should have a primary customer or sponsor.* Most projects have many interested parties or stakeholders, but someone must take the primary role of sponsorship. The **project sponsor** usually provides the direction and funding for the project. In this case, Anne Roberts would be the sponsor for the information technology collaboration project. Once further information technology projects are selected, however, the sponsors for those projects would be senior managers in charge of the main parts of the company affected by the projects. For example, if the vice president of sales initiates a project to improve direct product sales using the Internet, he or she might be the project sponsor. If several projects related to Internet technologies were undertaken, the organization might form a program. A **program** is "a group of related projects managed in a coordinated way to obtain benefits and control not available from managing them individually."[6] A program manager provides leadership for the project managers directing those projects, and the sponsors might come from several business areas.

■ *A project involves uncertainty.* Because every project is unique, it is sometimes difficult to define the project's objectives clearly, estimate how long it will take to complete, or determine how much it will cost. External factors also cause uncertainty, such as a supplier going out of business or a project team member needing unplanned time off. This uncertainty is one of the main reasons project management is so challenging, especially on projects involving new technologies.

A good **project manager** is crucial to a project's success. Project managers work with the project sponsors, the project team, and the other people involved in a project to meet project goals.

[6] Ibid, p. 16.

The Triple Constraint

Every project is constrained in different ways by its scope, time, and cost goals. These limitations are sometimes referred to in project management as the **triple constraint**. To create a successful project, a project manager must consider scope, time, and cost and balance these three often-competing goals. He or she must consider the following:

- *Scope*: What work will be done as part of the project? What unique product, service, or result does the customer or sponsor expect from the project?
- *Time:* How long should it take to complete the project? What is the project's schedule?
- *Cost*: What should it cost to complete the project? What is the project's budget?

Figure 1-1 illustrates the three dimensions of the triple constraint. Each area—scope, time, and cost—has a target at the beginning of the project. For example, the information technology collaboration project might have an initial scope of producing a forty- to fifty-page report and a one-hour presentation on thirty potential information technology projects. The project manager might further define project scope by providing a description of each potential project, an investigation of what other companies have implemented for similar projects, a rough time and cost estimate, and assessments of the risk and potential payoff as high, medium, or low. The initial time estimate for this project might be one month, and the cost estimate might be $50,000. These expectations provide the targets for the scope, time, and cost dimensions of the project.

Managing the triple constraint involves making trade-offs between scope, time, and cost goals for a project. For example, you might need to increase the budget for a project to meet scope and time goals. Alternatively, you might have to reduce the scope of a project to meet time and cost goals. Because projects involve uncertainty and limited resources, it is rare to complete many projects according to the exact scope, time, and cost plans originally predicted. Experienced project managers know that you must decide which aspect of the triple constraint is most important. If time is most important, you must often change the initial scope and/or cost goals to meet the schedule. If scope goals are most important, you may need to adjust time and/or cost goals.

For example, to generate project ideas, suppose the project manager for the information technology collaboration project sent an e-mail survey to all employees, as planned. The initial time and cost estimate may have been one week and $5,000 to collect ideas based on this e-mail survey. Now, suppose the e-mail survey generated only a few good project ideas, and the scope goal was to collect at least thirty good ideas. Should the project team use a different method like focus groups or interviews to collect ideas? Even though it was not in the initial scope, time, or cost estimates, it would really help the project.

Since good ideas are crucial to project success, it would make sense to inform the project sponsor that you want to make cost and/or schedule adjustments.

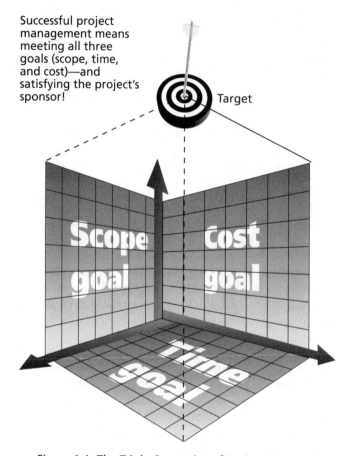

Successful project management means meeting all three goals (scope, time, and cost)—and satisfying the project's sponsor!

Target

Figure 1-1. The Triple Constraint of Project Management

Although the triple constraint describes how the basic elements of a project—scope, time, and cost—interrelate, other elements can also play significant roles. Quality is often a key factor in projects, as is customer or sponsor satisfaction. Some people, in fact, refer to the *"quadruple constraint"* of project management, including quality along with scope, time, and cost. Others believe that quality considerations, including customer satisfaction, must be inherent in setting the scope, time, and cost goals of a project. A project team may meet scope, time, and cost goals but fail to meet quality standards or satisfy their sponsor, if they have not adequately addressed these concerns. For example, Anne Roberts may receive a fifty-page report describing thirty potential information technology projects and hear a presentation on the findings of the report. The project team may have completed the work on time and within the cost constraint, but the quality may have been unacceptable. Anne's view of an executive presentation may be very different from the project team's view. The project manager should be communicating with the sponsor throughout the project to make sure the project meets his or her expectations.

How can you avoid the problems that occur when you meet scope, time, and cost goals, but lose sight of quality or customer satisfaction? The answer is *good project management, which includes more than meeting the triple constraint.*

WHAT IS PROJECT MANAGEMENT?

Project management is "the application of knowledge, skills, tools and techniques to project activities to meet project requirements."[7] Project managers must not only strive to meet specific scope, time, cost, and quality goals of projects, they must also facilitate the entire process to meet the needs and expectations of the people involved in or affected by project activities.

Figure 1-2 illustrates a framework to help you understand project management. Key elements of this framework include the project stakeholders, project management knowledge areas, project management tools and techniques, and the contribution of successful projects to the enterprise.

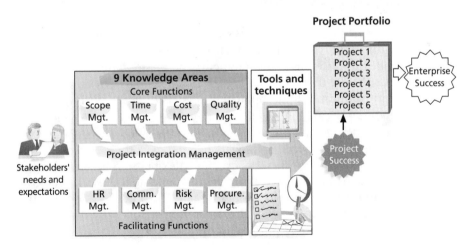

Figure 1-2. Project Management Framework

7 Ibid, p. 8.

Project Stakeholders

Stakeholders are the people involved in or affected by project activities and include the project sponsor, project team, support staff, customers, users, suppliers, and even opponents of the project. These stakeholders often have very different needs and expectations. For example, building a new house is a well-known example of a project. There are several stakeholders involved in a home construction project.

- The project sponsors would be the potential new homeowners. They would be the people paying for the house and could be on a very tight budget, so they would expect the contractor to provide accurate estimates of the costs involved in building the house. They would also need a realistic idea of when they could move in and what type of home they could afford given their budget constraints. The new homeowners would have to make important decisions to keep the costs of the house within their budget. Can they afford to finish the basement right away? If they can afford to finish the basement, will it affect the projected move-in date? In this example, the project sponsors are also the customers and users for the product, which is the house.

- The project manager in this example would normally be the general contractor responsible for building the house. He or she needs to work with all the project stakeholders to meet their needs and expectations.

- The project team for building the house would include several construction workers, electricians, carpenters, and so on. These stakeholders would need to know exactly what work they must do and when they need to do it. They would need to know if the required materials and equipment will be at the construction site or if they are expected to provide the materials and equipment. Their work would need to be coordinated since there are many interrelated factors involved. For example, the carpenter cannot put in kitchen cabinets until the walls are completed.

- Support staff might include the buyers' employers, the general contractor's administrative assistant, and other people who support other stakeholders. The buyers' employers might expect their employees to still complete their work but allow some flexibility so they can visit the building site or take phone calls related to building the house. The contractor's administrative assistant would support the project by coordinating meetings between the buyers, the contractor, suppliers, and so on.

- Building a house requires many suppliers. The suppliers would provide the wood, windows, flooring materials, appliances, and so on. Suppliers would expect exact details on what items they need to provide, where and when to deliver those items, and so on.

- There may or may not be opponents of a project. In this example, there might be a neighbor who opposes the project because the workers are making so much noise that she cannot concentrate on her work at home, or the noise might wake her sleeping children. She might interrupt the workers to voice her complaints or even file a formal complaint. Or the neighborhood might have association rules concerning new home design and construction. If the homeowners did not follow these rules, they might have to halt construction due to legal issues.

As you can see from this example, there are many different stakeholders on projects, and they all have different interests. Stakeholders' needs and expectations are important in the beginning and throughout the life of a project. Successful project managers develop good relationships with project stakeholders to understand and meet their needs and expectations.

Project Management Knowledge Areas

Project management knowledge areas describe the key competencies that project managers must develop. The center of Figure 1-2 shows the nine knowledge areas of project management. The four core knowledge areas of project management include project scope, time, cost, and quality management. These are core knowledge areas because they lead to specific project objectives. Brief descriptions of each core knowledge area are as follows:

- Project scope management involves defining and managing all the work required to complete the project successfully.
- Project time management includes estimating how long it will take to complete the work, developing an acceptable project schedule, and ensuring timely completion of the project.
- Project cost management consists of preparing and managing the budget for the project.
- Project quality management ensures that the project will satisfy the stated or implied needs for which it was undertaken.

The four facilitating knowledge areas of project management are human resource, communications, risk, and procurement management. These are called facilitating areas because they are the processes through which the project objectives are achieved. Brief descriptions of each facilitating knowledge area are as follows:

- Project human resource management is concerned with making effective use of the people involved with the project.
- Project communications management involves generating, collecting, disseminating, and storing project information.
- Project risk management includes identifying, analyzing, and responding to risks related to the project.
- Project procurement management involves acquiring or procuring goods and services for a project from outside the performing organization.

Project integration management, the ninth knowledge area, is an overarching function that affects and is affected by all of the other knowledge areas. Project managers must have knowledge and skills in all nine of these areas. This text includes an entire chapter on each of these knowledge areas, as all of them are crucial to project success.

Project Management Tools and Techniques

Thomas Carlyle, a famous historian and author, stated, "Man is a tool-using animal. Without tools he is nothing, with tools he is all." As the world continues to become more complex, it is even more important for people to develop and use tools, especially for managing important projects. **Project management tools and techniques** assist project managers and their teams in carrying out work in all nine knowledge areas. For example, some popular time-management tools and techniques include Gantt charts, project network diagrams, and critical path analysis. Table 1-1 lists some commonly used tools and techniques by knowledge area. You will learn more about these and other tools and techniques throughout this text.

Table 1-1: Common Project Management Tools and Techniques by Knowledge Area

KNOWLEDGE AREA/CATEGORY	TOOLS AND TECHNIQUES
Integration management	Project selection methods, project management methodology, stakeholder analysis, project charters, project management plans, project management software, change control boards, configuration management, project review meetings, work authorization systems
Scope management	Project scope statements, work breakdown structures, statements of work, scope management plan, requirements analysis, scope change control
Time management	Gantt charts, project network diagrams, critical path analysis, program evaluation review technique (PERT), critical chain scheduling, crashing, fast tracking, milestone reviews
Cost management	Net present value, return on investment, payback analysis, business cases, earned value management, project portfolio management, cost estimates, cost management plan, financial software
Quality management	Six Sigma, quality control charts, Pareto diagrams, fishbone or Ishikawa diagrams, quality audits, maturity models, statistical methods
Human resource management	Motivation techniques, empathic listening, team contracts, responsibility assignment matrices, resource histograms, resource leveling, team building exercises

Table 1-1: Common Project Management Tools and Techniques by Knowledge Area (continued)

KNOWLEDGE AREA/CATEGORY	TOOLS AND TECHNIQUES
Communications management	Communications management plan, conflict management, communications media selection, communications infrastructure, status reports, virtual communications, templates, project Web sites
Procurement management	Make-or-buy analysis, contracts, requests for proposals or quotes, source selection, negotiating, e-procurement
Risk management	Risk management plan, probability/impact matrix, risk ranking, Monte Carlo simulation, top-ten risk item tracking

Project managers must work with key stakeholders to define what constitutes success for a particular project and strive to complete their projects successfully by applying appropriate tools and techniques. In many organizations, project managers also support an emerging business strategy of **project portfolio management**, in which organizations group and manage projects as a portfolio of investments that contribute to the entire enterprise's success. You will learn more about project portfolio management in Chapter 7, Project Cost Management.

✓ *What Went Right?*

Follow-up studies done by the Standish Group showed some improvement in the statistics for information technology projects in the past decade:

■ The number of successful IT projects has more than doubled, from 16 percent in 1994 to 34 percent in 2002.

■ The number of failed projects has been cut in half, from 31 percent in 1994 to 15 percent in 2002.

■ The U.S. spent about the same amount of money on IT projects in 1994 and 2002 ($250 billion and $255 billion, respectively), but the amount of money wasted on challenged and failed projects was down to $55 billion in 2002 compared to $140 billion in 1994.[8]

Even though there have been significant improvements in managing information technology projects, there is still much room for improvement. The best news is that project managers are learning how to succeed more often. "The reasons for the increase in successful projects vary. First, the average cost of a project has been more than cut in half. Better tools have been created to monitor and control progress and better skilled project managers with better management processes are being used. The fact that there are processes is significant in itself."[9]

[8] The Standish Group, "Latest Standish Group CHAOS Report Shows Project Success Rates Have Improved by 50%," (March 25, 2003).

[9] The Standish Group, "CHAOS 2001: A Recipe for Success" (2001).

Despite its advantages, project management is not a silver bullet that guarantees success on all projects. Project management is a very broad, often complex discipline. What works on one project may not work on another, so it is essential for project managers to continue to develop their knowledge and skills in managing projects. It is also important to learn from the mistakes and successes of others.

Project Success Factors

Why do some projects succeed and others fail? Can organizations provide a better environment to help improve project success rates? There are no easy answers to any of these questions, but many people are contributing to our knowledge base to continue to improve the theory and practice of project management.

Table 1-2 summarizes the results of the 2001 Standish Group study describing, in order of importance, what factors contribute most to the success of information technology projects. Note that the study lists executive support as the most important factor, overtaking user involvement, which was most important in the earlier studies. Also note that several other success factors can be strongly influenced by executives, such as encouraging user involvement, providing clear business objectives, assigning an experienced project manager, using a standard software infrastructure, and following a formal methodology. Other success factors are related to good project scope and time management, such as having a minimized scope, firm basic requirements, and reliable estimates. In fact, *97 percent of successful projects were led by experienced project managers*, who can often help influence all of these factors to improve the probability of project success.

Table 1-2: What Helps Projects Succeed?

1. Executive support
2. User involvement
3. Experienced project manager
4. Clear business objectives
5. Minimized scope
6. Standard software infrastructure
7. Firm basic requirements
8. Formal methodology
9. Reliable estimates
10. Other criteria, such as small milestones, proper planning, competent staff, and ownership

The Standish Group, "Extreme CHAOS," (2001).

It is interesting to compare success factors for information technology projects in the U.S. with those in other countries. A 2004 study summarizes results of a survey of 247 information systems project practitioners in the Mainland of China. One of the study's key findings is that relationship management is viewed as a top success factor for information systems in China, while it is not mentioned in U.S. studies. The study also suggested that having competent team members is less important in China than in the U.S. The Chinese, like the U.S., included top management support, user involvement, and a competent project manager as vital to project success.[10]

It is also important to look beyond individual project success rates and focus on how organizations as a whole can improve project performance. Research comparing companies that excel in project delivery (the "winners") from those who do not found four significant best practices. The winners:

1. *Use an integrated toolbox.* Companies that consistently succeed in managing projects clearly define what needs to be done in a project, by whom, when, and how. They use an integrated toolbox, including project management tools, methods, and techniques. They carefully select tools, align them with project and business goals, link them to metrics, and provide them to project managers to deliver positive results.

2. *Grow project leaders.* The winners know that strong project managers—referred to as project leaders—are crucial to project success. They also know that a good project leader needs to be a business leader as well, with strong interpersonal and intrapersonal skills. Companies that excel in project management often grow their project leaders internally, providing them with career opportunities, training, and mentoring.

3. *Develop a streamlined project delivery process.* Winning companies have examined every step in the project delivery process, analyzed fluctuations in workloads, searched for ways to reduce variation, and eliminated bottlenecks to create a repeatable delivery process. All projects go through clear stages and clearly define key milestones. All project leaders use a shared road map, focusing on key business aspects of their projects while integrating goals across all parts of the organization.

4. *Measure project health using metrics.* Companies that excel in project delivery use performance metrics to quantify progress. They focus on a handful of important measurements and apply them to all projects. Metrics often include customer satisfaction, return on investment, and percentage of schedule buffer consumed. You will learn how to use these metrics in later chapters.[11]

Project managers play an important role in making projects, and therefore organizations, successful.

---

[10] Dong, Chang, K.B. Chuah, and Li Zhai, "A Study of Critical Success Factor of Information System Projects in China," Proceedings of PMI Research Conference (2004).

[11] Milosevic, Dragan and And Ozbay, "Delivering Projects: What the Winners Do," Proceedings of the Project Management Institute Annual Seminars & Symposium (November 2001).

THE ROLE OF THE PROJECT MANAGER

You have already read that project managers must work closely with the other stakeholders on a project, especially the sponsor and project team. They have to be familiar with the nine project management knowledge areas and the various tools and techniques related to project management. Experienced project managers help projects succeed. But what do project managers do, exactly? What skills do they really need to do a good job? The following section provides brief answers to these questions, and the rest of this book gives more insight into the role of the project manager. Even if you never become a project manager, you will probably be part of a project team, and it is important for team members to help their project managers.

Project Manager Job Description

A project manager can have many different job descriptions, which can vary tremendously based on the organization and the project. For example, Monster.com includes thousands of job listings for project managers. Below are a few edited postings from June 2004:

- Project manager for a consulting firm: Plans, schedules, and controls activities to fulfill identified objectives applying technical, theoretical, and managerial skills to satisfy project requirements. Coordinates and integrates team and individual efforts and builds positive professional relationships with clients and associates.

- IT project manager for a financial services firm: Manages, prioritizes, develops, and implements information technology solutions to meet business needs. Prepares and executes project plans using project management software following a standard methodology. Establishes cross-functional end-user teams defining and implementing projects on time and within budget. Acts as a liaison between third party service providers and end-users to develop and implement technology solutions. Participates in vendor contract development and budget management. Provides post implementation support.

- IT project manager for a non-profit consulting firm: Responsibilities include business analysis, requirements gathering, project planning, budget estimating, development, testing, and implementation. Responsible for working with various resource providers to ensure development is completed in a timely, high-quality, and cost-effective manner.

As you can see, the job description for a project manager can vary by industry and by organization, but there are similar tasks that most project managers perform regardless of these differences. A National Science Foundation study found that project management is a skill needed in every major information technology field, from database administrator to network specialist to technical writer. Table 1-3 lists fifteen project management job functions that are essential for good project management.

Table 1-3: Fifteen Project Management Job Functions

1. Define scope of project
2. Identify stakeholders, decision-makers, and escalation procedures
3. Develop detailed task list (work breakdown structure)
4. Estimate time requirements
5. Develop initial project management flowchart
6. Identify required resources and budget
7. Evaluate project requirements
8. Identify and evaluate risks
9. Prepare contingency plan
10. Identify interdependencies
11. Identify and track critical milestones
12. Participate in project phase review
13. Secure needed resources
14. Manage the change control process
15. Report project status

"Building a Foundation for Tomorrow: Skills Standards for Information Technology, Millennium Edition," Northwest Center for Emerging Technologies (NWCET), Belleview, WA, 1999.

Each of the job functions listed in Table 1-3 requires different skills, as do the duties described in the different job descriptions. So what skills do project managers need? Can they all be learned, or are some innate? How technical do IT project managers need to be?

Suggested Skills for Project Managers

As you can imagine, good project managers should have many skills. The Project Management Body of Knowledge (*PMBOK*® *Guide 2004)* suggests that the project management team understand and use expertise in the following areas:

- The project management body of knowledge
- Application area knowledge, standards, and regulations
- Project environment knowledge
- General management knowledge and skills
- Soft skills or human relations skills

The previous section introduced the nine project management knowledge areas, as well as some tools and techniques project managers use, in general. The following section focuses on the IT application area, including skills required in the project environment, general management, and soft skills.

The project environment differs from organization to organization and project to project, but there are some skills that will help in most project environments. These skills include understanding change, and understanding how organizations work within their social, political, and physical environments. Project managers must be comfortable leading and handling change, since most projects introduce changes in organizations and involve changes within the projects themselves. Project managers need to understand the organizations they work in and how products are developed and services are provided. And they must also understand the social, physical, and political environment. It takes very different skills and behavior to manage a project for a Fortune 100 company in the United States than it does to manage a government project for a new business in Poland. Chapter 2, The Project Management and Information Technology Context, provides detailed information on these topics.

Project managers should also possess general management knowledge and skills. They should understand important topics related to financial management, accounting, procurement, sales, marketing, contracts, manufacturing, distribution, logistics, the supply chain, strategic planning, tactical planning, operations management, organizational structures and behavior, personnel administration, compensation, benefits, career paths, and health and safety practices. On some projects, it will be critical for the project manager to have a lot of experience in one or several of these general management areas. On other projects, the project manager can delegate detailed responsibility for some of these areas to a team member, support staff, or even a supplier. Even so, the project manager must be intelligent and experienced enough to know which of these areas are most important and who is qualified to do the work. He or she must also make and/or take responsibility for all key project decisions.

Achieving high performance on projects requires soft skills or human relations skills. Some of these soft skills include effective communication, influencing the organization to get things done, leadership, motivation, negotiation, conflict management, and problem solving. Why do project managers need good soft skills? One reason is that to understand, navigate, and meet stakeholders' needs and expectations, project managers need to lead, communicate, negotiate, solve problems, and influence the organization at large. They need to be able to listen actively to what others are saying, help develop new approaches for solving problems, and then persuade others to work toward achieving project goals. Project managers must lead their project teams by providing vision, delegating work, creating an energetic and positive environment, and setting an example of appropriate and effective behavior. Project managers must focus on teamwork skills in order to use their people effectively. They need to be able to motivate different types of people and develop *esprit de corps* within the project team and with other project stakeholders. Since most projects involve changes and trade-offs between competing goals, it is important for project managers to have strong coping skills as well. It helps project managers maintain their sanity and reduce their stress levels if they cope with criticism and constant change. Project managers must be flexible, creative, and sometimes patient in working toward project goals; they must also be persistent in making project needs known. Lastly, project managers must be able to make effective use of technology as it relates to the specific project. Making effective use of technology often includes special product knowledge or experience with a particular industry. Project managers must make many decisions and deal with people in a wide variety of disciplines, so it helps tremendously to have a project manager who is confident in using the special tools or technologies that are the most effective in particular settings.

🎥 *Media Snapshot*

In 2004, millions of people watched the first season of the U.S. reality television show called *The Apprentice*, in which contestants vied for a high-level position working for Donald Trump. Each week, Trump fired one contestant and told the contestants bluntly why they were fired or why they were spared. Trump's reasons provide insight to improving project management skills, as follows:

1. Leadership and professionalism are crucial. No matter how smart you are (the first candidate fired had degrees in medicine and business), you must be professional in how you deal with people and display some leadership potential.

2. Know what your sponsor expects from the project, and learn from your mistakes. Jason, the second person and first project manager fired, decided not to take the time to meet with his project sponsors, causing his team to fail their assignment. Trump wanted everyone to remember that crucial mistake.

3. Trust your team and delegate decisions. Sam had several problems as a team member and project manager, but his lack of trust and respect for and from his teammates led to his downfall.

4. Know the business. Restaurants often have the highest profit margins on certain items, like drinks. Find out what's most important to your business when running projects. One team focused on increasing bar sales and easily won the competition that week.

5. Stand up for yourself. When Trump fired Kristi over the other women he explained his decision by saying that Kristi didn't fight for herself, while the other two did.

6. Be a team player. Tammy clearly did not get along with her team, and no one supported her in the boardroom when her team lost.

7. Don't be overly emotional and stay organized. Erika had a difficult time leading her team in selling Trump Ice, and she became flustered when they didn't get credit for sales because paperwork wasn't done correctly. Her emotions were evident in the boardroom when she was fired.

8. Work on projects and for people you believe in. Kwame's team selected an artist based on her profit potential, even though he and other teammates disliked her work. The other team picked an artist they liked, and they easily outsold Kwame's team.

9. Think outside the box. Troy led his team in trying to make the most money selling rickshaw rides. The other team brainstormed ideas and decided to sell advertising space on the rickshaws, which was a huge success.

10. There is some luck involved in project management, and you should always aim high. Nick and Amy were teamed against Bill, Troy, and Kwame to rent out a party room for the highest price. Troy's team seemed very organized and did get a couple of good bids, but Nick and Amy didn't seem to have any real prospects. They got lucky when one potential client came back at the last minute and agreed to a much higher than normal price.

Valuable Skills for IT Project Managers

What other skills do information technology project managers need to succeed in today's competitive market? People in industry and academia often debate the answer to this question. In an interview with two CIOs, John Oliver of True North Communications, Inc. and George Nassef of Hotjobs.com, both men provided very different responses when asked what skills good information technology project managers need. They could not agree on which skills were more important, such as can-do optimism versus assume-the-worst realism, or detail-oriented versus visionary. However, both agree that the most important skills seem to depend on the uniqueness of the project and the people involved.[12] Project managers need to have a wide variety of skills and be able to decide which particular skills are more important in different situations.

Some people believe that it is important for information technology project managers to understand the technologies they use on the projects they are managing. They do not normally have to be experts on any specific technology, but they have to know enough to build a strong team and ask the right questions to keep things on track. On small projects, the project manager might be the manager on a part-time basis and also be expected to produce some of the work. For example, on a small Web development project, the project manager might lead the team and also create some of the site. Other people say it is more important for information technology project managers to have strong business and behavioral skills so they can lead a project team and deliver a solution that will meet business needs.

It would be very difficult for someone with little or no background in information technology to become the project manager for a large information technology project. Not only would it be difficult to work with other managers and suppliers, it would also be difficult to earn the respect of the project team.

[12] Brandel, Mary, "The Perfect Project Manager," ComputerWorld (August 6, 2001).

However, project managers for large information technology projects do not have to be experts in the field of information technology. They should have some working knowledge of various technologies, but more importantly, they should understand how the project they are managing would enhance the business. Many companies have found that a good business manager can be a very good information technology project manager because they focus on meeting business needs and rely on key project members to handle the technical details. Although information technology project managers need to draw on their information technology expertise or the expertise of key team members, they must spend more time becoming better project managers and less time becoming information technology experts to lead their teams to success.

Lively debates continue on the differences between managing information technology projects and managing other types of projects. There are several differences, but there are even more similarities. Several articles and speakers joke about the differences between construction projects and software development projects. No, you cannot blow up an old information system like you can an old building and start from scratch. No, there are often no specific engineering principles and building codes that everyone knows and follows. Nevertheless, information technology project managers, like all project managers, still have the responsibility for working with their sponsors, project teams, and other stakeholders to achieve specific project and organizational goals. All project managers should continue to develop their knowledge and experience in project management, general management, and the industries they support.

Unfortunately, many people in information technology do not want to develop anything but their technical skills. They do not see how soft skills or business skills are going to improve their performance or help them earn higher salaries. Most managers disagree with that mindset because they see the need to improve communications between information technology professionals and their customers. Business people are now more savvy with information technology, but few information technology professionals have spent the time developing their business savvy.[13] Individuals must be willing to develop more than just their technical skills to be more productive team members and potential project managers. Everyone, no matter how technical they are, can develop business and soft skills.

Importance of Leadership Skills

In a recent study, one hundred project managers listed the characteristics they believed were critical for effective project management and the characteristics that made project managers ineffective. Table 1-4 lists the results. The study found that effective project managers provide leadership by example, are

[13] Thomsen-Moore, Lauren, "No 'soft skills' for us, we're techies," Computerworld Today (December 16, 2002).

visionary, technically competent, decisive, good communicators, and good motivators. They also stand up to top management when necessary, support team members, and encourage new ideas. The study also found that respondents believed *positive leadership contributes the most to project success.* The most important characteristics and behaviors of positive leaders include being a team builder and communicator, having high self-esteem, focusing on results, demonstrating trust and respect, and setting goals.

Table 1-4: Most Significant Characteristics of Effective and Ineffective Project Managers

EFFECTIVE PROJECT MANAGERS	INEFFECTIVE PROJECT MANAGERS
Lead by example	Set bad examples
Are visionaries	Are not self-assured
Are technically competent	Lack technical expertise
Are decisive	Are poor communicators
Are good communicators	Are poor motivators
Are good motivators	
Stand up to top management when necessary	
Support team members	
Encourage new ideas	

Zimmerer, Thomas W. and Mahmoud M. Yasin, "A Leadership Profile of American Project Managers," Project Management Journal (March 1998), 31-38.

Leadership and *management* are terms often used interchangeably, although there are differences. Generally, a **leader** focuses on long-term goals and big-picture objectives, while inspiring people to reach those goals. A **manager** often deals with the day-to-day details of meeting specific goals. Some people say that, "Managers do things right, and leaders do the right things." "Leaders determine the vision, and managers achieve the vision." "You lead people and manage things."

However, project managers often take on the role of both leader and manager. Good project managers know that people make or break projects, so they must set a good example to lead their team to success. They are aware of the greater needs of their stakeholders and organizations, so they are visionary in guiding their current projects and in suggesting future ones. As mentioned earlier, companies that excel in project management grow project "leaders," emphasizing development of business and communication skills. Yet good project managers must also focus on getting the job done by paying attention to the details and daily operations of each task. Instead of thinking of leaders and managers as specific people, it is better to think of people as having leadership skills, such as being visionary and inspiring, and management skills, such as being organized and effective. Therefore, the best project managers have leadership and management characteristics; they are visionary yet focused on the bottom line. Above all else, good project managers focus on achieving positive results!

Careers for Information Technology Project Managers

Paul Ziv, a recruitment strategist at ComputerJobs.com, described why project management made the top-ten list of information technology skills in demand in late 2002. He explained that information technology project managers are expected to do much more today. They have to understand the field and possess an executive skill set in order to lead teams in developing products and services that improve the bottom line. Table 1-5 summarizes the top-ten information technology skills and average salaries based on job postings in 2002.

Table 1-5: Top Ten Most In Demand Information Technology Skills

RANK	IT SKILL/JOB	AVERAGE ANNUAL SALARY
1	SQL Database Analyst	$80,664
2	Oracle Database Analyst	$87,144
3	C/C++ Programmer	$95,829
4	Visual Basic Programmer	$76,903
5	E-commerce/Java Developer	$89,163
6	Windows NT/2000 Expert	$80,639
7	Windows/Java Developer	$93,785
8	Security Architect	$86,881
9	**Project Manager**	**$95,719**
10	Network Engineer	$82,906

Ziv, Paul "The Top 10 IT Skills in Demand," Global Knowledge Webcast (*www.globalknowledge.com*) (11/20/2002).

Many people recognize the need to fill these other information technology careers, but most are unaware of the need for project managers or do not really understand what project management is all about. Each of the positions listed above is carried out in conjunction with the role of project management as the specific technologies are deployed. For example, an SQL database analyst would be a team member on a project that requires development or support of SQL databases.

Recent articles also emphasize the need for good project managers in the information technology field. In a 2004 survey by CIO.com, IT executives listed the IT skills that were most in demand: application development, project management, database management, and networking. Figure 1-3 shows these results in graphical format. Note that 58 percent of survey respondents included project management as a top IT skill in demand, the second most mentioned skill. Even if you choose to stay in a technical role, you still need project management knowledge and skills to help your team and your organization succeed.

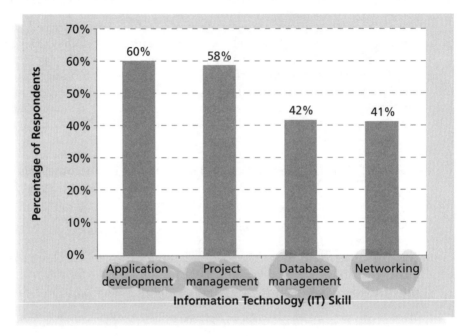

Figure 1-3. Top Information Technology Skills

Cosgrove, Lorraine, "January 2004 IT Staffing Update," CIO Research Reports, (February 3, 2004).

THE PROJECT MANAGEMENT PROFESSION

The profession of project management is growing at a very rapid pace. To understand this line of work, it is helpful to briefly review the history of project management, introduce you to the Project Management Institute (PMI) and some of its services (such as certification), and discuss the growth in project management software.

History of Project Management

Some people might argue that building the Egyptian pyramids or building the Great Wall of China was a project. Most people agree, however, that the modern concept of project management began with the Manhattan Project, which the U.S. military led to develop the atomic bomb. The Manhattan Project involved many people with different skills at several different locations. It also clearly separated the overall management of the project's mission, schedule, and budget under General Leslie R. Groves and the technical management of the project under Dr. Robert Oppenheimer. The Manhattan Project lasted about three years and cost almost $2 billion in 1946.

In developing the project, the military realized that scientists and other technical specialists often did not have the desire or the necessary skills to manage large projects. For example, after being asked several times for each team member's responsibilities at the new Los Alamos laboratory in 1943, Dr. Oppenheimer threw a piece of paper with an organization chart on it at his director and said, "Here's your damn organization chart."[14] Project management was recognized as a distinct discipline requiring people with special skills and, even more importantly, the desire to lead project teams.

The military was the key industry behind the development of several project management techniques. In 1917, Henry Gantt developed the famous Gantt chart as a tool for scheduling work in factories. A **Gantt chart** is a standard format for displaying project schedule information by listing project activities and their corresponding start and finish dates in a calendar format. Managers drew Gantt charts by hand to show project tasks and schedule information, and this tool provided a standard format for planning and reviewing all the work on early military projects.

Today's project managers still use the Gantt chart as the primary tool to communicate project schedule information, but with the aid of computers, it is no longer necessary to draw the charts by hand. Figure 1-4 displays a Gantt chart for developing an intranet. This version of the chart was created with Microsoft Project, the most widely used project management software today. You will learn more about using Project 2003, the most recent version of Microsoft Project, in Appendix A.

[14] The Regents of the University of California, Manhattan Project History, "Here's Your Damned Organization Chart," (1998-2001).

	Task Name		January	February	March	April	Ma
			4 11 18 25	1 8 15 22	1 8 15 22 29	5 12 19 26	3 10
1	⊟ *1 Concept*						
2	1.1 Evaluate current systems						
3	⊟ **1.2 Define Requirements**						
4	1.2.1 Define user requirements						
5	1.2.2 Define content requirements						
6	1.2.3 Define system requirements						
7	1.2.4 Define server owner requirements						
8	1.3 Define specific functionality						
9	1.4 Define risks and risk management approach						
10	1.5 Develop project plan						
11	1.6 Brief web development team						
12	⊞ *2 Web Site Design*						
30	⊞ *3 Web Site Development*						
50	⊞ *4 Roll Out*						
57	⊞ *5 Support*						

Figure 1-4. Sample Gantt Chart in Microsoft Project

Members of the Navy Polaris missile/submarine project first used network diagrams in 1958. These diagrams helped managers model the relationships among project tasks, which allowed them to create schedules that were more realistic. Figure 1-5 displays a network diagram created using Microsoft Project. Note that the diagram includes arrows that show which tasks are related and the sequence in which team members must perform the tasks. The concept of determining relationships among tasks is essential in helping to improve project scheduling. This concept allows you to find and monitor the **critical path**—the longest path through a network diagram that determines the earliest completion of a project. You will learn more about Gantt charts, network diagrams, critical path analysis, and other time management concepts in Chapter 6, Project Time Management.

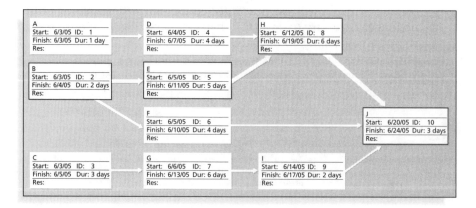

Figure 1-5. Sample Network Diagram in Microsoft Project

By the 1970s, the military had begun to use software to help manage large projects. Early project management software products were very expensive and ran on mainframe computers. For example, Artemis was an early project management software product that helped managers analyze complex schedules for designing aircraft. A full-time employee was often required to run the complicated software, and expensive pen plotters were used to draw network diagrams and Gantt charts. As computer hardware became smaller and more affordable and software became graphical and easy to use, project management software became less expensive and more widely used. Today, many industries use project management software on all types and sizes of projects. New software makes basic tools, such as Gantt charts and network diagrams, inexpensive, easy to create, and available for anyone to update. See the section in this chapter on project management software for more information.

In the 1990s, many companies began creating project management offices to help them handle the increasing number and complexity of projects. A **Project Management Office (PMO)** is an organizational group responsible for coordinating the project management function throughout an organization. There are different ways to structure a PMO, and they can have various roles and responsibilities. Below are possible goals of a PMO:

- Collect, organize, and integrate project data for the entire organization.
- Develop and maintain templates for project documents.
- Develop or coordinate training in various project management topics.
- Develop and provide a formal career path for project managers.
- Provide project management consulting services.
- Provide a structure to house project managers while they are acting in those roles or between projects.

By the end of the twentieth century, people in virtually every industry around the globe began to investigate and apply different aspects of project management to their projects. The sophistication and effectiveness with which project management tools are being applied and used today is influencing the way companies do business, use resources, and respond to market requirements with speed and accuracy. Web-based project management tools are now available to help organizations manage individual projects as well as portfolios of projects. Figure 1-6 provides a sample chart that could be displayed from an enterprise project management solution. **Enterprise project management software** integrates information from multiple projects to show the status of active, approved, and future projects across an entire organization and provides links to more detailed information. Summary charts often show status as green to indicate things are going well, yellow to indicate that there are some problems, and red to indicate major problems. Appendix A provides more information on Microsoft's Enterprise Project Management Solution, and the last section of this chapter describes recent advances in project management software.

Company ABC Project Portfolio				
Project Name	**Scope**	**Schedule**	**Budget**	**Links**
Active Projects				
Project 1	○	◉	◉	
Project 2	●	●	◉	
Project 3	○	○	○	
Project 4	○	◉	●	
Approved Projects				
Project 10	○	○	○	
Project 11	○	○	○	
Project 12	○	○	○	
Project 13	○	○	○	
Project 14	○	○	○	
Opportunities				
Project 100				
Project 200				
○	White = going well			
◉	Gray = some problems			
●	Black = major problems			

Figure 1-6. Sample Enterprise Project Management Tool

Many colleges, universities, and companies now offer courses related to various aspects of project management. You can even earn Bachelor's, Master's, and doctoral degrees in project management. The problems in managing projects, the publicity about project management, and the belief that it really can make a difference continue to contribute to the growth of this field.

The Project Management Institute

Although many professional societies are suffering from declining memberships, the **Project Management Institute (PMI)**, an international professional society for project managers founded in 1969, has continued to attract and retain members, reporting more than 133,000 members worldwide by May 2004. A large percentage of PMI members work in the information technology field and more than 15,000 pay additional dues to join the Information Systems Specific Interest Group. Because there are so many people working on projects in various industries, PMI has created Specific Interest Groups (SIGs) that enable members to share ideas about project management in their particular application areas, such as information systems. PMI also has SIGs for aerospace/defense, financial services, healthcare, hospitality management, manufacturing, new product development, retail, and urban development, to name a few. As a student, you can join PMI for a reduced fee. Consult PMI's Web site (*www.pmi.org*) or the Information Systems SIG site (*www.pmi-issig.org*) for more information. You can also network with other students studying project management by joining the Students of Project Management SIG at *www.studentsofpm.org*.

Project Management Certification

Professional certification is an important factor in recognizing and ensuring quality in a profession. PMI provides certification as a **Project Management Professional (PMP)**—someone who has documented sufficient project experience, agreed to follow the PMI code of professional conduct, and demonstrated knowledge of the field of project management by passing a comprehensive examination. Appendix B provides more information on PMP certification as well as other certification programs, such as CompTIA's Project+ certification.

The number of people earning PMP certification continues to increase. In 1993, there were about 1,000 certified project management professionals. By the end of May 2004, there were 81,913 certified project management professionals.[15] Figure 1-7 shows the rapid growth in the number of people earning project management professional certification from 1993 to 2003.

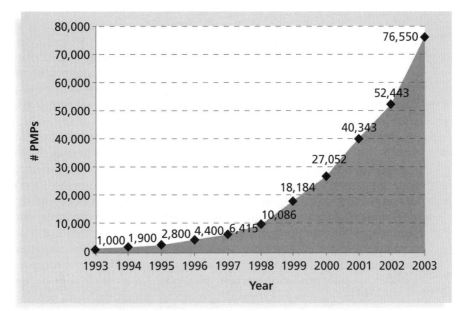

Figure 1-7. Growth in PMP Certification, 1993–2003

[15] The Project Management Institute, "PMI Today," (August 2004).

Several studies show that organizations supporting technical certification programs tend to operate in more complex information technology environments and are more efficient than companies that do not support certification. Likewise, organizations that support PMP certification see the value of investing in programs to improve their employees' knowledge in project management. Many employers today require specific certifications to ensure their workers have current skills, and job seekers find that they often have an advantage when they earn and maintain marketable certifications.

As information technology projects become more complex and global in nature, the need for people with demonstrated knowledge and skills in project management will continue. Just as passing the CPA exam is a standard for accountants, passing the PMP exam is becoming a standard for project managers. Some companies are requiring that all project managers be PMP certified. Project management certification is also enabling professionals in the field to share a common base of knowledge. For example, any person with PMP certification can list, describe, and use the nine project management knowledge areas. Sharing a common base of knowledge is important because it helps advance the theory and practice of project management. See Appendix B of this text for detailed information on certification.

Ethics in Project Management

Ethics is an important part of all professions. Project managers often face ethical dilemmas. For example, several projects involve different payment methods. If a project manager can make more money by doing a job poorly, should he or she do the job poorly? If a project manager is personally opposed to the development of nuclear weapons, should he or she refuse to manage a project that helps produce them? It is important for project managers to make decisions in an ethical manner.

PMI developed a PMP code of professional conduct that all applicants must sign in order to become certified project management professionals (PMPs). PMI states that it is vital for all PMPs to conduct their work in an ethical manner. Conducting work in an ethical manner helps the profession earn the confidence of the public, employers, employees, and project team members. The PMP code of professional conduct lists responsibilities to the profession, such as compliance with all organizational rules and policies, professional practice, and advancement of the profession. It also lists responsibilities to customers and the public, such as qualifications, experience, and performance of professional services, as well as conflict-of-interest situations. For example, one of the responsibilities to customers and the public includes refraining "from offering or accepting inappropriate payments, gifts or other forms of compensation for personal gain, unless in conformity with applicable laws or customs of the country where project management services are being provided."[16]

PMI even added a new series of questions to the PMP certification exam in March 2002 to emphasize the importance of professional responsibility. See Appendix B for more information.

Project Management Software

Unlike the cobbler neglecting to make shoes for his own children, the project management and software development communities have definitely responded to the need to provide more software to assist in managing projects. PMI published a Project Management Software Survey in 1999 that describes, compares, and contrasts more than 200 project management software tools. The Project Management Center, a Web site for people involved in project management, provides an alphabetical directory of more than 300 project management software solutions (*www.infogoal.com/pmc*). This site and others demonstrate the growth in available project management software products, especially Web-based tools. Deciding which project management software to use has become a project in itself. This section provides a summary of the basic types of project management software available and references for finding more information. In Appendix A, you will learn how to use Microsoft Project 2003, the most widely used project management software tool today.

Many people still use basic productivity software, such as Microsoft Word or Excel, to perform many project management functions, such as determining project scope, time, and cost, assigning resources, preparing project documentation, and so on. People often use productivity software instead of specialized project management software because they already have it and know how to use it. However, there are hundreds of project management software tools that provide specific functionality for managing projects. These project management software tools can be divided into three general categories based on functionality and price:

- *Low-end tools*: These tools provide basic project management features and generally cost less than $200 per user. They are often recommended for small projects and single users. Most of these tools allow users to create Gantt charts, which cannot be done easily using current productivity software. For example, Milestones Simplicity by KIDASA Software, Inc., has a Schedule Setup Wizard that walks users through simple steps to produce a Gantt chart. For $49 per user, this tool also includes a large assortment of symbols, flexible formatting, an outlining utility, and an Internet Publishing Wizard. Several companies provide add-on features to Excel or Access to provide basic project management functions using familiar software products.

[16] The Project Management Institute, Project Management Institute Certification Handbook, (January 2003) (22).

- *Midrange tools*: A step up from low-end tools, midrange tools are designed to handle larger projects, multiple users, and multiple projects. All of these tools can produce Gantt charts and network diagrams, and can assist in critical path analysis, resource allocation, project tracking, status reporting, and so on. Prices range from about $200 to $500 per user, and several tools require additional server software for using workgroup features. Microsoft Project is still the most widely used project management software today, and Project 2003 includes an enterprise version, as described briefly below and in more detail in Appendix A. Other companies that sell midrange project management tools include Artemis, PlanView, Primavera, and Welcom, to name just a few.

- *High-end tools*: Another category of project management software is high-end tools, sometimes referred to as enterprise project management software. These tools provide robust capabilities to handle very large projects, dispersed workgroups, and enterprise functions that summarize and combine individual project information to provide an enterprise view of all projects. These products are generally licensed on a per-user basis, integrate with enterprise database management software, and are accessible via the Internet. In mid 2002, Microsoft introduced the first version of their Enterprise Project Management software, and in 2003, they introduced the Microsoft Enterprise Project Management solution, as described in Appendix A. Several companies that provide midrange tools now offer enterprise versions of their software. There are also several inexpensive, Web-based products on the market. For example, VPMi Enterprise Online (*www.vcsonline.com*) is available for only $12 per user per month. See the Project Management Center Web site (*www.infogoal.com/pmc*) or similar sites for links to many companies that provide project management software.

As mentioned earlier, there are many reasons to study project management, particularly as it relates to information technology projects. The number of information technology projects continues to grow, the complexity of these projects continues to increase, and the profession of project management continues to expand and mature. As more people study and work in this important field, the success rate of information technology projects should continue to improve.

CASE WRAP-UP

Anne Roberts worked with the VPs and the CEO to begin reorganizing several parts of the company to support their new emphasis on projects. They formed a project team to implement an enterprise project management software tool across the organization. They formed another team to develop project-based reward systems for all employees. They also authorized funds for a project to educate all employees in project management and to develop a mentoring program. Anne had successfully convinced everyone that effectively managing projects was crucial to their company's future.

CHAPTER SUMMARY

There is a new or renewed interest in project management today as the number of projects continues to grow and their complexity continues to increase. The success rate of information technology projects has more than doubled since 1995, but still only about a third are successful in meeting scope, time, and cost goals. Using a more disciplined approach to managing projects can help projects and organizations succeed.

A project is a temporary endeavor undertaken to create a unique product, service, or result. An information technology project involves the use of hardware, software, and/or networks. Projects are unique, temporary, and developed incrementally; they require resources, have a sponsor, and involve uncertainty. The triple constraint of project management refers to managing the scope, time, and cost dimensions of a project.

Project management is the application of knowledge, skills, tools, and techniques to project activities to meet project requirements. Stakeholders are the people involved in or affected by project activities. A framework for project management includes the project stakeholders, project management knowledge areas, and project management tools and techniques. The nine knowledge areas are project integration management, scope, time, cost, quality, human resources, communications, risk, and procurement management. Project portfolio management involves organizing and managing projects as a portfolio of investments that contribute to the entire enterprise's success. Studies show that executive support is crucial to project success, as are other factors like user involvement, an experienced project manager, and clear business objectives.

Project managers play a key role in helping projects and organizations succeed. They must perform various job duties, possess many skills, and continue to develop skills in project management, general management, and their application area, such as information technology. Soft skills, especially leadership, are particularly important for project managers.

The profession of project management continues to grow and mature. In the U.S., the military took the lead in project management and developed many tools, such as Gantt charts and network diagrams, but today people in virtually every industry around the globe use project management. The Project Management Institute (PMI) is an international professional society that provides certification as a Project Management Professional (PMP) and upholds a code of ethics. Today, hundreds of project management software products are available to assist people in managing projects.

DISCUSSION QUESTIONS

1. Why is there a new or renewed interest in the field of project management?
2. What is a project, and what are its main attributes? How is a project different from what most people do in their day-to-day jobs? What is the triple constraint?
3. What is project management? Briefly describe the project management framework, providing examples of stakeholders, knowledge areas, tools and techniques, and project success factors.
4. What is the role of the project manager? What are suggested skills for all project managers and for information technology project managers? Why is leadership so important for project managers? How is the job market for information technology project managers?
5. Briefly describe some key events in the history of project management. What role does the Project Management Institute play in helping the profession? What functions can you perform with project management software? What are some popular names of low-end, midrange, and high-end project management tools?

EXERCISES

1. Visit the Standish Group's Web site at *www.standishgroup.com*. Read one of the CHAOS articles or a similar report on information technology project management on their site or a similar site. Write a one-page summary of the report, its key conclusions, and your opinion of the report.
2. Find someone who works as a project manager or is a member of a project team. Prepare several interview questions, and then ask them your questions in person, via the phone, or via the Internet. Write a one- to two-page summary of your findings.
3. Search the Internet for the terms *project management*, *project management careers*, *project portfolio management*, and *information technology project management*. Write down the number of hits that you received for each of these phrases. Find at least three Web sites that provide interesting information on one of the topics. In a one- to two-page paper, summarize key information about these three Web sites as well as the Project Management Institute's Web site (*www.pmi.org*).

4. Find any example of a real project with a real project manager. Feel free to use projects in the media (the Olympics, television shows, movies, etc.) or a project from your work, if applicable. Write a one-page paper describing the project in terms of its scope, time, and cost goals. Discuss what went right and wrong on the project and the role of the project manager and sponsor. Also describe if the project was a success or not and why. Include at least one reference and cite it on the last page.

5. Skim through Appendix A on Microsoft Project 2003. Review information about Project 2003 from Microsoft's Web site (*www.microsoft.com/project*). Also, visit The Project Management Center (*www.infogoal.com/pmc*) and visit two other sites for project management software providers. Write a two-page paper answering the following questions:

a) What functions does project management software provide that you cannot do easily using other tools such as a spreadsheet or database?

b) How do the three different tools you reviewed compare, based on cost of the tool, key features, and other relevant criteria?

c) How can organizations justify investing in enterprise project management software?

SUGGESTED READINGS

1. Harvard Business Review. "Your Best Managers Lead *and* Manage," Harvard Business School Publishing Corporation (2003).

   This publication includes three articles: Abraham Zaleznik's "Managers and Leaders: Are They Different?"; Henry Mintzberg's "The Manager's Job: Folklore and Fact'; and Jonathan Gosling and Henry Mintzberg's "The Five Minds of a Manager." The main conclusion of these articles is that organizations need people who can manage and lead, as suggested in this chapter.

2. Ibbs, C. William and Young, H. Kwak. "Calculating Project Management's Return on Investment," *Project Management Journal* (*www.ce.berkeley.edu/pmroi/calculating-PMROI.pdf*) (June 2000).

   Many companies want proof that project management is a good investment. This paper describes a procedure to measure the return on investment for project management. Information is based on thirty-eight companies and government agencies in four different industries. William Ibbs also published a book called Quantifying the Value of Project Management *in 2002.*

3. Milosevic, Dragan. "Project Management Toolbox: Tools and Techniques for the Practicing Project Manager," John Wiley & Sons, Inc. (2003).

   This text includes detailed descriptions of more than fifty tools and techniques used in project management. Tools are organized to assist in project initiation, planning, implementation, and closure.

4. Musser, John. "Project Management Resources," Columbia University, (*www.columbia.edu/~jm2217/#SampleSDP*) (2004).

> *Professor John Musser at Columbia University has created an excellent Web site with links to hundreds of online references on various project management topics. You can also link to this site from the online companion for this text.*

5. Project Management Institute Web site (*www.pmi.org*).

> *PMI's Web site changes often, but it provides useful information on project management as well as links to other sites, several articles and publications, and their online bookstore. Students can join PMI at reduced rates and receive copies of* PM Network *magazine monthly and the* Project Management Journal *quarterly.*

6. Shenhar, Aaron J. and Dov Dvir. "Project Management Evolution: Past History and Future Research Directions," Proceedings of PMI Research Conference (2004).

> *PMI sponsored its first international research conference in Paris, France in June 2000, the second research conference in Seattle in July 2002, and the third in London in 2004. This article is one of many excellent articles from these proceedings. The authors discuss project management research in its historical perspective and offer insights into its evolution.*

7. Yourdon, Ed. "Surviving a Death March Project," *Software Development* (July 1997).

> *Ed Yourdon is famous for his books on software development, including a 1999 and 2004 text titled* Death March. *Yourdon offers practical advice on how to handle software projects that are plagued with problems from the start. He describes the importance of a project manager being able to negotiate schedules, budgets, and other aspects of a project with users, managers, and other stakeholders. Yourdon's latest book,* Brain Drain, *describes the effects of outsourcing on the software development industry.*

KEY TERMS

- **critical path** — the longest path through a network diagram that determines the earliest completion of a project
- **enterprise project management software** — software that integrates information from multiple projects to show the status of active, approved, and future projects across an entire organization
- **Gantt chart** — a standard format for displaying project schedule information by listing project activities and their corresponding start and finish dates in a calendar format

- **leader** — a person who focuses on long-term goals and big-picture objectives, while inspiring people to reach those goals
- **manager** — a person who deals with the day-to-day details of meeting specific goals
- **program** — a group of projects managed in a coordinated way to obtain benefits and control not available from managing them individually
- **project** — a temporary endeavor undertaken to create a unique product, service, or result
- **project management** — the application of knowledge, skills, tools, and techniques to project activities to meet project requirements
- **project manager** — the person responsible for working with the project sponsor, the project team, and the other people involved in a project to meet project goals
- **Project Management Institute (PMI)** — international professional society for project managers
- **project management knowledge areas** — project integration management, scope, time, cost, quality, human resource, communications, risk, and procurement management
- **Project Management Office (PMO)** — an organizational group responsible for coordinating the project management functions throughout an organization
- **Project Management Professional (PMP)** — certification provided by PMI that requires documenting project experience, agreeing to follow the PMI code of ethics, and passing a comprehensive exam
- **project management tools and techniques** — methods available to assist project managers and their teams; some popular tools in the time management knowledge area include Gantt charts, network diagrams, and critical path analysis
- **project portfolio management** — when organizations group and manage projects as a portfolio of investments that contribute to the entire enterprise's success
- **project sponsor** — the person who provides the direction and funding for a project
- **stakeholders** — people involved in or affected by project activities
- **triple constraint** — balancing scope, time, and cost goals

2

The Project Management and Information Technology Context

┌─ Objectives ┐

After reading this chapter, you will be able to:

1. *Describe the systems view of project management and how it applies to information technology projects*
2. *Understand organizations, including the four frames, organizational structures, and organizational culture*
3. *Explain why stakeholder management and top management commitment are critical for a project's success*
4. *Understand the concept of a project phase and the project life cycle and distinguish between project development and product development*
5. *Discuss the unique attributes and diverse nature of information technology projects*

OPENING CASE

om Walters recently accepted a new position at his college as the Director of Information Technology. Tom had been a respected faculty member at the college for the past fifteen years. The college—a small, private college in the Southwest—offered a variety of programs in the liberal arts and professional areas. Enrollment included 1500 full-time traditional students and about 1000 working-adult students attending an evening program. Many instructors supplemented their courses with information on the Internet and course Web sites, but they did not offer any distance-learning programs. The college's niche was serving students in that region who liked the setting of a small liberal arts college. Like most colleges, its use of information technology had grown tremendously in the past five years. There were a few classrooms on campus with computers for the instructors and students, and a few more with just instructor stations and projection systems. Tom knew that several colleges throughout the country required that all students lease laptops and that these colleges incorporated technology components into most courses. This idea fascinated him. He and two other members of the Information Technology department visited a local college that had required all students to lease laptops for the past three years, and they were very impressed with what they saw and heard. Tom and his staff developed plans to start requiring students to lease laptops at their college the next year.

Tom sent an e-mail to all faculty and staff in September, which briefly described this and other plans. He did not get much response, however, until the February faculty meeting when, as he described some of the details of his plan, the chairs of the History, English, Philosophy, and Economics departments all voiced their opposition to the idea. They eloquently stated that the college was not a technical training school, and that they thought the idea was ludicrous. Members of the Computer Science department voiced their concern that all of their students already had state-of-the art desktop computers and would not want to pay a mandatory fee to lease less-powerful laptops. The director of the adult education program expressed her concern that many adult-education students would balk at an increase in fees. Tom was in shock to hear his colleagues' responses, especially after he and his staff had spent a lot of time planning details of how to implement laptops at their campus. Now what should he do?

Many of the theories and concepts of project management are not difficult to understand. What *is* difficult is implementing them in various environments. Project managers must consider many different issues when managing projects. Just as each project is unique, so is its environment. This chapter discusses some of the components involved in understanding the project environment, such as using a systems approach, understanding organizations, managing stakeholders, matching product life cycles to the project environment, and understanding the context of information technology projects.

A SYSTEMS VIEW OF PROJECT MANAGEMENT

Even though projects are temporary and intended to provide a unique product or service, you cannot run projects in isolation. If project managers lead projects in isolation, it is unlikely that those projects will ever truly serve the needs of the organization. Therefore, projects must operate in a broad organizational environment, and project managers need to consider projects within the greater organizational context. To handle complex situations effectively, project managers need to take a holistic view of a project and understand how it relates to the larger organization. **Systems thinking** describes this holistic view of carrying out projects within the context of the organization.

What Is a Systems Approach?

The term **systems approach** emerged in the 1950s to describe a holistic and analytical approach to solving complex problems that includes using a systems philosophy, systems analysis, and systems management. A **systems philosophy** is an overall model for thinking about things as systems. **Systems** are sets of interacting components working within an environment to fulfill some purpose. For example, the human body is a system composed of many subsystems—the nervous system, the skeletal system, the circulatory system, the digestive system, and so on. **Systems analysis** is a problem-solving approach that requires defining the scope of the system, dividing it into its components, and then identifying and evaluating its problems, opportunities, constraints, and needs. Once this is completed, the systems analyst then examines alternative solutions for improving the current situation, identifies an optimum, or at least satisfactory, solution or action plan, and examines that plan against the entire system. **Systems management** addresses the business, technological, and organizational issues associated with creating, maintaining, and making a change to a system.

🎥 *Media Snapshot*

The Press Association Ltd., the largest news agency in the United Kingdom, appeared to have it all as it continued to grow and dominate its industry. However, management noticed its profit margins were sliding, so they hired a consulting firm to help turn things around. The consultants found the organization's rapid growth rate helped extend its reach throughout the country with the acquisition of several news organizations, but these disparate groups were not assimilated into overall operations. People were not sharing work processes and technologies across the organization, so operating costs were high.

To solve this problem, the consultants suggested using a top-down strategy to make sure projects were initiated to support key business goals. Richard Stoneham, principal for

Stratege Consulting of Sydney, Australia, stressed the importance of using a holistic view of the organization and its projects: *"If you approach everything from systems thinking, you see everything as connected...Successful change has to be planned and managed at the organizational level so the whole system is taken into account."* Stoneham also suggested releasing short-term results to accrue benefits on an incremental basis and reviewing projects on a regular basis to ensure strategic alignment.[1]

Using a systems approach is critical to successful project management. Top management and project managers must follow a systems philosophy to understand how projects relate to the whole organization. They must use systems analysis to address needs with a problem-solving approach. They must use systems management to identify key business, technological, and organizational issues related to each project in order to identify and satisfy key stakeholders and do what is best for the entire organization.

In the opening case, when Tom Walters planned the laptop project, he did not use a systems approach. Members of the Information Technology department did all of the planning. Even though Tom sent an e-mail describing the laptop project to all faculty and staff, he did not address many of the organizational issues involved in such a complex project. Most faculty and staff are very busy at the beginning of fall term and many may not have read the entire message. Others may have been too busy to communicate their concerns to the Information Technology department. Tom was unaware of the effects the laptop project would have on other parts of the college. He did not clearly define the business, technological, and organizational issues associated with the project. Tom and the Information Technology department began work on the laptop project in isolation. If they had taken a systems approach, considering other dimensions of the project, and involving key stakeholders, they could have identified and addressed many of the issues raised at the February faculty meeting *before* the meeting.

The Three-Sphere Model for Systems Management

Many business and information technology students understand the concepts of systems and performing a systems analysis. However, they often gloss over the topic of systems management. The simple idea of addressing the three spheres of systems management—business, organization, and technology—can have a huge impact on selecting and managing projects successfully.

Figure 2-1 provides a sample of some of the business, organizational, and technological issues that could be factors in the laptop project. In this case, technological issues, though not simple by any means, are probably the least difficult to identify and resolve. However, projects must address issues in all three spheres of

---

[1] Jackson, Lynne, "Forge Ahead," PM Network (April 2004), p.48.

the systems management model. Although it is easier to focus on the immediate and sometimes narrow concerns of a particular project, project managers and other staff must keep in mind the effects of any project on the interests and needs of the entire system or organization.

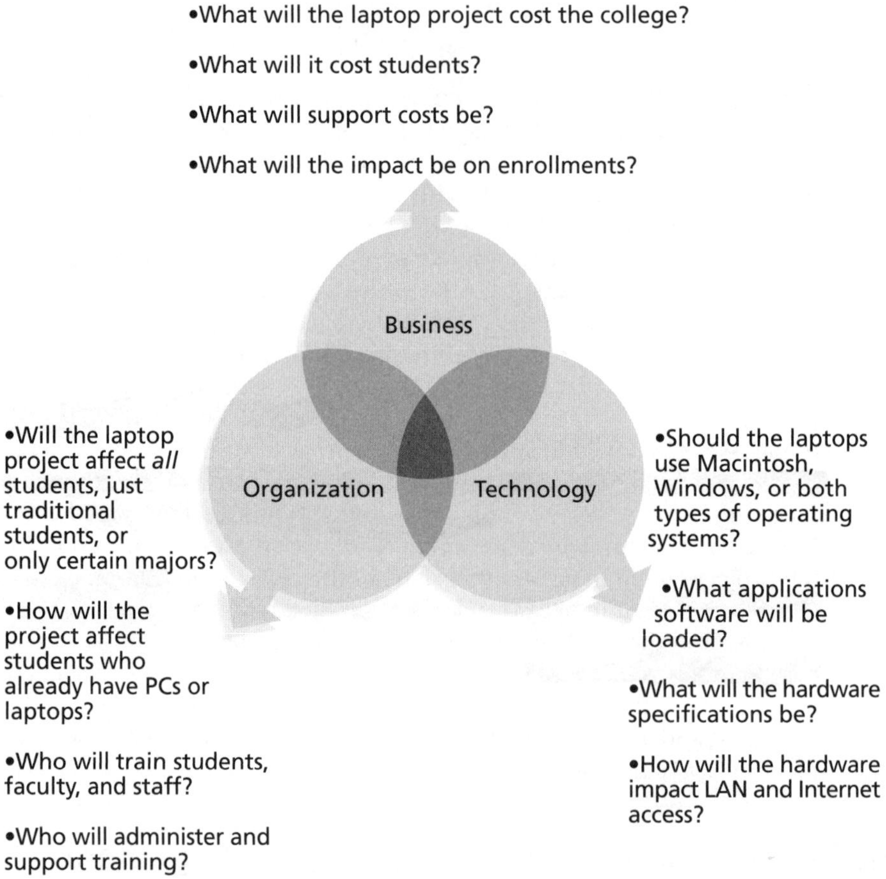

•What will the laptop project cost the college?

•What will it cost students?

•What will support costs be?

•What will the impact be on enrollments?

Business

Organization   Technology

•Will the laptop project affect *all* students, just traditional students, or only certain majors?

•How will the project affect students who already have PCs or laptops?

•Who will train students, faculty, and staff?

•Who will administer and support training?

•Should the laptops use Macintosh, Windows, or both types of operating systems?

•What applications software will be loaded?

•What will the hardware specifications be?

•How will the hardware impact LAN and Internet access?

Figure 2-1. Three-Sphere Model for Systems Management

Many information technology professionals become captivated with the technology and day-to-day problem solving involved in working with information systems. They tend to become frustrated with many of the "people problems" or politics involved in most organizations. In addition, many information technology professionals ignore important business issues—such as, "Does it make financial sense to pursue this new technology?" or, "Should the company develop this software in-house or purchase it off-the-shelf?" Using a more holistic approach helps project managers integrate business and organizational issues into their planning. It also helps them look at projects as a series of interrelated phases. When you integrate business and organizational issues into project planning and look at projects as a series of interrelated phases, you do a better job of ensuring project success.

UNDERSTANDING ORGANIZATIONS

The systems approach requires that project managers always view their projects in the context of the larger organization. Organizational issues are often the most difficult part of working on and managing projects. For example, many people believe that most projects fail because of company politics. Project managers often do not spend enough time identifying all the stakeholders involved in projects, especially the people opposed to the projects. Similarly, they often do not consider the political context of a project or the culture of the organization. To improve the success rate of information technology projects, it is important for project managers to develop a better understanding of people as well as organizations.

The Four Frames of Organizations

Organizations can be viewed as having four different frames: structural, human resources, political, and symbolic:[2]

- The **structural frame** deals with how the organization is structured (usually depicted in an organizational chart) and focuses on different groups' roles and responsibilities in order to meet the goals and policies set by top management. This frame is very rational and focuses on coordination and control. For example, within the structural frame, a key information technology issue is whether a company should centralize the information technology personnel in one department or decentralize across several departments. You will learn more about organizational structures in the next section.

- The **human resources frame** focuses on producing harmony between the needs of the organization and the needs of the people. It recognizes that there are often mismatches between the needs of the organization and the needs of individuals and groups and works to resolve any potential problems. For example, many projects might be more efficient for the organization if personnel worked 80 or more hours a week for several months. This work schedule would probably conflict with the personal lives of those people. Important issues in information technology related to the human resources frame are the shortage of skilled information technology workers within the organization and unrealistic schedules imposed on many projects.

- The **political frame** addresses organizational and personal politics. **Politics** in organizations take the form of competition among groups or individuals for power and leadership. The political frame assumes that organizations are

[2] Bolman, Lee G. and Deal, Terrence E., *Reframing Organizations*, Jossey-Bass Publishers, 1991.

coalitions composed of varied individuals and interest groups. Often, important decisions need to be made based on the allocation of scarce resources. Competition for scarce resources makes conflict a central issue in organizations, and power improves the ability to obtain scarce resources. Project managers must pay attention to politics and power if they are to be effective. It is important to know who opposes your projects as well as who supports them. Important issues in information technology related to the political frame are the power shifts from central functions to operating units or from functional managers to project managers.

■ The **symbolic frame** focuses on symbols and meanings. What is most important about any event in an organization is not what actually happened, but what it means. Was it a good sign that the CEO came to a kick-off meeting for a project, or was it a threat? The symbolic frame also relates to the company's culture. How do people dress? How many hours do they work? How do they run meetings? Many information technology projects are international and include stakeholders from various cultures. Understanding those cultures is also a crucial part of the symbolic frame.

⊗ *What Went Wrong?*

Several large organizations have installed or tried to install enterprise resource planning (ERP) systems to integrate business functions such as ordering, inventory, delivery, accounting, and human resource management. They understand the potential benefits of an ERP system and can analyze its various technical issues, but many companies do not realize how important the organizational issues are to ERP implementations.

For example, in early 2001, Sobey's, Canada's second largest grocery store chain with 1400 stores, abandoned its two-year, $90-million investment in an ERP system. The system was developed by SAP, the largest enterprise software company and the third-largest software supplier. Unfortunately, the system did not work properly due to several organizational challenges. Every department has to work together to implement an ERP system, and it is often difficult to get departments to communicate their needs. As Dalhousie University Associate Professor Sunny Marche states, "The problem of building an integrated system that can accommodate different people is a very serious challenge. You can't divorce technology from the sociocultural issues. They have an equal role." Sobey's ERP system shut down for five days and employees were scrambling to stock potentially empty shelves in several stores for weeks. The system failure cost Sobey's more than $90 million and caused shareholders to take an 82-cent after-tax hit per share.[3]

---

[3] Hoare, Eva. "Software hardships," The Herald, Halifax, Nova Scotia, Canada (2001).

Project managers must learn to work within all four organizational frames to function well in organizations. Chapter 9, Project Human Resource Management, and Chapter 10, Communications Management, further develop some of the organizational issues. The following sections on organizational structures, stakeholder management, and the need for top management commitment provide additional information related to the structural and political frames.

Organizational Structures

Many discussions of organizations focus on organizational structure. Three general classifications of organizational structures are functional, project, and matrix. Most companies today involve all three structures somewhere in the organization, but one is usually most common. Figure 2-2 portrays these three organizational structures. A **functional organizational structure** is the hierarchy most people think of when picturing an organizational chart. Functional managers or vice presidents in specialties such as engineering, manufacturing, information technology (IT), and human resources (HR) report to the chief executive officer (CEO). Their staffs have specialized skills in their respective disciplines. For example, most colleges and universities have very strong functional organizations. Only faculty in the Business department teach business courses; faculty in the History department teach history; faculty in the Art department teach art, and so on.

A **project organizational structure** also has a hierarchical structure, but instead of functional managers or vice presidents reporting to the CEO, program managers report to the CEO. Their staffs have a variety of skills needed to complete the projects within their programs. An organization that uses this structure earns their revenue primarily from performing projects for other groups under contract. For example, many defense, architectural, engineering, and consulting companies use a project organizational structure. These companies often hire people specifically to work on particular projects.

A **matrix organizational structure** represents the middle ground between functional and project structures. Personnel often report to both a functional manager and one or more project managers. For example, information technology personnel at many companies often split their time between two or more projects, but they report to their manager in the Information Technology department. Project managers in matrix organizations have staff from various functional areas working on their projects, as shown in Figure 2-2. Matrix organizational structures can be strong, weak, or balanced, based on the amount of control exerted by the project managers.

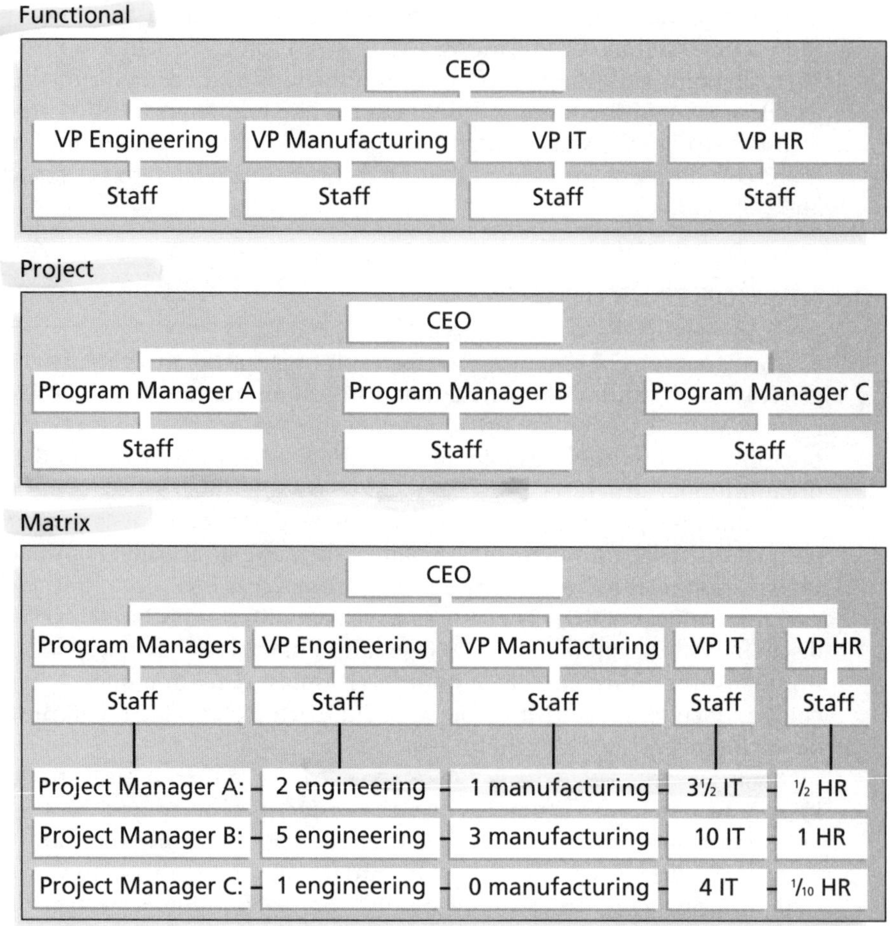

Functional

CEO			
VP Engineering	VP Manufacturing	VP IT	VP HR
Staff	Staff	Staff	Staff

Project

CEO		
Program Manager A	Program Manager B	Program Manager C
Staff	Staff	Staff

Matrix

CEO				
Program Managers	VP Engineering	VP Manufacturing	VP IT	VP HR
Staff	Staff	Staff	Staff	Staff

Project Manager A:	2 engineering	1 manufacturing	3½ IT	½ HR
Project Manager B:	5 engineering	3 manufacturing	10 IT	1 HR
Project Manager C:	1 engineering	0 manufacturing	4 IT	1/10 HR

Figure 2-2. Functional, Project, and Matrix Organizational Structures

Table 2-1 summarizes how organizational structures influence projects and project managers. Project managers have the most authority in a pure project organizational structure and the least amount of authority in a pure functional organizational structure. It is important that project managers understand the current organizational structure under which they are working. For example, if someone in a functional organization is asked to lead a project that requires strong support from several different functional areas, he or she should ask for top management sponsorship. This sponsor should solicit support from all relevant functional managers to ensure that they cooperate on the project and that qualified people are available to work as needed. The project manager might also ask for a separate budget to pay for project-related trips, meetings, and training or to provide financial incentives to the people supporting the project.

Table 2-1: Organizational Structure Influences on Projects

PROJECT CHARACTERISTICS	ORGANIZATIONAL STRUCTURE TYPE				
	FUNCTIONAL	MATRIX			PROJECT
		WEAK MATRIX	BALANCED MATRIX	STRONG MATRIX	
Project manager's authority	Little or none	Limited	Low to Moderate	Moderate to high	High to almost total
Percent of performing organization's personnel assigned full-time to project work	Virtually none	0–25%	15–60%	50–95%	85–100%
Who controls the project budget	Functional manager	Functional manager	Mixed	Project manager	Project manager
Project manager's role	Part-time	Part-time	Full-time	Full-time	Full-time
Common title for project manager's role	Project Coordinator/ Project Leader	Project Coordinator/ Project Leader	Project Manager/ Project Officer	Project Manager/ Program Manager	Project Manager/ Program Manager
Project management administrative staff	Part-time	Part-time	Part-time	Full-time	Full-time

PMBOK® Guide, 2000, p.19, and *PMBOK® Guide 2004,* p.28.

Even though project managers have the most authority in the project organizational structure, this type of organization is often inefficient for the company as a whole. Assigning staff full-time to the project often creates underutilization and/or misallocation of staff resources. For example, if a technical writer is assigned full-time to a project but there is no work for him or her on a particular day, the organization is wasting money by paying that person a full-time wage. Project organizations may also miss economies of scale available through the pooling of requests for materials with other projects.

Disadvantages such as these illustrate the benefit of using a systems approach to managing projects. For example, the project manager might suggest hiring an independent contractor to do the technical writing work instead of using a full-time employee. This approach would save the organization money while still meeting the needs of the project. When project managers use a systems approach, they are better able to make decisions that address the needs of the entire organization.

Organizational Culture

Just as an organization's structure affects its ability to manage projects, so does an organization's culture. **Organizational culture** is a set of shared assumptions, values, and behaviors that characterize the functioning of an organization. It often includes elements of all four frames described above. Organizational culture is very powerful, and many people believe the underlying causes of many companies' problems are not in the organizational structure or staff; they are in the culture. It is also important to note that the same organization can have different sub-cultures. The information technology department may have a different organizational culture than the finance department, for example. Some organizational cultures make it easier to manage projects.

According to Stephen P. Robbins, author of a popular textbook on organizational behavior (see the Suggested Readings at the end of this chapter), there are ten characteristics of organizational culture:

1. Member identity: The degree to which employees identify with the organization as a whole rather than with their type of job or profession. For example, a project manager or team member might feel more dedicated to his or her company or project team than to their job or profession, or they might not have any loyalty to a particular company or team. As you can guess, an organizational culture where employees identify more with the whole organization are more conducive to a good project culture.

2. Group emphasis: The degree to which work activities are organized around groups or teams, rather than individuals. An organizational culture that emphasizes group work is best for managing projects.

3. People focus: The degree to which management's decisions take into account the effect of outcomes on people within the organization. A project manager might assign tasks to certain people without considering their individual needs, or the project manager might know each person very well and focus on individual needs when assigning work or making other decisions. Good project managers often balance the needs of individuals and the organization.

4. Unit integration: The degree to which units or departments within an organization are encouraged to coordinate with each other. Most project managers strive for strong unit integration to deliver a successful product, service, or result. An organizational culture with strong unit integration makes the project manager's job easier.

5. Control: The degree to which rules, policies, and direct supervision are used to oversee and control employee behavior. Experienced project managers know it is often best to balance the degree of control to get good project results.

6. Risk tolerance: The degree to which employees are encouraged to be aggressive, innovative, and risk seeking. An organizational culture with a higher risk tolerance is often best for project management since projects often involve new technologies, ideas, and processes.

7. Reward criteria: The degree to which rewards, such as promotions and salary increases, are allocated according to employee performance rather than seniority, favoritism, or other nonperformance factors. Project managers and their teams often perform best when rewards are based mostly on performance.

8. Conflict tolerance: The degree to which employees are encouraged to air conflicts and criticism openly. It is very important for all project stakeholders to have good communications, so it is best to work in an organization where people feel comfortable discussing conflict openly.

9. Means-ends orientation: The degree to which management focuses on outcomes rather than on techniques and processes used to achieve results. An organization with a balanced approach in this area is often best for project work.

10. Open-systems focus: The degree to which the organization monitors and responds to changes in the external environment. As discussed earlier in this chapter, projects are part of a larger organizational environment, so it is best to have a strong open-systems focus.

As you can see, there is a definite relationship between organizational culture and successful project management. Project work is most successful in an organizational culture where employees identify more with the organization, where work activities emphasize groups, and where there is strong unit integration, high risk tolerance, performance-based rewards, high conflict tolerance, an open-systems focus, and a balanced focus on people, control, and means-orientation.

STAKEHOLDER MANAGEMENT

Recall from Chapter 1 that project stakeholders are the people involved in or affected by project activities. Stakeholders can be internal to the organization, external to the organization, directly involved in the project, or simply affected by the project. Internal project stakeholders generally include the project sponsor, project team, support staff, and internal customers for the project. Other internal stakeholders include top management, other functional managers, and other project managers. Since organizations have limited resources, projects affect top management, other functional managers, and other project managers by using some of the organization's limited resources. Thus, while additional internal stakeholders may not be directly involved in the project, they are still stakeholders because the project affects them in some way. External project stakeholders include the project's customers (if they are external to the organization), competitors, suppliers, and other external groups potentially involved in or affected by the project, such as government officials or concerned citizens. Since the purpose of project management is to meet project requirements and satisfy stakeholders, it is critical that project managers take adequate time to identify, understand, and manage relationships with all project stakeholders. Using the four frames of organizations to think about project stakeholders can help you meet their expectations.

Consider again the laptop project from the opening case. Tom Walters seemed to focus on just a few internal project stakeholders. He viewed only part of the structural frame of the college. Since his department would do most of the work in administering the laptop project, he focused on those stakeholders. Tom did not even involve the main customers for this project—the students at the college. Even though Tom sent an e-mail to faculty and staff, he did not hold meetings with senior administration or faculty at the college. Tom's view of who the stakeholders were for the laptop project was very limited.

During the faculty meeting, it became evident that the laptop project had many stakeholders in addition to the Information Technology department and students. If Tom had expanded his view of the structural frame of his organization by reviewing an organizational chart for the entire college, he could have identified other key stakeholders. He would have been able to see that the laptop project would affect academic department heads and members of different administrative areas. If Tom had focused on the human resources frame, he would have been able to tap his knowledge of the college and identify individuals who would most support or oppose requiring laptops. By using the political frame, Tom could have considered the main interest groups that would be most affected by this project's outcome. Had he used the symbolic frame, Tom could have tried to address what moving to a laptop environment would really mean for the college. He then could have anticipated some of the opposition from people who were not in favor of increasing the use of technology on campus. He also could have solicited a strong endorsement from the college president or dean before talking at the faculty meeting.

Tom Walters, like many new project managers, learned the hard way that his technical and analytical skills were not enough to guarantee success in project management. To be more effective, he had to identify and address the needs of different stakeholders and understand how his project related to the entire organization.

The Importance of Top Management Commitment

People in top management positions, of course, are key stakeholders in projects. A very important factor in helping project managers successfully lead projects is the level of commitment and support they receive from top management. In fact, without top management commitment, many projects will fail. Some projects have a senior manager called a **champion** who acts as a key proponent for a project. As described earlier, projects are part of the larger organizational environment, and many factors that might affect a project are out of the project manager's control. Several studies cite executive support as one of the key factors associated with virtually all project success.

Top management commitment is crucial to project managers for the following reasons:

- Project managers need adequate resources. The best way to kill a project is to withhold the required money, human resources, and visibility for the project. If project managers have top management commitment, they will also have adequate resources and not be distracted by events that do not affect their specific projects.

- Project managers often require approval for unique project needs in a timely manner. For example, on large information technology projects, top management must understand that unexpected problems may result from the nature of the products being produced and the specific skills of the people on the project team. For example, the team might need additional hardware and software halfway through the project for proper testing, or the project manager might need to offer special pay and benefits to attract and retain key project personnel. With top management commitment, project managers can meet these specific needs in a timely manner.

- Project managers must have cooperation from people in other parts of the organization. Since most information technology projects cut across functional areas, top management must help project managers deal with the political issues that often arise in these types of situations. If certain functional managers are not responding to project managers' requests for necessary information, top management must step in to encourage functional managers to cooperate.

- Project managers often need someone to mentor and coach them on leadership issues. Many information technology project managers come from technical positions and are inexperienced as managers. Senior managers should take the time to pass on advice on how to be good leaders. They should encourage new project managers to take classes to develop leadership skills and allocate the time and funds for them to do so.

Information technology project managers work best in an environment in which top management values information technology. Working in an organization that values good project management and sets standards for its use also helps project managers succeed.

The Need for Organizational Commitment to Information Technology

Another factor affecting the success of information technology projects is the organization's commitment to information technology in general. It is very difficult for a large information technology project (or a small one, for that matter) to be successful if the organization itself does not value information technology. Many companies have realized that information technology is integral to their business and have created a vice president or equivalent-level position for the head of information technology, often called the Chief Information Officer (CIO). Some companies assign people from non-information technology areas to work on large projects full-time to increase involvement from end users of the systems. Some CEOs even take a strong leadership role in promoting the use of information technology in their organizations.

The Gartner Group, Inc., a well-respected information technology consulting firm, awarded Boston's State Street Bank and Trust Company's CEO, Marshall Carter, the 1998 Excellence in Technology Award. Carter provided the vision and leadership for his organization to implement new information technology that successfully expanded the bank's business. They had to gather, coordinate, and analyze vast amounts of data from around the globe to provide new asset management services to their customers. It took six years to transform State Street Bank and Trust into a company providing state-of-the-art tools and services to its customers. The bank's revenues, profits, and earnings per share more than doubled during Carter's first five years as CEO. One key to Carter's success was his vision that technology was an integral part of the business and not just a means of automating old banking services. Carter used a highly personal style to keep his people motivated, and he often showed up at project review meetings to support his managers on information technology projects [4]

Recent finalists and winners of Gartner Group awards include Hard Rock Café, Boise Cascade Office Products, Telecom Italia Mobile (TIM), and Grupo Financiero Bital, a financial institution in Mexico. Whether a business is providing food and entertainment, office supplies, mobile technology, or financial services, information technology plays an important role in helping organizations compete in today's market.

[4] Melymuka, Kathleen, "Old Bank, New Ideas," *ComputerWorld* (February 15, 1999).

The Need for Organizational Standards

Another problem in most organizations is not having standards or guidelines to follow that could help in performing project management. These standards or guidelines might be as simple as providing standard forms or templates for common project documents, examples of good project plans, or guidelines on how the project manager should provide status information to top management. The content of a project plan and how to provide status information might seem like common sense to senior managers, but many new information technology project managers have never created plans or given a non-technical status report. Top management must support the development of these standards and guidelines and encourage or even enforce their use. For example, an organization might require all potential project information in a standard format to make project portfolio management decisions. If a project manager does not submit a potential project in the proper format, it could be rejected.

Some organizations invest heavily in project management by creating a project management office or center of excellence, as described in Chapter 1. A project management office or center of excellence is an organizational entity created to assist project managers in achieving project goals. Rachel Hollstadt, founder and CEO of a project management consulting firm, suggests that organizations consider adding a new position, a Chief Project Officer (CPO). Some organizations develop career paths for project managers. Some require that all project managers have Project Management Professional (PMP) certification and that all employees have some type of project management training. The implementation of all of these standards demonstrates an organization's commitment to project management.

PROJECT PHASES AND THE PROJECT LIFE CYCLE

Since projects operate as part of a system and involve uncertainty, it is good practice to divide projects into several phases. A **project life cycle** is a collection of project phases. Some organizations specify a set of life cycles for use on all of their projects, while others follow common industry practices based on the types of projects involved. In general, project life cycles define what work will be performed in each phase, what deliverables will be produced and when, who is involved in each phase, and how management will control and approve work produced in each phase. A **deliverable** is a product or service, such as a report, a training session, a piece of hardware, or a segment of software code, produced or provided as part of a project. See Chapter 5, Project Scope Management, for detailed information on deliverables.

In early phases of a project life cycle, resource needs are usually lowest and the level of uncertainty is highest. Project stakeholders have the greatest opportunity to influence the final characteristics of the project's products, services, or results during the early phases of a project life cycle. It is much more expensive to make major changes to a project during latter phases. During the middle phases of a project life cycle, the certainty of completing a project improves as a project continues, and more resources are usually needed than during the initial or final phase. The final phase of a project focuses on ensuring that project requirements were met and that the project sponsor approves completion of the project.

Project phases vary by project or industry, but some general phases in traditional project management are often called the concept, development, implementation, and close-out phases. The first two phases (concept and development) focus on planning and are often referred to as **project feasibility**. The last two phases (implementation and close-out) focus on delivering the actual work and are often referred to as **project acquisition**. A project should successfully complete each phase before moving on to the next. This project life cycle approach provides better management control and appropriate links to the ongoing operations of the organization.

Figure 2-3 provides a summary framework for the general phases of the traditional project life cycle. In the concept phase of a project, managers usually briefly describe the project—they develop a very high-level or summary plan for the project, which describes the need for the project and basic underlying concepts. A preliminary or rough cost estimate is developed in this first phase, and an overview of the work involved is created. A work breakdown structure (WBS) outlines project work and is a deliverable-oriented document that defines the total scope of the project. (You will learn more about the work breakdown structure in Chapter 5, Project Scope Management.) For example, if Tom Walters (from the opening case) had followed the project life cycle instead of moving full-steam ahead with the laptop project, he could have created a committee of faculty and staff to study the concept of increasing the use of technology on campus. This committee might have developed a management plan that included an initial, smaller project to investigate alternative ways of increasing the use of technology. They might have estimated that it would take six months and $20,000 to conduct a detailed technology study. The WBS at this phase of the study might have three levels and partition the work to include a competitive analysis of what five similar campuses were doing, a survey of local students, staff, and faculty, and a rough assessment of how using more technology would affect costs and enrollments. At the end of the concept phase, the committee would be able to deliver a report and presentation on its findings. The report and presentation would be an example of a deliverable.

After the concept phase is completed, the next project phase—development—begins. In the development phase, the project team creates more detailed project plans, a more accurate cost estimate, and a more thorough WBS. In the example under discussion, suppose the concept phase report suggested that requiring students to have laptops was one means of increasing the use of technology on campus. The project team could then further expand this idea in the development phase. They would have to decide if students would purchase or lease the laptops, what type of hardware and software the laptops would require, how much to charge students, how to handle training and maintenance, how to integrate the use of the new technology with current courses, and so on. If, however, the concept phase report showed that the laptop idea was not a good idea for the college, then the project team would no longer consider increasing the use of technology by requiring laptops in the development phase. This phased approach minimizes the time and money spent developing inappropriate projects. A project idea must pass the concept phase before evolving into the development phase.

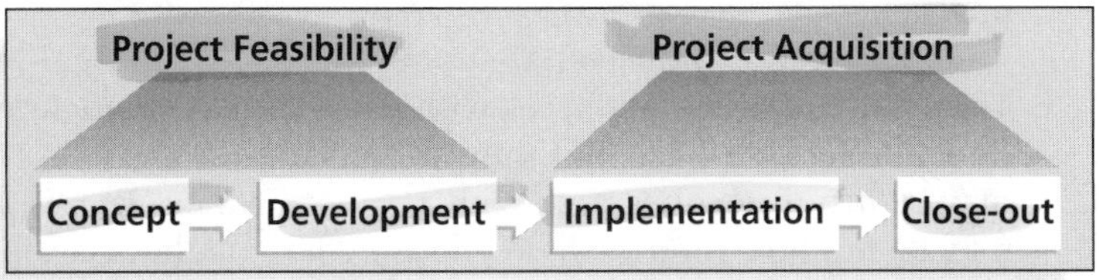

	Project Feasibility		Project Acquisition	
	Concept	**Development**	**Implementation**	**Close-out**
Sample deliverables for each phase	Management plan	Project plans	Last work package	Completed work
	Preliminary cost estimate	Budgetary cost estimate	Definitive cost estimate	Lessons learned
	2-level WBS	3+-level WBS	Performance reports	Customer acceptance

Figure 2-3. Phases of the Traditional Project Life Cycle

The third phase of the traditional project life cycle is implementation. In this phase, the project team creates a definitive or very accurate cost estimate, delivers the required work, and provides performance reports to stakeholders. Suppose Tom Walters' college took the idea of requiring students to have laptops through the development phase. During the implementation phase, the project team would need to obtain the required hardware and software, install the necessary network equipment, deliver the laptops to the students, create a process for collecting fees, provide training to students, faculty, and staff, and so on. Other people on campus would also be involved in the implementation phase. Faculty would need to consider how best to take advantage of the new technology. The recruiting staff would have to update their materials to reflect this new feature of the college. Security would need to address new problems that might result from having students carry around expensive equipment. The project team usually spends the bulk of their efforts and money during the implementation phase of projects.

The last phase of the traditional project life cycle is close-out. In the close-out phase, all of the work is completed, and there should be some sort of customer acceptance of the entire project. The project team should document their experiences on the project in a lessons-learned report. If the laptop idea made it all the way through the implementation phase and all students received laptops, the project team would then complete the project by closing out any related activities. They might administer a survey to students, faculty, and staff in order to gather opinions on how the project fared. They would ensure that any contracts with suppliers were completed and appropriate payments made. They would transition future work related to the laptop project to other parts of the organization. The project team could also share their lessons-learned report with other college campuses that are considering implementing a similar program.

Many projects, however, do not follow this traditional project life cycle. They still have general phases with some similar characteristics as the traditional project life cycle, but they are much more flexible. For example, there may be just three phases, the initial, intermediate, and final phase. Or there may be multiple intermediate phases. There might be a separate project just to complete a feasibility study. Regardless of the project life cycle's specific phases, it is good practice to think of projects as having phases that connect the beginning and the end of the project, so that people can measure progress toward achieving project goals during each phase.

Just as a *project* has a life cycle, so does a *product*. Information technology projects help produce products and services such as new software, hardware, networks, research reports, and training on new systems. Understanding the product life cycle is just as important to good project management as understanding the phases of the traditional project life cycle.

Product Life Cycles

All products follow some type of life cycle—cars, buildings, even amusement parks. The Disney Corporation, for example, follows a rigorous process to design, build, and test new products. They assign project managers to oversee the development of all new products, such as rides, parks, and cruise lines. Likewise, major automotive companies follow product life cycles to produce new cars, trucks, and other products. Most information technology professionals are familiar with the concept of a product life cycle, especially for developing software.

Software development projects are one subset of information technology projects. In general, information technology projects involve researching, analyzing, and then purchasing and installing new hardware and software with little or no actual software development required. However, some projects involve minor software modifications to enhance existing software or to integrate one application with another. Other projects involve a major amount of software development. Many argue that developing software requires project managers to modify traditional project management methods, depending on a particular product's life cycle.

A **systems development life cycle (SDLC)** is a framework for describing the phases involved in developing information systems. Some popular models of a systems development life cycle include the waterfall model, the spiral model, the incremental build model, the prototyping model, and the Rapid Application Development (RAD) model. These life cycle models are examples of a **predictive life cycle**, meaning that the scope of the project can be clearly articulated and the schedule and cost can be accurately predicted. The project team spends a large portion of the project effort attempting to clarify the requirements of the entire system and then producing a design. Users are often unable to see any tangible results in terms of working software for an extended period. Below are brief descriptions of several predictive SDLC models: [5]

- The waterfall life cycle model has well-defined, linear stages of systems development and support. This life cycle model assumes that requirements will remain stable after they are defined.

- The spiral life cycle model was developed based on experience with various refinements of the waterfall model as applied to large government software projects. It recognizes the fact that most software is developed using an iterative or spiral approach rather than a linear approach.

- The incremental build life cycle model provides for progressive development of operational software, with each release providing added capabilities.

- The prototyping life cycle model is used for developing software prototypes to clarify user requirements for operational software. It requires heavy user involvement, and developers use a model to generate functional requirements and physical design specifications simultaneously. Developers can throw away or keep prototypes, depending on the project.

- The Rapid Application Development (RAD) life cycle model uses an approach in which developers work with an evolving prototype. This life cycle model also requires heavy user involvement and helps produce systems quickly without sacrificing quality. Developers use RAD tools such as CASE (Computer Aided Software Engineering), JRP (Joint Requirements Planning), and JAD (Joint Application Design) to facilitate rapid prototyping and code generation.

In contrast to the predictive life cycle models, the **Adaptive Software Development (ASD)** life cycle model assumes that software development follows an adaptive approach because the requirements cannot be clearly expressed early in the life cycle. Important attributes of this approach are that the projects are mission driven and component based, using time-based cycles

---

[5] Desaulniers, Douglas H. and Robert J. Anderson, "Matching Software Development Life Cycles to the Project Environment," Proceedings of the Project Management Institute Annual Seminars & Symposium, (November 1–10, 2001).

to meet target dates. Requirements are developed using an iterative approach, and development is risk driven and change tolerant to address and incorporate rather than mitigate risks. More recently, the term **agile software development** has become popular to describe new approaches that focus on close collaboration between programming teams and business experts. Two popular ASD life cycle models are extreme programming and Scrum.

■ **Extreme programming (XP)** has become a popular buzzword, although some people use the term agile development to describe extreme programming. This life cycle model meets the needs of people developing software in rapidly changing environments. An XP development team includes software developers, managers, and user representatives. One unique feature of the XP life cycle model is that developers program in pairs to promote synergy and increase productivity. Another unique feature of XP is that software developers must write the tests for their own code. XP development teams emphasize user satisfaction by producing tangible results in a short time while promoting teamwork and collective code ownership. Some disadvantages of the XP life cycle model are that it does not work for large projects (more than twenty people), it requires full-time user representatives, it assumes developers are highly disciplined, and it will not work if the team must predict the cost and schedule at the project's onset and then try to meet those commitments. The term extreme project management loosely describes how to manage these types of projects.

■ **Scrum** is similar to other life cycle models that use iterative development to address changing requirements, but in the Scrum life cycle model, the repetitions are referred to as sprints, which normally last thirty days. Each day the entire team meets for a short meeting, called a scrum, where they decide what to accomplish that day. Team members identify any obstacles they might have in accomplishing their goals, and the project manager must resolve and remove those obstacles. Project managers can use Scrum on large development projects because each team focuses on a discrete piece of work or object, with clear interfaces and behaviors. Scrum works best for projects using object-oriented technology and requires strong leadership to coordinate the work of each smaller team.

These life cycle models are all examples of systems development life cycles. Many Web sites and introductory management information systems texts describe each of them in detail. The type of software and complexity of the information system in development determines which life cycle model to use. It is important to understand the product life cycle to meet the needs of the project environment.

Most large information technology products are developed as a series of projects. For example, the systems planning phase for a new information system can include a project to hire an outside consulting firm to help identify and evaluate potential strategies for developing a particular business application, such as a new order processing system or general ledger system. It can also include a project to develop, administer, and evaluate a survey of users to get their opinions on the current information systems used for performing that business function in the organization. The systems analysis phase might include a project to create process models for certain business functions in the organization. It can also include a project to create data models of existing databases in the company related to the business function and application. The implementation phase might include a project to hire contract programmers to code a part of the system. The close-out phase might include a project to develop and run several training sessions for users of the new application. All of these examples show that large information technology projects are usually composed of several smaller projects. It is often good practice to view large projects as a series of smaller, more manageable ones, especially when there is a lot of uncertainty involved. Successfully completing one small project at a time will help the project team succeed in completing the larger project.

Because some aspects of project management need to occur during each phase of the product life cycle, it is critical for information technology professionals to understand and practice good project management throughout the product life cycle.

The Importance of Project Phases and Management Reviews

Due to the complexity and importance of many information technology projects and their resulting products, it is important to take time to review the status of a project at each phase. A project should successfully pass through each of the main project or product phases before continuing to the next. Since the organization usually commits more money as a project continues, a management review should occur after each phase to evaluate progress, potential success, and continued compatibility with organizational goals. These management reviews, called **phase exits** or **kill points**, are very important for keeping projects on track and determining if they should be continued, redirected, or terminated. Recall that projects are just one part of the entire system of an organization. Changes in other parts of the organization might affect a project's status, and a project's status might likewise affect what is happening in other parts of the organization. By breaking projects into phases, top management can make sure that the projects are still compatible with the needs of the rest of the company.

Let's take another look at the opening case. Suppose Tom Walters' college did a study on increasing the use of technology that was sponsored by the college president. At the end of the concept phase, the project team could have presented information to the faculty, president, and other staff members that described different options for increasing the use of technology, an analysis of what competing colleges were doing, and results of a survey of local stakeholders' opinions on the subject. This presentation at the end of the concept phase represents one form of a management review. Suppose the study reported that 90 percent of students, faculty, and staff surveyed strongly opposed the idea of requiring all students to have laptops and that many adult students said they would attend other colleges if they were required to pay for the additional technology. The college would probably decide not to pursue this idea any further. Had Tom taken a phased approach, he and his staff would not have wasted the time and money it took to develop detailed plans.

In addition to formal management reviews, it is important to have top management involvement throughout the life cycle of most projects. It is unwise to wait for the end of project or product phases to have management inputs. Many projects are reviewed by management on a regular basis, such as weekly or even daily, to make sure they are progressing well. Everyone wants to be successful in accomplishing goals at work, and having management involvement ensures that they are on track in accomplishing both project and organizational goals.

☑ *What Went Right?*

Having specific deliverables and kill points at the end of project or product phases helps managers make better decisions about whether to proceed, redefine, or kill a project. Improvement in information technology project success rates reported by the Standish Group has been due, in part, to an increased ability to know when to cancel failing projects. Standish Group Chairman Jim Johnson made the following observation:"The real improvement that I saw was in our ability to—in the words of Thomas Edison—know when to stop beating a dead horse . . . Edison's key to success was that he failed fairly often; but as he said, he could recognize a dead horse before it started to smell . . . In information technology we ride dead horses—failing projects—a long time before we give up. But what we are seeing now is that we are able to get off them; able to reduce cost overrun and time overrun. That's where the major impact came on the success rate."[6]

Another example of the power of management oversight comes from Huntington Bancshares, Inc. This company, like many others, had an **executive steering committee**, a group of senior executives from various parts of the organization, who regularly reviewed important corporate projects and issues. This Ohio-based, $26 billion bank holding company completed a year-long Web site redesign effort using XML technology to give its online customers access to real-time account information as well as other banking services. The CIO, Joe Gottron, said there were "four or five very intense moments" when the whole project was

---

[6] Cabanis, Jeannette, "A Major Import: The Standish Group's Jim Johnson On Project Management and IT Project Success," *PM Network*, PMI, (September 1998), p.7.

almost stopped due to its complexity. An executive steering committee met weekly to review the project's progress and discuss work planned for the following week. Gottron said the meetings ensured that "if we were missing a beat on the project, no matter which company [was responsible], we were on top of it and adding additional resources to make up for it."[7]

THE CONTEXT OF INFORMATION TECHNOLOGY PROJECTS

As described earlier, software development projects can follow several different product life cycles based on the project context. There are several other issues related to managing information technology projects. In fact, several groups have recognized the additional knowledge that information technology project managers must possess by creating separate certification programs specifically for them. PMI launched a new Certificate of Added Qualification (CAQ) exam in 2002 that PMPs can take to demonstrate their knowledge of managing projects in information technology. CompTIA's Project+ exam includes project scenarios and information specific to working in the information technology industry. Appendix B includes more information on these exams. This section highlights some of the issues unique to the information technology industry that affect project management, including the nature of projects, the characteristics of project team members, and the diverse nature of technologies involved.

The Nature of Information Technology Projects

Unlike projects in many other industries, projects labeled as information technology projects can be very diverse. Some involve a small number of people installing off-the-shelf hardware and associated software. Others involve hundreds of people analyzing several organizations' business processes and then developing new software in a collaborative effort with users to meet business needs. Even for small hardware-oriented projects, there is a wide diversity in the types of hardware that could be involved—personal computers, mainframe computers, network equipment, kiosks, or small mobile devices, to name a few. The network equipment might be wireless, phone-based, cable-based, or require a satellite connection. The nature of software development projects is even more diverse than hardware-oriented projects. A software development project might include developing a simple, stand-alone Excel or Access application or a sophisticated, global e-commerce system using state-of-the-art programming languages.

[7] Mearian, Lucas, "Bank Hones Project Management Skills With Redesign," *ComputerWorld*, (April 29, 2002).

Information technology projects also support every possible industry and business function. Managing an information technology project for a film company's animation department would require different knowledge and skills of the project manager and team members than a project to improve a federal tax collection system or install a communication infrastructure in a third-world country. Because of the diversity of information technology projects and the newness of the field, it is important to develop and follow best practices in managing these varied projects. That way, information technology project managers will have a common starting point and method to follow with every project.

Characteristics of Information Technology Project Team Members

Because of the nature of information technology projects, the people involved come from very diverse backgrounds and possess different skill sets. Most trade schools, colleges, and universities did not start offering degrees in computer technology, computer science, management information systems, or other information technology areas until the 1970s. Therefore, many people in the field do not have a common educational background. Many companies purposely hire graduates with degrees in other fields such as business, mathematics, or the liberal arts to provide different perspectives on information technology projects. Even with these different educational backgrounds, there are some common job titles for people working on most information technology projects, such as business analyst, programmer, network specialist, database analyst, quality assurance expert, technical writer, security specialist, hardware engineer, software engineer, and system architect. Within the category of programmer, there are several other job titles used to describe the specific technologies the programmer uses, such as Java programmer, XML programmer, C/C++ programmer, and so on.

Some information technology projects require the skills of people in just a few of these job functions, but many require inputs from many or all of them. Occasionally, information technology professionals move around between these job functions, but more often people become technical experts in one area or they decide to move into a management position. It is also rare for technical specialists or project managers to remain with the same company for a long time, and in fact, many information technology projects include a large number of contract workers. Working with this "army of free agents," as Rob Thomsett, author and consultant for the Cutter Consortium, calls them, creates special challenges. (See the Suggested Readings at the end of this chapter for an article on this topic by Thomsett.)

Diverse Technologies

Many of the job titles for information technology professionals reflect the different technologies required to hold that position. Unfortunately, hardware specialists might not understand the language of database analysts, and vice versa. Security specialists may have a hard time communicating with business analysts. It is also unfortunate that people within the same information technology job function often do not understand each other because each uses different technology. For example, someone with the title of programmer can often use several different programming languages. However, a COBOL programmer cannot be of much help on a Java project. These highly specialized positions also make it difficult for project managers to form and lead project teams.

Another problem with diverse technologies is that many of them change rapidly. A project team might be close to finishing a project when they discover a new technology that can greatly enhance the project and better meet long-term business needs. New technologies have also shortened the timeframe many businesses have to develop, produce, and distribute new products and services. This fast-paced environment requires equally fast-paced processes to manage and produce information technology projects and products.

As you can see, working as an information technology project manager or team member is an exciting and challenging job. It's important to focus on successfully completing projects that will have a positive impact on the organization as a whole.

CASE WRAP-UP

After several people voiced concerns about the laptop idea at the faculty meeting, the president of the college directed that a committee be formed to formally review the concept of requiring students to have laptops in the near future. Because the college was dealing with several other important enrollment-related issues, the president named the vice president of enrollment to head the committee. Other people soon volunteered or were assigned to the committee, including Tom Walters as head of Information Technology, the director of the adult education program, the chair of the Computer Science department, and the chair of the History department. The president also insisted that the committee include at least two members of the student body. The president knew everyone was busy, and he questioned whether the laptop idea was a high-priority issue for the college. He directed the committee to present a proposal at the next month's faculty meeting, either to recommend the creation of a formal project team (of which these committee members would commit to be a part) to fully investigate requiring laptops, or to recommend terminating the concept. At the next faculty meeting, few people were surprised to hear the recommendation to terminate the concept. Tom Walters learned that he had to pay much more attention to the needs of the entire college before proceeding with detailed information technology plans.

CHAPTER SUMMARY

Projects operate in an environment broader than the project itself. Project managers need to take a systems approach when working on projects; they need to consider projects within the greater organizational context.

Organizations have four different frames: structural, human resources, political, and symbolic. Project managers need to understand all of these aspects of organizations to be successful. The structural frame focuses on different groups' roles and responsibilities to meet the goals and policies set by top management. The human resources frame focuses on producing harmony between the needs of the organization and the needs of people. The political frame addresses organizational and personal politics. The symbolic frame focuses on symbols and meanings.

The structure of an organization has strong implications for project managers, especially in terms of the amount of authority the project manager has. The three basic organizational structures include functional, matrix, and project. Project managers have the most authority in a pure project organization, an intermediate amount of authority in a matrix organization, and the least amount of authority in a pure functional organization.

Organizational culture also affects project management. A culture where employees have a strong identity with the organization, where work activities emphasize groups, where there is strong unit integration, high risk tolerance, performance-based rewards, high conflict tolerance, an open-systems focus, and a balance on the dimensions of people focus, control, and means-orientation is more conducive to project work.

Project stakeholders are individuals and organizations who are actively involved in the project or whose interests may be positively or negatively affected because of project execution or successful project completion. Project managers must identify and understand the different needs of all stakeholders on their projects.

Top management commitment is crucial for project success. Since projects often affect many areas in an organization, top management must assist project managers if they are to do a good job of project integration. Organizational commitment to information technology is also important to the success of information technology projects. Development standards and guidelines assist most organizations in managing projects.

A project life cycle is a collection of project phases. Traditional project phases include concept, development, implementation, and close-out phases. Projects often produce products, which follow product life cycles. Examples of product life cycles for software development include the waterfall, spiral, incremental build, prototyping, RAD, extreme programming, and Scrum models. Project managers must understand the specific life cycle of the products they are producing as well as the general project life cycle model.

A project should successfully pass through each of the project phases in order to continue to the next phase. A management review should occur at the end of each project phase, and more frequent management inputs are often needed. These management reviews and inputs are important for keeping projects on track and determining if projects should be continued, redirected, or terminated.

Project managers need to consider several factors due to the unique context of information technology projects. The diverse nature of these projects and the wide range of business areas and technologies involved make information technology projects especially challenging to manage. Leading project team members with a wide variety of specialized skills and understanding rapidly changing technologies are also important considerations.

DISCUSSION QUESTIONS

1. What does it mean to take a systems view of a project? How does taking a systems view of a project apply to project management?

2. Explain the four frames of organizations. How can they help project managers understand the organizational context for their projects?

3. Briefly explain the differences between functional, matrix, and project organizations. Describe how each structure affects the management of the project.

4. Describe how organizational culture is related to project management. What type of culture promotes a strong project environment?

5. Discuss the importance of top management commitment and the development of standards for successful project management. Provide examples to illustrate the importance of these items based on your experience on any type of project.

6. What are the phases in a traditional project life cycle? How does a project life cycle differ from a product life cycle? Why does a project manager need to understand both?

7. What makes information technology projects different from other types of projects? How should project managers adjust to these differences?

EXERCISES

1. Summarize the three-sphere model of systems management in your own words. Then use your own experience or interview someone who recently completed an information technology project and list several business, technology, and organizational issues addressed during the project. Which issues were most important to the project and why? Summarize your answers in a two-page paper.

2. Apply the four frames of organizations to an information technology project with which you are familiar. If you cannot think of a good information technology project, use your personal experience in deciding where to attend college to apply this framework. Write a one- to two-page paper describing key issues related to the structural, human resources, political, and symbolic frames. Which frame seemed to be the most important and why? For example, did you decide where to attend college primarily because of the curriculum and structure of the program? Did you follow your friends? Did your parents have a lot of influence in your decision? Did you like the culture of the campus?

3. Read one of the suggested readings or search the Internet for an interesting article about software development life cycles, including agile software development. Also, find two informative Web sites related to this subject. What do these sources say about project management? Write a one- to two-page summary of your findings.

4. Search the Internet and scan information technology industry magazines or Web sites to find an example of an information technology project that had problems due to organizational issues. Write a one- to two-page paper summarizing who the key stakeholders were for the project and how they influenced the outcome.

5. Write a one- to two-page summary of an article about the importance of top management support for successful information technology projects.

SUGGESTED READINGS

1. Bolman, Lee G. and Deal, Terrence E. *Reframing Organizations*. Jossey-Bass Publishers (1991).

    This book describes the four frames of organizations: structural, human resources, political, and symbolic. It explains how to look at situations from more than one vantage point to help bring order out of confusion.

2. Desaulniers, Douglas H. and Anderson, Robert J. "Matching Software Development Life Cycles to the Project Environment," Proceedings of the Project Management Institute Annual Seminars & Symposium, (November 1–10, 2001).

    Software development projects require special attention to the product life cycle. This article provides an excellent summary of several software development life cycles and the importance of using the correct one for your project environment.

3. Highsmith, Jim. "Agile Project Management: Creating Innovative Products," Pearson Education (2004).

    This book describes how a variety of projects can be better managed by applying agile principles and practices. The author outlines the five phases in the agile process management framework: envision, speculate, explore, adapt, and close. He also identifies specific practices for each phase that align with agile values and guiding principles.

4. McConnell, Steve. "10 Myths of Rapid Software Development," Project Management Institute Information Systems Specific Interest Group, March 2002.

    McConnell has written several books and articles on rapid software development. McConnell developed this presentation to highlight common myths about the subject. Visit Construx's Web site (www.construx.com) for further information and access to free resources to advance the field of software engineering.

5. Robbins, Stephen P. "Organizational Behavior, 10th Edition," Prentice Hall (2002).

    This comprehensive textbook presents a reality-based review of organizational behavior. It helps readers to understand and predict behavior in organizations at three levels—the individual, the group, and the organization system. It includes detailed information on organizational culture.

6. Thomsett, Rob. "Extreme Project Management," *Cutter Consortium* (*www.cutter.com*) (2002).

    Rob Thomsett, an author and senior consultant for the Cutter Consortium, suggests the need for a new paradigm on project management called extreme project management. In this article, he discusses how three major forces—a power shift, the free-agent army, and the global e-economy— have changed the way projects should be managed. Thomsett also wrote a book in 2002 called Radical Project Management.

KEY TERMS

- **adaptive software development (ASD)** — a software development approach used when requirements cannot be clearly expressed early in the life cycle
- **agile software development** — a method for software development that uses new approaches, focusing on close collaboration between programming teams and business experts
- **champion** — a senior manager who acts as a key proponent for a project
- **deliverable** — a product or service, such as a report, a training session, a piece of hardware, or a segment of software code, produced or provided as part of a project
- **executive steering committee** — a group of senior executives from various parts of the organization who regularly review important corporate projects and issues
- **extreme programming (XP)** — an approach to programming for developing software in rapidly changing environments
- **functional organizational structure** — an organizational structure that groups people by functional areas such as information technology, manufacturing, engineering, and human resources
- **human resources frame** — focuses on producing harmony between the needs of the organization and the needs of people
- **matrix organizational structure** — an organizational structure in which employees are assigned to both functional and project managers
- **organizational culture**— a set of shared assumptions, values, and behaviors that characterize the functioning of an organization
- **phase exit** or **kill point** — management review that should occur after each project phase to determine if projects should be continued, redirected, or terminated
- **political frame** — addresses organizational and personal politics
- **politics** — competition between groups or individuals for power and leadership
- **predictive life cycle** — a software development approach used when the scope of the project can be clearly articulated and the schedule and cost can be accurately predicted
- **project acquisition** — the last two phases in a project (implementation and close-out) that focus on delivering the actual work
- **project feasibility** — the first two phases in a project (concept and development) that focus on planning

- **project life cycle** — a collection of project phases, such as concept, development, implementation, and close-out
- **project organizational structure** — an organizational structure that groups people by major projects, such as specific aircraft programs
- **Scrum** — an iterative development process for software where the iterations are referred to as sprints, which normally last thirty days
- **structural frame** — deals with how the organization is structured (usually depicted in an organizational chart) and focuses on different groups' roles and responsibilities to meet the goals and policies set by top management
- **symbolic frame** — focuses on the symbols, meanings, and culture of an organization
- **systems** — sets of interacting components working within an environment to fulfill some purpose
- **systems analysis** — a problem-solving approach that requires defining the scope of the system to be studied, and then dividing it into its component parts for identifying and evaluating its problems, opportunities, constraints, and needs
- **systems approach** — a holistic and analytical approach to solving complex problems that includes using a systems philosophy, systems analysis, and systems management
- **systems development life cycle (SDLC)** — a framework for describing the phases involved in developing and maintaining information systems
- **systems management** — addressing the business, technological, and organizational issues associated with creating, maintaining, and making a change to a system
- **systems philosophy** — an overall model for thinking about things as systems
- **systems thinking** — taking a holistic view of an organization to effectively handle complex situations

3

The Project Management Process Groups: A Case Study

Objectives

After reading this chapter, you will be able to:

1. *Describe the five project management process groups, the typical level of activity for each, and the interactions among them*
2. *Understand how the project management process groups relate to the project management knowledge areas*
3. *Discuss how organizations develop information technology project management methodologies to meet their needs*
4. *Review a case study of an organization applying the project management process groups to manage an information technology project, and understand the contribution that effective project initiation, project planning, project execution, project monitoring and controlling, and project closing make to project success*

OPENING CASE

*E*rica Bell was in charge of the Project Management Office (PMO) for her consulting firm. The firm, JWD (Job Well Done) Consulting, had grown to include more than 200 full-time consultants and even more part-time consultants. JWD Consulting provides a variety of consulting services to assist organizations in selecting and managing information technology projects. The firm focuses on finding and managing high-payoff projects and developing strong metrics to measure project performance and benefits to the organization after the project is implemented. The firm's emphasis on metrics and working collaboratively with its customers gives it an edge over many competitors.

Joe Fleming, the CEO, wanted his company to continue to grow and become a world-class consulting organization. Since the core of the business was helping other organizations with project management, he felt it was crucial for JWD Consulting to have an exemplary process for managing its own projects. He asked Erica to work with her team and other consultants in the firm to develop an intranet where they could share their project management knowledge. He also thought it would make sense to make some of the information available to the firm's clients. For example, the firm could provide project management templates, tools, articles, links to other sites, and an "Ask the Expert" feature to help build relationships with current and future clients. Since JWD Consulting emphasizes the importance of high-payoff projects, Joe also wanted to see a business case for this project before proceeding.

Recall from Chapter 1 that project management consists of nine knowledge areas: integration, scope, time, cost, quality, human resources, communications, risk, and procurement. Another important concept to understand is that projects involve five project management process groups: initiating, planning, executing, monitoring and controlling, and closing. Tailoring these process groups to meet individual project needs increases the chance of success in managing projects. This chapter describes each project management process group in detail through a simulated case study based on JWD Consulting. Although you will learn more about each knowledge area in Chapters 4 though 12, it's important first to learn how they fit into the big-picture of managing a project. Understanding how the knowledge areas and project management process groups function together will lend context to the remaining chapters.

PROJECT MANAGEMENT PROCESS GROUPS

Project management is an integrative endeavor; decisions and actions taken in one knowledge area at a certain time usually affect other knowledge areas. Managing these interactions often requires making trade-offs among the project's scope, time, and cost—the triple constraint of project management described in Chapter 1. A project manager may also need to make trade-offs between other knowledge areas, such as between risk and human resources. Consequently, you can view project management as a number of related processes.

- A **process** is a series of actions directed toward a particular result. **Project management process groups** progress from initiation activities to planning activities, executing activities, monitoring and controlling activities, and closing activities. **Initiating processes** include defining and authorizing a project or project phase. To initiate a project or just the concept phase of a project, someone must define the business need for the project, someone must sponsor the project, and someone must take on the role of project manager. Initiating processes take place during *each* phase of a project. Therefore, you cannot equate process groups with project phases. Recall that there can be different project phases, but all projects will include all five process groups. For example, project managers and teams should reexamine the business need for the project during every phase of the project life cycle to determine if the project is worth continuing. Initiating processes are also required to end a project. Someone must initiate activities to ensure that the project team completes all the work, documents lessons learned, reassigns project resources, and that the customer accepts the work.

- **Planning processes** include devising and maintaining a workable scheme to ensure that the project addresses the organization's needs. There normally is no single "project plan." There are several project plans, such as the scope management plan, schedule management plan, cost management plan, procurement management plan, and so on, defining each knowledge area as it relates to the project at that point in time. For example, a project team must develop a plan to define the work that needs to be done for the project, to schedule activities related to that work, to estimate costs for performing the work, to decide what resources to procure to accomplish the work, and so on. To account for changing conditions on the project and in the organization, project teams often revise project plans during each phase of the project life cycle. The project management plan, described in Chapter 4, coordinates information from all other plans.

- **Executing processes** include coordinating people and other resources to carry out the project plans and produce the products, services, or results of the project or phase. Examples of executing processes include developing the project team, directing and managing the project team, performing quality assurance, distributing information, and selecting sellers.

- **Monitoring and controlling processes** include regularly measuring and monitoring progress to ensure that the project team meets the project objectives. The project manager and staff monitor and measure progress against the plans and take corrective action when necessary. A common monitoring and controlling process is performance reporting, where project stakeholders can identify any necessary changes that may be required to keep the project on track.

- **Closing processes** include formalizing acceptance of the project or project phase and ending it efficiently. Administrative activities are often involved in this process group, such as archiving project files, closing out contracts, documenting lessons learned, and receiving formal acceptance of the delivered work as part of the phase or project.

Figure 3-1 shows the project management process groups and how they relate to each other in terms of typical level of activity, time frame, and overlap. Notice that the process groups are not isolated events. The level of activity and length of each process group varies for every project. Normally, the executing processes require the most resources and time, followed by the planning processes. The initiating and closing processes are usually the shortest and require the least amount of resources and time. However, every project is unique, so there can be exceptions. You can apply the process groups for each major phase of a project, or you can apply the process groups to an entire project, as the JWD Consulting case study does in this chapter.

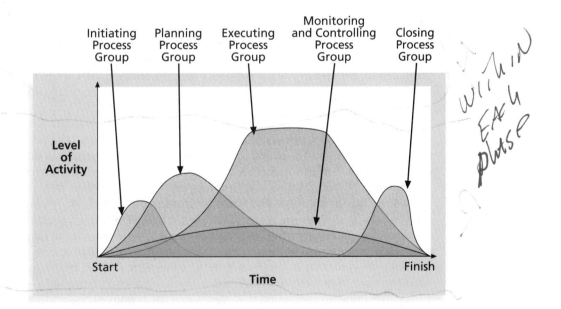

Figure 3-1. Level of Activity and Overlap of Process Groups Over Time

⊗ *What Went Wrong?*

Many readers of *CIO Magazine* commented on its cover story about problems with information systems at the U.S. Internal Revenue Service (IRS). The article described serious problems the IRS has had in managing information technology projects. Philip A. Pell, PMP, believes that having a good project manager and following a good project management process would help the IRS and many organizations tremendously. Mr. Pell provided the following feedback:

> Pure and simple, good, methodology-centric, predictable, and repeatable project management is the SINGLE greatest factor in the success (or in this case failure) of any project. When a key stakeholder says, 'I didn't know how bad things were,' it is a direct indictment of the project manager's communications management plan. When a critical deliverable like the middleware infrastructure that makes the whole thing work is left without assigned resources and progress tracking, the project manager has failed in his duty to the stakeholders. When key stakeholders (people and organizations that will be affected by the project, not just people who are directly working on the project) are not informed and their feedback incorporated into the project plan, disaster is sure to ensue. The project manager is ultimately responsible for the success or failure of the project.[1]

Each of the five project management process groups is characterized by the completion of certain tasks. During initiating processes for a new project, the organization recognizes that a new project exists. Often, completing a business case and project charter accomplishes this recognition (see Chapter 4). These documents identify the main stakeholders for a project, justify the project, and specify the high-level scope, time, and cost goals of the project. Usually, the project manager and key team members are also selected during the initiating process group, if they haven't been selected already in a process some organizations call pre-initiating.

Outcomes of the planning process group include completing the work breakdown structure and scope statement, the project schedule, and the project cost estimate (see Chapters 5, 6, and 7, respectively). Planning processes are especially important for information technology projects. Everyone who has ever worked on a large information technology project that involves new technology knows the saying, "A dollar spent up front in planning is worth one hundred dollars spent after the system is implemented." Planning is crucial in information technology projects because once a project team implements a new system, it takes a considerable amount of effort to change the system. Research suggests that companies working to implement best practices should spend at least 20 percent of project time in initiating and planning.[2]

The executing process group involves taking the actions necessary to complete

---

[1] Pell, Phillip A., Comments posted on CIO Magazine Web site on article "For the IRS, There's No EZ Fix," (April 1, 2004).

[2] PCI Group, "PM Best Practices Report," (October 2001).

the work described in the planning activities. The main outcome of this process group is the delivery of the actual work of the project. For example, if an information technology project involves providing new hardware, software, and training, the executing processes would include leading the project team and other stakeholders to purchase the hardware, develop and test the software, and deliver and participate in the training. The executing process group should overlap the other process groups and generally requires the most resources.

E

Monitoring and controlling processes measure progress toward the project objectives, monitor deviation from the plan, and take corrective action to match progress with the plan. The project manager should be monitoring progress closely to ensure that deliverables are being completed and objectives are being met. The project manager must work closely with the project team and other stakeholders and take appropriate actions to keep the project running smoothly. The ideal outcome of the monitoring and controlling process group is to complete a project successfully by delivering the agreed-upon project scope within time, cost, and quality constraints. If changes to project objectives or plans are required, monitoring and controlling processes ensure that these changes are made efficiently and effectively to meet stakeholder needs and expectations. Monitoring and controlling processes overlap all of the other project management process groups because changes can occur at any time.

M

During the closing processes, the project team works to gain acceptance of the end products, services, or results and bring the phase or project to an orderly end. Key outcomes of this process group are formal acceptance of the work and creation of closing documents, such as a final project report and lessons-learned report.

C

🎥 Media Snapshot

Just as information technology projects need to follow the project management process groups, so do other projects, such as the production of a movie. Processes involved in making movies might include screenwriting (initiating), producing (planning), acting and directing (executing), editing (monitoring and controlling), and releasing the movie to theaters (closing). Many people enjoy watching the extra features on a DVD that describe how these processes lead to the creation of a movie. For example, the DVD for Lord of the Rings: The Two Towers Extended Edition includes detailed descriptions of how the script was created, how huge structures were built, how special effects were made, and how talented professionals overcame numerous obstacles to complete the project. This acted "...not as promotional filler but as a serious and meticulously detailed examination of the entire filmmaking process."[3] Project managers in any field know how important it is to follow a good process.

[3] Jacks, Brian, "Lord of the Rings: The Two Towers Extended Edition (New Line)," *Underground Online* (accessed from *www.ugo.com* August 4, 2004).

MAPPING THE PROCESS GROUPS TO THE KNOWLEDGE AREAS

You can map the main activities of each project management process group into the nine project management knowledge areas. Table 3-1 provides a big-picture view of the relationships among the 44 project management activities, the process groups in which they are typically completed, and the knowledge areas into which they fit. The activities listed in the table are the main processes for each knowledge area listed in the *PMBOK® Guide 2004*. Note that these processes have changed slightly from the *PMBOK® Guide 2000*. Also note that some of the processes begin with verbs and some do not, as listed in the *PMBOK® Guide 2004*. The wording of these processes may be slightly different in other parts of this text.

Table 3-1: Project Management Process Groups and Knowledge Area Mapping

KNOWLEDGE AREA	PROJECT MANAGEMENT PROCESS GROUPS				
	INITIATING	PLANNING	EXECUTING	MONITORING & CONTROLLING	CLOSING
Project Integration Management	Develop project charter, Develop preliminary project scope statement	Develop project management plan	Direct and manage project execution	Monitor and control project work, Integrated change control	Close project
Project Scope Management		Scope planning, Scope definition, Create WBS		Scope verification, Scope control	
Project Time Management		Activity definition, Activity sequencing, Activity resource estimating, Activity duration estimating, Schedule development		Schedule control	
Project Cost Management		Cost estimating, Cost budgeting		Cost control	

Table 3-1: Project Management Process Groups and Knowledge Area Mapping (continued)

KNOWLEDGE AREA	PROJECT MANAGEMENT PROCESS GROUPS				
	INITIATING	PLANNING	EXECUTING	MONITORING & CONTROLLING	CLOSING
Project Quality Management		Quality planning	Perform quality assurance	Perform quality control	
Project Human Resource Management		Human resource planning	Acquire project team, Develop project team	Manage project team	
Project Communications Management		Communications planning	Information distribution	Performance reporting, Manage stakeholders	
Project Risk Management		Risk management planning, Risk identification, Qualitative risk analysis, Quantitative risk analysis, Risk response planning		Risk monitoring and control	
Project Procurement Management		Plan purchases and acquisitions, Plan contracting	Request seller responses, Select sellers	Contract administration	Contract closure

PMBOK® Guide 2004, p. 69

Several organizations use PMI's information as a foundation for developing their own project management methodologies, as described in the next section. Notice in Table 3-1 that the majority of project management processes occur as part of the planning process group. Since each project is unique, project teams are always trying to do something that has not been done before. To succeed at unique and new activities, project teams must do a fair amount of planning. Recall, however, that the most time and money is normally spent on executing. It is good practice for organizations to determine how project management will work best in their own organizations.

DEVELOPING AN INFORMATION TECHNOLOGY PROJECT MANAGEMENT METHODOLOGY

Many organizations spend a great deal of time and money on training efforts for general project management skills, but after the training, project managers may still not know how to tailor their project management skills to the organization's particular needs. Because of this problem, some organizations develop their own internal information technology project management methodologies. The *PMBOK® Guide* is a standard that describes best practices for *what* should be done to manage a project. A **methodology** describes *how* things should be done, and different organizations often have different ways of doing things.

For example, after implementing a systems development life cycle (SDLC) at BlueCross BlueShield of Michigan, the Methods department became aware that developers and project managers were often working on different information technology projects in different ways. Deliverables were often missing or looked different from project to project. They may have all had a project charter, status reports, technical documents (i.e., database design documents, user interface requirements, and so on), but how they were producing and delivering these deliverables was different. There was a general lack of consistency and a need for standards to guide both new and experienced project managers. Top management decided to authorize funds to develop a methodology for project managers that could also become the basis for information technology project management training within the organization. It was also part of an overall effort to help raise the company's Software Capability Maturity Model level. (See Chapter 8, Project Quality Management, for more information on maturity models.)

BlueCross BlueShield of Michigan launched a three-month project to develop its own project management methodology. Some of the project team members had already received PMP certification, so they decided to base their methodology on the *PMBOK® Guide 2000*, making adjustments as needed to best describe how their organization managed information technology projects. See a complete article on this project on the companion Web site for this text. Also see the Suggested Reading to review the State of Michigan Project Management Methodology, which provides another good example of an information technology project management methodology.

Many organizations include project management in their methodologies for managing Six Sigma projects, which you will learn about in Chapter 8. Other organizations include project management in their software development methodologies, such as the **Rational Unified Process (RUP)** framework. RUP is an iterative software development process that focuses on team productivity and delivers software best practices to all team members. According to RUP expert Bill Cottrell, "RUP embodies industry-standard management and technical methods and techniques to provide a software engineering process particularly suited to creating and maintaining component-based software system solutions."[4] Cottrell explains that you can tailor RUP to include the PMBOK process groups. Specifically, IBM Rational, the creators of RUP, found that it could adjust RUP input artifacts with PMBOK process inputs, RUP steps with PMBOK process tools and techniques, and RUP resulting artifacts with PMBOK process outputs.

☑ *What Went Right?*

Jordan Telecom (JT), Jordan's only telecom operator, introduced new customized project management processes to improve efficiency and reduce costs in its Information Technology department. The first step the organization took was to acquire International Standards Organization (ISO) 9000 certification in April 2004 to make sure JT had documented processes that everyone followed. It also wanted to make sure its processes were optimized and that projects followed a customized methodology. JT created three lines of processes based on the size of the project: high, medium, or low. JT used information from the *PMBOK® Guide* and PMI's Organizational Project Management Maturity Model to assess the maturity of JT's Information Technology department and design and develop a customized information technology project management process. Rula Ammuri, JT's Chief Information Officer, believes this new methodology will result in a 40-50 percent increase in productivity. Ammuri said, "I believe that optimizing our process and working according to a methodology will have an enhancement in the organization."[5]

The following section describes an example of applying the project management process groups to a project at JWD Consulting. It uses some of the ideas from the *PMBOK® Guides 2000* and *2004*, some ideas from the BlueCross BlueShield of Michigan methodology, some ideas from Six Sigma and RUP, and some new ideas to meet unique project needs. This example also includes the use of Microsoft Project to demonstrate how project management software can assist in several aspects of managing a project. Several templates will illustrate how project teams prepare various project management documents. Details on creating many of the documents shown are provided in later chapters, so do not worry if you do not understand everything right now. You might want to read this section again to enhance your learning. The following fictitious case provides an example of the elements involved in managing a project from start to finish.

[4] Cottrell, Bill, "Standards, compliance, and Rational Unified Process, Part I: Integrating RUP and the PMBOK," IBM Developerworks, (May 10, 2004).

[5] Al-Tamimi, Fairooz, "Jordanian Company Uses PMI Methods to 'Go Global,' Improve Productivity," *PMI Today* (August 2004).

CASE STUDY: JWD CONSULTING'S PROJECT MANAGEMENT INTRANET SITE PROJECT

Project Initiation

In project management, initiating includes recognizing and starting a new project. An organization should put considerable thought into project selection to ensure that it initiates the right kinds of projects for the right reasons. It is better to have a moderate or even small amount of success on an important project than huge success on one that is unimportant. The selection of projects for initiation, therefore, is crucial, as is the selection of project managers. Ideally, the project manager would be involved in initiating a project, but often the project manager is selected after many initiation decisions have already been made. You will learn more about project selection in Chapter 4, Project Integration Management. Organizations must also understand and plan for the ongoing support that is often required after implementing a new system or other product or service resulting from a project.

It is important to remember that strategic planning should serve as the foundation for deciding which projects to pursue. The organization's strategic plan expresses the vision, mission, goals, objectives, and strategies of the organization. It also provides the basis for information technology project planning. Information technology is usually a support function in an organization, so it is critical that the people initiating information technology projects understand how those projects relate to current and future needs of the organization. For example, JWD Consulting's main business is providing consulting services to other organizations, not developing its own intranet. Information systems, therefore, must support the firm's business goals, such as providing consulting services more effectively and efficiently.

An organization may initiate information technology projects for several reasons, but the most important reason is to support business objectives. Providing a good return on investment at a reasonable level of risk is also important, especially in tough economic times. As mentioned in the opening case, JWD Consulting wants to follow an exemplary process for managing its projects since its core business is helping other organizations manage projects. Developing an intranet to share its project management knowledge could help JWD Consulting reduce internal costs by working more effectively, and by allowing existing and potential customers to access some of the firm's information. JWD Consulting could also increase revenues by bringing in more business.

The *PMBOK® Guide 2004* includes only two items as outputs of the initiating process group: the project charter and the preliminary project scope statement, created as part of project integration management. This differs from the *PMBOK® Guide 2000*, which lists project scope management as the only knowledge area

2000

involved in initiation and identifies four outputs—the project charter, assignment of the project manager, clarification of constraints, and a list of assumptions.

JWD Consulting also believes it is important to identify key project stakeholders and to write a business case for the project during project initiation. Some organizations require an approved corporate project request and a more detailed business case before project initiation begins. Other organizations require these items in a phase called pre-initiation. Remember that it is important to tailor the project management process to meet unique project needs.

Erica, head of the Project Management Office for JWD Consulting, has reviewed possible outputs of project initiation and tailored them to meet the needs of the Project Management Intranet Site Project. Since the outputs are needed but have not yet been completed, she includes the tasks of identifying stakeholders and developing a business case under the initiating process. JWD Consulting does not require formal paperwork for internal project requests, so she decides not to include that item. The project's business case includes a clarification of constraints and a list of assumptions as well as several items often included in a preliminary scope statement, so she decides not to include those items as separate outputs of initiation. The project charter will include the preliminary budget and schedule information for the project. Table 3-2 summarizes the desired outputs of project initiation for this particular project. Note that Erica is tailoring information to meet the needs of her project and organization. She knows that her project requires a strong business case, and she knows that it is crucial to identify key stakeholders early in her organization.

Table 3-2: Outputs of Project Initiation

Project manager assigned
Key stakeholders identified   *& then private list*   *cover*
Business case completed
Project charter completed and signed

As described in the opening case, Joe Fleming, the CEO at JWD Consulting, has assigned Erica Bell as the project manager for the intranet project, so that particular initiating task is finished. Erica has managed several projects in the past, and as the head of the firm's Project Management Office, she has expertise that would be instrumental to this project's success.

Erica met with Joe Fleming, the project's sponsor, to help identify other key stakeholders for this project. They decided to contact one full-time consultant with an outstanding record, Michael Chen, one part-time consultant, Jessie Faue, who was new to the company and supported the Project Management Office, and two members of the Information Technology department who supported the current intranet, Kevin Dodge and Cindy Dawson. They also knew that client inputs would be important for this project, so Joe agreed to call the CEOs of two of the firm's largest clients to see if they would be willing to provide representatives to work on this project at their own expense. After Joe and Erica made the preliminary contacts, Erica documented the stakeholders' roles, names, organizations, and contact information. All of the internal staff Joe and Erica recommended agreed to work on the project, and the two client representatives would be Kim Phuong and Page Miller. Erica would have these stakeholders review and sign the project charter.

Erica drafted a business case for the project, getting input and feedback from Joe, one of her senior staff members in the Project Management Office, and a member of the Finance department. Table 3-3 provides the business case. (Note that this example and others are abbreviated examples. See the companion Web site and the author's Web site, *www.kathyschwalbe.com*, for additional examples of business cases and other project documents and to download a business case template and other templates. You will learn how to perform financial calculations in later chapters.) Notice that the following information is included in the business case:

- Introduction
- Business Objective
- Current situation and problem/opportunity statement
- Critical assumptions and constraints
- Analysis of options and recommendation
- Preliminary project requirements
- Budget estimate and financial analysis
- Schedule estimate
- Potential risks
- Exhibits

Since this project is relatively small and is for an internal sponsor, the business case is not as long as many other business cases. Erica reviewed the business case with Joe before proceeding to develop the project charter. Joe agreed that the project was worth pursuing, and he told Erica to proceed with developing the project charter to officially recognize the existence of this project.

Table 3-3: JWD Consulting's Business Case

1.0 INTRODUCTION/BACKGROUND

JWD Consulting's core business goal is to provide world-class project management consulting services to various organizations. The CEO, Joe Fleming, believes the firm can streamline operations and increase business by providing information related to project management on its intranet site, making some information and services accessible to current and potential clients.

Table 3-3: JWD Consulting's Business Case (continued)

2.0 BUSINESS OBJECTIVE

JWD Consulting's strategic goals include continuing growth and profitability. The Project Management Intranet Site Project will support these goals by increasing visibility of the firm's expertise to current and potential clients by allowing client and public access to some sections of the intranet. It will also improve profitability by reducing internal costs by providing standard tools, techniques, templates, and project management knowledge to all internal consultants. Since JWD Consulting focuses on identifying profitable projects and measuring their value after completion, this project must meet those criteria.

3.0 CURRENT SITUATION AND PROBLEM/OPPORTUNITY STATEMENT

JWD Consulting has a corporate Web site as well as an intranet. The firm currently uses the Web site for marketing information. The primary use of the intranet is for human resource information, such as where consultants enter their hours on various projects, change and view their benefits information, access an online directory and Web-based e-mail system, and so on. The firm also uses an enterprise-wide project management system to track all project information, focusing on the status of deliverables and meeting scope, time, and cost goals. There is an opportunity to provide a new section on the intranet dedicated to sharing consultants' project management knowledge across the organization. JWD Consulting only hires experienced consultants and gives them freedom to manage projects as they see fit. However, as the business grows and projects become more complex, even experienced project managers are looking for suggestions on how to work more effectively.

4.0 CRITICAL ASSUMPTION AND CONSTRAINTS

The proposed intranet site must be a valuable asset for JWD Consulting. Current consultants and clients must actively support the project, and it must pay for itself within one year by reducing internal operating costs and generating new business. The Project Management Office manager must lead the effort, and the project team must include participants from several parts of the company, as well as current client organizations. The new system must run on existing hardware and software, and it should require minimal technical support. It must be easily accessible by clients and the public yet secure from unauthorized users.

5.0 ANALYSIS OF OPTIONS AND RECOMMENDATION

There are three options for addressing this opportunity:

1. Do nothing. The business is doing well, and we can continue to operate without this new project.
2. Purchase access to specialized software to support this new capability with little in-house development.
3. Design and implement the new intranet capabilities in-house using mostly existing hardware and software.

Based on discussions with stakeholders, we believe that option 3 is the best option.

6.0 PRELIMINARY PROJECT REQUIREMENTS

The main features of the project management intranet site include the following:

1. Access to several project management templates and tools. Users must be able to search for templates and tools, read instructions on using these templates and tools, and see examples of how to apply them to real projects. Users must also be able to submit new templates and tools, which should be first screened or edited by the Project Management Office.

Table 3-3: JWD Consulting's Business Case (continued)

2. Access to relevant project management articles. Many consultants and clients feel as though there is an information overload when they research project management information. They often waste time they should be spending with their clients. The new intranet should include access to several important articles on various project management topics, which are searchable by topic, and allow users to request the Project Management Office staff to find additional articles to meet their needs.

3. Links to other, up-to-date Web sites, with brief descriptions of the main features of the external site.

4. An "Ask the Expert" feature to help build relationships with current and future clients and share knowledge with internal consultants.

5. Appropriate security to make the entire intranet site accessible to internal consultants and certain sections accessible to others.

6. The ability to charge money for access to some information. Some of the information and features of the intranet site should prompt external users to pay for the information or service. Payment options should include a credit card option or similar online payment transactions. After the system verifies payment, the user should be able to access or download the desired information.

7. Other features suggested by users, if they add value to the business.

7.0 BUDGET ESTIMATE AND FINANCIAL ANALYSIS

A preliminary estimate of costs for the entire project is $140,000. This estimate is based on the project manager working about 20 hours per week for six months and other internal staff working a total of about 60 hours per week for six months. The customer representatives would not be paid for their assistance. A staff project manager would earn $50 per hour. The hourly rate for the other project team members would be $70 per hour, since some hours normally billed to clients may be needed for this project. The initial cost estimate also includes $10,000 for purchasing software and services from suppliers. After the project is completed, maintenance costs of $40,000 are included for each year, primarily to update the information and coordinate the "Ask the Expert" feature and online articles.

Projected benefits are based on a reduction in hours consultants spend researching project management information, appropriate tools and templates, and so on. Projected benefits are also based on a small increase in profits due to new business generated by this project. If each of more than 400 consultants saved just 40 hours each year (less than one hour per week) and could bill that time to other projects that generate a conservative estimate of $10 per hour in *profits*, then the projected benefit would be $160,000 per year. If the new intranet increased business by just 1 percent, using past profit information, increased profits due to new business would be at least $40,000 each year. Total projected benefits, therefore, are about $200,000 per year.

Exhibit A summarizes the projected costs and benefits and shows the estimated net present value (NPV), return on investment (ROI), and year in which payback occurs. It also lists assumptions made in performing this preliminary financial analysis. All of the financial estimates are very encouraging. The estimated payback is within one year, as requested by the sponsor. The NPV is $272,800, and the discounted ROI based on a three-year system life is excellent at 112 percent.

8.0 SCHEDULE ESTIMATE

The sponsor would like to see the project completed within six months, but there is some flexibility in the schedule. We also assume that the new system will have a useful life of at least three years.

Table 3-3: JWD Consulting's Business Case (continued)

9.0 POTENTIAL RISKS

There are several risks involved with this project. The foremost risk is a lack of interest in the new system by our internal consultants and external clients. User inputs are crucial for populating information into this system and realizing the potential benefits from using the system. There are some technical risks in choosing the type of software used to search the system, check security, process payments, and so on, but the features of this system all use proven technologies. The main business risk is investing the time and money into this project and not realizing the projected benefits.

10.0 EXHIBITS

Exhibit A: Financial Analysis for Project Management Intranet Site Project

Discount rate	8%				
Assume the project is done in about 6 months		Year			
	0	1	2	3	Total
Costs	140,000	40,000	40,000	40,000	
Discount factor	1	0.93	0.86	0.79	
Discounted costs	140,000	37,037	34,294	31,753	243,084
Benefits	0	200,000	200,000	200,000	
Discount factor	1	0.93	0.86	0.79	
Discounted benefits	0	186,185	171,468	158,766	515,419
Discounted benefits - costs	(140,000)	148,148	137,174	127,013	
Cumulative benefits - costs	(140,000)	8,148	145,322	272,336	← NPV
	Payback in Year 1				
Discounted life cycle ROI----------->	112%				
Assumptions					
Costs	# hours				
PM (500 hours, $50/hour)	25,000				
Staff (1500 hours, $70/hour)	105,000				
Outsourced software and services	10,000				
Total project costs (all applied in year 0)	140,000				
Benefits					
# consultants	400				
Hours saved	40				
$/hour profit	10				
Benefits from saving time	160,000				
Benefits from 1% increase in profits	40,000				
Total annual projected benefits	200,000				

Erica drafted a project charter and had the project team members review it before showing it to Joe. Joe made a few minor changes, which Erica incorporated, and then all the key stakeholders signed the project charter. Table 3-4 shows the final project charter (see Chapter 4 for more information on project charters). Note the items included on the project charter and its short length. JWD Consulting believes that project charters should preferably be one or two pages long, and they may refer to other documents, such as a business case, as needed. Erica felt the most important parts of the project charter were the signatures of key stakeholders and their individual comments. It is hard to get stakeholders to agree on even a one-page project charter, so everyone has a chance to make their concerns known in the comments section. Note that Michael Chen, the senior consultant asked to work on the project, was concerned about working on this project when he felt that his other assignments with external clients might have a higher priority. He offered to have an assistant help as needed. The information technology staff members mentioned their concerns about testing and security issues. Erica knew that she would have to consider these concerns when managing the project.

Table 3-4: Project Charter

Project Title: Project Management Intranet Site Project

Project Start Date: May 2, 2005   **Projected Finish Date:** November 4, 2005

Budget Information: The firm has allocated $140,000 for this project. The majority of costs for this project will be internal labor. An initial estimate provides a total of 80 hours per week.

Project Manager: Erica Bell, (310) 555-5896, erica_bell@jwdconsulting.com

Project Objectives: Develop a new capability accessible on JWD Consulting's intranet site to help internal consultants and external customers manage projects more effectively. The intranet site will include several templates and tools that users can download, examples of completed templates and related project management documents used on real projects, important articles related to recent project management topics, an article retrieval service, links to other sites with useful information, and an "Ask the Expert" feature, where users can post questions they have about their projects and receive advice from experts in the field. Some parts of the intranet site will be accessible free to the public, other parts will only be accessible to current customers and/or internal consultants, and other parts of the intranet site will be accessible for a fee.

Approach:

- Develop a survey to determine critical features of the new intranet site and solicit input from consultants and customers.

- Review internal and external templates and examples of project management documents.

- Research software to provide security, manage user inputs, and facilitate the article retrieval and "Ask the Expert" features.

- Develop the intranet site using an iterative approach, soliciting a great deal of user feedback.

- Determine a way to measure the value of the intranet site in terms of reduced costs and new revenues, both during the project and one year after project completion.

ROLES AND RESPONSIBILITIES:

Name	Role	Position	Contact Information
Joe Fleming	Sponsor	JWD Consulting, CEO	joe_fleming@jwdconsulting.com
Erica Bell	Project Manager	JWD Consulting, manager	erica_bell@jwdconsulting.com
Michael Chen	Team Member	JWD Consulting, senior consultant	michael_chen@jwdconsulting.com

Table 3-4: Project Charter (continued)

NAME	ROLE	POSITION	CONTACT INFORMATION
Jessie Faue	Team Member	JWD Consulting, consultant	jessie_faue@jwdconsulting.com
Kevin Dodge	Team Member	JWD Consulting, IT department	kevin_dodge@jwdconsulting.com
Cindy Dawson	Team Member	JWD Consulting, IT department	cindy_dawson@jwdconsulting.com
Kim Phuong	Advisor	Client representative	kim_phuong@client1.com
Page Miller	Advisor	Client representative	page_miller@client2.com

Comments: (Handwritten or typed comments from above stakeholders, if applicable)
"I will support this project as time allows, but I believe my client projects take priority. I will have one of my assistants support the project as needed." —Michael Chen
"We need to be extremely careful testing this new system, especially the security in giving access to parts of the intranet site to the public and clients." —Kevin Dodge and Cindy Dawson

Project Planning

Planning is often the most difficult and unappreciated process in project management. Because planning is not always used to facilitate action, many people view planning negatively. The main purpose of project plans, however, is *to guide project execution*. To guide execution, plans must be realistic and useful, so a fair amount of time and effort must go into the planning process; people knowledgeable with the work need to plan the work.

Table 3-5 lists the project management knowledge areas, processes, and outputs of project planning according to the *PMBOK® Guide 2004*. There are many potential outputs from the planning process group, and every knowledge area is included. Just a few planning documents from JWD Consulting's Project Management Intranet Site Project are provided in this chapter as examples. Recall that the *PMBOK® Guide 2004* is only a guide, so many organizations may have different planning outputs based on their particular needs.

Table 3-5: Planning Processes and Outputs

KNOWLEDGE AREA	PLANNING PROCESS	OUTPUTS
Project Integration Management	Develop project management plan	■ Project management plan
Project Scope Management	Scope planning	■ Project scope management plan
	Scope definition	■ Project scope statement ■ Requested changes ■ Project scope management plan (updates)

Table 3-5: Planning Processes and Outputs (continued)

KNOWLEDGE AREA	PLANNING PROCESS	OUTPUTS
	Create WBS	■ Project scope statement (updates) ■ WBS and WBS dictionary ■ Scope baseline ■ Project scope management plan (updates) ■ Requested changes
Project Time Management	Activity definition	■ Activity list and attributes ■ Milestone list ■ Requested changes
	Activity sequencing	■ Project schedule network diagram ■ Activity list and attributes (updates) ■ Requested changes
	Activity resource estimating	■ Activity resource requirements ■ Activity attributes (updates) ■ Resource breakdown structure ■ Resource calendars (updates) ■ Requested changes
	Activity duration estimating	■ Activity duration estimates ■ Activity attributes (updates)
	Schedule development	■ Project schedule ■ Schedule model data ■ Schedule baseline ■ Updates to resource requirements, activity attributes, project calendar and project management plan ■ Requested changes
Project Cost Management	Cost estimating	■ Activity cost estimates and supporting detail ■ Requested change ■ Cost management plan (updates)
	Cost budgeting	■ Cost baseline ■ Project funding requirements ■ Cost management plan (updates) ■ Requested changes
Project Quality Management	Quality planning	■ Quality management plan ■ Quality metrics ■ Quality checklists ■ Process improvement plan ■ Quality baseline ■ Project management plan (updates)
Project Human Resource Management	Human resource planning	■ Roles and responsibilities ■ Project organization charts ■ Staffing management plan

Table 3-5: Planning Processes and Outputs (continued)

KNOWLEDGE AREA	PLANNING PROCESS	OUTPUTS
Project Communications Management	Communications planning	■ Communications management plan
Project Risk Management	Risk management planning	■ Risk management plan
	Risk identification	■ Risk register
	Qualitative risk analysis	■ Risk register (updates)
	Quantitative risk analysis	■ Risk register (updates)
	Risk response planning	■ Risk register (updates) ■ Project management plan (updates) ■ Risk-related contractual agreements
Project Procurement Management	Plan purchases and acquisitions	■ Procurement management plan ■ Contract statement of work ■ Make-or-buy decisions ■ Requested changes
	Plan contracting	■ Procurement documents ■ Evaluation criteria ■ Contract statement of work (updates)

Since the Project Management Intranet Site Project is relatively small, Erica believes some of the most important planning documents to focus on are the following:

- A team contract (not listed in Table 3-5, which is based on the *PMBOK® Guide 2004*)
- A project scope statement
- A work breakdown structure (WBS)
- A project schedule, in the form of a Gantt chart with all dependencies and resources entered
- A list of prioritized risks (part of a risk register)

All of these documents, as well as other project-related information, will be available to all team members on a project Web site. JWD Consulting has used project Web sites for several years, and has found that they really help facilitate communications and document project information. For larger projects, JWD Consulting also creates project organization charts, a formal communications management plan, a quality management plan, detailed cost estimates, procurement plans, and other planning documents. You will learn more about these documents by knowledge area in the following chapters. Appendix D provides templates for creating many of these documents.

Soon after the project team signed the project charter, Erica organized a kick-off meeting for the Project Management Intranet Site Project. An important part of the meeting was helping the project team get to know each other. Erica had met and talked to each member separately, but this was the first time the project team would all meet each other. Jessie Faue worked in the Project Management Office with Erica, so they knew each other well, but Jessie was new to the company and did not know any of the other team members. Michael Chen was a senior consultant and often worked on the highest priority projects for external clients. He attended the kick-off meeting with his assistant, Jill Anderson, who would also support the project when Michael was too busy. Everyone valued Michael's expertise, and he was extremely straightforward in dealing with people. He also knew both of the client representatives from past projects. Kevin Dodge was JWD Consulting's intranet guru who tended to focus on technical details. Cindy Dawson was also from the Information Technology department and had experience working as a business consultant and negotiating with outside suppliers. Kim Phuong and Page Miller, the two client representatives, were excited about the project, but they were wary of sharing sensitive information about their company.

Erica had everyone introduce him or herself, and then she facilitated an ice-breaker activity so everyone would be more relaxed. She asked everyone to describe his or her dream vacation, assuming cost was no issue. This activity helped everyone get to know each other and show different aspects of their personalities. Erica knew that it was important to build a strong team and have everyone work well together.

Erica then explained the importance of the project, reviewing the signed project charter. She explained that an important tool to help a project team work together was to have members develop a team contract that everyone felt comfortable signing. JWD Consulting believed in using team contracts for all projects to help promote teamwork and clarify team communications. She explained the main topics covered in a team contract and showed them a team contract template. She then had the team members form two smaller groups, with one consultant, one Information Technology department member, and one client representative in each group. These smaller groups made it easier for everyone to contribute ideas. Each group shared their ideas for what should go into the contract, and then they worked together to form one project team contract. Table 3-6 shows the resulting team contract, which took about ninety minutes to create. Erica could see that there were different personalities on this team, but she felt they all could work together well.

Table 3-6: Team Contract

Code of Conduct: As a project team, we will:
- Work proactively, anticipating potential problems and working to prevent them.
- Keep other team members informed of information related to the project.
- Focus on what is best for the entire project team.

Table 3-6: Team Contract (continued)

Participation: We will:

- Be honest and open during all project activities.
- Encourage diversity in team work.
- Provide the opportunity for equal participation.
- Be open to new approaches and consider new ideas.
- Have one discussion at a time.
- Let the project manager know well in advance if a team member has to miss a meeting or may have trouble meeting a deadline for a given task.

Communication: We will:

- Decide as a team on the best way to communicate. Since a few team members cannot meet often for face-to-face meetings, we will use e-mail, a project Web site, and other technology to assist in communicating.
- Have the project manager facilitate all meetings and arrange for phone and video-conferences, as needed.
- Work together to create the project schedule and enter actuals into our enterprise-wide project management system by 4 p.m. every Friday.
- Present ideas clearly and concisely.
- Keep discussions on track and have one discussion at a time.

Problem Solving: We will:

- Encourage everyone to participate in solving problems.
- Only use constructive criticism and focus on solving problems, not blaming people.
- Strive to build on each other's ideas.

Meeting Guidelines: We will:

- Plan to have a face-to-face meeting the first and third Tuesday morning of every month.
- Meet more frequently the first month.
- Arrange for telephone or videoconferencing for participants as needed.
- Hold other meetings as needed.
- Record meeting minutes and send them out via e-mail within 24 hours of all project meetings, focusing on decisions made and action items from each meeting.
- Develop an agenda before all meetings with our project sponsor and client advisors.
- Document major issues and decisions related to the project and send them out via e-mail to all team members and the project sponsor.

Erica wanted to keep their meeting to its two-hour time limit. Their next task would be to clarify the scope of the project by developing a scope statement and WBS. She knew it took time to develop these documents, but she wanted to get a feel for what everyone thought were the main deliverables for this project, their roles in producing those deliverables, and what areas of the project scope needed clarification. She reminded everyone what their budget and schedule goals were so they would keep that in mind as they discussed the scope of the project. She also asked each person to provide the number of hours he or she would be available to work on this project each month for the next six months. She then had each person write down his or her answers to the following questions:

1. List one item that is most unclear to you about the scope of this project.
2. What other questions do you have or issues do you foresee about the scope of the project?

3. List what you believe to be the main deliverables for this project.

4. Which deliverables do you think you will help create or review?

Erica collected everyone's inputs. She explained that she would take this information and work with Jessie to develop the first draft of the scope statement that she would e-mail to everyone by the end of the week. She also suggested that they all meet again in one week to develop the scope statement further and to start creating the WBS for the project.

Erica and Jessie reviewed all the information and created the first draft of the scope statement. At their next team meeting, they discussed the scope statement and got a good start on the WBS. Table 3-7 shows a portion of the scope statement that Erica created after a few more e-mails and another team meeting. Note that the scope statement lists the importance of documenting the product characteristics and requirements, summarizes the deliverables, and describes project success criteria.

Table 3-7: Scope Statement (Draft Version)

Project Title: Project Management Intranet Site Project

Date: May 18, 2005 **Prepared by:** Erica Bell, Project Manager, (310) 555-5896, erica_bell@jwdconsulting.com

Project Justification: Joe Fleming, CEO of JWD Consulting, requested this project to assist the company in meeting its strategic goals. The new intranet site will increase visibility of the company's expertise to current and potential clients though the sections of the intranet to which they will be accessible. It will also help reduce internal costs and improve profitability by providing standard tools, techniques, templates, and project management knowledge to all internal consultants. The budget for the project is $140,000. An additional $40,000 per year will be required for operational expenses after the project is completed. Estimated benefits are $200,000 each year. It is important to focus on the system paying for itself within one year.

Product Characteristics and Requirements:

1. Templates and tools: The intranet site will allow authorized users to download files they can use to create project management documents and to help them use project management tools. These files will be in Microsoft Word, Excel, Access, Project, or in HTML or PDF format, as appropriate.

2. User submissions: Users will be encouraged to e-mail files with sample templates and tools to the Webmaster. The Webmaster will forward the files to the appropriate person for review and then post the files to the intranet site, if desired.

3. Articles: Articles posted on the intranet site will have appropriate copyright permission. The preferred format for articles will be PDF. The project manager may approve other formats.

4. Requests for articles: The intranet site will include a section for users to request someone from the Project Management Office (PMO) at JWD Consulting to research appropriate articles for them. The PMO manager must first approve the request and negotiate payments, if appropriate.

5. Links: All links to external sites will be tested on a weekly basis. Broken links will be fixed or removed within five working days of discovery.

Table 3-7: Scope Statement (Draft Version) (continued)

6. The "Ask the Expert" feature must be user-friendly and capable of soliciting questions and immediately acknowledging that the question has been received in the proper format. The feature must also be capable of forwarding the question to the appropriate expert (as maintained in the system's expert database) and capable of providing the status of questions that are answered. The system must also allow for payment for advice, if appropriate.

7. Security: The intranet site must provide several levels of security. All internal employees will have access to the entire intranet site when they enter their security information to access the main, corporate intranet. Part of the intranet will be available to the public from the corporate Web site. Other portions of the intranet will be available to current clients based on verification with the current client database. Other portions of the intranet will be available after negotiating a fee or entering a fixed payment using preauthorized payment methods.

8. Search feature: The intranet site must include a search feature for users to search by topic, key words, etc.

9. The intranet site must be accessible using a standard Internet browser. Users must have appropriate application software to open several of the templates and tools.

10. The intranet site must be available 24 hours a day, 7 days a week, with one hour per week for system maintenance and other periodic maintenance, as appropriate.

Summary of Project Deliverables

Project management-related deliverables: Business case, charter, team contract, scope statement, WBS, schedule, cost baseline, status reports, final project presentation, final project report, lessons-learned report, and any other documents required to manage the project.

Product-related deliverables:

1. Survey: Survey current consultants and clients to help determine desired content and features for the intranet site.

2. Files for templates: The intranet site will include templates for at least twenty documents when the system is first implemented, and it will have the capacity to store up to one hundred documents. The project team will decide on the initial twenty templates based on survey results.

3. Examples of completed templates: The intranet site will include examples of projects that have used the templates available on the intranet site. For example, if there is a template for a business case, there will also be an example of a real business case that uses the template.

4. Files for tools: The intranet site will include information on how to use several project management tools, including the following as a minimum: work breakdown structures, Gantt charts, network diagrams, cost estimates, and earned value management. Where appropriate, sample files will be provided in the application software appropriate for the tool. For example, Microsoft Project files will be available to show sample work breakdown structures, Gantt charts, network diagrams, cost estimates, and applications of earned value management. Excel files will be available for sample cost estimates and earned value management charts.

5. Example applications of tools: The intranet site will include examples of real projects that have applied the tools listed in number 4 above.

6. Articles: The intranet site will include at least ten useful articles about relevant topics in project management. The intranet site will have the capacity to store at least 1000 articles in PDF format with an average length of ten pages each.

Table 3-7: Scope Statement (Draft Version) (continued)

7. Links: The intranet site will include links with brief descriptions for at least twenty useful sites. The links will be categorized into meaningful groups.

8. Expert database: In order to deliver an "Ask the Expert" feature, the system must include and access a database of approved experts and their contact information. Users will be able to search for experts by pre-defined topics.

9. User Requests feature: The intranet site will include an application to solicit and process requests from users.

10. Intranet site design: An initial design of the new intranet site will include a site map, suggested formats, appropriate graphics, etc. The final design will incorporate comments from users on the initial design.

11. Intranet site content: The intranet site will include content for the templates and tools section, articles section, article retrieval section, links section, "Ask the Expert" section, User Requests feature, security, and payment features.

12. Test plan: The test plan will document how the intranet site will be tested, who will do the testing, and how bugs will be reported.

13. Promotion: A plan for promoting the intranet site will describe various approaches for soliciting inputs during design. The promotion plan will also announce the availability of the new intranet site.

14. Project benefit measurement plan: A project benefit plan will measure the financial value of the intranet site.

Project Success Criteria: Our goal is to complete this project within six months for no more than $140,000. The project sponsor, Joe Fleming, has emphasized the importance of the project paying for itself within one year after the intranet site is complete. To meet this financial goal, the intranet site must have strong user inputs. We must also develop a method for capturing the benefits while the intranet site is being developed and tested, and after it is rolled out. If the project takes a little longer to complete or costs a little more than planned, the firm will still view it as a success if it has a good payback and helps promote the firm's image as an excellent consulting organization.

Time Budget

As the project team worked on the scope statement, they also developed the work breakdown structure (WBS) for the project. The WBS is a very important tool in project management because it provides the basis for deciding how to do the work. The WBS also provides a basis for creating the project schedule and performing earned value management for measuring and forecasting project performance. Erica and her team decided to use the project management process groups as the main categories for the WBS (Figure 3-2). They included completed work from the initiating process to provide a complete picture of the project's scope. The group also wanted to list several milestones on their schedule, such as the completion of key deliverables, so they prepared a separate list of milestones that they would include on the Gantt chart. You will learn more about creating a WBS in Chapter 5, Project Scope Management.

1.0 Initiating
    1.1 Determine/assign project manager
    1.2 Identify key stakeholders
    1.3 Prepare business case
    1.4 Prepare project charter
2.0 Planning
    2.1 Hold project kickoff meeting
    2.2 Prepare team contract
    2.3 Prepare scope statement
    2.4 Prepare WBS
    2.5 Prepare schedule and cost baseline
        2.5.1 Determine task resources
        2.5.2 Determine task durations
        2.5.3 Determine task dependencies
        2.5.4 Create draft Gantt chart
        2.5.5 Review and finalize Gantt chart
    2.6 Identify, discuss, and prioritize risks
3.0 Executing
    3.1 Survey
    3.2 User inputs
    3.3 Intranet site content
        3.3.1 Templates and tools
        3.3.2 Articles
        3.3.3 Links
        3.3.4 Ask the Expert
        3.3.5 User requests feature
    3.4 Intranet site design
    3.5 Intranet site construction
    3.6 Intranet site testing
    3.7 Intranet site promotion
    3.8 Intranet site roll-out
    3.9 Project benefits measurement
4.0 Controlling
    4.1 Status reports
5.0 Closing
    5.1 Prepare final project report
    5.2 Prepare final project presentation
    5.3 Lessons learned

Figure 3-2. JWD Consulting Intranet Project Work Breakdown Structure (WBS)

After preparing the WBS, the project team held another face-to-face meeting to develop the project schedule, following the steps outlined in section 2.5 of the WBS. Several of the project schedule tasks are dependent on one another. For example, the intranet site testing was dependent on the construction and completion of the content tasks. Everyone participated in the development of the schedule, especially the tasks on which each would be working. Some of the tasks were broken down further so the team members had a better understanding of what they had to do and when. They also kept their workloads and cost constraints in mind when developing the duration estimates. For example, Erica was scheduled to work twenty hours per week on this project, and the other project team members combined should not spend more than sixty hours per week on average for the project. As team members provided duration estimates, they also estimated how many work hours they would spend on each task.

After the meeting, Erica worked with Jessie to enter all of the information into Microsoft Project. Erica was using the intranet site project to train Jessie in applying several project management tools and templates. They entered all of the tasks, duration estimates, and dependencies to develop the Gantt chart. Erica decided to enter the resource and cost information after reviewing the schedule. Their initial inputs resulted in a completion date a few weeks later than planned. Erica and Jessie reviewed the critical path for the project, and Erica had to shorten the duration estimates for a few critical tasks in order to meet their schedule goal of completing the project within six months. She talked to the team members working on those tasks, and they agreed that they could plan to work more hours each week on those tasks in order to complete them on time. Erica considered adding a project buffer at the end of the project, but she was confident they could meet the schedule goal. You will learn more about project buffers and developing project schedules in Chapter 6, Project Time Management.

Figure 3-3 shows the resulting Gantt chart created in Microsoft Project. Only the executing tasks are expanded to show the subtasks under that category. Figure 3-4 shows part of the network diagram, again emphasizing the executing tasks. The highlighted boxes represent tasks on the critical path. (You will learn how to use Project 2003 in Appendix A. Chapter 6, Project Time Management, explains how to use network diagrams to find the critical path and explains Gantt chart symbols.) The baseline schedule projects a completion date of November 1, 2005. The project charter had a planned completion date of November 4, 2005. Erica wanted to complete the project on time, and although three extra days was not much of a buffer, she felt the baseline schedule was very realistic. She would do her best to help everyone meet their deadlines.

6 MOS

M 6MNS

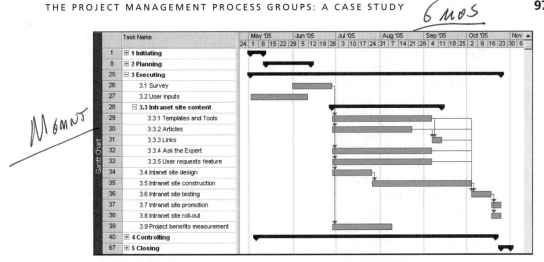

Figure 3-3. JWD Consulting Intranet Site Project Baseline Gantt Chart

PJCT
TEAM

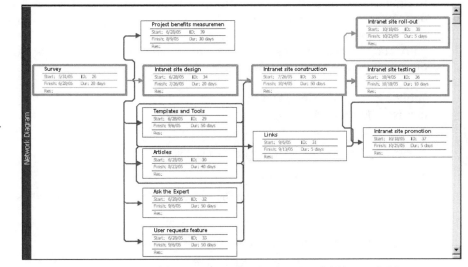

Figure 3-4. JWD Consulting Intranet Site Project Partial Network Diagram

The majority of the costs for this project were internal labor, and the team kept their labor hour constraints in mind when developing task duration estimates. Erica and Jessie entered each project team member's name and labor rate in the resource sheet for their Microsoft Project file. The client representatives were not being paid for their time, so she left their labor rates at the default value of zero. Erica had also included $10,000 for procurement in the financial analysis she prepared for the business case, and she showed Jessie how to enter that amount as a fixed cost split equally between the "Ask the Expert" and User Requests features, where she thought they would have to purchase some external software and/or services. Erica then helped Jessie assign resources to tasks, entering the projected number of hours everyone planned to work each week on each task. They then ran several cost reports and made a few minor adjustments to resource assignments to make their planned total cost meet their budget constraints. Their cost baseline was very close to their planned budget of $140,000.

The last deliverable her team needed to create within the planning process group was a list of prioritized risks. This information will be updated and expanded as the project progresses to include information on root causes of the risks, warning signs that potential risks might occur, and response strategies for the risks. Erica reviewed the risks she had mentioned in the business case as well as the comments team members made on the project charter and in their team meetings. She held a special meeting for everyone to brainstorm and discuss potential risks. They posted all of the risks they identified on a probability/impact matrix, and then they grouped some of the ideas. There was only one risk in the high probability and high impact category, and several with medium impact in one or both categories. They chose not to list the low probability and low impact risks. After some discussion, the team developed the list of prioritized risks shown in Table 3-8.

Table 3-8: List of Prioritized Risks

RANKING	POTENTIAL RISK
1	Lack of inputs from internal consultants
2	Lack of inputs from client representatives
3	Security of new system
4	Outsourcing/purchasing for the article retrieval and "Ask the Expert" features
5	Outsourcing/purchasing for processing online payment transactions
6	Organizing the templates and examples in a useful fashion
7	Providing an efficient search feature
8	Getting good feedback from Michael Chen and other senior consultants
9	Effectively promoting the new system
10	Realizing the benefits of the new system within one year

Project Executing

Executing the project involves taking the actions necessary to ensure that activities in the project plan are completed. It also includes work required to introduce any new hardware, software, and procedures into normal operations. The products of the project are produced during project execution, and it usually takes the most resources to accomplish this process. Table 3-9 lists the knowledge areas, executing processes, and outputs of project execution listed in the *PMBOK® Guide 2004*. Many project sponsors and customers focus on outputs or deliverables related to providing the products, services, or results desired from the project. Implementing solutions to problems, such as change requests or corrective actions, are also important outputs, as are the other outputs listed below.

Table 3-9: Executing Processes and Outputs

KNOWLEDGE AREA	EXECUTING PROCESS	OUTPUTS
Project Integration Management	Direct and manage project execution	■ Deliverables ■ Requested changes ■ Implemented solutions to problems (i.e., change requests, corrective actions, preventive actions, and defect repairs) ■ Work performance information
Project Quality Management	Perform quality assurance	■ Requested changes ■ Recommended corrective actions ■ Organizational process assets (updates) ■ Project management plan (updates)
Project Human Resource Management	Acquire project team	■ Project staff assignments ■ Resource availability ■ Staffing management plan (updates)
	Develop project team	■ Team performance assessment
Project Communications Management	Information distribution	■ Organizational process assets (updates) ■ Requested changes
Project Procurement Management	Request seller responses	■ Qualified sellers list ■ Procurement document package ■ Proposals
	Select sellers	■ Selected sellers ■ Contract ■ Contract management plan ■ Resource availability ■ Procurement management plan (updates) ■ Requested changes

For this relatively small project, Erica could work closely with all the team members to make sure they were producing the desired work results. She also used her networking skills to get input from other people in the firm and from external sources at no additional cost to the project. She knew that providing strong leadership and using good communication skills were crucial to good project execution. The firm did have a formal change request form, but primarily used it for external projects. The firm also had contract specialists and templates for several procurement documents that the project team would use for the portions of the project they planned to outsource. See Appendix D for sample templates related to change requests and contracts. Chapter 4, Project Integration Management, discusses integrated change control, and Chapter 12, Project Procurement Management, describes various types of contracts and other procurement documents.

As mentioned earlier, Erica knew that Joe, the CEO and project sponsor, liked to see progress on projects through milestone reports, which are one type of project report. He also wanted Erica to alert him to any potential issues or problems. Table 3-10 shows a sample of a milestone report for the Project Management Intranet Site Project that Erica reviewed with Joe in mid-June. Erica met with most of her project team members often, and she talked to Joe about once a week to review progress on completing milestones and to discuss any other project issues. Although Erica could have used Project 2003 to create milestone reports, she instead used word processing software because this project was small, and she could more easily manipulate report format.

Table 3-10: Milestone Report as of 6/17/05

MILESTONE	DATE	STATUS	RESPONSIBLE	ISSUES/COMMENTS
Initiating				
Project manager determined/assigned	5/2/05	Completed	Joe	
Business case created	5/6/05	Completed	Erica	
Project charter signed	5/10/05	Completed	Erica	
Planning				
Project kickoff meeting held	5/13/05	Completed	Erica	Went well
Team contract signed	5/13/05	Completed	Erica	
Scope statement completed	5/27/05	Completed	Erica	
WBS completed	5/31/05	Completed	Erica	
List of prioritized risks completed	6/3/05	Completed	Erica	Reviewed with sponsor and team
Schedule and cost baseline completed	6/13/05	Completed	Erica	
Executing				
Survey completed	6/28/05		Erica	Poor response so far!
Intranet site design completed	7/26/05		Kevin	
Project benefits measurement completed	8/9/05		Erica	

Table 3-10: Milestone Report as of 6/17/05 (continued)

MILESTONE	DATE	STATUS	RESPONSIBLE	ISSUES/COMMENTS
User inputs collected	8/9/05		Jessie	
Articles completed	8/23/05		Jessie	
Templates and tools completed	9/6/05		Erica	
Ask the Expert completed	9/6/05		Michael	
User Requests feature completed	9/6/05		Cindy	
Links completed	9/13/05		Kevin	
Intranet site construction completed	10/4/05		Kevin	
Intranet site testing completed	10/18/05		Cindy	
Intranet site promotion completed	10/25/05		Erica	
Intranet site roll-out completed	10/25/05		Kevin	
Monitoring and Controlling				
Status reports	Every Friday		All	
Closing				
Final project presentation completed	10/27/05		Erica	
Sponsor sign-off on project completion	10/27/05		Joe	
Final project report completed	10/28/05		Erica	
Lessons-learned reports submitted	11/1/05		All	

   Human resource issues often occur during project execution, especially conflicts. At several of the team meetings, Erica could see that Michael seemed to be bored and often left the room to make phone calls to clients. She talked to Michael about the situation, and she discovered that Michael was supportive of the project, but he knew he could only spend a minimal amount of time on it. He was much more productive outside of meetings, so Erica agreed to have Michael attend a minimal amount of project team meetings. She could see that Michael was contributing to the team by the feedback he provided and his leadership on the "Ask the Expert" feature for the intranet site. Erica adjusted her communication style to meet his specific needs.

Another problem occurred when Cindy was contacting potential suppliers for software to help with the "Ask the Expert" and User Requests features. Kevin wanted to write all of the software for the project himself, but Cindy knew it made better business sense to purchase these new software capabilities from a reliable source. Cindy had to convince Kevin that it was worth buying some software from other sources.

Cindy also discovered that their estimate of $10,000 was only about half the amount they needed. She discussed the problem with Erica, explaining the need for some custom development no matter which supplier they chose. Erica agreed that they should go with an outside source, and she asked their sponsor to approve the additional funds. Joe agreed, but he stressed the importance of still having the system pay for itself within a year.

Erica also had to ask Joe for help when the project team received a low response rate to their survey and requests for user inputs. Joe sent out an e-mail to all of JWD Consulting's consultants describing the importance of the project. He also offered a free family vacation and five extra vacation days to the person who provided the best examples of how they used tools and templates to manage their projects. Erica then received informative input from the consultants. Having effective communication skills and strong top management support are essential to good project execution.

Project Monitoring and Controlling

Monitoring and controlling is the process of measuring progress toward project objectives, monitoring deviation from the plan, and taking corrective action to match progress with the plan. Monitoring and controlling affects all other phases of the project life cycle. It also involves seven of the nine project management knowledge areas. Table 3-11 lists the knowledge areas, monitoring and controlling processes, and outputs, according to the *PMBOK® Guide 2004*.

Table 3-11: Monitoring and Controlling Processes and Outputs

KNOWLEDGE AREA	MONITORING AND CONTROLLING PROCESS	OUTPUTS
Project Integration Management	Monitor and control project work	■ Recommended corrective actions, preventive actions, and defect repair ■ Forecasts ■ Requested changes
	Integrated change control	■ Approved change requests, corrective actions, preventive actions, and defect repair ■ Rejected change requests ■ Project management plan (updates) ■ Project scope statement (updates) ■ Validated defect repair ■ Deliverables

Table 3-11: Monitoring and Controlling Processes and Outputs (continued)

KNOWLEDGE AREA	MONITORING AND CONTROLLING PROCESS	OUTPUTS
Project Scope Management	Scope verification	■ Accepted deliverables ■ Requested changes ■ Recommended corrective actions
	Scope change control	■ Project scope statement (updates) ■ WBS and dictionary (updates) ■ Scope baseline (updates) ■ Requested changes ■ Recommended corrective actions ■ Organizational process assets (updates) ■ Project management plan (updates)
Project Time Management	Schedule control	■ Schedule model data (updates) ■ Schedule baseline (updates) ■ Performance measurements ■ Requested changes ■ Recommended corrective actions ■ Organizational process assets (updates) ■ Activity list and attributes (updates) ■ Project management plan (updates)
Project Cost Management	Cost control	■ Cost estimate (updates) ■ Cost baseline (updates) ■ Performance measurements ■ Forecasted completion ■ Requested changes ■ Recommended corrective action ■ Organizational process assets (updates) ■ Project management plan (updates)
Project Quality Management	Perform quality control	■ Quality control measurements ■ Validated defect repair and deliverables ■ Quality baseline (updates) ■ Recommended corrective actions, preventive actions, and defect repair ■ Requested changes ■ Organizational process assets (updates) ■ Project management plan (updates)
Project Human Resource Management	Manage project team	■ Requested changes ■ Recommended corrective and preventive actions ■ Organizational process assets (updates) ■ Project management plan (updates)
Project Communications Management	Performance reporting	■ Performance reports ■ Forecasts ■ Requested changes ■ Recommended corrective actions ■ Organizational process assets (updates)

Table 3-11: Monitoring and Controlling Processes and Outputs (continued)

KNOWLEDGE AREA	MONITORING AND CONTROLLING PROCESS	OUTPUTS
	Manage stakeholders	■ Resolved issues ■ Approved change requests and corrective actions ■ Organizational process assets (updates) ■ Project management plan (updates)
Project Risk Management	Risk monitoring and control	■ Risk register (updates) ■ Requested changes ■ Recommended corrective and preventive actions ■ Organizational process assets (updates) ■ Project management plan (updates)
Project Procurement Management	Contract administration	■ Contract documentation ■ Requested changes ■ Recommended corrective actions ■ Organizational process assets (updates) ■ Project management plan (updates)

On the Project Management Intranet Site Project, there were several updates to the project management plan to reflect changes made to the project scope, schedule, and budget. Erica and other project team members took corrective action when necessary. For example, when they were not getting many responses to their survey, Erica asked Joe for help. When Cindy had trouble negotiating with a supplier, she got help from another senior consultant who had worked with that supplier in the past. Erica also had to request more funds for that part of the project.

Project team members submitted a brief status report every Friday. They were originally using a company template for status reports, but Erica found that by modifying the old template, she received better information to help her team work more effectively. She wanted team members not only to report what they did but also to focus on what was going well or not going well and why. This extra information helped team members reflect on the project's progress and identify areas in need of improvement. Table 3-12 is an example of one of Cindy's status reports.

Table 3-12: Sample Weekly Status Report

Project Name: Project Management Intranet Project
Team Member Name: Cindy Dawson, cindy_dawson@jwdconsulting.com
Date: August 5, 2005

Work completed this week:
-Worked with Kevin to start the intranet site construction
-Organized all the content files
-Started developing a file naming scheme for content files
-Continued work on "Ask the Expert" and User Requests features

Table 3-12: Sample Weekly Status Report (continued)

-Met with preferred supplier
-Verified that their software would meet our needs
-Discovered the need for some customization

Work to complete next week:
-Continue work on intranet site construction
-Prepare draft contract for preferred supplier
-Develop new cost estimate for outsourced work

What's going well and why:
The intranet site construction started well. The design was very clear and easy to follow. Kevin really knows what he's doing.

What's not going well and why:
It is difficult to decide how to organize the templates and examples. Need more input from senior consultants and clients.

Suggestions/Issues:
-Hold a special meeting to decide how to organize the templates and examples on the intranet site.
-Get some sample contracts and help in negotiating with the preferred supplier.

Project changes:
I think we can stay on schedule, but it looks like we'll need about $10,000 more for out-sourcing. That's doubling our budget in that area.

In addition to status reports, an important tool for monitoring and controlling the project was using project management software. Each team member submitted his or her actual hours worked on tasks each Friday afternoon by 4 p.m. via the firm's enterprise-wide project management software. They were using the enterprise version of Microsoft Project 2003, so they could easily update their task information via the Web. Erica worked with Jessie to analyze the information, paying special attention to the critical path and earned value data. (See Chapter 6 on Project Time Management for more information on critical path analysis; Chapter 7 on Project Cost Management for a description of earned value management; and Appendix A for more information on using Project 2003 to help control projects.) Erica wanted to finish the project on time, even if it meant spending more money. Joe agreed with that approach, and approved the additional funding Erica projected they would need based on the earned value projections and the need to make up a little time on critical tasks.

Joe again emphasized the importance of the new system paying for itself within a year. Erica was confident that they could exceed the projected financial benefits, and she decided to begin capturing benefits as soon as the project team began testing the system. When she was not working on this project, Erica was managing JWD Consulting's Project Management Office (PMO), and she could already see how the intranet site would help her staff save time and make their consultants more productive. One of her staff members wanted to move into the consulting group, and she believed the PMO could continue to provide its current services with one less person due to this new system—a benefit she had not considered before. Several of the firm's client contracts were based on performance and not hours billed, so she was excited to start measuring the value of the new intranet site to their consultants as well.

Project Closing

The closing process involves gaining stakeholder and customer acceptance of the final products and services and bringing the project, or project phase, to an orderly end. It includes verifying that all of the deliverables are complete, and it often includes a final presentation. Even though many information technology projects are canceled before completion, it is still important to formally close any project and reflect on what can be learned to improve future projects. As philosopher George Santayana said, "Those who cannot remember the past are condemned to repeat it."

It is also important to plan for and execute a smooth transition of the project into the normal operations of the company. Most projects produce results that are integrated into the existing organizational structure. For example, JWD Consulting's Project Management Intranet Site Project will require staff to support the intranet site after it is operational. Erica did include support costs of $40,000 per year for the projected three-year life of the new system. She also created a transition plan as part of the final report to provide for a smooth transition of the system into the firm's operations. The plan included a list of issues that had to be resolved before the firm could put the new intranet site into production. For example, Michael Chen would not be available to work on the intranet site after the six-month project was complete, so they had to know who would support the "Ask the Expert" feature and plan some time for Michael to work with him or her.

Table 3-13 lists the knowledge areas, processes, and outputs of project closing based on the *PMBOK® Guide 2004*. During the closing processes of any project, project team members should take the time to develop appropriate closing procedures, deliver the final product, service, or result of the project, and update organizational process assets, such as project files and a lessons-learned report. If the project team procured items during the project, they must formally complete or close out all contracts.

Table 3-13: Closing Processes and Output

KNOWLEDGE AREA	PROCESS	OUTPUTS
Project Integration Management	Close project	■ Administrative and contract closure procedures ■ Final product, service, or result ■ Organizational process assets (updates)
Project Procurement Management	Contract closure	■ Closed contracts ■ Organizational process assets (updates)

Erica and her team prepared a final report, final presentation, contract files, and lessons-learned report in closing the project. Erica reviewed the confidential, individual lessons-learned report from each team member and wrote one summary lessons-learned report to include in the final documentation, part of which is provided in Table 3-14. Notice the highlighted items, such as the importance of having a good kick-off meeting, working together to develop a team contract, using project management software, and communicating well with the project team and sponsor.

Table 3-14: Lessons-Learned Report (abbreviated)

Project Name:	JWD Consulting Project Management Intranet Site Project
Project Sponsor:	Joe Fleming
Project Manager:	Erica Bell
Project Dates:	May 2, 2005 – November 4, 2005
Final Budget:	$150,000

1. Did the project meet scope, time, and cost goals?

   We did meet scope and time goals, but we had to request an additional $10,000, which the sponsor did approve.

2. What were the success criteria listed in the project scope statement?

   Below is what we put in our project scope statement under project success criteria:

   "Our goal is to complete this project within six months for no more than $140,000. The project sponsor, Joe Fleming, has emphasized the importance of the project paying for itself within one year after the intranet site is complete. To meet this financial goal, the intranet site must have strong user input. We must also develop a method for capturing the benefits while the intranet site is being developed and tested, and after it is rolled out. If the project takes a little longer to complete or costs a little more than planned, the firm will still view it as a success if it has a good payback and helps promote the firm's image as an excellent consulting organization."

3. Reflect on whether or not you met the project success criteria.

   As stated above, the sponsor was not too concerned about going over budget as long as the system would have a good payback period and help promote our firm's image. We have already documented some financial and image benefits of the new intranet site. For example, we have decided that we can staff the PMO with one less person, resulting in substantial cost savings. We have also received excellent feedback from several of our clients about the new intranet site.

4. In terms of managing the project, what were the main lessons your team learned from this project?

   The main lessons we learned include the following:

   - Having a good project sponsor was instrumental to project success. We ran into a couple of difficult situations, and Joe was very creative in helping us solve problems.
   - Teamwork was essential. It really helped to take time for everyone to get to know each other at the kick-off meeting. It was also helpful to develop and follow a team contract.
   - Good planning paid off in execution. We spent a fair amount of time developing a good project charter, scope statement, WBS, schedules, and so on. Everyone worked together to develop these planning documents, and there was strong buy-in.
   - Project management software was very helpful throughout the project.

Table 3-14: Lessons-Learned Report (abbreviated) (continued)

5. Describe one example of what went right on this project.
6. Describe one example of what went wrong on this project.
7. What will you do differently on the next project based on your experience working on this project?

Erica also had Joe sign a client acceptance form, one of the sample templates on the new intranet site that the project team suggested all consultants use when closing their projects.

Table 3-15 provides the table of contents for the final project report. The cover page included the project title, date, and team member names. (Templates for lessons-learned reports, client acceptance forms, and final project reports are included in Appendix D.) Notice the inclusion of a transition plan and a plan to analyze the benefits of the system each year in the final report. Also, notice that the final report includes attachments for all the project management and product-related documents. Erica knew how important it was to provide good final documentation on projects. The project team produced a hard copy of the final documentation and an electronic copy to store on the new intranet site for other consultants to use as desired.

Table 3-15: Final Project Report Table of Contents

1. Project Objectives
2. Summary of Project Results
3. Original and Actual Start and End Dates
4. Original and Actual Budget
5. Project Assessment (Why did you do this project? What did you produce? Was the project a success? What went right and wrong on the project?)
6. Transition Plan
7. Annual Project Benefits Measurement Approach

Attachments:

A. Project Management Documentation
   - Business case
   - Project charter
   - Team contract
   - Scope statement
   - WBS
   - Baseline and actual Gantt chart
   - List of prioritized risks
   - Milestone reports
   - Status reports
   - Contract files
   - Lessons-learned reports
   - Final presentation
   - Client acceptance form

Table 3-15: Final Project Report Table of Contents (continued)

B. Product-Related Documentation
- Survey and results
- Summary of user inputs
- Intranet site content
- Intranet site design documents
- Test plans and reports
- Intranet site promotion information
- Intranet site roll-out information
- Project benefits measurement information

Erica also organized a project closure luncheon for the project team right after their final project presentation. She used the luncheon to share lessons learned and celebrate a job well done!

The project management process groups—initiating, planning, executing, monitoring and controlling, and closing—provide a useful framework for understanding project management. They apply to most projects (information technology and non-information technology) and, along with the project management knowledge areas, help project managers see the big picture of managing a project in their particular organization.

CASE WRAP-UP

Erica Bell and her team finished the Project Management Intranet Site Project on Friday, November 4, 2005, as planned in their project charter. They did go over budget, however, but Joe had approved Erica's request for additional funds, primarily for purchasing external software and customization. Like any project, they had a few challenges, but they worked together as a team and used good project management to meet their sponsor's and users' needs. They received positive initial feedback from internal consultants and some of their clients on the new intranet site. People were asking for templates, examples, and expert advice even before the system was ready. About a year after the project was completed, Erica worked with a member of the Finance department to review the benefits of the new system. The Project Management Office did lose one of its staff members, but it did not request a replacement since the new system helped reduce the PMO's workload. This saved the firm about $70,000 a year for the salary and benefits of that staff position. They also had data to show that the firm saved more than $180,000 on contracts with clients due to the new system, while they had projected just $160,000. The firm was breaking even with the "Ask the Expert" feature the first year, and Erica estimated that the system provided $30,000 in additional profits the first year by generating new business, not the $40,000 they had projected. However, savings from the PMO staff position salary and the extra savings on contracts more than made up for the $10,000 difference. Joe was proud of the project team and the system they produced to help make JWD Consulting a world-class organization.

CHAPTER SUMMARY

Project management is often a number of interlinked processes. The five project management process groups are initiating, planning, executing, monitoring and controlling, and closing. These processes occur at varying levels of intensity throughout each phase of a project, and specific outcomes are produced as a result of each process. Normally the executing processes require the most resources and time, followed by the planning processes.

Mapping the main activities of each project management process group into the nine project management knowledge areas provides a big picture of what activities are involved in project management.

Some organizations develop their own information technology project management methodologies, often using the standards found in the *PMBOK® Guide 2000* or *2004* as a foundation. It is important to tailor project management methodologies to meet the organization's particular needs. Popular methodologies like RUP and Six Sigma include project management processes.

The JWD Consulting case study demonstrates how one organization managed an information technology project from its initiation through its closure. The case study provides several samples of outputs produced for initiating, planning, executing, monitoring and controlling, and closing as follows:

- Business case
- Project charter
- Team contract
- Work breakdown structure
- Gantt chart
- Network diagram

- List of prioritized risks
- Milestone report
- Status reports
- Lessons-learned report
- Final project report

Later chapters in this text provide detailed information on creating these documents and using several of the tools and techniques described in this case study.

DISCUSSION QUESTIONS

1. Briefly describe what happens in each of the five project management process groups (initiating, planning, executing, monitoring and controlling, and closing). On which process should team members spend the most time? Why?

2. Which process group includes information from every single knowledge area? Why?

3. Why do organizations need to tailor project management information found in the *PMBOK® Guide* to create their own methodologies?

4. What are some of the key outputs of each process group?

5. What are some of the typical challenges project teams face during each of the five process groups?

EXERCISES

1. Study the WBS and Gantt charts provided in Figures 3-3 and 3-4. Enter the WBS into Project 2003, indenting tasks as shown to create the WBS hierarchy. Then enter durations and dependencies to try to reproduce the Gantt chart. Check your work with the files available on the companion Web site for this text.

2. Read the article by William Munroe regarding BlueCross BlueShield of Michigan's information technology project management methodology (available on the companion Web site for this text). Write a two-page summary of the article, its key conclusions, and your opinion of it. Do you think many other organizations could apply this methodology, or does each organization need to create its own methodology?

3. Read the "ResNet Case Study" available from the companion Web site for this text at *www.course.com/mis/schwalbe4e* under Chapter 3. This real case study about Northwest Airlines' reservation system illustrates another application of the project management process groups. Write a three-page paper summarizing the main outputs produced during each project process group in this case. Also, include your opinion of whether or not Peeter Kivestu was an effective project manager.

4. JWD Consulting wrote a business case during project initiation. Review the contents of this document (Table 3-3) and find another example of a business case for a project. Write a two-page paper comparing the contents of these documents. In addition, describe whether you think all projects should include a business case before the project sponsors officially approve the project.

5. Read an article about a recipient of PMI's Project of the Year award, such as the Salt Lake City Organizing committee mentioned on PMI's Web site. Write a one-page paper summarizing the project, focusing on how the project manager and team used good project management practices.

RUNNING CASE

Manage Your Health, Inc. (MYH) is a Fortune 500 company that provides a variety of health care services across the globe. MYH has more than 20,000 full-time employees and more than 5,000 part-time employees. Before deciding which projects to pursue, MYH wants to develop a methodology for managing all information technology projects. Management has decided to develop an approach where suggested project management outputs will be similar based on the size of the project. You are part of a team charged with developing this methodology. For example, a communications management plan might be very short and produced by one or two people for small projects, while one for a large project might be fifty or more pages long and created by a committee of key stakeholders. Projects under a certain dollar value might not be tracked in the company's enterprise project management system, while others would.

Tasks

1. After reviewing information technology projects completed in the past few years, your team has found that MYH has a very large volume of projects with varying schedules, budgets, and complexities. List at least five key characteristics your team can use to group projects into three basic categories, which you will later use to determine the project management methodology that projects should follow. Prepare a one-page document that provides clear guidelines for placing projects into one of those three categories.

2. Review the outputs by project management process group based on the *PMBOK® Guide 2004* summarized in this chapter and the outputs used for the JWD Consulting project. Using the three categories you created in Task 1 above, list key recommended outputs by project management process group. Also discuss any other important guidelines that MYH should include in its project management methodology, such as how these outputs should be created. Prepare a two- to four-page document with your suggestions.

ADDITIONAL RUNNING CASES AND OTHER APPENDICES

Appendix C provides additional running cases and questions you can use to practice applying the concepts, tools, and techniques you are learning throughout this and subsequent chapters. Review the running cases provided in Appendix C and on the companion Web site (*www.course.com/mis/schwalbe4e*). Appendix D includes information on templates for various project management documents. For additional sample documents based on real projects, visit the author's Web site (*www.kathyschwalbe.com*). Appendix E and the CD-ROM included with this text include a computer simulation where you can also practice applying the project management process groups and knowledge areas.

SUGGESTED READINGS

1. Charbonneau, Serge, "Software Project Management – A Mapping between RUP and the PMBOK," IBM Developerworks (May 3, 2004).

   In this article, the author explains Rational Unified Process (RUP), including its lifecycle of four phases: inception, elaboration, construction, and transition. He describes the main documents created using a RUP methodology and then presents a detailed mapping comparing it to the project management process groups based on the PMBOK® Guide 2000. He concludes by saying that there are no fundamental incompatibilities between the two standards.

2. Doloi, Hemanta, David Gunaratnam, and Ali Jaafari, "Life Cycle Project Management: A Platform for Strategic Project Management," PM Research Conference Proceedings (July 2004).

   In this article, three researchers from the University of Sydney present a Life Cycle Project Management model to address the need for a more strategic approach toward managing projects. The concepts adopted in their model require a shift in project management culture away from some of the traditional process-based and activity-driven models to emphasize newer approaches embodied in dynamic project organizations.

3. Hartnett, Ed, "Procedures for Professional Software Engineers," (*www.projectmagazine.com*) (January 15, 2003).

   Every professional follows standard procedures when doing his or her work. In this article, Harnett describes the most important standard practices to follow when developing software.

4. Munroe, William, "Developing and Implementing an IT Project Management Process," *ISSGI Review* (First Quarter 2001).

   A copy of this article is available on the companion Web site for this text under the information for Chapter 3. BlueCross BlueShield developed its own information technology project management methodology. They based their methodology on standards found in the PMBOK® Guide 2000, making adjustments as needed to best describe how their organization managed information technology projects.

5. Schwalbe, Kathy, "ResNet Case Study," *Information Technology Project Management, Second Edition* (2001).

This case study is available on the companion Web site for this text under the information for Chapter 3. ResNet is the software that Northwest Airlines uses for making reservations. This case study documents the initiating, planning, executing, monitoring and controlling, and closing process of the ResNet system, providing many examples of real project documents.

6. State of Michigan, "State of Michigan Project Management Methodology" (*http://michigan.gov/dit*) (May 2001).

This document provides an excellent example of how an organization developed its own methodology to manage its information technology projects. The Web site also provides other valuable resources, such as templates for various documents.

KEY TERMS

- **closing processes** — formalizing acceptance of the project or project phase and ending it efficiently.
- **executing processes** — coordinating people and other resources to carry out the project plans and produce the products, services, or results of the project or project phase.
- **initiating processes** — defining and authorizing a project or project phase
- **methodology** — describes *how* things should be done
- **monitoring and controlling processes** — regularly measuring and monitoring progress to ensure that the project team meets the project objectives
- **planning processes** — devising and maintaining a workable scheme to ensure that the project addresses the organization's needs
- **process** — a series of actions directed toward a particular result
- **project management process groups** — the progression of project activities from initiation to planning, executing, monitoring and controlling, and closing
- **Rational Unified Process (RUP)** — an iterative software development process that focuses on team productivity and delivers software best practices to all team members

4

Project Integration Management

Objectives

After reading this chapter, you will be able to:

1. *Describe an overall framework for project integration management as it relates to the other project management knowledge areas and the project life cycle*
2. *Explain the strategic planning process and apply different project selection methods*
3. *Explain the importance of creating a project charter to formally initiate projects*
4. *Discuss the process of creating a preliminary project scope statement*
5. *Describe project management plan development, including content, using guidelines and templates for developing plans, and performing a stakeholder analysis to help manage relationships*
6. *Explain project execution, its relationship to project planning, the factors related to successful results, and tools and techniques to assist in project execution*
7. *Describe the process of monitoring and controlling project work*
8. *Understand the integrated change control process, planning for and managing changes on information technology projects, and developing and using a change control system*
9. *Explain the importance of developing and following good procedures for closing projects*
10. *Describe how software can assist in project integration management*

OPENING CASE

Nick Carson recently became the project manager of a critical biotech project for his company in Silicon Valley. This project involved creating the hardware and software for a DNA-sequencing instrument used in assembling and analyzing the human genome.

Each instrument sold for approximately $200,000, and various clients would purchase several instruments. One hundred instruments running twenty-four hours per day could decipher the entire human genome in less than two years. The biotech project was the company's largest project, and it had tremendous potential for future growth and revenues. Unfortunately, there were problems managing this large project. It had been underway for three years and had already gone through three different project managers. Nick had been the lead software developer on the project before top management made him the project manager. Top management told him to do whatever it took to deliver the first version of the software for the DNA-sequencing instrument in four months and a production version in nine months. Negotiations for a potential corporate buyout with a larger company influenced top management's sense of urgency to complete the project.

Nick was highly energetic and intelligent and had the technical background to make the project a success. He delved into the technical problems and found some critical flaws that were keeping the DNA-sequencing instrument from working. He was having difficulty, however, in his new role as project manager. Although Nick and his team got the product out on time, top management was upset because Nick did not focus on managing all aspects of the project. He never provided top management with accurate schedules or detailed plans of what was happening on the project. Instead of performing the work of project manager, Nick had taken on the role of software integrator and troubleshooter. Nick, however, did not understand top management's problem—he delivered the product, didn't he? Didn't they realize how valuable he was?

WHAT IS PROJECT INTEGRATION MANAGEMENT?

Project integration management involves coordinating all of the other project management knowledge areas throughout a project's life cycle. This integration ensures that all the elements of a project come together at the right times to complete a project successfully. According to the *PMBOK® Guide 2004*, there are seven main processes involved in project integration management:

1. **Develop the project charter**, which involves working with stakeholders to create the document that formally authorizes a project—the charter.

2. **Develop the preliminary project scope statement**, which involves further work with stakeholders, especially users of the project's products, services, or results, to develop the high-level scope requirements. The output of this process is the preliminary project scope statement.

3. **Develop the project management plan**, which involves coordinating all planning efforts to create a consistent, coherent document—the project management plan.

4. **Direct and manage project execution,** which involves carrying out the project management plan by performing the activities included in it. The outputs of this process are deliverables, requested changes, work performance information, implemented change requests, corrective actions, preventive actions, and defect repair.

5. **Monitor and control the project work**, which involves overseeing project work to meet the performance objectives of the project. The outputs of this process are recommended corrective and preventive actions, forecasts, recommended defect repair, and requested changes.

6. **Perform integrated change control,** which involves coordinating changes that affect the project's deliverables and organizational process assets. The outputs of this process include approved and rejected change requests, approved corrective and preventive actions, approved and validated defect repair, deliverables, and updates to the project management plan and project scope statement.

7. **Close the project,** which involves finalizing all project activities to formally close the project. Outputs of this process include final products, services, or results, administrative and contract closure procedures, and updates to organizational process assets.

Many people consider project integration management the key to overall project success. Someone must take responsibility for coordinating all of the people, plans, and work required to complete a project. Someone must focus on the big picture of the project and steer the project team toward successful completion. Someone must make the final decisions when there are conflicts among project goals or people involved. Someone must communicate key project information to top management. This someone is the project manager, and the project manager's chief means for accomplishing all these tasks is project integration management.

Good project integration management is critical to providing stakeholder satisfaction. Project integration management includes interface management.

Interface management involves identifying and managing the points of interaction between various elements of the project. The number of interfaces can increase exponentially as the number of people involved in a project increases. Thus, one of the most important jobs of a project manager is to establish and maintain good communication and relationships across organizational interfaces. The project manager must communicate well with all project stakeholders, including customers, the project team, top management, other project managers, and opponents of the project.

What happens when a project manager does not communicate well with all stakeholders? In the opening case, Nick Carson seemed to ignore a key stakeholder for the DNA-sequencing instrument project—his top management. Nick was comfortable working with other members of the project team, but he was not familiar with his new job as project manager or the needs of the company's top management. Nick continued to do his old job of software developer and took on the added role of software integrator. He mistakenly thought project integration management meant software integration management and focused on the project's technical problems. He totally ignored what project integration management is really about—integrating the work of all of the people involved in the project by focusing on good communication and relationship management. Recall that project management is applying knowledge, skills, tools, and techniques to meet project requirements, while also meeting or exceeding stakeholder needs and expectations. Nick did not take the time to find out what top management expected from him as the project manager; he assumed that completing the project on time and within budget was sufficient to make top management happy.

In addition to not understanding project integration management, Nick did not use holistic or systems thinking (see Chapter 2). He burrowed into the technical details of his particular project. He did not stop to think about what it meant to be the project manager, how this project related to other projects in the organization, or what top management's expectations were of him and his team.

Project integration management must occur within the context of the entire organization, not just within a particular project. The project manager must integrate the work of the project with the ongoing operations of the performing organization. In the opening case, Nick's company was negotiating a potential buyout with a larger company. Consequently, top management needed to know when the DNA-sequencing instrument would be ready, how big the market was for the product, and if they had enough in-house staff to continue to manage projects like this one in the future. They wanted to see a project management plan and a schedule to help them monitor the project's progress and show their potential buyer what was happening. When top managers tried to talk to Nick about these issues, Nick soon returned to discussing the technical details of the project. Even though Nick was very bright, he had no experience or real interest in many of the business aspects of how the company operated. Project managers must always view their projects in the context of the changing needs of their organizations and respond to requests from top management.

Following a standard process for managing projects can help prevent some of the typical problems new and experienced project managers face, including communicating with managing stakeholders. Before organizations begin projects, however, they should go through a formal process to decide what projects to pursue.

STRATEGIC PLANNING AND PROJECT SELECTION

Successful leaders look at the big picture or strategic plan of the organization to determine what types of projects will provide the most value. Therefore, project initiation starts with identifying potential projects, using realistic methods to select which projects to work on, and then formalizing their initiation by issuing some sort of project charter.

Identifying Potential Projects

The first step in project management is deciding what projects to do in the first place. Some may argue that project managers should not be involved in strategic planning and project selection, but successful organizations know that current project managers can provide valuable insight into the project selection process. Figure 4-1 shows a four-stage planning process for selecting information technology projects. Note the hierarchical structure of this model and the results produced from each stage. The first step in this process, starting at the top of the hierarchy, is to tie the information technology strategic plan to the organization's overall strategic plan. **Strategic planning** involves determining long-term objectives by analyzing the strengths and weaknesses of an organization, studying opportunities and threats in the business environment, predicting future trends, and projecting the need for new products and services. Many people are familiar with "SWOT" analysis—analyzing **S**trengths, **W**eaknesses, **O**pportunities, and **T**hreats—that is used to aid in strategic planning. It is very important to have managers from outside the information technology department assist in the information technology planning process, as they can help information technology personnel understand organizational strategies and identify the business areas that support them.

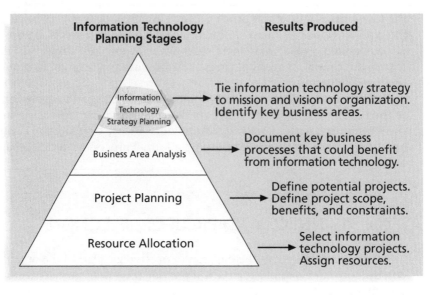

Figure 4-1. Planning Process for Selecting Information Technology Projects

After identifying strategic goals, the next step in the planning process for selecting information technology projects is to perform a business area analysis. This analysis outlines business processes that are central to achieving strategic goals and helps determine which ones could most benefit from information technology. Then, the next step is to start defining potential information technology projects, their scope, benefits, and constraints. The last step in the planning process for selecting information technology projects is choosing which projects to do and assigning resources for working on them.

Aligning Information Technology with Business Strategy

It is often easier to identify and select appropriate information technology projects when organizations align their information technology departments more closely with the business. Even though many information technology projects do not produce "strategic information systems" or receive great publicity, it is critical that the information technology planning process start by analyzing the overall strategy of the organization. Most organizations face thousands of problems and opportunities for improvement. Therefore, an organization's strategic plan should guide the information technology project selection process. An organization must develop a strategy for using information technology to define how it will support the organization's objectives. This information technology strategy must align with the organization's strategic plans and strategy. In fact, research shows that supporting explicit business objectives is the number one reason cited for why organizations invest in information technology projects. Other top criteria for investing in information technology projects include supporting implicit business objectives and providing financial incentives, such as a good internal rate of return (IRR) or net present value (NPV).[1] You will learn more about these financial criteria later in this section.

A 2002 study found a direct correlation between closer business and information technology alignment and reporting structure. Hacket Best Practices found that companies in which the chief information officer (CIO) reports directly to the chief executive officer (CEO) lowered their operational support costs per end user by 17 percent. Many CIOs now report to the CEO, increasing from 24 percent in 1999 to 37 percent in 2002. The study also found that companies with centralized or consolidated information technology operations had 24 percent lower operational support costs per end user. The consistent use of information technology standards also lowered companies' application development costs by 41 percent per user.[2]

Information systems can be and often are central to business strategy. Author Michael Porter, who developed the concept of the strategic value of competitive advantage, has written several books and articles on strategic planning and competition. He and many other experts have emphasized the importance of using information technology to support strategic plans and provide a competitive advantage. Many information systems are classified as "strategic" because they directly support key business strategies. For example, information systems can help an organization support a strategy of being a low-cost producer. As one of the largest retailers in the U.S., Wal-Mart's inventory control system is a classic example of such a system. Information systems can support a strategy of providing specialized products or services that set a company apart from others in the industry. Consider the classic example of Federal Express's introduction of online package tracking systems. They were the first company to provide this type of service, which gave them a competitive advantage until others developed similar systems. Information systems can also support a strategy of selling to a particular market or occupying a specific project niche. Owens-Corning developed a strategic information system that boosted the sales of its home-insulation products by providing its customers with a system for evaluating the energy efficiency of building designs.

[1] Bacon, James. "The Use of Decision Criteria in Selecting Information Systems/Technology Investments," *MIS Quarterly*, Vol. 16, No. 3 (September 1992).

[2] Cosgrove Ware, Lorraine. "By the Numbers," *CIO Magazine* (*www.cio.com*) (September 1, 2002).

Methods for Selecting Projects

Organizations identify many potential projects as part of their strategic planning processes, and often rely on experienced project managers to help them make project selection decisions. However, organizations need to narrow down the list of potential projects to those projects that will be of most benefit. Selecting projects is not an exact science, but it is a critical part of project management. Many methods exist for selecting from among possible projects. Five common techniques are:

1. Focusing on broad organizational needs
2. Categorizing information technology projects
3. Performing net present value or other financial analyses
4. Using a weighted scoring model
5. Implementing a balanced scorecard

In practice, organizations usually use a combination of these approaches to select projects. Each approach has advantages and disadvantages, and it is up to management to decide the best approach for selecting projects based on their particular organization.

Focusing on Broad Organizational Needs

Top managers must focus on meeting their organization's many needs when deciding what projects to undertake, when to undertake them, and to what level. Projects that address broad organizational needs are much more likely to be successful because they will be important to the organization. For example, a broad organizational need might be to improve safety, increase morale, provide better communications, or improve customer service. However, it is often difficult to provide a strong justification for many information technology projects related to these broad organizational needs. For example, it is often impossible to estimate the financial value of such projects, but everyone agrees that they do have a high value. As the old proverb says, "It is better to measure gold roughly than to count pennies precisely."

One method for selecting projects based on broad organizational needs is to determine whether they first meet three important criteria: need, funding, and will. Do people in the organization agree that the project needs to be done? Does the organization have the will and the capacity to provide adequate funds to perform the project? Is there a strong will to make the project succeed? For example, many visionary CEOs can describe a broad need to improve certain aspects of their organizations, such as communications. Although they cannot specifically describe how to improve communications, they might allocate funds to projects that address this need. As projects progress, the organization must reevaluate the need, funding, and will for each project to determine if the project should be continued, redefined, or terminated.

Categorizing Information Technology Projects

Another method for selecting projects is based on various categorizations. One type of categorization assesses whether projects provide a response to a problem, an opportunity, or a directive.

- **Problems** are undesirable situations that prevent an organization from achieving its goals. These problems can be current or anticipated. For example, users of an information system may be having trouble logging onto the system or getting information in a timely manner because the system has reached its capacity. In response, the company could initiate a project to enhance the current system by adding more access lines or upgrading the hardware with a faster processor, more memory, or more storage space.

- **Opportunities** are chances to improve the organization. For example, the project described in the opening case involves creating a new product that can make or break the entire company.

- **Directives** are new requirements imposed by management, government, or some external influence. For example, many projects involving medical technologies must meet rigorous government requirements.

Organizations select projects for any of these reasons. It is often easier to get approval and funding for projects that address problems or directives because the organization must respond to these categories of projects to avoid hurting their business. Many problems and directives must be resolved quickly, but managers must also apply systems thinking and seek opportunities for improving the organization through information technology projects.

Another categorization for information technology projects is based on the time it will take to complete a project or the date by which it must be done. For example, some potential projects must be finished within a specific time window. If they cannot be finished by this set date, they are no longer valid projects. Some projects can be completed very quickly—within a few weeks, days, or even minutes. Many organizations have an end user support function to handle very small projects that can be completed quickly. Even though many information technology projects can be completed quickly, it is still important to prioritize them.

A third categorization for project selection is the overall priority of the project. Many organizations prioritize information technology projects as being high, medium, or low priority based on the current business environment. For example, if it is crucial to cut operating costs quickly, projects that have the most potential to do so would be given a high priority. The organization should always complete high-priority projects first, even if a low- or medium-priority project could be finished in less time. Usually there are many more potential information technology projects than an organization can undertake at any one time, so it is very important to work on the most important ones first.

Performing Net Present Value Analysis, Return on Investment, and Payback Analysis

Financial considerations are often an important aspect of the project selection process, especially during tough economic times. As authors Dennis Cohen and Robert Graham put it, "Projects are never ends in themselves. Financially they are always a means to an end, cash."[3] Many organizations require an approved business case before pursuing projects, and financial projections are a critical component of the business case. (See Chapter 3 for a sample business case.) Three primary methods for determining the projected financial value of projects include net present value analysis, return on investment, and payback analysis. Because project managers often deal with business executives, they must understand how to speak their language, which often boils down to these important financial concepts.

Net Present Value Analysis

Everyone knows that a dollar earned today is worth more than a dollar earned five years from now. **Net present value (NPV) analysis** is a method of calculating the expected net monetary gain or loss from a project by discounting all expected future cash inflows and outflows to the present point in time. An organization should consider only projects with a positive NPV if financial value is a key criterion for project selection. This is because a positive NPV means the return from a project exceeds the **cost of capital**—the return available by investing the capital elsewhere. Projects with higher NPVs are preferred to projects with lower NPVs, if all other factors are equal.

Figure 4-2 illustrates this concept for two different projects. Note that this example starts discounting right away in Year 1 and uses a 10 percent discount rate. You can use the NPV function in Microsoft Excel to calculate the NPV quickly. Detailed steps on performing this calculation manually are provided later in this section. Note that Figure 4-2 lists the projected benefits first, followed by the costs, and then the calculated cash flow amount. Note that the sum of the **cash flow**—benefits minus costs or income minus expenses—is the same for both projects at $5,000. The net present values are different, however, because they account for the time value of money. Project 1 had a negative cash flow of $5,000 in the first year, while Project 2 had a negative cash flow of only $1,000 in the first year. Although both projects had the same total cash flows without discounting, these cash flows are not of comparable financial value. Project 2's NPV of $3,201 is better than Project 1's NPV of $2,316. NPV analysis, therefore, is a method for making equal comparisons between cash flows for multi-year projects.

[3] Cohen, Dennis J. and Robert J. Graham, *The Project Manager's MBA*, San Francisco, Jossey-Bass (2001) p. 31.

	A	B	C	D	E	F	G
1	Discount rate	10%					
2							
3	**PROJECT 1**	YEAR 1	YEAR 2	YEAR 3	YEAR 4	YEAR 5	**TOTAL**
4	Benefits	$0	$2,000	$3,000	$4,000	$5,000	$14,000
5	Costs	$5,000	$1,000	$1,000	$1,000	$1,000	$9,000
6	Cash flow	($5,000)	$1,000	$2,000	$3,000	$4,000	**$5,000**
7	NPV ➔	$2,316					
8		Formula =npv(b1,b6:f6)					
9							
10	**PROJECT 2**	YEAR 1	YEAR 2	YEAR 3	YEAR 4	YEAR 5	**TOTAL**
11	Benefits	$1,000	$2,000	$4,000	$4,000	$4,000	$15,000
12	Costs	$2,000	$2,000	$2,000	$2,000	$2,000	$10,000
13	Cash flow	($1,000)	$0	$2,000	$2,000	$2,000	**$5,000**
14	NPV ➔	$3,201					
15		Formula =npv(b1,b13:f13)					
16							
17							

Figure 4-2. Net Present Value Example

There are some items to consider when calculating NPV. Some organizations refer to the investment year(s) for project costs as Year 0 instead of Year 1 and do not discount costs in Year 0. Other organizations start discounting immediately based on their financial procedures; it's simply a matter of preference for the organization. The discount rate can also vary, often based on the prime rate and other economic considerations. You can enter costs as negative numbers instead of positive numbers, and you can list costs first and then benefits. For example, Figure 4-3 shows the financial calculations JWD Consulting provided in the business case for the Project Management Intranet Site Project described in Chapter 3. Note that the discount rate is 8 percent, benefits are listed first, and costs are entered as positive numbers. The NPV and other calculations are still the same; only the format is different. A project manager must be sure to check with his or her organization to find out its guidelines for when discounting starts, what discount rate to use, and what format the organization prefers.

Discount rate	8%					
Assume the project is completed in Year 0			Year			
	0	**1**	**2**	**3**	**Total**	
Costs	140,000	40,000	40,000	40,000		
Discount factor	1	0.93	0.86	0.79		
Discounted costs	140,000	37,200	34,400	31,600	243,200	
Benefits	0	200,000	200,000	200,000		
Discount factor	1	0.93	0.86	0.79		
Discounted benefits	0	186,000	172,000	158,000	516,000	
Discounted benefits - costs	(140,000)	148,800	137,600	126,400	272,800	◄—NPV
Cumulative benefits - costs	(140,000)	8,800	146,400	272,800		
ROI ————————————►	112%					
		Payback In Year 1				

Figure 4-3. JWD Consulting Net Present Value Example

To determine NPV, follow these steps:

1. Determine the estimated costs and benefits for the life of the project and the products it produces. For example, JWD Consulting assumed the project would produce a system in about six months that would be used for three years, so costs are included in Year 0, when the system is developed, and ongoing system costs and projected benefits are included for Years 1, 2, and 3.

2. Determine the discount rate. A **discount rate** is the rate used in discounting future cash flow. It is also called the capitalization rate or opportunity cost of capital. In Figure 4-2, the discount rate is 10 percent per year, and in Figure 4-3, the discount rate is 8 percent per year.

3. Calculate the net present value. There are several ways to calculate NPV. Most spreadsheet software has a built-in function to calculate NPV. For example, Figure 4-2 shows the formula that Microsoft Excel uses: =npv(discount rate, range of cash flows), where the discount rate is in cell B1 and the range of cash flows for Project 1 are in cells B6 through F6. (See Chapter 7, Project Cost Management, for more information on cash flow and other cost-related terms.) To use the NPV function, there must be a row in the spreadsheet (or column, depending how it is organized) for the cash flow each year, which is the benefit amount for that year minus the cost amount. The result of the formula yields an NPV of

$2316 for Project 1 and $3201 for Project 2. Since both projects have positive NPVs, they are both good candidates for selection. However, since Project 2 has a higher NPV than Project 1 (38 percent higher), it would be the better choice between the two. If the two numbers are close, then other methods should be used to help decide which project to select.

The mathematical formula for calculating NPV is:

$$NPV = \sum_{t=1...n} A/(1+r)^t$$

where t equals the year of the cash flow, n is the last year of the cash flow, A is the amount of cash flow each year, and r is the discount rate. If you cannot enter the data into spreadsheet software, you can perform the calculations by hand or with a simple calculator. First, determine the annual **discount factor**—a multiplier for each year based on the discount rate and year—and then apply it to the costs and benefits for each year. The formula for the discount factor is $1/(1+r)^t$ where r is the discount rate, such as 8 percent, and t is the year. For example, the discount factors used in Figure 4-3 are calculated as follows:

Year 0: discount factor = $1/(1+0.08)^0$ = 1
Year 1: discount factor = $1/(1+0.08)^1$ = .93
Year 2: discount factor = $1/(1+0.08)^2$ = .86
Year 3: discount factor = $1/(1+0.08)^3$ = .79

After determining the discount factor each year, multiply the costs and benefits each year by the appropriate discount factor. For example, in Figure 4-3, the discounted cost for Year 1 is $40,000 * .93 = $37,200. Next, sum all of the discounted costs and benefits each year to get a total. For example, the total discounted costs in Figure 4-3 are $243,200. To calculate the NPV, take the total discounted benefits and subtract the total discounted costs. In this example, the NPV is $516,000 – $243,200 = $272,800.

Return on Investment

Another important financial consideration is return on investment. **Return on investment (ROI)** is the result of subtracting the project costs from the benefits and then dividing by the costs. For example, if you invest $100 today and next year it is worth $110, your ROI is ($110 -100)/100 or 0.10 (10 percent). Note that the ROI is always a percentage. It can be positive or negative. It is best to consider discounted costs and benefits for multi-year projects when calculating ROI. Figure 4-3 shows an ROI of 112 percent. You calculate this number as follows:

ROI = (total discounted benefits – total discounted costs)/discounted costs
ROI = (516,000 – 243,200) / 243,200 = 112%

The higher the ROI, the better. An ROI of 112 percent is outstanding. Many organizations have a required rate of return for projects. The **required rate of return** is the minimum acceptable rate of return on an investment. For example, an organization might have a required rate of return of at least 10 percent for projects. The organization bases the required rate of return on what it could expect to receive elsewhere for an investment of comparable risk. You can also determine a project's **internal rate of return (IRR)** by finding what discount rate results in an NPV of zero for the project. You can use the Goal Seek function in Excel (use Excel's Help function for more information on Goal Seek) to determine the IRR quickly. Simply set the cell containing the NPV calculation to zero while changing the cell containing the discount rate. For example, in Figure 4-2, you could set cell B7 to zero while changing cell B1 to find that the IRR for Project 1 is 27 percent.

Many organizations use ROI in the project selection process. The magazine *Information Week* asked 200 information technology and business professionals how often they measure formal ROI for information technology investments. The poll found that 41 percent of organizations required ROI calculations for all information technology investments, and 59 percent required it for specific initiatives.[4]

Payback Analysis

Payback analysis is another important financial tool to use when selecting projects. **Payback period** is the amount of time it will take to recoup, in the form of net cash inflows, the total dollars invested in a project. In other words, payback analysis determines how much time will lapse before accrued benefits overtake accrued and continuing costs. Payback occurs when the net discounted cumulative benefits and costs reach zero. Figure 4-3 shows how to find the payback period. The cumulative benefits minus costs for Year 0 are ($140,000). Adding that number to the discounted benefits minus costs for Year 1 results in $8,800. Since that number is positive, the payback occurs in Year 1.

Creating a chart helps illustrate more precisely when the payback period occurs. Figure 4-4 charts the cumulative discounted costs and cumulative discounted benefits each year using the numbers from Figure 4-3. Note that the lines cross a little after Year 1 starts. The cumulative discounted benefits and costs are greater than zero where the lines cross. A payback early in Year 1 is normally considered very good.

---

[4] Klein, Paula, "Fickle Yet Focused," *Information Week* (August 3, 2001).

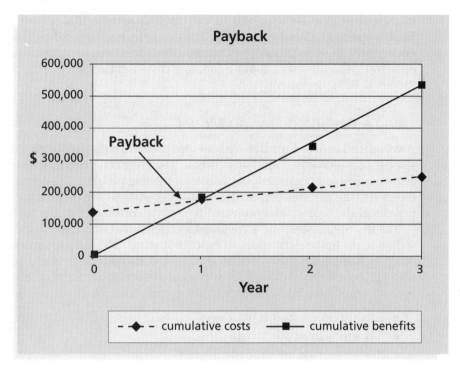

Figure 4-4. Charting the Payback Period

Many organizations have certain recommendations for the length of the payback period of an investment. They might require all information technology projects to have a payback period of less than two years or even one year, regardless of the estimated NPV or ROI. Rhonda Hocker, CIO at San Jose-based BEA Systems Inc., said the middleware vendor has a general rule that its information technology projects should have a payback period of less than one year. The company also tries to limit project teams to no more than twelve people doing a job within four months. Given the economic climate and rapid pace of change in businesses and technology, the company has to focus on delivering positive financial results quickly.[5] However, organizations must also consider long-range goals when making technology investments. Many crucial projects cannot achieve a payback so quickly or be completed in such a short time period.

To aid in project selection, it is important for project managers to understand the organization's financial expectations for projects. It is also important for top

[5] Songini, Marc L, "Tight Budgets Put More Pressure on IT," *Computer World* (December 2, 2002).

management to understand the limitations of financial estimates, particularly for information technology projects. For example, it is very difficult to develop good estimates of projected costs and benefits for information technology projects. You will learn more about estimating costs and benefits in Chapter 7, Project Cost Management.

Using a Weighted Scoring Model

A **weighted scoring model** is a tool that provides a systematic process for selecting projects based on many criteria. These criteria can include factors such as meeting broad organizational needs; addressing problems, opportunities, or directives; the amount of time it will take to complete the project; the overall priority of the project; and projected financial performance of the project.

The first step in creating a weighted scoring model is to identify criteria important to the project selection process. It often takes time to develop and reach agreement on these criteria. Holding facilitated brainstorming sessions or using groupware to exchange ideas can aid in developing these criteria. Some possible criteria for information technology projects include:

- Supports key business objectives
- Has strong internal sponsor
- Has strong customer support
- Uses realistic level of technology
- Can be implemented in one year or less
- Provides positive NPV
- Has low risk in meeting scope, time, and cost goals

Next, you assign a weight to each criterion. Once again, determining weights requires consultation and final agreement. These weights indicate how much you value each criterion or how important each criterion is. You can assign weights based on percentages, and the sum of all of the criteria's weights must total 100 percent. You then assign numerical scores to each criterion (for example, 0 to 100) for each project. The scores indicate how much each project meets each criterion. At this point, you can use a spreadsheet application to create a matrix of projects, criteria, weights, and scores. Figure 4-5 provides an example of a weighted scoring model to evaluate four different projects. After assigning weights for the criteria and scores for each project, you calculate a weighted score for each project by multiplying the weight for each criterion by its score and adding the resulting values.

For example, you calculate the weighted score for Project 1 in Figure 4-5 as:

$$25\% * 90 + 15\% * 70 + 15\% * 50 + 10\% * 25 + 5\% * 20 + 20\% * 50 + 10\% * 20 = 56$$

	A	B	C	D	E	F
1	Criteria	Weight	Project 1	Project 2	Project 3	Project 4
2	Supports key business objectives	25%	90	90	50	20
3	Has strong internal sponsor	15%	70	90	50	20
4	Has strong customer support	15%	50	90	50	20
5	Uses realistic level of technology	10%	25	90	50	70
6	Can be implemented in one year or less	5%	20	20	50	90
7	Provides positive NPV	20%	50	70	50	50
8	Has low risk in meeting scope, time, and cost goals	10%	20	50	50	90
9	**Weighted Project Scores**	**100%**	**56**	**78.5**	**50**	**41.5**

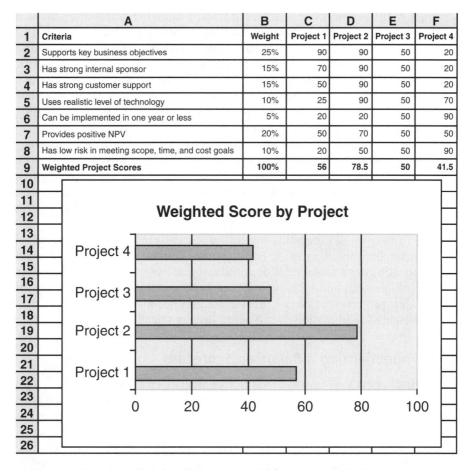

Figure 4-5. Sample Weighted Scoring Model for Project Selection

Note that in this example, Project 2 would be the obvious choice for selection because it has the highest weighted score. Creating a bar chart to graph the weighted scores for each project allows you to see the results at a glance. If you create the weighted scoring model in a spreadsheet, you can enter the data, create and copy formulas, and perform a "what-if" analysis. For example, suppose you change the weights for the criteria. By having the weighted scoring model in a spreadsheet, you can easily change the weights to update the weighted scores and charts automatically. This capability allows you to investigate various options for different stakeholders quickly. Ideally, the result should be reflective of the group's consensus, and any major disagreements should be documented.

Many teachers use a weighted scoring model to determine grades. Suppose grades for a class are based on two homework assignments and two exams. To calculate final grades, the teacher would assign a weight to each of these items. Suppose Homework One is worth 10 percent of the grade, Homework Two is worth 20 percent of the grade, Test One is worth 20 percent of the grade, and Test Two is worth 50 percent of the grade. Students would want to do well on each of these items, but they should focus on performing well on Test Two since it is 50 percent of the grade.

You can also establish weights by assigning points. For example, a project might receive 10 points if it definitely supports key business objectives, 5 points if it somewhat supports them, and 0 points if it is totally unrelated to key business objectives. With a point model, you can simply add all the points to determine the best projects for selection, without having to multiply weights and scores and sum the results.

You can also determine minimum scores or thresholds for specific criteria in a weighted scoring model. For example, suppose an organization really should not consider a project if it does not score at least 50 out of 100 on every criterion. You can build this type of threshold into the weighted scoring model to reject projects that do not meet these minimum standards. As you can see, weighted scoring models can aid in project selection decisions.

Implementing a Balanced Scorecard

Drs. Robert Kaplan and David Norton developed another approach to help select and manage projects that align with business strategy. A **balanced scorecard** is a methodology that converts an organization's value drivers, such as customer service, innovation, operational efficiency, and financial performance, to a series of defined metrics. Organizations record and analyze these metrics to determine how well projects help them achieve strategic goals. Using a balanced scorecard involves several detailed steps. You can learn more about how balanced scorecards work from the Balanced Scorecard Institute (*www.balancedscorecard.org*) or other sources. Although this concept can work within an information technology department specifically, it is best to implement a balanced scorecard throughout an organization because it helps foster alignment between business and information technology.[6] The Balanced Scorecard Institute, which provides training and guidance to organizations using this methodology, quotes Kaplan and Norton's description of the balanced scorecard as follows:

> "The balanced scorecard retains traditional financial measures. But financial measures tell the story of past events, an adequate story for industrial age companies

---

[6] Berkman, Eric, "How to Use the Balanced Scorecard," *CIO Magazine* (May 15, 2002).

for which investments in long-term capabilities and customer relationships were not critical for success. These financial measures are inadequate, however, for guiding and evaluating the journey that information age companies must make to create future value through investment in customers, suppliers, employees, processes, technology, and innovation."[7]

As you can see, organizations can use many approaches to select projects. Many project managers have some say in which projects their organization selects for implementation. Even if they do not, they need to understand the motive and overall business strategy for the projects they are managing. Project managers and team members are often asked to justify their projects, and understanding many of these project selection methods can help them to do so.

Project Charters

After top management decides on which projects to pursue, it is important to let the rest of the organization know about these projects. Management needs to create and distribute documentation to authorize project initiation. This documentation can take many different forms, but one common form is a project charter. A **project charter** is a document that formally recognizes the existence of a project and provides direction on the project's objectives and management. It authorizes the project manager to use organizational resources to complete the project. Ideally, the project manager will provide a major role in developing the project charter. Instead of project charters, some organizations initiate projects using a simple letter of agreement, while others use much longer documents or formal contracts. Key project stakeholders should sign a project charter to acknowledge agreement on the need for and intent of the project. A project charter is a key output of the initiation process, as described in Chapter 3.

The *PMBOK® Guide 2004* lists inputs, tools and techniques, and outputs of the seven project integration management processes, as shown in Figure 4-6. For example, inputs that are helpful in developing a project charter include the following:

- A contract: If you are working on a project under contract, the contract should include much of the information needed for creating a good project charter. Some people might use a contract in place of a charter; however, many contracts are difficult to read and can often change, so it is still a good idea to create a project charter.

---

[7] The Balanced Scorecard Institute. "What Is a Balanced Scorecard," (*www.balancedscorecard.org*) (February 2003).

- A statement of work: A statement of work is a document that describes the products or services to be created by the project team. It usually includes a description of the business need for the project, a summary of the requirements and characteristics of the products or services, and organizational information, such as appropriate parts of the strategic plan, showing the alignment of the project with strategic goals.

- Enterprise environmental factors: These factors include the organization's structure, culture, infrastructure, human resources, personnel policies, marketplace conditions, stakeholder risk tolerances, industry risk information, and project management information systems.

- Organizational process assets information: **Organizational process assets** include formal and informal plans, policies, procedures, guidelines, information systems, financial systems, management systems, lessons learned, and historical information that help people understand, follow, and improve business processes in a specific organization. How an organization manages its business processes, promotes learning, and shares knowledge can also provide important information when creating a project charter. Managers should review formal and informal enterprise plans, policies, procedures, guidelines, and management systems when developing a project charter.

PROJECT INTEGRATION MANAGEMENT

4.1 Develop Project Charter

1. **Inputs**
   .1 Contract
   .2 Statement of work
   .3 Enterprise environmental factors
   .4 Organizational process assets
2. **Tools and Techniques**
   .1 Project selection methods
   .2 Project management methodology
   .3 Project management information system
   .4 Expert judgment
3. **Outputs**
   .1 Project charter

4.2 Develop Preliminary Project Scope Statement

1. **Inputs**
   .1 Project charter
   .2 Statement of work
   .3 Enterprise environmental factors
   .4 Organizational process assets
2. **Tools and Techniques**
   .1 Project management methodology
   .2 Project management information system
   .3 Expert judgment
3. **Outputs**
   .1 Preliminary project scope statement

4.3 Develop Project Management Plan

1. **Inputs**
   .1 Preliminary project scope statement
   .2 Project management processes
   .3 Enterprise environmental factors
   .4 Organizational process assets
2. **Tools and Techniques**
   .1 Project management methodology
   .2 Project management information system
   .3 Expert judgment
3. **Outputs**
   .1 Project management plan

4.4 Direct and Manage Project Execution

1. **Inputs**
   .1 Project management plan
   .2 Approved corrective actions
   .3 Approved preventive actions
   .4 Approved change requests
   .5 Approved defect repair
   .6 Validated defect repair
   .7 Administrative closure procedures
2. **Tools and Techniques**
   .1 Project management methodology
   .2 Project management information system
3. **Outputs**
   .1 Deliverables
   .2 Requested changes
   .3 Implemented change requests
   .4 Implemented corrective actions
   .5 Implemented preventive actions
   .6 Implemented defect repair
   .7 Work performance information

4.5 Monitor and Control Project Work

1. **Inputs**
   .1 Project management plan
   .2 Work performance information
   .3 Rejected change requests
2. **Tools and Techniques**
   .1 Project management methodology
   .2 Project management information system
   .3 Earned value technique
   .4 Expert judgment
3. **Outputs**
   .1 Recommended corrective actions
   .2 Recommended preventive actions
   .3 Forecasts
   .4 Recommended defect repair
   .5 Requested changes

4.6 Integrated Change Control

1. **Inputs**
   .1 Project management plan
   .2 Requested changes
   .3 Work performance information
   .4 Recommended preventive actions
   .5 Recommended corrective actions
   .6 Recommended defect repair
   .7 Deliverables
2. **Tools and Techniques**
   .1 Project management methodology
   .2 Project management information system
   .3 Expert judgment
3. **Outputs**
   .1 Approved change requests
   .2 Rejected change requests
   .3 Project management plan (updates)
   .4 Project scope statement (updates)
   .5 Approved corrective actions
   .6 Approved preventive actions
   .7 Approved defect repair
   .8 Validated defect repair
   .9 Deliverables

4.7 Close Project

1. **Inputs**
   .1 Project management plan
   .2 Contract documentation
   .3 Enterprise environmental factors
   .4 Organizational process assets
   .5 Work performance information
   .6 Deliverables
2. **Tools and Techniques**
   .1 Project management methodology
   .2 Project management information system
   .3 Expert judgment
3. **Outputs**
   .1 Administrative closure procedure
   .2 Contract closure procedure
   .3 Final product, service, or result
   .4 Organizational process assets (updates)

Figure 4-6. Project Integration Management Overview
PMBOK® Guide 2004, p. 79.

Tools and techniques for developing a project charter listed in the *PMBOK® Guide 2004* include project selection methods, a project management methodology, a project management information system, and expert judgment. The previous section described several project selection methods, and Chapter 3 discussed using a project management methodology. A project management information system is a set of automated tools that help to organize and integrate project management information throughout an organization. In addition to project management software, such as Project 2003 or similar software, a project management information system can include additional software that helps people in an organization create, modify, track, and communicate project information. See the sections in Chapters 4–12 about using software to assist in each knowledge area. Expert judgment is also an important technique for creating a project charter as well as for performing many other project management processes.

The only output of the process to develop a project charter is a project charter. Although the format of project charters can vary tremendously, they should include the following basic information:

- The project's title and date of authorization
- The project manager's name and contact information
- A summary schedule, including the planned start and finish dates; if a summary milestone schedule is available, it should also be included or referenced
- A summary of the project's budget or reference to budgetary documents
- A brief description of the project objectives, including the business need or other justification for authorizing the project
- A summary of the planned approach for managing the project, which should describe stakeholder needs and expectations, important assumptions, and constraints, and refer to related documents, such as a communications management plan, as available
- A roles and responsibilities matrix
- A sign-off section for signatures of key project stakeholders
- A comments section in which stakeholders can provide important comments related to the project

Chapters 3 and 5 include samples of project charters. Unfortunately, many internal projects, like the one in the opening case of this chapter, do not have project charters. They often have a budget and general guidelines, but no formal, signed documentation. If Nick had a project charter to refer to—especially if it included information on the approach for managing the project—his job would have been easier. Project charters are usually not difficult to write. What is difficult is getting people with the proper knowledge and authority to write and sign the project charters. Top management should have reviewed the charter with Nick, since he was the project manager. In their initial meeting, they should have discussed roles and responsibilities, as well as their expectations of how Nick should work with them. If there is no project charter, the project manager should work with key stakeholders, including top management, to create one.

Since many projects fail because of unclear requirements and expectations, starting with a project charter makes a lot of sense. If project managers are having difficulty obtaining support from project stakeholders, for example, they can refer to what everyone agreed to in the project charter. After formally recognizing the existence of a project, the next step in project integration management is preparing a preliminary scope statement.

PRELIMINARY SCOPE STATEMENTS

A **scope statement** is a document used to develop and confirm a common understanding of the project scope. It describes in detail the work to accomplish on the project and is an important tool for preventing **scope creep**—the tendency for project scope to keep getting bigger, as described in more detail in Chapter 5. It is helpful to create a *preliminary* or initial scope statement during project initiation so that the entire project team can start important discussions and work related to the project scope. A more detailed scope statement is prepared as part of project scope management. There are usually several versions of the scope statement, and each one becomes more detailed as the project progresses and more information becomes available.

Scope statements, like project charters, also vary by type of project. Complex projects have very long scope statements, whereas smaller projects have shorter scope statements. See Chapter 3 for a sample scope statement for JWD Consulting's Project Management Intranet Site Project. Items often described in a preliminary scope statement include the project objectives, product or service requirements and characteristics, project boundaries, deliverables, product acceptance criteria, project assumptions and constraints, the organizational structure for the project, an initial list of defined risks, a summary of schedule milestones, a rough order of magnitude cost estimate, configuration management requirements, and a description of approval requirements.

Referring again to the opening case, we can assume that Nick and his project team had a good handle on the project scope since they did produce the desired product and results. Most of the team members had been working on the project for a long time, and they were experts in various aspects of biotechnology. However, Nick and senior management should have discussed and agreed on what information they needed to review, such as schedule milestones and product acceptance criteria, since those items had the most effect on the corporate buyout. A preliminary and more detailed scope statement, updated as the project progressed, would have been another tool for improving communications between Nick and senior management.

PROJECT MANAGEMENT PLANS

To coordinate and integrate information across project management knowledge areas and across the organization, there must be a good project management plan. A **project management plan** is a document used to coordinate all project planning documents and help guide a project's execution and control. Plans created in the other knowledge areas are considered subsidiary parts of the overall project management plan. Project management plans also document project planning assumptions and decisions regarding choices, facilitate communication among stakeholders, define the content, extent, and timing of key management reviews, and provide a baseline for progress measurement and project control. Project management plans should be dynamic, flexible, and subject to change when the environment or project changes. These plans should greatly assist the project manager in leading the project team and assessing project status.

To create and assemble a good project management plan, the project manager must practice the art of project integration management, since information is required from all of the project management knowledge areas. Working with the project team and other stakeholders to create a project management plan will help the project manager guide the project's execution and understand the overall project.

Project Management Plan Contents

Just as projects are unique, so are project management plans. A small project involving a few people over a couple of months might have a project management plan consisting of only a project charter, scope statement, and Gantt chart. A large project involving a hundred people over three years would have a much more detailed project management plan. It is important to tailor project management plans to fit the needs of specific projects. The project management plans should guide the work, so they should be only as detailed as needed for each project. (See the first Suggested Reading for information on accessing a real example of a complete project management plan.)

There are, however, common elements to most project management plans. Parts of a project management plan include an introduction or overview of the project, a description of how the project is organized, the management and technical processes used on the project, and sections describing the work to be performed, the schedule, and the budget.

The introduction or overview of the project should include, as a minimum, the following information:

- The project name: Every project should have a unique name. Unique names help distinguish each project and avoid confusion among related projects.

- A brief description of the project and the need it addresses: This description should clearly outline the goals of the project and reason for the project. It should be written in layperson's terms, avoid technical jargon, and include a rough time and cost estimate.

- The sponsor's name: Every project needs a sponsor. Include the name, title, and contact information of the sponsor in the introduction.

- The names of the project manager and key team members: The project manager should always be the contact for project information. Depending on the size and nature of the project, names of key team members may also be included.

- Deliverables of the project: This section should briefly list and describe the products that will be produced as part of the project. Software packages, pieces of hardware, technical reports, and training materials are examples of deliverables.

- A list of important reference materials: Many projects have a history preceding them. Listing important documents or meetings related to a project helps project stakeholders understand that history. This section should reference the plans produced for other knowledge areas. (Recall from Chapter 3 that every single knowledge area includes some planning processes.) Therefore, the project management plan should reference and summarize important parts of the scope management, schedule management, cost management, quality management, human resource management, communications management, risk management, and procurement management plans.

- A list of definitions and acronyms, if appropriate: Many projects, especially information technology projects, involve terminology unique to a particular industry or technology. Providing a list of definitions and acronyms will help avoid confusion.

The description of how the project is organized should include the following information:

- Organizational charts: In addition to an organizational chart for the company sponsoring the project and for the customer's company (if it is an external customer), there should be a project organizational chart to show the lines of authority, responsibilities, and communication for the project. For example, the Manhattan Project introduced in Chapter 1 had a very detailed organizational chart to show all the people working on the project.

- Project responsibilities: This section of the project plan should describe the major project functions and activities and identify those individuals who are responsible for them. A responsibility assignment matrix (described in Chapter 9) is a tool often used for displaying this information.

- Other organizational or process-related information: Depending on the nature of the project, there may be a need to document major processes followed on the project. For example, if the project involves releasing a major software upgrade, it might help everyone involved in the project to see a diagram or timeline of the major steps involved in this process.

The section of the project management plan describing management and technical approaches should include the following information:

- Management objectives: It is important to understand top management's view of the project, what the priorities are for the project, and any major assumptions or constraints.
- Project controls: This section describes how to monitor project progress and handle changes. Will there be monthly status reviews and quarterly progress reviews? Will there be specific forms or charts to monitor progress? Will the project use earned value management (described in Chapter 7) to assess and track performance? What is the process for change control? What level of management is required to approve different types of changes? (You will learn more about change control later in this chapter.)
- Risk management: This section briefly addresses how the project team will identify, manage, and control risks. It should refer to the risk management plan, if one is required for the project.
- Project staffing: This section describes the number and types of people required for the project. It should refer to the staffing management plan, if one is required for the project.
- Technical processes: This section describes specific methodologies a project might use and explains how to document information. For example, many information technology projects follow specific software development methodologies or use particular Computer Aided Software Engineering (CASE) tools. Many companies or customers also have specific formats for technical documentation. It is important to clarify these technical processes in the project management plan.

The next section of the project management plan should describe the work to perform and reference the scope management plan. It should summarize the following:

- Major work packages: A project manager usually organizes the project work into several work packages using a work breakdown structure (WBS), and produces a scope statement to describe the work in more detail. This section should briefly summarize the main work packages for the project and refer to appropriate sections of the scope management plan.
- Key deliverables: This section lists and describes the key products produced as part of the project. It should also describe the quality expectations for the product deliverables.

- Other work-related information: This section highlights key information related to the work performed on the project. For example, it might list specific hardware or software to use on the project or certain specifications to follow. It should document major assumptions made in defining the project work.

The project schedule information section should include the following:

- Summary schedule: It is helpful to see a one-page summary of the overall project schedule. Depending on the size and complexity of the project, the summary schedule might list only key deliverables and their planned completion dates. For smaller projects, it might include all of the work and associated dates for the entire project in a Gantt chart. For example, the Gantt chart and milestone schedule provided in Chapter 3 for JWD Consulting were fairly short and simple.

- Detailed schedule: This section provides information on the project schedule that is more detailed. It should reference the schedule management plan and discuss dependencies among project activities that could affect the project schedule. For example, it might explain that a major part of the work cannot start until an external agency provides funding. A network diagram can show these dependencies (see Chapter 6, Project Time Management).

- Other schedule-related information: Many assumptions are often made in preparing project schedules. This section should document major assumptions and highlight other important information related to the project schedule.

The budget section of the project management plan should include the following:

- Summary budget: The summary budget includes the total estimate of the overall project's budget. It could also include the budget estimate for each month or year by certain budget categories. It is important to provide some explanation of what these numbers mean. For example, is the total budget estimate a firm number that cannot change, or is it a rough estimate based on projected costs over the next three years?

- Detailed budget: This section summarizes what is in the cost management plan and includes more detailed budget information. For example, what are the fixed and recurring cost estimates for the project each year? What are the projected financial benefits of the project? What types of people are needed to do the work, and how are the labor costs calculated? See Chapter 7, Project Cost Management, for more information on creating cost estimates and budgets.

■ Other budget-related information: This section documents major assumptions and highlights other important information related to financial aspects of the project.

Using Guidelines to Create Project Management Plans

Many organizations use guidelines to create project management plans. Project 2003 and other project management software packages come with several template files to use as guidelines. However, do not confuse a project management plan with a Gantt chart. The project management plan includes all project-planning documents. Plans created in the other knowledge areas can be considered subsidiary parts of the overall project management plan.

Many government agencies also provide guidelines for creating project management plans. For example, the U.S. Department of Defense (DOD) Standard 2167, Software Development Plan, describes the format for contractors to use in creating a plan for software development for DOD projects. The Institute of Electrical and Electronics Engineers (IEEE) Standard 1058-1998 describes the contents of a Software Project Management Plan (SPMP). Table 4-1 provides some of the categories for the IEEE SPMP. Companies working on software development projects for the Department of Defense must follow this or a similar standard.

Table 4-1: Sample Contents for a Software Project Management Plan (SPMP)

| | | MAJOR PROJECT MANAGEMENT PLAN SECTIONS | | | |
	OVERVIEW	PROJECT ORGANIZATION	MANAGERIAL PROCESS PLANS	TECHNICAL PROCESS PLANS	SUPPORTING PROCESS PLANS
Section Topics	Purpose, scope, and objectives; assumptions and constraints; project deliverables; schedule and budget summary; evolution of the plan	External interfaces; internal structure; roles and responsibilities	Start-up plans (estimation, staffing, resource acquisition, and project staff training plans); work plan (work activities, schedule, resource, and budget allocation); control plan; risk management plan; closeout plan	Process model; methods, tools, and techniques; infrastructure plan; product acceptance plan	Configuration management plan; verification and validation plan; documentation plan; quality assurance plan; reviews and audits; problem resolution plan; subcontractor management plan; process improvement plan

IEEE Standard 1058-1998.

In many private organizations, specific documentation standards are not as rigorous; however, there are usually guidelines for developing project management plans. It is good practice to follow standards or guidelines for developing project management plans in an organization to facilitate the development and execution of those plans. The organization can work more efficiently if all project management plans follow a similar format. Recall from Chapter 1 that companies that excel in project management develop and deploy standardized project delivery systems.

> The winners clearly spell out what needs to be done in a project, by whom, when, and how. For this they use an integrated toolbox, including PM tools, methods, and techniques...If a scheduling template is developed and used over and over, it becomes a repeatable action that leads to higher productivity and lower uncertainty. Sure, using scheduling templates is neither a breakthrough nor a feat. But laggards exhibited almost no use of the templates. Rather, in constructing schedules their project managers started with a clean sheet, a clear waste of time.[8]

For example, in the opening case, Nick Carson's top managers were disappointed because he did not provide them with the project planning information they needed to make important business decisions. They wanted to see detailed project management plans, including schedules and a means for tracking progress. Nick had never created a project management plan or a status report before, and the organization did not provide templates or examples to follow. If it had, Nick might have been able to deliver the information top management was expecting. See the companion Web site for templates and samples or real project documents. The first Suggested Reading includes an entire project management plan, which is available on the author's Web site.

Stakeholder Analysis and Top Management Support

Because the ultimate goal of project management is to meet or exceed stakeholder needs and expectations from a project, it is important to include a stakeholder analysis as part of project planning. A **stakeholder analysis** documents information such as key stakeholders' names and organizations, their roles on the project, unique facts about each stakeholder, their level of interest in the project, their influence on the project, and suggestions for managing relationships with each stakeholder. Since a stakeholder analysis often includes sensitive information, it should not be part of the overall project plan, which would be available for all stakeholders to review. In many cases, only project managers and other key project team members should see the stakeholder analysis. Table 4-2 provides an example of a stakeholder analysis that

---

[8] Milosevic, Fragan and And Ozbay. "Delivering Projects: What the Winners Do." Proceedings of the Project Management Institute Annual Seminars & Symposium (November 2001).

Nick Carson could have used to help him manage the DNA-sequencing instrument project described in the opening case. It is important for project managers to take the time to perform a stakeholder analysis of some sort to help understand and meet stakeholder needs and expectations.

Table 4-2: Sample Stakeholder Analysis

	AHMED	SUSAN	ERIK	MARK	DAVID
	KEY STAKEHOLDERS				
Organization	Internal senior management	Project team	Project team	Hardware vendor	Project manager for other internal projects
Role on project	Project sponsor and one of the company's founders	DNA sequencing expert	Lead programmer	Supplier of some instrument hardware	Competitor for company resources
Unique facts	Quiet, demanding, likes details, business-focused, Stanford MBA	Ph.D. in biology, easy to work with, has toddler	Very smart, best programmer I know, weird sense of humor	Head of a start-up company, he knows we can make him rich if this works	Nice guy, one of the oldest people at company, has three kids in college
Level of interest	Very high	Very high	High	Very high	Low to medium
Level of influence	Very high; can call the shots	Subject matter expert; critical to success	High; hard to replace	Low; other vendors available	Low to medium
Suggestions on managing relationship	Keep informed, let him lead conversations, do as he says and quickly	Make sure she reviews specifications and leads testing; can do some work from home	Keep him happy so he stays; emphasize stock options; likes Mexican food	Give him enough lead time to deliver hardware	He knows his project takes a back seat to this one, but I can learn from him

A stakeholder analysis will also help the project manager lead the execution of the project plan. For example, Nick Carson should have taken the time to get to know key stakeholders, especially the project sponsor, so he could understand the project sponsor's expectations and working style. Because Nick had a tendency to talk at great lengths about technical details and Ahmed, the project sponsor, was very quiet by nature, Nick should have let him lead conversations about the project to determine his concerns and interests. Once Ahmed told Nick what to do, Nick should have focused on meeting his sponsor's expectations. If Nick did not know how to perform certain project management functions, he should have asked for help. Nick should have realized his own limitations and perhaps hired an assistant with a project management background.

Likewise, Ahmed and other top managers at the company should have taken the time to understand Nick's needs and provided him with more support. Nick heard, "Do whatever it takes to get the job done," so he delved into what he loved to do: technical problem solving. Ahmed did not hire someone to fill Nick's previous technical position, so how could he expect Nick to do both the management work and technical work required for the project? He should have seen that Nick needed help and worked with him to plan the best course of action. Recall from Chapter 2 that executive support contributed the most to successful information technology projects. Top management must work with project managers to help them plan and execute projects successfully. Since Nick was the fourth project manager on this project, top management at his company obviously had some problems choosing and working with project managers in general.

PROJECT EXECUTION

Directing and managing project execution involves managing and performing the work described in the project management plan. The majority of time on a project is usually spent on execution, as is most of the project's budget. The application area of the project directly affects project execution because the products of the project are produced during project execution. For example, the DNA-sequencing instrument project from the opening case and all associated software and documentation would be produced during project execution. The project team would need to use their expertise in biology, hardware and software development, and testing to produce the product successfully.

The project manager would also need to focus on leading the project team and managing stakeholder relationships to execute the project management plan successfully. Project human resource management and project communications management are crucial to a project's success. See Chapters 9 and 10 respectively for more information on those knowledge areas. If the project involves a significant amount of risk or outside resources, the project manager also needs to be well versed in project risk management and project procurement management. See Chapters 11 and 12 for details on those knowledge areas. Many unique situations occur during project execution, so project managers must be flexible and creative in dealing with them. Review the situation Erica Bell faced during project execution in Chapter 3. Also review the ResNet case study (available on the companion Web site for this text) to understand the execution challenges project manager Peeter Kivestu and his project team faced.

Coordinating Planning and Execution

Project integration management views project planning and execution as intertwined and inseparable activities. The main function of creating a project management plan and all other project plans (e.g., the scope management plan, schedule management plan, and so on) is to guide project execution. A good plan should help produce good products or work results. Plans should document what good work results consist of. Updates to plans should reflect knowledge gained from completing work earlier in the project. Anyone who has tried to write a computer program from poor specifications appreciates the importance of a good plan. Anyone who has had to document a poorly programmed system appreciates the importance of good execution.

A commonsense approach to improving the coordination between project plan development and execution is to follow this simple rule: Those who will do the work should plan the work. All project personnel need to develop both planning and executing skills and need experience in these areas. In information technology projects, programmers who have had to write detailed specifications and then create the code from their own specifications become better at writing specifications. Likewise, most systems analysts begin their careers as programmers, so they understand what type of analysis and documentation they need to write good code. Although project managers are responsible for developing the overall project management plan, they must solicit input from the project team members who are developing plans in each knowledge area.

Providing Strong Leadership and a Supportive Culture

Strong leadership and a supportive organizational culture are crucial during project execution. Project managers must lead by example to demonstrate the importance of creating good project plans and then following them in project execution. Project managers often create plans for things they need to do themselves. If project managers follow through on their own plans, their team members are more likely to do the same.

⊗ What Went Wrong?

Many people have a poor view of plans based on their experiences. Top managers often require a project management plan, but then no one follows up on whether the plan was followed. For example, one project manager said he would meet with each project team leader within two months to review their project plans. The project manager created a detailed schedule for these reviews. He cancelled the first meeting due to another business commitment. He rescheduled the next meeting for unexplained personal reasons. Two months later, the project manager had still not met with over half of the project team leaders. Why should project members feel obligated to follow their own plans when the project manager obviously did not follow his?

Another example of misusing plans comes from a college dean who refused to authorize hiring a new faculty member, even though adjunct faculty staffed half of the classes in that academic discipline. The dean said the department's strategic plan was outdated, and they could not hire anyone else until they had a new strategic plan. The college itself had not updated its strategic plan in over six years, and members of the department suspected the situation was related to the dean wanting more political power within the department. A faculty member who taught strategic planning quickly wrote a new department strategic plan, got approval from other department members, and sent it to the dean. The department did hire a new faculty member, but no one ever looked at that strategic plan again. Remember that project plans, unlike many other plans, should guide execution, not satisfy personal or political goals.

Good project execution also requires a supportive organizational culture. For example, organizational procedures can help or hinder project execution. If an organization has useful guidelines and templates for project management that everyone in the organization follows, it will be easier for project managers and their teams to plan and do their work. If the organization uses the project plans as the basis for performing and monitoring progress during execution, the culture will promote the relationship between good planning and execution. On the other hand, if organizations have confusing or bureaucratic project management guidelines that hinder getting work done or do not measure progress against plans, project managers and their teams will be frustrated.

Even with a supportive organizational culture, project managers may sometimes find it necessary to break the rules to produce project results in a timely manner. When project managers break the rules, politics will play a role in the results. For example, if a particular project requires use of nonstandard software, the project manager must use his or her political skills to convince concerned stakeholders of the need to break the rules on using only standard software. Breaking organizational rules—and getting away with it—requires excellent leadership, communication, and political skills.

Capitalizing on Product, Business, and Application Area Knowledge

In addition to possessing strong leadership, communication, and political skills, project managers also need to possess product, business, and application area knowledge to execute projects successfully. It is often helpful for information technology project managers to have prior technical experience or at least a working knowledge of information technology products. For example, if the project manager were leading a Joint Application Design (JAD) team to help define user requirements, it would be helpful for him or her to understand the language of the business and technical experts on the team. See Chapter 5, Project Scope Management, for more information on JAD.

Many information technology projects are small, so project managers may be required to perform some technical work or mentor team members to complete the project. For example, a three-month project to develop a Web-based application with only three team members would benefit most from a project manager who can complete some of the technical work. On larger projects, however, the project manager's primary responsibility is to lead the team and communicate with key project stakeholders. He or she would not have time to do any of the technical work. In this case, it is usually best that the project manager understand the business and application area of the project more than the technology involved.

However, it is very important on large projects for the project manager to understand the business and application area of their project. For example, Northwest Airlines completed a series of projects in the last ten years to develop and upgrade its reservation systems. The company spent millions of dollars and had more than seventy full-time people working on the projects at peak periods. The project manager, Peeter Kivestu, had never worked in an information technology department, but he had extensive knowledge of the airline industry and the reservations process. He carefully picked his team leaders, making sure they had the required technical and product knowledge. ResNet was the first large information technology project at Northwest Airlines led by a business manager instead of a technical expert, and it was a roaring success. Many organizations have found that large information technology projects require experienced general managers who understand the business and application area of the technology, not the technology itself. (You can read the entire ResNet case study from the companion Web site listed in the second Suggested Reading for Chapter 3.)

Project Execution Tools and Techniques

Directing and managing project execution requires specialized tools and techniques, some of which are unique to project management. Project managers can use specific tools and techniques to perform activities that are part of execution processes. These include:

- **Project management methodology**: As mentioned earlier, many experienced project managers believe the most effective way to improve project management is to follow a methodology that describes not only what to do in managing a project, but how to do it. For example, if Nick and his senior management had agreed that they would have weekly status review meetings in Ahmed's office where Nick and his key project team members would provide short presentations on specific topics, the execution of the project would have gone much more smoothly.

- **Project management information systems**: As described in Chapter 1, there are hundreds of project management software products available on the market today. Many large organizations are moving toward powerful enterprise project management systems that are accessible via the Internet. Project managers or other team members can create Gantt charts that include links to other planning documents. For example, Nick or one of his teammates could have created a detailed Gantt chart for the biotech project and created a link to other key planning documents. Nick could have shown the summary tasks during the status review meetings, and if top management had questions, they could link to supporting detail. They could also have set baselines for completing the project and tracked their progress toward achieving those goals.

Although these tools and techniques can aid in project execution, project managers must remember that positive leadership and strong teamwork are critical to successful project management. Project managers should delegate the detailed work involved in using these tools to other team members and focus on providing leadership for the whole project to ensure project success.

☑ *What Went Right?*

Malaysia's capital, Kuala Lumpur, has become one of Asia's busiest, most exciting cities. With growth, however, came traffic. To help alleviate this problem, the city hired a local firm in mid-2003 to manage a MYR $400 million (U.S. $105 million) project to develop a state-of-the-art Integrated Transport Information System (ITIS). The Deputy Project Director, Lawrence Liew, explained that they broke the project into four key phases and focused on several key milestones. They deliberately kept the work loosely structured to allow the team to be more flexible and creative in handling uncertainties. They based the entire project team within a single project office to streamline communications and facilitate quick problem solving through ad hoc working groups. They also used a dedicated project intranet to exchange information between the project team and sub-contractors. The new system is planned for completion in mid-2005.[9]

Project managers and their teams are most often remembered for how well they executed a project and handled difficult situations. As John Madden often says when announcing Super Bowl events, the key to winning is good execution. When one *losing* coach was asked what he thought about his team's execution, he responded, "I'm all for it!"

[9] Parkes, Sarah, "Crosstown Traffic," *PM Network* (August 2004).

MONITORING AND CONTROLLING PROJECT WORK

On large projects, many project managers say that 90 percent of the job is communicating and managing changes. Changes are inevitable on most projects, so it's important to develop and follow a process to monitor and control changes.

Monitoring project work includes collecting, measuring, and disseminating performance information. It also involves assessing measurements and analyzing trends to determine what process improvements can be made. The project team should continuously monitor project performance to assess the overall health of the project and identify areas that require special attention.

The project management plan, work performance information, performance reports, and change requests are all important inputs for monitoring and controlling project work. Key tools and techniques for performing this process include using a project management methodology and a project management information system, as described for project execution. Additional tools and techniques include expert judgment and earned value management, described in detail in Chapter 7, Project Cost Management.

Media Snapshot

Few events get more media attention than the Olympic Games. Imagine all the work involved in planning and executing an event that involves thousands of athletes from around the world with millions of spectators. The 2002 Olympic Winter Games and Paralympics took five years to plan and cost more than $1.9 billion. PMI awarded the Salt Lake Organizing Committee (SLOC) the Project of the Year award for delivering world-class games that, according to the International Olympic Committee, "made a profound impact upon the people of the world."[10]

Four years before the Games began, the SLOC used a Primavera software-based system with a cascading color-coded WBS to integrate planning. A year before the Games, they added a Venue Integrated Planning Schedule to help the team integrate resource needs, budgets, and plans. For example, this software helped the team coordinate different areas involved in controlling access into and around a venue, such as roads, pedestrian pathways, seating and safety provisions, and hospitality areas, saving nearly $10 million.

When the team experienced a budget deficit three years before the games, they separated "must-have" from "nice-to-have" items and implemented a rigorous expense approval process. According to Matthew Lehman, SLOC managing director, using classic project management tools turned a $400 million deficit into a $100 million surplus.

The SLOC also used an Executive Roadmap, a one-page list of the top 100 Games-wide activities, to keep executives apprised of progress. Activities were tied to detailed project

[10] Foti, Ross, "The Best Winter Olympics, Period," *PM Network* (January 2004) p. 23.

information within each department's schedule. A 90-day highlighter showed which managers were accountable for each integrated activity. Fraser Bullock, SLOC Chief Operating Officer and Chief, said, "We knew when we were on and off schedule and where we had to apply additional resources. The interrelation of the functions meant they could not run in isolation—it was a smoothly running machine."[11]

Two important outputs of monitoring and controlling project work include recommended corrective and preventive actions. Corrective actions should result in improvements in project performance. Preventive actions reduce the probability of negative consequences associated with project risks. For example, if project team members have not been reporting hours that they worked, a corrective action would be to show them how to enter the information and let them know that they need to do it. A preventive action might be modifying a time-tracking system screen to avoid common errors people made in the past. Forecasts are also important outputs of monitoring and controlling. Forecasts estimate conditions and events in the project's future based on past information. For example, project managers often provide forecasts of the amount of money they will need to complete a project based on past performance (see Chapter 7, Project Cost Management). Two other outputs of monitoring and controlling project work are recommended defect repairs and requested changes.

INTEGRATED CHANGE CONTROL

Integrated change control involves identifying, evaluating, and managing changes throughout the project life cycle. The three main objectives of integrated change control are:

1. Influencing the factors that create changes to ensure that changes are beneficial: To ensure that changes are beneficial and that a project is successful, project managers and their teams must make trade-offs among key project dimensions, such as scope, time, cost, and quality.
2. Determining that a change has occurred: To determine that a change has occurred, the project manager must know the status of key project areas at all times. In addition, the project manager must communicate significant changes to top management and key stakeholders. Top management and other key stakeholders do not like surprises, especially ones that mean the project might produce less, take longer to complete, cost more than planned, or be of lower quality than desired.

---

[11] Ibid. 23.

3. Managing actual changes as they occur: Managing change is a key role of project managers and their teams. It is important that project managers exercise discipline in managing the project to help minimize the number of changes that occur.

Important inputs to the integrated change control process include the project management plan, work performance information (usually in the form of performance reports), requested changes, recommended preventive and corrective actions, recommended defect repair, and deliverables. Important outputs include approved and rejected change requests, approved corrective and preventive actions, approved and validated defect repair, deliverables, and updates to the project management plan and project scope statement.

The project management plan provides the baseline for identifying and controlling project changes. A **baseline** is the approved project management plan plus approved changes. For example, the project management plan includes a section describing the work to perform on a project. This section of the plan describes the key deliverables for the project, the products of the project, and quality requirements. The schedule section of the project management plan lists the planned dates for completing key deliverables, and the budget section of the project management plan provides the planned cost for these deliverables. The project team must focus on delivering the work as planned. If the project team or someone else causes changes during project execution, they must revise the project management plan and have it approved by the project sponsor. Many people refer to different types of baselines, such as a cost baseline or schedule baseline, to describe different project goals more clearly and performance toward meeting them.

Performance reports provide information on how project execution is going. The main purpose of these reports is to alert the project manager and project team of issues that might cause problems in the future. The project manager and project team must decide if corrective or preventive actions are needed, what the best course of action is, and when to act. For example, suppose one of the key deliverables in a project management plan for the opening case is installation of a new server for the project. If one of the project team members reports that he or she is having problems coordinating the purchase and installation of this server, the project manager should assess what will happen to the project if this deliverable is a little late. If late installation will cause problems in other areas of the project, the project manager should take necessary actions to help the team member complete this task on time. Perhaps the purchase request is held up because one of the purchasing clerks is on vacation. The project manager could talk to the head of the purchasing department to make sure someone processes the order. If nothing can be done to meet the planned installation date, the project manager should alert the other people who would be affected by this schedule change. The project manager should also look at the big picture of how the project is going. If there is a consistent trend that schedules are slipping, the project manager should alert key stakeholders and negotiate a later completion date for the project.

Change requests are common on projects and occur in many different forms. They can be oral or written, formal or informal. For example, the project team member responsible for installing the server might ask the project manager at a progress review meeting if it is all right to order a server with a faster processor than planned, from the same manufacturer for the same approximate cost. Since this change is positive and should have no negative effects on the project, the project manager might give a verbal approval at the progress review meeting. Nevertheless, it is still important that the project manager document this change to avoid any potential problems in the future. The appropriate team member should update the section of the scope statement with the new specifications for the server. Still, keep in mind that many change requests can have a major impact on a project. For example, customers changing their minds about the number of pieces of hardware they want as part of a project will have a definite impact on the scope and cost of the project. Such a change might also affect the project's schedule. The project team must present such significant changes in written form, and there should be a formal review process for analyzing and deciding whether to approve these changes.

Change is unavoidable and often expected on most information technology projects. Technologies change, personnel change, organizational priorities change, and so on. Careful change control on information technology projects is a critical success factor. A good change control system is also important for project success.

Change Control on Information Technology Projects

From the 1950s to the 1980s, a widely held view of information technology (then often referred to as data automation) project management was that the project team should strive to do exactly what they planned on time and within budget. The problem with this view was that project teams could rarely meet original project goals, especially on projects involving new technologies. Stakeholders rarely agreed up front on what the scope of the project really was or what the finished product should really look like. Time and cost estimates created early in a project were rarely accurate.

Beginning in the 1990s, most project managers and top management realized that project management is a process of constant communication and negotiation about project objectives and stakeholder expectations. This view assumes that changes happen throughout the project life cycle and recognizes that changes are often beneficial to some projects. For example, if a project team member discovers a new hardware or software technology that could satisfy the customers' needs for less time and money, the project team and key stakeholders should be open to making major changes in the project.

All projects will have some changes, and managing them is a key issue in project management, especially for information technology projects. Many information technology projects involve the use of hardware and software that is updated frequently. For example, the initial plan for the specifications of the server described earlier may have been cutting-edge technology at that time. If the actual ordering of the server occurred six months later, it is quite possible that a more powerful server could be ordered at the same cost. This example illustrates a positive change. On the other hand, the manufacturer of the server specified in the project plan could go out of business, which would result in a negative change. Information technology project managers should be accustomed to changes such as these and build some flexibility into their project plans and execution. Customers for information technology projects should also be open to meeting project objectives in different ways.

Even if project managers, project teams, and customers are flexible, it is important that projects have a formal change control system. This formal change control system is necessary to plan for managing change.

Change Control System

A **change control system** is a formal, documented process that describes when and how official project documents may be changed. It also describes the people authorized to make changes, the paperwork required for this change, and any automated or manual tracking systems the project will use. A change control system often includes a change control board (CCB), configuration management, and a process for communicating changes.

A **change control board (CCB)** is a formal group of people responsible for approving or rejecting changes to a project. The primary functions of a change control board are to provide guidelines for preparing change requests, evaluating change requests, and managing the implementation of approved changes. An organization could have key stakeholders for the entire organization on this board, and a few members could rotate based on the unique needs of each project. By creating a formal board and process for managing changes, better overall change control should result.

However, CCBs can have some drawbacks. One drawback is the time it takes to make decisions on proposed changes. CCBs often meet only once a week or once a month and may not make decisions in one meeting. Some organizations have streamlined processes for making quick decisions on smaller project changes. One company created a "48-hour policy," in which task leaders on a large information technology project would reach agreements on key decisions or changes within their expertise and authority. The person in the area most affected by this decision or change then had 48 hours to go to his or her top management to seek approval. If for some reason the project team's decision could not be implemented, the top manager consulted would have 48 hours to reverse a decision; otherwise, the project team's decision was approved. This type of process is a great way to deal with the many time-sensitive decisions or changes that project teams must make on many information technology projects.

Configuration management is another important part of integrated change control. **Configuration management** ensures that the descriptions of the project's products are correct and complete. It involves identifying and controlling the functional and physical design characteristics of products and their support documentation. Members of the project team, frequently called configuration management specialists, are often assigned to perform configuration management for large projects. Their job is to identify and document the functional and physical characteristics of the project's products, control any changes to such characteristics, record and report the changes, and audit the products to verify conformance to requirements.

Another factor in change control is communication. Project managers should use written and oral performance reports to help identify and manage project changes. For example, on software development projects, most programmers must make their edits to one master file in a database that requires the programmers to "check out" the file to edit it. If two programmers check out the same file, they must coordinate their work before they can check the file back in to the database. In addition to written or formal communication methods, oral and informal communications are also important. Some project managers have stand-up meetings once a week or even every morning, depending on the nature of the project. The goal of a stand-up meeting is to communicate what is most important on the project quickly. For example, the project manager might have an early morning stand-up meeting every day with all of his or her team leaders. There might be a weekly stand-up meeting every Monday morning with all interested stakeholders. Requiring participants to stand keeps meetings short and forces everyone to focus on the most important project events.

Why is good communication so critical to success? One of the most frustrating aspects of project change is not having everyone coordinated and informed about the latest project information. Again, it is the project manager's responsibility to integrate all project changes so that the project stays on track. The project manager and his or her staff must develop a system for notifying everyone affected by a change in a timely manner. E-mail, real-time databases, cell phones, and the Web make it easier to disseminate the most current project information. You will learn more about good communication in Chapter 10, Project Communications Management.

Table 4-3 lists suggestions for performing integrated change control. As described earlier, project management is a process of constant communication and negotiation. Project managers should plan for changes and use appropriate tools and techniques such as a change control board, configuration management, and good communication. It is helpful to define procedures for making timely decisions on small changes, use written and oral performance reports to help identify and manage changes, and use software to assist in planning, updating, and controlling projects.

Table 4-3: Suggestions for Performing Integrated Change Control

View project management as a process of constant communication and negotiation.

Plan for change.

Establish a formal change control system, including a change control board (CCB).

Use good configuration management.

Define procedures for making timely decisions on smaller changes.

Use written and oral performance reports to help identify and manage change.

Use project management and other software to help manage and communicate changes.

Focus on leading the project team and meeting overall project goals and expectations.

Project managers must also provide strong leadership to steer the project to successful completion. They must not get too involved in managing project changes. Project managers should delegate much of the detailed work to project team members and focus on providing overall leadership for the project in general. Remember, project managers must focus on the big picture and perform project integration management well to lead their team and organization to success.

CLOSING PROJECTS

The last process in project integration management is closing the project. In order to close a project, you must finalize all activities and transfer the completed or cancelled work to the appropriate people. The main outputs of closing projects are:

- Administrative closure procedures: It is important for project teams and other stakeholders to develop and follow a step-by-step process for closing projects. In particular, administrative closure procedures should define the approval process for all project deliverables.

- Contract closure procedures: Many projects involve contracts, which are legally binding agreements. Contract closure procedures describe the methodology for making sure the contract has been completed, including delivery of goods and services and payment for them.

- Final products, services, or results: Project sponsors are usually most interested in making sure they receive delivery of the final products, services, or results they expected when they authorized the project.

- Organizational process asset updates: The project team should provide a list of project documentation, project closure documents, and historical information produced by the project in a useful format. This information is considered a process asset. For example, project teams often write a lessons-learned report at the end of a project, and this information can be a tremendous asset for future projects. (See Chapter 10, Project Communications Management for more information on creating lessons-learned reports and Appendix D on the companion Web site for a template for documenting lessons learned.)

USING SOFTWARE TO ASSIST IN PROJECT INTEGRATION MANAGEMENT

As described throughout this chapter, project teams can use various types of software to assist in project integration management. Project teams can create documents with word processing software, give presentations with presentation software, track information with spreadsheets, databases, or customized software, and transmit information using various types of communication software.

Project management software is also an important tool for developing and integrating project planning documents, executing the project management plan and related project plans, monitoring and controlling project activities, and performing integrated change control. Small project teams can use low-end or mid-range project management software products to coordinate their work. For large projects, however, like managing the Olympic Games described in the Media Snapshot, organizations may benefit most from high-end tools that provide enterprise project management capabilities and integrate all aspects of project management. All projects can benefit from using some type of project management information system to coordinate and communicate project information.

As you can see, there is a lot of work involved in project integration management. Project managers and their teams must focus on pulling all the elements of a project together to successfully complete projects.

CASE WRAP-UP

Without consulting Nick Carson or his team, Nick's CEO hired a new person, Jim, to act as a middle manager between himself and the people in Nick's department. The CEO and other top managers really liked Jim, the new middle manager. He met with them often, shared ideas, and had a great sense of humor. He started developing standards the company could use to help manage projects in the future.

For example, he developed templates for creating plans and progress reports and put them on the company's intranet. However, Jim and Nick did not get along. Jim accidentally sent an e-mail to Nick that was supposed to go to the CEO. In this e-mail, Jim said that Nick was hard to work with and preoccupied with the birth of his son.

Nick was furious when he read the e-mail and stormed into the CEO's office. The CEO suggested that Nick move to another department, but Nick did not like that option. Without considering the repercussions, the CEO offered Nick a severance package to leave the company. Because of the planned corporate buyout, the CEO knew they might have to let some people go anyway. Nick talked the CEO into giving him a two-month sabbatical he had not yet taken plus a higher percentage on his stock options. After discussing the situation with his wife and realizing that he would get over $70,000 if he resigned, Nick took the severance package. He had such a bad experience as a project manager that he decided to stick with being a technical expert. Jim, however, thrived in his position and helped the company improve their project management practices and ensure success in a highly competitive market.

CHAPTER SUMMARY

Project integration management is usually the most important project management knowledge area, since it ties together all the other areas of project management. A project manager's primary focus should be on project integration management.

Before selecting projects to pursue, it is important for organizations to follow a strategic planning process. Information technology projects should support the organization's overall business strategy. Common techniques for selecting projects include focusing on broad organizational needs, categorizing projects, performing financial analyses, developing weighted scoring models, and using balanced scorecards.

Project integration management includes the following processes:

- Developing the project charter involves working with stakeholders to create the document that formally authorizes a project. Project charters can have different formats, but they should include basic project information and signatures of key stakeholders.

■ Developing the preliminary project scope statement involves further work with stakeholders, especially users of the project's products, services, or results, to develop the high-level scope requirements. A good preliminary project scope statement can help prevent scope creep.

■ Developing the project management plan involves coordinating all planning efforts to create a consistent, coherent document—the project management plan. The main purpose of project plans is to facilitate action.

■ Directing and managing project execution involves carrying out the project plans by performing the activities included in it. Project plan execution should require the majority of a project's budget.

■ Monitoring and controlling project work is needed to meet the performance objectives of the project. The project team should continuously monitor project performance to assess the overall health of the project.

■ Integrated change control involves coordinating changes that affect the project's deliverables and organizational process assets. A change control system often includes a change control board (CCB), configuration management, and a process for communicating changes.

■ Closing the project involves finalizing all project activities. It is important to follow good procedures to ensure that all project activities are completed and that the project sponsor accepts delivery of the final products, services, or results of the project.

DISCUSSION QUESTIONS

1. Describe project integration management. How does project integration management relate to the project life cycle, stakeholders, and the other project management knowledge areas?

2. Briefly describe the strategic planning process. Which project selection method(s) do you think organizations use most often for justifying information technology projects?

3. Summarize key work involved in each of the seven processes for project integration management.

4. Describe a well-planned and executed project with which you are familiar. Describe a disastrous project. What were some of the main differences between these projects?

5. Discuss the importance of following a well-integrated change control process on information technology projects. What do you think of the suggestions made in this chapter? Think of three additional suggestions for integrated change control on information technology projects.

EXERCISES

1. Write a two-page paper based on the opening case. Answer the following questions:

   a. What do you think the real problem was in this case?

   b. Does the case present a realistic scenario? Why or why not?

   c. Was Nick Carson a good project manager? Why or why not?

   d. What could Nick have done to be a better project manager?

   e. What should top management have done to help Nick?

2. Use spreadsheet software to create Figures 4-2 through 4-5 in this text. Make sure your formulas work correctly.

3. Perform a financial analysis for a project using the format provided in Figure 4-3. Assume the projected costs and benefits for this project are spread over four years as follows: Estimated costs are $100,000 in Year 1 and $25,000 each year in Years 2, 3, and 4. Estimated benefits are $0 in Year 1 and $80,000 each year in Years 2, 3, and 4. Use an 8 percent discount rate. Create a spreadsheet (or use the business case financials template provided on the companion Web site) to calculate and clearly display the NPV, ROI, and year in which payback occurs. In addition, write a paragraph explaining whether you would recommend investing in this project, based on your financial analysis.

4. Create a weighted scoring model to determine grades for a course. Final grades are based on three exams worth 15%, 20%, and 25%, respectively; homework is worth 20%; and a group project is worth 20%. Enter scores for three students. Assume Student 1 earns 100% (or 100) on every item. Assume Student 2 earns 80% on each of the exams, 90% on the homework, and 95% on the group project. Assume Student 3 earns 90% on Exam 1, 75% on Exam 2, 80% on Exam 3, 90% on the homework, and 70% on the group project. You can use the weighted scoring model template, create your own spreadsheet, or make the matrix by hand.

5. Develop an outline (major headings and subheadings only) for a project management plan to create a Web site for your class, and then fill in the details for the introduction or overview section. Assume that this Web site would include a home page with links to a syllabus for the class, lecture notes or other instructional information, links to the Web site for this textbook, links to other Web sites with project management information, and links to personal pages for each member of your class and future classes. Also, include a bulletin board and chat room feature where students and the instructor can exchange information. Assume your instructor is the project's sponsor, you are the project manager, your classmates are your project team, and you have one year to complete the project.

6. Critique the sample project management plan mentioned in the first Suggested Reading. Write a two-page paper summarizing your opinion of this reading. What do you like and dislike about the plan? How could it be improved?

RUNNING CASE

Manage Your Health, Inc. (MYH) is a Fortune 500 company that provides a variety of health care services across the globe. MYH has more than 20,000 full-time employees and more than 5,000 part-time employees. MYH recently updated its strategic plan; key goals include reducing internal costs, increasing cross-selling of products, and exploiting new Web-based technologies to help employees, customers, and suppliers work together to improve the development and delivery of its health care products and services. Below are some ideas the Information Technology department has developed for supporting these strategic goals:

1. Recreation and Wellness Intranet Project: Provide an application on the current intranet to help employees improve their health. A recent study found that MYH, Inc. pays 20 percent more than the industry average for employee health care premiums, primarily due to the poor health of its employees. You believe that this application will help improve employee health within one year of its rollout so that you can negotiate lower health insurance premiums, providing net savings of at least $30/employee/year for full-time employees over the next four years. This application would include the following capabilities:

   - Allow employees to register for company-sponsored recreational programs, such as soccer, softball, bowling, jogging, walking, and other sports.

   - Allow employees to register for company-sponsored classes and programs to help them manage their weight, reduce stress, stop smoking, and manage other health-related issues.

   - Track data on employee involvement in these recreational and health-management programs.

   - Offer incentives for people to join the programs and do well in them (e.g., incentives for achieving weight goals, winning sports team competitions, etc.).

2. Health Coverage Costs Business Model: Develop an application to track employee health care expenses and company health care costs. Health care premiums continue to increase, and the company has changed insurance carriers several times in the past ten years. This application should allow

business modeling of various scenarios as well as tracking and analyzing current and past employee health care expenses and company health care costs. This application must be secure and run on the current intranet so several managers and analysts could access it and download selective data for further analysis. The new application must also import data from the current systems that track employee expenses submitted to the company and the company's costs to the insurance provider. You believe that having this data will help you revise policies concerning employee contributions to health care premiums and help you negotiate for lower premiums with insurance companies. You estimate that this application would save your company about $20/employee/year for full-time employees over the next four years and cost about $100,000 to develop.

3. Cross-Selling System: Develop an application to improve cross-selling to current customers. The current sales management system has separate sections for major product/service categories and different sales reps based on those products and services. You see great opportunities to increase sales to current customers by providing them discounts when they purchase multiple products/services. You estimate that this system would increase profits by $1 million each year for the next three years and cost about $800,000 each year for development and maintenance.

4. Web-Enhanced Communications System: Develop a Web-based application to improve development and delivery of products and services. There are currently several incompatible systems related to the development and delivery of products and services to customers. This application would allow customers and suppliers to provide suggestions, enter orders, view the status and history of orders, and use electronic commerce capabilities to purchase and sell their products. You estimate that this system would save your company about $2 million each year for three years after implementation. You estimate it will take one year and $3 million to develop and require 20 percent of development costs each year to maintain.

TASKS

1. Summarize each of the above-proposed projects in a simple table format suitable for presentation to top management. Include the name for each project, identify how each one supports business strategies, assess the potential financial benefits and other benefits of each project, and provide your initial assessment of the value of each project. Write your results in a one- to two-page memo to top management, including appropriate back-up information and calculations.

2. Prepare a weighted scoring model using the template provided under Appendix D on the companion Web site for this text to evaluate these

four projects. Develop at least four criteria, assign weights to each criterion, assign scores, and then calculate the weighted scores. Print the spreadsheet and bar chart with the results. Also write a one-page paper describing this weighted scoring model and what the results show.

3. Prepare a business case for the Recreation and Wellness Intranet Project. Assume the project will take six months to complete and cost about $200,000. Use the business case template provided under Appendix D on the companion Web site for this text

4. Prepare a project charter for the Recreation and Wellness Intranet Project. Assume the project will take six months to complete and cost about $200,000. Use the project charter template provided in this text and the sample project charter provided in Chapter 3 (Table 3-4) as a guide.

5. Prepare a stakeholder analysis for this project, using the template provided under Appendix D on the companion Web site for this text and example in this chapter as a guide. Be creative in making up information about stakeholders.

ADDITIONAL RUNNING CASES AND OTHER APPENDICES

Appendix C provides additional case studies and questions you can use to practice applying the concepts, tools, and techniques you are learning throughout this and subsequent chapters. Review the running cases provided in Appendix C and on the companion Web site (*www.course.com/mis/schwalbe4e*). Appendix D includes templates for various project management documents. For additional sample documents based on real projects, visit the author's Web site at *www.kathyschwalbe.com*. Appendix E and the CD-ROM included with this text include a computer simulation where you can also practice applying the project management process groups and knowledge areas.

SUGGESTED READINGS

1. ABC I-Net Project Team. ABC I-Net Sample Project Management Plan, (October 9, 2001).

   This 44-page sample project management plan is from a real project to upgrade the intranet and Internet infrastructures for a major corporation. It describes the benefits of the project, scope, critical success factors, cost estimate, management approach, and other important items. This sample project management plan is not perfect, but it does provide a real example of a plan created by a real project team. You can download this entire sample plan and other sample documents from the author's Web site at www.kathyschwalbe.com or from the companion Web site for the text. The name of this document is ABC I-Net.doc.

2. Ashurst, Colin and Niel F. Dohert. "Towards the Formulation of a 'Best Practice' Framework for Benefits Realisation in IT Projects," Electronic Journal of Information Systems Evaluation (EJISE), (December 2003).

> *This paper introduces a framework for establishing how explicitly and proactively organizations practice benefits realization management for information technology projects. The initial empirical results have shown that there is a very significant gap between the best practice framework and the practice of benefits realization management. The majority of organizations and projects studied adopted the traditional measures of project success, namely delivery on time and on budget, and there was little evidence of any explicit focus on benefits delivery or business change.*

3. Badir, Yuosre F., Remi Founou, Claude Stricker, and Vincent Bourquin. "Management of Global Large-scale Projects Through a Federation of Multiple Web-based Workflow Management Systems," Project Management Journal (September 2003).

> *A special challenge in project management is in dealing with global large-scale projects (GLSPs). This article by members of the Swiss Federal Institute of Technology describes a conceptual framework for Web-based information and project management that helps GLSP management teams monitor, control, and coordinate multiple workflows between organizations.*

4. Clay, Melanie, Helen M. Edwards, and John Maguire. "Establishing the Strategic Context of IT Projects — A Case Study from the Automotive Industry," Electronic Journal of Information Systems Evaluation (December 2003).

> *This case study demonstrates that many factors can affect an information technology project's evaluation and procurement. One of the most critical, however, is a sound basis from which to make judgments on the need for a system. Many new information systems require changes in current business processes. In this case study, benefits from a new bar-coding system could only be achieved after implementing business process improvements.*

5. IEEE Computer Society. "Std 1058-1998." *IEEE Standard for Software Project Management Plans* (1998).

> *This standard describes the format and content of software project management plans. It identifies the minimum set of elements that should appear in all software project management plans.*

Key Terms

- **balanced scorecard** — a methodology that converts an organization's value drivers to a series of defined metrics
- **baseline** — the approved project management plan plus approved changes

- **cash flow** — benefits minus costs or income minus expenses
- **change control board (CCB)** — a formal group of people responsible for approving or rejecting changes on a project
- **change control system** — a formal, documented process that describes when and how official project documents may be changed
- **configuration management** — a process that ensures that the descriptions of the project's products are correct and complete
- **cost of capital** — the return available by investing the capital elsewhere
- **directives** — new requirements imposed by management, government, or some external influence
- **discount factor** — a multiplier for each year based on the discount rate and year
- **discount rate** — the rate used in discounting future cash flow; also called the capitalization rate or opportunity cost of capital
- **integrated change control** — identifying, evaluating, and managing changes throughout the project life cycle
- **interface management** — identifying and managing the points of interaction between various elements of a project
- **internal rate of return (IRR)** — the discount rate that results in an NPV of zero for a project
- **net present value (NPV) analysis** — a method of calculating the expected net monetary gain or loss from a project by discounting all expected future cash inflows and outflows to the present point in time
- **opportunities** — chances to improve the organization
- **organizational process assets** — formal and informal plans, policies, procedures, guidelines, information systems, financial systems, management systems, lessons learned, and historical information that help people understand, follow, and improve business processes in a specific organization
- **payback period** — the amount of time it will take to recoup, in the form of net cash inflows, the total dollars invested in a project
- **problems** — undesirable situations that prevent the organization from achieving its goals
- **project charter** — a document that formally recognizes the existence of a project and provides direction on the project's objectives and management
- **project integration management** — coordinating all of the other project management knowledge areas throughout a project's life. These processes include developing the project charter, developing the preliminary project scope statement, developing the project management plan, directing and managing the project, monitoring and controlling the project, providing integrated change control, and closing the project
- **project management plan** — a document used to coordinate all project planning documents and guide project execution and control

- **required rate of return** — the minimum acceptable rate of return on an investment
- **return on investment (ROI)** — (benefits minus costs) divided by costs
- **scope creep** — the tendency for project scope to keep getting bigger
- **scope statement** — a document used to develop and confirm a common understanding of the project scope; the first version is often called a preliminary scope statement
- **stakeholder analysis** — an analysis of information such as key stakeholders' names and organizations, their roles on the project, unique facts about each stakeholder, their level of interest in the project, their influence on the project, and suggestions for managing relationships with each stakeholder
- **strategic planning** — determining long-term objectives by analyzing the strengths and weaknesses of an organization, studying opportunities and threats in the business environment, predicting future trends, and projecting the need for new products and services
- **weighted scoring model** — a technique that provides a systematic process for basing project selection on numerous criteria

5

Project Scope Management

OPENING CASE

Kim Nguyen was leading an initial project team meeting to create the work breakdown structure (WBS) for the IT Upgrade Project. This project was necessary for implementing several high-priority, Internet-based applications the company was developing. The IT Upgrade Project involved creating and implementing a plan to get all employees' information technology assets to meet new corporate standards within nine months. These standards specified the minimum equipment required for each desktop or laptop computer, namely the type of processor, amount of memory, hard disk size, type of network connection, and software. Kim knew that to perform the upgrades, they would first have to create a detailed inventory of all of the current hardware, networks, and software in the entire company of 2000 employees.

Kim had worked with other stakeholders to develop a project charter and preliminary scope statement. The project charter included rough cost and schedule estimates for the project and signatures of key stakeholders; the preliminary scope statement provided a start in defining the hardware, software, and network requirements as well as other information related to the project scope. Kim called a meeting with her project team and other stakeholders to define the scope of the project further. She wanted to get everyone's ideas on what was involved in doing the project, who would do what, and how they could avoid potential scope creep. The company's new CEO, Walter Schmidt, was known for keeping a close eye on major projects like this one. They had started using a new project management information system that let everyone know the status of projects at a detailed and high level. Kim knew that a good WBS was the foundation for scope, time, and cost performance, but she had never led a team in creating one or allocating costs based on a WBS. Where should she begin?

WHAT IS PROJECT SCOPE MANAGEMENT?

Recall from Chapter 1 that several factors are associated with project success. Many of these factors, such as user involvement, clear business objectives, minimized scope, and firm basic requirements, are elements of project scope management. William V. Leban, program manager at Keller Graduate School of Management, cites the lack of proper project definition and scope as the main reason why projects fail.[1]

One of the most important and most difficult aspects of project management, therefore, is defining the scope of a project. **Scope** refers to *all* the work involved in creating the products of the project and the processes used to create them.

[1] Chalfin, Natalie, "Four Reasons Why Projects Fail," *PM Network* (June 1998): p. 7.

Recall from Chapter 2 that the term **deliverable** describes a product produced as part of a project. Deliverables can be product-related, such as a piece of hardware or software, or process-related, such as a planning document or meeting minutes. Project stakeholders must agree on what the products of the project are and, to some extent, how they should produce them to define all of the deliverables.

Project scope management includes the processes involved in defining and controlling what is or is not included in a project. It ensures that the project team and stakeholders have the same understanding of what products the project will produce and what processes the project team will use to produce them. There are five main processes involved in project scope management:

1. **Scope planning** involves deciding how the scope will be defined, verified, and controlled and how the WBS will be created. The project team creates a scope management plan as the main output of the project scope planning process.

2. **Scope definition** involves reviewing the project charter and preliminary scope statement created during the initiation process and adding more information during the planning process as requirements are developed and change requests are approved. The main outputs of scope definition are the project scope statement, requested changes to the project, and updates to the project scope management plan.

3. **Creating the WBS** involves subdividing the major project deliverables into smaller, more manageable components. The main outputs include a work breakdown structure (WBS), a WBS dictionary, a scope baseline, requested changes to the project, and updates to the project scope statement and project scope management plan.

4. **Scope verification** involves formalizing acceptance of the project scope. Key project stakeholders, such as the customer and sponsor for the project, inspect and then formally accept the deliverables of the project during this process. If the deliverables are not acceptable, the customer or sponsor usually requests changes, which result in recommendations for taking corrective actions. The main outputs of this process, therefore, are accepted deliverables, requested changes, and recommended corrective actions.

5. **Scope control** involves controlling changes to project scope, which is a challenge on many information technology projects. Scope control includes identifying, evaluating, and implementing changes to project scope as the project progresses. Scope changes often influence the team's ability to meet project time and cost goals, so project managers must carefully weigh the costs and benefits of scope changes. The main outputs of this process are requested changes, recommended corrective actions, and updates to the project scope statement, WBS and WBS dictionary, scope baseline, project management plan, and organizational process assets.

SCOPE PLANNING AND THE SCOPE MANAGEMENT PLAN

The first step in project scope management is scope planning. The project's size, complexity, importance, and other factors will affect how much effort is spent on scope planning. For example, a team working on a project to upgrade the entire corporate accounting system for a multibillion dollar company with more than 50 geographic locations should spend a fair amount of time on scope planning. A project to upgrade the hardware and software for a small accounting firm with only five employees, on the other hand, would need a much smaller scope planning effort. In any case, it is important for a project team to decide how they will define the scope, develop the detailed scope statement, create the work breakdown structure, verify the scope, and control the scope for every project they undertake.

☑ *What Went Right?*

Many financial service companies use customer relationship management (CRM) systems to improve their understanding of and responsiveness to customers. A senior management team at the Canadian money management company Dynamic Mutual Funds (DMF) launched an enterprise-wide, national program to build and manage its customer relationships. It soon became clear that the way the company developed its project scope planning and definition in the past would not work for this important program. They needed a more organized, highly participative approach that could be accomplished quickly.

The team proposed a new concept, project scope design, which consisted of seven non-sequential steps:

1. Analyze the project atmosphere, stakeholders, and centers of influence.

2. Align the project scope with the organization's strategic objectives and business challenges.

3. Determine where to add value to the business.

4. Study the process flow between the business units.

5. Develop an efficient communication strategy.

6. Develop the project approach.

7. Coordinate the new project with the other initiatives already under way.

DMF's program to manage customer relationships successfully completed its first phase in June 2001, and that October DMF won an eCustomer World Golden Award for world-class innovation in the Canadian marketplace.[2]

---

[2] Kayed, Omar, "Seven Steps to Dynamic Scope Design," *PM Network* (December 2003).

The main output of scope planning is a scope management plan. The **scope management plan** is a document that includes descriptions of how the team will prepare the project scope statement, create the WBS, verify completion of the project deliverables, and control requests for changes to the project scope. Kim, the project manager in the opening case, should work with her project team to write the scope management plan. After they have a good draft, she should review it with the project sponsor to make sure their approach meets expectations.

Key inputs of the scope management plan include the project charter, preliminary scope statement, and project management plan, as described in Chapter 4. Table 5-1 shows the project charter for the project described in the opening case. Notice how information from the project charter provides a basis for making scope management decisions. It describes the high-level scope goals for the project, a general approach to accomplishing the project's goals, and the main roles and responsibilities of important project stakeholders.

Table 5-1: Sample Project Charter

Project Title: Information Technology (IT) Upgrade Project

Project Start Date: March 4, 2007   **Projected Finish Date:** December 4, 2007

Project Manager: Kim Nguyen, 691-2784, *knguyen@course.com*

Project Objectives: Upgrade hardware and software for all employees (approximately 2,000) within nine months based on new corporate standards. See attached sheet describing the new standards. Upgrades may affect servers, as well as associated network hardware and software. Budgeted $1,000,000 for hardware and software costs and $500,000 for labor costs.

Approach:

- Update the information technology inventory database to determine upgrade needs

- Develop detailed cost estimate for project and report to CIO

- Issue a request for quote to obtain hardware and software

- Use internal staff as much as possible for planning, analysis, and installation

ROLES AND RESPONSIBILITIES:

NAME	ROLE	RESPONSIBILITY
Walter Schmidt	CEO	Project sponsor, monitor project
Mike Zwack	CIO	Monitor project, provide staff
Kim Nguyen	Project Manager	Plan and execute project
Jeff Johnson	Director of Information, Technology Operations	Mentor Kim
Nancy Reynolds	VP, Human Resources	Provide staff, issue memo to all employees about project
Steve McCann	Director of Purchasing	Assist in purchasing hardware and software

Table 5-1: Sample Project Charter (continued)

Sign-off: (Signatures of all the above stakeholders)

Walter Schmidt	*Steve McCann*
Mike Zwack	*Nancy Reynolds*
Kim Nguyen	*Jeff Johnson*

Comments: (Handwritten or typed comments from above stakeholders, if applicable)

"This project must be done within ten months at the absolute latest." Mike Zwack, CIO

"We are assuming that adequate staff will be available and committed to supporting this project. Some work must be done after hours to avoid work disruptions, and overtime will be provided." Jeff Johnson and Kim Nguyen, Information Technology department

This short document includes important information that will help Kim guide her project team in developing the scope management plan. Additional input that assist in scope planning include information related to organizational process assets, such as policies and procedures related to scope planning and management, and historical information about previous projects. Environmental factors, such as the organization's infrastructure or marketplace conditions, also affect how scope should be managed on a project.

The main tools and techniques available for scope planning include templates, forms, and standards, as well as expert judgment. For example, if a project involves developing a database, project team members can decide to use common systems analysis and design standards, such as creating entity relationship diagrams, use cases, data flow diagrams, and so on to help document the scope. Many software tools include online templates and forms for creating these and similar items. Expert judgment should also be used to help decide the best way to manage scope for particular projects. For example, organizations often hire experts from outside companies to evaluate and recommend off-the-shelf software and then assist in managing the purchase and installation of the new software.

SCOPE DEFINITION AND THE PROJECT SCOPE STATEMENT

The next step in project scope management is to define further the work required for the project. Good scope definition is very important to project success because it helps improve the accuracy of time, cost, and resource estimates, it defines a baseline for performance measurement and project control, and it aides in communicating clear work responsibilities. The main tools and techniques used in scope definition include analyzing products, identifying alternative approaches to doing the work, understanding and analyzing stakeholder needs, and using expert judgment. The main output of scope definition is the project scope statement.

As described in Chapter 4, the project team develops a preliminary scope statement in initiating a project as part of the project integration management knowledge area. This document, as well as the project charter, organizational process assets, and approved change requests provide a basis for creating the **project scope statement**. Table 3-7 in Chapter 3 includes a sample project scope statement. The preliminary project scope statement should provide basic scope information, and the project scope statement should continue to clarify and provide information that is more specific.

p992

Although contents vary, project scope statements should include, at a minimum, a description of the project, including its overall objectives and justification, detailed descriptions of all project deliverables, and the characteristics and requirements of products and services produced as part of the project. It is also helpful to document project success criteria in the project scope statement, as well as provide other scope-related information, such as the project boundaries, product acceptance criteria, project constraints and assumptions, project organization, defined risks, schedule milestones, order of magnitude cost estimate, configuration management requirements, and approval requirements. It should also reference supporting documents, such as product specifications that will affect what products are produced or purchased, or corporate policies, which might affect how products or services are produced. Many information technology projects also require detailed functional and design specifications for developing software, which also should be referenced in the detailed scope statement.

As time progresses, the scope of a project should become more clear and specific. For example, the project charter for the IT Upgrade Project shown in Table 5-1 includes a short statement about the servers and other computers and software that the IT Upgrade Project may affect. Table 5-2 provides an example of how the scope becomes progressively more detailed in the preliminary scope statement, and then the project scope statement.

Table 5-2: Further Defining Project Scope

Project Charter:

Upgrades may affect servers...

Preliminary Scope Statement:

Servers: If additional servers are required to support this project, they must be compatible with existing servers. If it is more economical to enhance existing servers, a detailed description of enhancements must be submitted to the CIO for approval. See current server specifications provided in Atch 6. The CEO must approve a detailed plan describing the servers and their location at least two weeks before installation.

Project Scope Statement, Version 1:

Servers: This project will require purchasing ten new servers to support Web, network, database, application, and printing functions. Two of each type of server will be purchased and dedicated to this project. Detailed descriptions of the servers are provided in a product brochure in Appendix 8 along with a plan describing where they will be located.

Notice in Table 5-2 that the preliminary and project scope statements often refer to related documents, such as product specifications, product brochures, or other plans. As more information becomes available and decisions are made related to project scope, such as specific products that will be purchased or changes that have been approved, the project team should update the project scope statement. They might name different iterations of the scope statement Version 1, Version 2, and so on. These updates may also require changes to be made to the project scope management plan. For example, if the company must purchase servers for the project from a supplier they have never worked with before, the scope management plan should include information on working with that new supplier.

As IT Changes

An up-to-date project scope statement is an important document for developing and confirming a common understanding of the project scope. It describes in detail the work to be accomplished on the project and is an important tool for ensuring customer satisfaction and preventing scope creep, as described later in this chapter.

Recall from Chapter 1 the importance of addressing the triple constraint of project management—meeting scope, time, and cost goals for a project. Time and cost goals are normally straightforward. For example, the time goal for the IT Upgrade Project is nine months, and the cost goal is $1.5 million. It is much more difficult to describe, agree upon, and meet the scope goal of many projects.

🎥 *Media Snapshot*

Many people enjoy watching television shows like *Changing Rooms* or *Trading Spaces*, where participants have two days and $1,000 to update a room in their neighbor's house. Since the time and cost are set, it's the scope that has the most flexibility. Examples of some of the work completed include new flooring, light fixtures, paint, new shelves, artwork, etc. to brighten up a dull room.

Designers on these shows often have to change initial scope goals due to budget or time constraints. For example, designers often go back to local stores to exchange items, such as lights, artwork, or fabric, for less expensive items to meet budget constraints. Or they might describe a new piece of furniture they'd like the carpenter to build, but the carpenter changes the design or materials to meet time constraints. Occasionally designers can buy more expensive items or have more elaborate furniture built because they underestimated costs and schedules.

Another important issue related to project scope management is meeting customer expectations. Who wouldn't be happy with a professionally designed room at no cost to them? Although most homeowners are very happy with work done on the show, some are obviously disappointed. Unlike most projects where the project team works closely with the customer, homeowners have little say in what gets done and cannot inspect the work along the way. They walk into their newly decorated room with their eyes closed. Modernizing a room can mean something totally different to a homeowner and the interior designer. For example, one woman was obviously shocked when she saw her bright orange kitchen with black appliances. Another couple couldn't believe there was moss on their bedroom walls. What happens when the homeowners don't like the work that's been done? The FAQ section of tlc.com says, "Everyone on our show is told upfront that there's a chance they won't like the final design of the room. Each applicant signs a release acknowledging that the show is not responsible for redecorating a room that isn't to the owner's taste." Too bad you can't get sponsors for most projects to sign a similar release form. It would make project scope management much easier!

CREATING THE WORK BREAKDOWN STRUCTURE

After completing scope planning and definition processes, the next step in project scope management is to create a work breakdown structure. A **work breakdown structure (WBS)** is a deliverable-oriented grouping of the work involved in a project that defines the total scope of the project. Because most projects involve many people and many different deliverables, it is important to organize and divide the work into logical parts based on how the work will be performed. The WBS is a foundation document in project management because it provides the basis for planning and managing project schedules, costs, resources, and changes. Since the WBS defines the total scope of the project, some project management experts believe that work should not be done on a project if it is not included in the WBS. Therefore, it is crucial to develop a good WBS.

The project scope statement and project management plan are the primary input for creating a WBS. The main tools and techniques include using WBS templates, as described below, and using **decomposition** or subdividing project deliverables into smaller pieces. The output of the process to create the WBS are the WBS itself, the WBS dictionary, a scope baseline, and updates to the project scope statement and scope management plan.

What does a WBS look like? A WBS is often depicted as a task-oriented family tree of activities, similar to an organizational chart. A project team often organizes the WBS around project products, project phases, or using the project management process groups. Many people like to create a WBS in chart form first to help them visualize the whole project and all of its main parts. For example, Figure 5-1 shows a WBS for an intranet project. Notice that product areas provide the basis for its organization. In this case, there are main boxes or groupings on the WBS for developing the Web site design, the home page for the intranet, the marketing department's pages, and the sales department's pages.

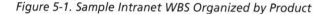

Figure 5-1. Sample Intranet WBS Organized by Product

In contrast, a WBS for the same intranet project can be organized around project phases, as shown in Figure 5-2.[3] Notice that project phases of concept, Web site design, Web site development, roll out, and support provide the basis for its organization.

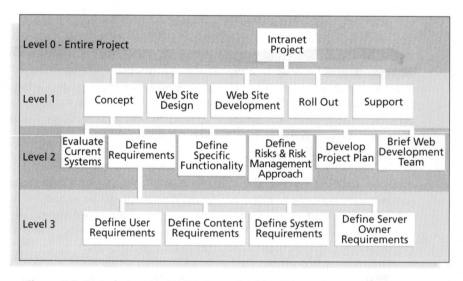

Figure 5-2. Sample Intranet WBS Organized by Phase

A WBS can also be shown in tabular form as an indented list of tasks that shows the same groupings of the work. For example, Table 5-3 shows the WBS from Figure 5-2 in tabular form. The items on the WBS are the same, but the numbering scheme and indentation of tasks show the structure. Many documents, such as contracts, use this tabular form. Project management software also uses this form. The WBS is actually the contents of the Task Name column in Project 2003. See Appendix A for detailed information on creating a WBS using Project 2003.

---

[3] This particular structure is based on a sample Project 98 file.

Table 5-3: Intranet WBS in Tabular Form

1.0 Concept

      1.1 Evaluate current systems

      1.2 Define requirements

            1.2.1 Define user requirements

            1.2.2 Define content requirements

            1.2.3 Define system requirements

            1.2.4 Define server owner requirements

      1.3 Define specific functionality

      1.4 Define risks and risk management approach

      1.5 Develop project plan

      1.6 Brief Web development team

2.0 Web Site Design

3.0 Web Site Development

4.0 Roll Out

5.0 Support

Figure 5-3 shows the phase-oriented intranet WBS, using the same numbering scheme from Table 5-3, in the form of a Gantt chart created in Project 2003. You can see from this figure that the WBS is the basis for project schedules. Notice that the WBS is in the left part of the figure under the Task Name column. The resulting schedule is in the right part of the figure. You will learn more about Gantt charts in Chapter 6, Project Time Management.

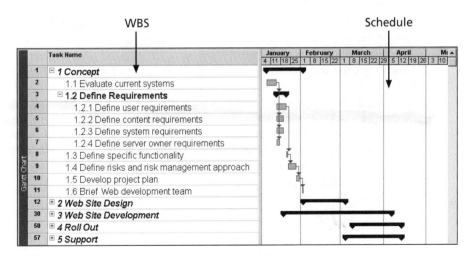

Figure 5-3. Intranet Gantt Chart in Microsoft Project

The work breakdown structures in Figures 5-1, 5-2, and 5-3 and in Table 5-3 present information in hierarchical form. The 0 level of a WBS represents the entire project and is the top level. (Note the labels on the left side of Figure 5-2. Also, note that some sources call the entire project level 1 instead of level 0.) The next level down is level 1, which represents the major products or phases of the project. Level 2 includes the major subsets of level 1. For example, in Figure 5-2 the level 2 items under the level 1 item called "Concept" include: evaluate current systems, define requirements, define specific functionality, define risks and risk management approach, develop project plan, and brief Web development team. Under the level 2 item called "Define Requirements" are four level 3 items on the WBS: define user requirements, define content requirements, define server requirements, and define server owner requirements.

In Figure 5-2, the lowest level is level 3. A **work package** is a task at the lowest level of the WBS. It also represents the lowest level of work that the project manager is using to monitor and control the project. Generally, each work package in a WBS should represent roughly 80 hours of effort. You can also think of work packages in terms of accountability and reporting. If a project has a relatively short time frame and requires weekly progress reports, a work package might represent 40 hours of work or less. On the other hand, if a project has a very long time frame and requires quarterly progress reports, a work package might represent more than 100 hours of work. A work package might also be a specific piece of hardware or equipment, such as a specific server.

Another way to think of work packages relates to entering data into project management software. *You can only enter duration estimates for work packages.* The rest of the WBS items are just groupings or summary tasks for the work packages. The software automatically calculates duration estimates for various WBS levels based on data entered for each work package and the WBS hierarchy.

The sample WBSs shown here seem somewhat easy to construct and understand. *Nevertheless, it is very difficult to create a good WBS.* To create a good WBS, you must understand both the project and its scope and incorporate the needs and knowledge of the stakeholders. The project manager and the project team must decide as a group how to organize the work and how many levels to include in the WBS. Many project managers have found that it is better to focus on getting the top levels done well before getting too bogged down in more detail.

Many people confuse tasks on a WBS with specifications. Tasks on a WBS represent work that needs to be done to complete the project. For example, if you are creating a WBS to redesign a kitchen, you might have level 1 categories called design, purchasing, flooring, walls, cabinets, and appliances. Under flooring, you might have tasks to remove the old flooring, install the new flooring, and install the trim. You would not have tasks like "12' by 14' of light oak" or "flooring must be durable."

Another concern when creating a WBS is how to organize it so that it provides the basis for the project schedule. You should focus on what work needs to be done and how it will be done, not when it will be done. In other words, the tasks do not have to be developed as a sequential list of steps. If you do want some time-based flow for the work, you can create a WBS using the project management process groups of initiating, planning, executing, controlling, and closing as level 1 in the WBS. By doing this, not only does the project team follow good project management practice, but the WBS tasks can also be mapped more easily against time. For example, Figure 5-4 shows a WBS and Gantt chart for the intranet project, organized by the five project management process groups. Tasks under initiating include selecting a project manager, forming the project team, and developing the project charter. Tasks under planning include developing a scope statement, creating a WBS, and developing and refining other plans, which would be broken down in more detail for a real project. The tasks of concept, Web site design, Web site development, and roll out, which were WBS level 1 items in Figure 5-2, now become WBS level 2 items under executing. The executing tasks vary the most from project to project, but many of the tasks under the other project management process groups would be similar for all projects. If you do not use the project management process groups in the WBS, you can have a level 1 category called project management to make sure that tasks related to managing the project are accounted for. Remember that all work should be included in the WBS, including project management.

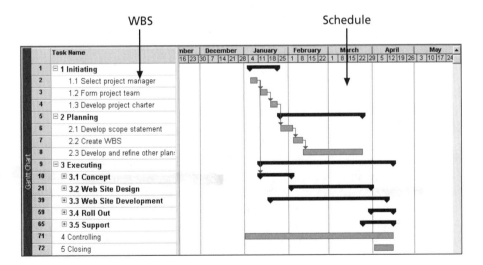

Figure 5-4. Intranet Gantt Chart Organized by Project Management Process Groups

JWD Consulting used the project management process groups for the level 1 items in its WBS for the Project Management Intranet Site Project in Chapter 3. The project team focused on the product deliverables they had to produce for the project in breaking down the executing task. Table 5-4 shows the categories they used for that part of the WBS. Some project teams like to list every deliverable they need to produce and then use those as the basis for creating all or part of their WBS. Recall that the scope statement should list and describe all of the deliverables required for the project. It is very important to ensure consistency between the project charter, scope statement, WBS, and Gantt chart to define the scope of the project accurately.

Table 5-4: Executing Tasks for JWD Consulting's WBS

3.0 Executing

    3.1 Survey

    3.2 User inputs

    3.3 Intranet site content

        3.3.1 Templates and tools

        3.3.2 Articles

        3.3.3 Links

        3.3.4 Ask the expert

        3.3.5 User requests

    3.4 Intranet site design

    3.5 Intranet site construction

    3.6 Site testing

    3.7 Site promotion

    3.8 Site roll-out

    3.9 Project benefits measurement

It is also very important to involve the entire project team and the customer in creating and reviewing the WBS. *People who will do the work should help to plan the work* by creating the WBS. Having group meetings to develop a WBS helps everyone understand *what* work must be done for the entire project and *how* it should be done, given the people involved. It also helps to identify where coordination between different work packages will be required.

Approaches to Developing Work Breakdown Structures

There are several approaches you can use to develop a work breakdown structure. These approaches include:

- Using guidelines
- The analogy approach
- The top-down approach
- The bottom-up approach
- The mind-mapping approach

Using Guidelines

If guidelines for developing a WBS exist, it is very important to follow them. Some organizations—for example, the U.S. Department of Defense (DOD)—prescribe the form and content for WBSs for particular projects. Many DOD projects require contractors to prepare their proposals based on the DOD-provided WBS. These proposals must include cost estimates for each task in the WBS at a detailed and summary level. The cost for the entire project must be calculated by summing the costs of all of the lower-level WBS tasks. When DOD personnel evaluate cost proposals, they must compare the contractors' costs with the DOD's estimates. A large variation in costs for a certain WBS task often indicates confusion as to what work must be done.

Consider a large automation project for the U.S. Air Force. In the mid-1980s, the Air Force developed a request for proposals for the Local On-Line Network System (LONS) to automate 15 Air Force Systems Command bases. This $250 million project involved providing the hardware and developing software for sharing documents such as contracts, specifications, requests for proposals, and so on. The Air Force proposal guidelines included a WBS that contractors were required to follow in preparing their cost proposals. Level 1 WBS items included hardware, software development, training, project management, and the like. The hardware item was composed of several level 2 items, such as servers, workstations, printers, network hardware, and so on. Air Force personnel reviewed the contractors' cost proposals against their internal cost estimate, which was also based on this WBS. Having a prescribed WBS helped contractors to prepare their cost proposals and the Air Force to evaluate them.

Many organizations provide guidelines and templates for developing WBSs, as well as examples of WBSs from past projects. At the request of many of its members, the Project Management Institute recently developed a WBS

Practice Standard to provide guidance for developing and applying the WBS to project management (see the third Suggested Reading at the end of this chapter). This document includes sample WBSs for a wide variety of projects in various industries, including projects for Web design, telecom, service industry outsourcing, and software implementation.

Project managers and their teams should review appropriate information to develop their unique project WBSs more efficiently. For example, Kim Nguyen and key team members from the opening case should review their company's WBS guidelines, templates, and other related information before and during the team meetings to create their WBS.

The Analogy Approach

Another approach for constructing a WBS is the analogy approach. In the **analogy approach**, you use a similar project's WBS as a starting point. For example, Kim Nguyen from the opening case might learn that one of her organization's suppliers did a similar information technology upgrade project last year. She could ask them to share their WBS for that project to provide a starting point for her own project.

McDonnell Aircraft Company, now part of Boeing, provides an example of using an analogy approach when creating WBSs. McDonnell Aircraft Company designed and manufactured several different fighter aircraft. When creating a WBS for a new aircraft design, it started by using 74 predefined subsystems for building a fighter aircraft based on past experience. There was a level 1 WBS item for the airframe that was composed of level 2 items, such as a forward fuselage, center fuselage, aft fuselage, and wings. This generic product-oriented WBS provided a starting point for defining the scope of new aircraft projects and developing cost estimates for new aircraft designs.

Some organizations keep a repository of WBSs and other project documentation on file to assist people working on projects. Project 2003 and many other software tools include sample files to assist users in creating a WBS and Gantt chart. Viewing examples of other similar projects' WBSs allows you to understand different ways to create a WBS. See the companion Web site for this text and the author's Web site for sample WBSs.

The Top-down and Bottom-up Approaches

Two other approaches for creating WBSs are the top-down and bottom-up approaches. Most project managers consider the top-down approach of WBS construction to be conventional. To use the **top-down approach**, start with the largest items of the project and break them into their subordinate items. This process involves refining the work into greater and greater levels of detail. For example, Figure 5-2 shows how work was broken down to level 3 for part of the intranet project. After finishing the process, all resources should be assigned at the work package level. The top-down approach is best suited to project managers who have vast technical insight and a big-picture perspective.

In the **bottom-up approach**, team members first identify as many specific tasks related to the project as possible. They then aggregate the specific tasks and organize them into summary activities, or higher levels in the WBS. For example, a group of people might be responsible for creating a WBS to create an e-commerce application. Instead of looking for guidelines on how to create a WBS or viewing similar projects' WBSs, they could begin by listing detailed tasks they think they would need to do in order to create the application. After listing detailed tasks, they would group the tasks into categories. Then, they would group these categories into higher-level categories. Some people have found that writing all possible tasks down on notes and then placing them on a wall helps them see all the work required for the project and develop logical groupings for performing the work. For example, a business analyst on the project team might know that they had to define user requirements and content requirements for the e-commerce application. These tasks might be part of the requirements documents they would have to create as one of the project deliverables. A hardware specialist might know they had to define system requirements and server requirements, which would also be part of a requirements document. As a group, they might decide to put all four of these tasks under a higher-level item called "define requirements" that would result in the delivery of a requirements document. Later, they might realize that defining requirements should fall under a broader category of concept design for the e-commerce application, along with other groups of tasks related to the concept design. The bottom-up approach can be very time-consuming, but it can also be a very effective way to create a WBS. Project managers often use the bottom-up approach for projects that represent entirely new systems or approaches to doing a job, or to help create buy-in and synergy with a project team.

Mind Mapping

Some project managers like to use mind mapping to help develop WBSs. **Mind mapping** is a technique that uses branches radiating out from a core idea to structure thoughts and ideas. Instead of writing tasks down in a list or immediately trying to create a structure for tasks, mind mapping allows people to write and even draw pictures of ideas in a nonlinear format. This more visual, less structured approach to defining and then grouping tasks can unlock creativity among individuals and increase participation and morale among teams.[4]

---

4 Mindjet Visual Thinking, "About Mind Maps," (*www.mindjet.com*) (2002).

Figure 5-5 shows a diagram that uses mind mapping to create a WBS for the IT Upgrade Project. The circle in the center represents the entire project. Each of the four main branches radiating out from the center represents the main tasks or level 1 items for the WBS. Different people at the meeting creating this mind map might have different roles in the project, which could help in deciding the tasks and WBS structure. For example, Kim would want to focus on all of the project management tasks, and she might also know that they will be tracked in a separate budget category. People who are familiar with acquiring or installing hardware and software might focus on that work, and so on. Branching off from the main task called "Update inventory" are two subtasks, "Perform physical inventory" and "Update database." Branching off from the "Perform physical inventory" subtask are three further subdivisions, labeled Building A, Building B, and Building C, and so on. The team would continue to add branches and items until they have exhausted ideas on what work needs to be performed.

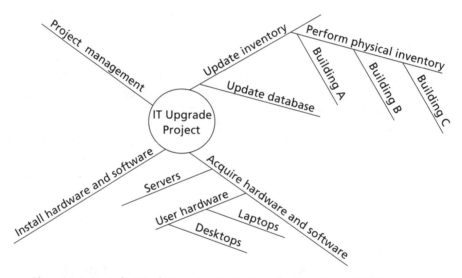

Figure 5-5. Sample Mind-Mapping Technique for Creating a WBS

After discovering WBS items and structure using the mind-mapping technique, you could then translate the information into chart or tabular form, as described earlier. Figure 5-6 shows a chart of the WBS items from the mind map in Figure 5-5.

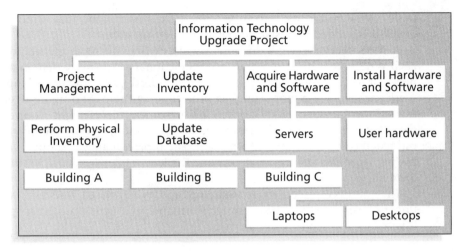

Figure 5-6. Resulting WBS in Chart Form

Mind mapping can be used for developing WBSs using the top-down or bottom-up approach. For example, you could conduct mind mapping for an entire project by listing the whole project in the center of a document, adding the main categories on branches radiating out from the center, and then adding branches for appropriate subcategories. You could also develop a separate mind-mapping document for each deliverable of a project by putting the deliverable in the center of a document and then surrounding it with other mind-mapping documents to create one for the entire project. You can also add items anywhere on a mind-mapping document without following a strict top-down or bottom-up approach. After the mind-mapping documents are complete, you can convert them into a chart or tabular WBS form.

The WBS Dictionary and Scope Baseline

As you can see from these sample WBSs, many of the items listed on them are rather vague. What exactly does "Update database" mean, for example? The person responsible for this task might think that it does not need to be broken down any further, which could be fine. However, the task should be described in more detail so everyone has the same understanding of what it involves. What if someone else has to perform the task? What would you tell him/her to do? What will it cost to complete the task? Information that is more detailed is needed to answer these and other questions.

A **WBS dictionary** is a document that describes detailed information about each WBS item. The format of the WBS dictionary can vary based on project needs. It might be appropriate to have just a short paragraph describing each work package. For a more complex project, an entire page or more might be needed for the work package descriptions. Some projects might require that each WBS item describe the responsible organization, resource requirements, estimated costs, and other information. The project manager should work with his/her team and sponsor to determine the level of detail needed in the WBS dictionary.

The approved project scope statement and its associated WBS and WBS dictionary form the **scope baseline**. Performance in meeting project scope goals is based on this scope baseline.

Advice for Creating a WBS and WBS Dictionary

As stated previously, creating a good WBS is no easy task and usually requires several iterations. Often, it is best to use a combination of approaches to create a project WBS. There are some basic principles, however, that apply to creating any good WBS and its WBS dictionary.

- A unit of work should appear at only one place in the WBS.
- The work content of a WBS item is the sum of the WBS items below it.
- A WBS item is the responsibility of only one individual, even though many people may be working on it.
- The WBS must be consistent with the way in which work is actually going to be performed; it should serve the project team first, and other purposes only if practical.
- Project team members should be involved in developing the WBS to ensure consistency and buy-in.
- Each WBS item must be documented in a WBS dictionary to ensure accurate understanding of the scope of work included and not included in that item.
- The WBS must be a flexible tool to accommodate inevitable changes while properly maintaining control of the work content in the project according to the scope statement.[5]

---

[5] Cleland, David I. *Project Management: Strategic Design and Implementation,* 2nd ed. (New York: McGraw-Hill, 1994.)

SCOPE VERIFICATION

It is difficult to create a good project scope statement and WBS for a project. It is even more difficult, especially on information technology projects, to verify the project scope and minimize scope changes. Some project teams know from the start that the scope is very unclear and that they must work closely with the project customer to design and produce various deliverables. In this case, the project team must develop a process for scope verification that meets unique project needs. Careful procedures must be developed to ensure the customer is getting what they want and the project team has enough time and money to produce the desired products and services.

Even when the project scope is fairly well defined, many information technology projects suffer from **scope creep**—the tendency for project scope to keep getting bigger and bigger. There are many horror stories about information technology projects failing due to scope problems such as scope creep. For this reason, it is very important to verify the project scope and develop a process for controlling scope changes.

(X) *What Went Wrong?*

A project scope that is too broad and grandiose can cause severe problems. Scope creep and an overemphasis on technology for technology's sake resulted in the bankruptcy of a large pharmaceutical firm, Texas-based FoxMeyer Drug. In 1994, the CIO was pushing for a $65 million system to manage the company's critical operations. He did not believe in keeping things simple, however. The company spent nearly $10 million on state-of-the-art hardware and software and contracted the management of the project to a prestigious (and expensive) consulting firm. The project included building an $18 million robotic warehouse, which looked like something out of a science fiction movie, according to insiders. The scope of the project kept getting bigger and more impractical. The elaborate warehouse was not ready on time, and the new system generated erroneous orders that cost FoxMeyer Drug more than $15 million in unrecovered excess shipments. In July of 1996, the company took a $34 million charge for its fourth fiscal quarter, and by August of that year, FoxMeyer Drug filed for bankruptcy.[6]

Another major source of information technology project scope problems is a lack of user involvement. For example, in the late 1980s at Northrop Grumman, which specializes in defense electronics, information technology, and advanced aircraft, shipbuilding, and space technology, an information technology project team became convinced that it could, and should, automate the review and approval process of government proposals. The team implemented a powerful workflow system to manage the whole process. Unfortunately, the end users for the system were aerospace engineers who preferred to work in a more

---

[6] James, Geoffrey, "Information Technology fiascoes . . . and how to avoid them," *Datamation* (November 1997).

casual, ad hoc fashion. They dubbed the system "Naziware" and refused to use it. This example illustrates an information technology project that wasted millions of dollars developing a system that was not in touch with the way end users did their work. [7]

Failing to follow good project management processes and use off-the-shelf software often results in scope problems. 21st Century Insurance Group in Woodland Hills, California, paid Computer Sciences Corporation $100 million on a project to develop a system for managing business applications, including managing insurance policies, billing, claims, and customer service. After five years, the system was still in development and used to support less than 2 percent of the company's business in 2002. Joshua Greenbaum, an analyst at Enterprise Applications Consulting, called the project a "huge disaster" and questioned the insurance company's ability "to manage a process that is pretty well known these days...I'm surprised that there wasn't some way to build what they needed using off-the-shelf components and lower their risk." [8]

Scope verification involves formal acceptance of the completed project scope by the stakeholders. This acceptance is often achieved by a customer inspection and then sign-off on key deliverables. To receive formal acceptance of the project scope, the project team must develop clear documentation of the project's products and procedures to evaluate if they were completed correctly and satisfactorily. To minimize scope changes, it is crucial to do a good job of verifying project scope.

The project scope statement, WBS dictionary, project scope management plan, and deliverables are the main input for scope verification. The main tool for performing scope verification is inspection. The customer, sponsor, or user inspects the work after it is delivered. The main outputs of scope verification are accepted deliverables, requested changes, and recommended corrective actions. For example, suppose Kim's team members deliver upgraded computers to users as part of the IT Upgrade Project. Several users might complain because the computers did not include special keyboards they need for medical reasons. Appropriate people would review this change request and take appropriate corrective action, such as getting sponsor approval for purchasing the special keyboards.

SCOPE CONTROL

As discussed in the section of Chapter 4 on integrated change control, change is inevitable on projects, especially changes to the scope of information technology projects. **Scope control** involves controlling changes to the project scope. Users often are not exactly sure how they want screens to look or what functionality

---

[7] Ibid.

[8] Songini, Marc L., "21st Century Insurance apps in limbo despite $100M investment," *ComputerWorld* (December 6, 2002).

they will really need to improve business performance. Developers are not exactly sure how to interpret user requirements, and they also have to deal with constantly changing technologies.

The goal of scope control is to influence the factors that cause scope changes, assure changes are processed according to procedures developed as part of integrated change control, and manage changes when they occur. You cannot do a good job of scope control if you do not first do a good job of scope definition and verification. How can you prevent scope creep when you have not agreed on the work to be performed and your sponsor hasn't verified that the proposed work was acceptable? You also need to develop a process for soliciting and monitoring changes to project scope. Stakeholders should be encouraged to suggest changes that will benefit the overall project and discouraged from suggesting unnecessary changes.

The project scope statement, project scope management plan, WBS and WBS dictionary, performance reports, work performance information, and approved change requests are the main input to scope control. Two important tools for performing scope control include a change control system and configuration management, as described in Chapter 4. Other tools include replanning project scope and performing variance analysis. **Variance** is the difference between planned and actual performance. For example, you can measure cost and schedule variance, as described in more detail in Chapter 7, Project Cost Management. The outputs of scope control include requested changes, recommended corrective action, and updates to the project scope statement, WBS and WBS dictionary, scope baseline, organizational process assets, and project management plan.

Table 1-2 in Chapter 1 lists the top ten factors that help information technology projects succeed. Four of these ten factors are related to scope verification and control: user involvement, clear business objectives, minimized scope, and firm basic requirements. To avoid project failures, therefore, it is crucial for information technology project managers and their teams to work on improving user input and reducing incomplete and changing requirements and specifications.

Suggestions for Improving User Input

Lack of user input leads to problems with managing scope creep and controlling change. How can you manage this important issue? Following are suggestions for improving user input:

- Develop a good project selection process for information technology projects. Insist that all projects have a sponsor from the user organization. The sponsor should not be someone in the information technology department, nor should the sponsor be the project manager. Make project information, including the project charter, project management plan, project scope statement, WBS, and WBS dictionary, easily available in the organization. Making basic project information available will help avoid duplication of effort and ensure that the most important projects are the ones on which people are working.

- Have users on the project team. Some organizations require project managers to come from the business area of the project instead of the information technology group. Some organizations assign co-project managers to information technology projects, one from information technology and one from the main business group. Users should be assigned full-time to large information technology projects and part-time to smaller projects. A key success factor in Northwest Airline's ResNet project (see the companion Web site for this text for an entire case study on this project) was training reservation agents—the users—in how to write programming code for their new reservation system. Because the sales agents had intimate knowledge of the business, they provided excellent input and actually created most of the software.

- Have regular meetings with defined agendas. Meeting regularly sounds obvious, but many information technology projects fail because the project team members do not have regular interaction with users. They assume they understand what users need without getting direct feedback. To encourage this interaction, users should sign off on key deliverables presented at meetings.

- Deliver something to project users and sponsors on a regular basis. If it is some sort of hardware or software, make sure it works first.

- Do not promise to deliver what cannot be delivered in a particular timeframe. Make sure the project schedule allows enough time to produce the deliverables.

- Co-locate users with the developers. People often get to know each other better by being in close proximity. If the users cannot be physically moved to be near developers during the entire project, they could set aside certain days for co-location.

Suggestions for Reducing Incomplete and Changing Requirements

Some requirement changes are expected on information technology projects, but many projects have too many changes to their requirements, especially during later stages of the project life cycle when it is more difficult to implement them. The following are suggestions for improving the requirements process:

- Develop and follow a requirements management process that includes procedures for initial requirements determination. (See the Suggested Reading by Robertson for detailed information on managing requirements.)

- Employ techniques such as prototyping, use case modeling, and Joint Application Design to understand user requirements thoroughly. **Prototyping** involves developing a working replica of the system or some aspect of the system. These working replicas may be throwaways or an incremental component of the deliverable system. Prototyping is an effective tool for gaining an understanding of requirements, determining the feasibility of requirements, and resolving user interface uncertainties. **Use case modeling** is a process for identifying and modeling business events, who initiated them, and how the system should respond to them. It is an effective tool for understanding requirements for information systems. **Joint Application Design (JAD)** uses highly organized and intensive workshops to bring together project stakeholders—the sponsor, users, business analysts, programmers, and so on—to jointly define and design information systems. These techniques also help users become more active in defining system requirements.

- Put all requirements in writing and keep them current and readily available. Several tools are available to automate this function. For example, a type of software called a requirements management tool aids in capturing and maintaining requirements information, provides immediate access to the information, and assists in establishing necessary relationships between requirements and information created by other tools.

- Create a requirements management database for documenting and controlling requirements. Computer Aided Software Engineering (CASE) tools or other technologies can assist in maintaining a repository for project data.

- Provide adequate testing to verify that the project's products perform as expected. Conduct testing throughout the project life cycle. Chapter 8, Project Quality Management, includes more information on testing.

- Use a process for reviewing requested requirements changes from a systems perspective. For example, ensure that project scope changes include associated cost and schedule changes. Require approval by appropriate stakeholders. For example, at PanEnergy Corporation in Houston, Bruce Woodland completed a $10 million electronic commerce project on time and under budget. He says, "Whenever someone wanted to add or change something, we'd tell them how long it would take and then ask them how they wanted to deal with it. . . . The first few times, the users almost took us out behind the dumpster and hanged us. But when we could match specific functionalities they wanted with the time and dollars they required, we didn't have a problem."[9]

- Emphasize completion dates. For example, a project manager at Farmland Industries, Inc. in Kansas City, Missouri, kept her 15-month, $7 million integrated supply-chain project on track by setting the project deadline.

[9] King, Julia, "IS reins in runaway projects," *ComputerWorld* (September 24, 1997).

She says, "May 1 was the drop-dead date, and everything else was backed into it. Users would come to us and say they wanted something, and we'd ask them what they wanted to give up to get it. Sticking to the date is how we managed scope creep."[10]

■ Allocate resources specifically for handling change requests. For example, Peeter Kivestu and his ResNet team at Northwest Airlines knew that users would request enhancements to the reservations system they were developing. They provided a special function key on the ResNet screen for users to submit their requests, and the project included three full-time programmers to handle these requests. Users made over 11,000 enhancement requests. The managers who sponsored the four main software applications had to prioritize the software enhancement requests and decide as a group what changes to approve. The three programmers then implemented as many items as they could, in priority order, given the time they had. Although they only implemented 38 percent of the requested enhancements, they were the most important ones, and the users were very satisfied with the system and process.

USING SOFTWARE TO ASSIST IN PROJECT SCOPE MANAGEMENT

Project managers and their teams can use several types of software to assist in project scope management. As shown in several of the figures and tables in this chapter, you can use word processing software to create scope-related documents, and most people use spreadsheet or presentation software to develop various charts and graphs related to scope management. Project stakeholders also transmit project scope management information using various types of communication software such as e-mail and assorted Web-based applications.

Project management software helps you develop a WBS, which serves as a basis for creating Gantt charts, assigning resources, allocating costs, and so on. You can also use the templates that come with various project management software products to help you create a WBS for your project. See the section on project scope management in Appendix A for detailed information on using Project 2003 and Appendix D for information on templates related to project scope management.

You can also use many types of specialized software to assist in project scope management. Many information technology projects use special software for requirements management, prototyping, modeling, and other scope-related work. Because scope is such a crucial part of project management, there are many software products available to assist in managing project scope.

[10] Ibid.

Project scope management is very important, especially on information technology projects. After selecting projects, organizations must plan what is involved in performing the work of the project, break down the work into manageable pieces, verify the scope with project stakeholders, and manage changes to project scope. Using the basic project management concepts, tools, and techniques discussed in this chapter can help you successfully perform project scope management.

CASE WRAP-UP

Kim Nguyen reviewed guidelines for creating WBSs provided by her company and other sources. She had a meeting with the three team leaders for her project to get their input on how to proceed. They reviewed several sample documents and decided to have major groupings for their project based on updating the inventory database, acquiring the necessary hardware and software, installing the hardware and software, and performing project management. After they decided on a basic approach, Kim led a meeting with the entire project team of twelve people. She reviewed the project charter and preliminary scope statement, described the basic approach they would use to manage the project scope, and reviewed sample WBSs. Kim opened the floor for questions, which she answered confidently. She then let each team leader work with his or her people to start writing the detailed scope statement and their sections of the WBS and WBS dictionary. Everyone participated in the meeting, sharing their individual expertise and openly asking questions. Kim could see that the project was off to a good start.

CHAPTER SUMMARY

Project scope management includes the processes required to ensure that the project addresses all the work required, and only the work required, to complete the project successfully. The main processes include scope planning, scope definition, WBS creation, scope verification, and scope control.

The first step in project scope management is scope planning, where a scope management plan is created. This plan should include descriptions of how the team will prepare the detailed scope statement, create the WBS, verify completion of the project deliverables, and control requests for changes to the project scope.

A project scope statement is created in the scope definition process. This document often includes a project justification, a brief description of the project's products, a summary of all project deliverables, and a statement of what determines project success. There are often several versions of the project scope statement to keep scope information up-to-date.

A work breakdown structure (WBS) is a deliverable-oriented grouping of the work involved in a project that defines the total scope of the project. The WBS forms the basis for planning and managing project schedules, costs, resources, and changes. You cannot use project management software without first creating a good WBS. A WBS dictionary is a document that describes detailed information about each WBS item. A good WBS is often difficult to create because of the complexity of the project. There are several approaches for developing a WBS, including using guidelines, the analogy approach, the top-down approach, the bottom-up approach, and mind mapping.

Scope verification involves formal acceptance of the project scope by the stakeholders. Scope control involves controlling changes to the project scope.

Poor project scope management is one of the key reasons projects fail. For information technology projects, it is important for good project scope management to have strong user involvement, a clear statement of requirements, and a process for managing scope changes.

There are many software products available to assist in project scope management. The WBS is a key concept in properly using project management software since it provides the basis for entering tasks.

DISCUSSION QUESTIONS

1. What is involved in project scope management, and why is good project scope management so important on information technology projects?
2. Discuss the process of further defining project scope, going from information in a project charter to a preliminary scope statement, project scope statement, WBS, and WBS dictionary.
3. Describe different ways to develop a WBS and explain why it is often so difficult to do.
4. Describe a project that suffered from scope creep. Could it have been avoided? How? Can scope creep be a good thing? When?
5. Why do you need a good WBS to use project management software? What other types of software can you use to assist in project scope management?

EXERCISES

1. Use PowerPoint, Visio, or similar software to create a WBS in chart form (similar to an organizational chart—see the sample in Figure 5-2). Assume the level 1 categories are initiating, planning, executing, controlling, and closing. Under the executing section, include level 2 categories of analysis, design, prototyping, testing, implementation, and support. Assume the support category includes level 3 items called training, documentation, user support, and enhancements.

2. Create the same WBS described in Exercise 1 using Project 2003, indenting categories appropriately. Use the outline numbering feature to display the outline numbers (click Tools on the menu bar, click Options, and then click Show outline number). For example, your WBS should start with 1.0 Initiating. Do not enter any durations or dependencies. See Appendix A or Project 2003's Help for assistance. Print the resulting Gantt chart on one page, being sure to display the entire Task Name column.

3. Create a WBS for one of the following projects:

   ■ Introducing self-checkout registers at your school's bookstore
   ■ Updating fifty laptops from Project 2002 to Project 2003
   ■ Creating a new information system for your school or company

   Decide on all of the level 1 categories for the WBS. Then break down the work to at least the third level for one of the level 1 items. Enter the WBS into Project 2003 and print out the Gantt chart. Do not enter any durations or dependencies. Make notes of questions you had while completing this exercise.

4. Review a template file in the Microsoft Project 2003 templates folder, from Microsoft's "templates on Office online" (*http://office.microsoft.com/templates*), or from another source. What do you think about the WBS? Write a one-page paper summarizing your analysis, providing at least three suggestions for improving the WBS.

5. Read one of the suggested readings or find an article related to project scope management. Write a one-page summary of the article, its key conclusions, and your opinion.

RUNNING CASE

Managers at Manage Your Health, Inc. (MYH) selected Tony Prince as the project manager for the Recreation and Wellness Intranet Project. The schedule goal is six months, and the budget is $200,000. Tony had previous project management and systems analysis experience within the company, and he was an avid sports enthusiast. Tony was starting to put the project team together. Tony knew he would have to develop a survey to solicit input from all employees about this new system and make sure it was very user-friendly.

Recall from Chapter 4 that this application would include the following capabilities:

- Allow employees to register for company-sponsored recreational programs, such as soccer, softball, bowling, jogging, walking, and other sports.

- Allow employees to register for company-sponsored classes and programs to help them manage their weight, reduce stress, stop smoking, and manage other health-related issues.

- Track data on employee involvement in these recreational and health-management programs.

- Offer incentives for people to join the programs and do well in them (e.g., incentives for achieving weight goals, winning sports team competitions, etc.).

Assume that MYH would not need to purchase any additional hardware or software for the project.

TASKS

1. Develop a first version of a project scope statement for the project. Use the template provided on the companion Web site for this text under Appendix D and the example in Chapter 3 as guides. Be as specific as possible in describing product characteristics and requirements, as well as all of the project's deliverables. Be sure to include testing and training as part of the project scope.

2. Develop a work breakdown structure (WBS) for the project. Break down the work to level 2 or level 3, as appropriate. Use the template on the companion Web site for this text under Appendix D and samples in Chapters 3 and 5 as guides. Print the WBS in list form as a Word file. Be sure the WBS is based on the project charter (created in the Chapter 4 Running Case), the project scope statement created in Task 1 above, and other relevant information.

3. Use the WBS you developed in Task 2 above to begin creating a Gantt chart in Project 2003 for the project. Use the outline numbering feature to display the outline numbers (click Tools on the menu bar, click Options, and then click Show outline number). Do not enter any durations or

dependencies. Print the resulting Gantt chart on one page, being sure to display the entire Task Name column.

4. Develop a strategy for scope verification and change control for this project. Write a two-page paper summarizing key points of the strategy.

ADDITIONAL RUNNING CASES AND OTHER APPENDICES

Appendix C provides additional case studies and questions you can use to practice applying the concepts, tools, and techniques you are learning throughout this and subsequent chapters. Review the running cases provided in Appendix C and on the companion Web site (*www.course.com/mis/schwalbe4e*). Appendix D includes templates for various project management documents. For additional sample documents based on real projects, visit the author's Web site at *www.kathyschwalbe.com*. Appendix E and the CD-ROM included with this text include a computer simulation where you can also practice applying the project management process groups and knowledge areas.

SUGGESTED READINGS

1. Abramovici, Adrian. "Controlling Scope Creep," *PM Network* (January 2000).

    This article states that scope creep is one of the most common problems that project managers face. The author illustrates the problems with scope creep in a short case study and provides practical advice on how to control it.

2. Levinson, Meridith. "Home Improvement," *CIO Magazine* (*www.cio.com*) (August 1, 2004).

    Home Depot, the second largest retailer in the world (trailing only Wal-Mart), is investing nearly $1 billion in information technology infrastructure and modernization efforts. This article describes several components involved in the massive scope of this effort, business issues driving the changes, and results to date.

3. Project Management Institute. "Project Management Institute Practice Standard for Work Breakdown Structures" (2001).

    The WBS Practice Standard is intended to provide guidance for developing and applying the WBS to project management. PMI members can download a free copy of this document from PMI's Web site (www.pmi.org). Nonmembers can purchase the document from PMI. As well as providing guidelines, this document includes several sample WBSs for a variety of projects, from various industries.

4. Robertson, Susan and James. *Mastering the Requirements Process*. Addison-Wesley (1999).

This book does an outstanding job of explaining the requirements management process. The heart of this book is the Volere Requirements Process Model, an industry-tested and adaptable template for gathering and verifying requirements for software products.

5. Turbit, Neville. "Defining the Scope in IT," The Project Perfect White Paper Collection (*www.projectperfect.com.au*) (October 29, 2003).

This is one of a series of white papers available from the Project Perfect Web site. The author provides a detailed, step-by-step process for defining many information technology deliverables by defining the functionality, the data, and the technical structure.

KEY TERMS

- **analogy approach** — creating a WBS by using a similar project's WBS as a starting point
- **bottom-up approach** — creating a WBS by having team members identify as many specific tasks related to the project as possible and then grouping them into higher level categories
- **decomposition** — subdividing project deliverables into smaller pieces
- **deliverable** — a product, such as a report or segment of software code, produced as part of a project
- **Joint Application Design (JAD)** — using highly organized and intensive workshops to bring together project stakeholders—the sponsor, users, business analysts, programmers, and so on—to jointly define and design information systems
- **mind mapping** — a technique that can be used to develop WBSs by using branches radiating out from a central core idea to structure thoughts and ideas
- **project scope management** — the processes involved in defining and controlling what is or is not included in a project
- **project scope statement** — a document that includes, at a minimum, a description of the project, including its overall objectives and justification, detailed descriptions of all project deliverables, and the characteristics and requirements of products and services produced as part of the project
- **prototyping** — developing a working replica of the system or some aspect of the system to help define user requirements
- **scope** — all the work involved in creating the products of the project and the processes used to create them
- **scope baseline** — the approved project scope statement and its associated WBS and WBS dictionary
- **scope control** — controlling changes to the project scope

- **scope creep** — the tendency for project scope to keep getting bigger
- **scope management plan** — document that includes descriptions of how the project team will prepare the project scope statement, create the WBS, verify completion of the project deliverables, and control requests for changes to the project scope
- **scope verification** — formalizing acceptance of the project scope, sometimes by customer sign-off
- **top-down approach** — creating a WBS by starting with the largest items of the project and breaking them into their subordinate items
- **use case modeling** — a process for identifying and modeling business events, who initiated them, and how the system should respond to them
- **variance** — the difference between planned and actual performance
- **WBS dictionary** — a document that describes detailed information about each WBS item
- **work breakdown structure (WBS)** — a deliverable-oriented grouping of the work involved in a project that defines the total scope of the project
- **work package** — a task at the lowest level of the WBS

6

Project Time Management

Objectives

After reading this chapter, you will be able to:

1. *Understand the importance of project schedules and good project time management*
2. *Define activities as the basis for developing project schedules*
3. *Describe how project managers use network diagrams and dependencies to assist in activity sequencing*
4. *Understand the relationship between estimating resources and project schedules*
5. *Explain how various tools and techniques help project managers perform activity duration estimating*
6. *Use a Gantt chart for planning and tracking schedule information, find the critical path for a project, and describe how critical chain scheduling and the Program Evaluation and Review Technique (PERT) affect schedule development*
7. *Discuss how reality checks and people issues are involved in controlling and managing changes to the project schedule*
8. *Describe how project management software can assist in project time management and review words of caution before using this software*

OPENING CASE

Sue Johnson was the project manager for a consulting company contracted to provide a new online registration system at a local college. This system absolutely had to be operational by May 1st so students could use it to register for the fall semester. Her company's contract had a stiff penalty clause if the system was not ready by then, and Sue and her team would get nice bonuses for doing a good job on this project and meeting the schedule. Sue knew that it was her responsibility to meet the schedule and manage scope, cost, and quality expectations. She and her team developed a detailed schedule and network diagram to help organize the project.

Developing the schedule turned out to be the easy part; keeping the project on track was more difficult. Managing people issues and resolving schedule conflicts were two of the bigger challenges. Many of the customers' employees took unplanned vacations and missed or rescheduled project review meetings. These changes made it difficult for the project team to follow their planned schedule for the system because they had to have customer sign-off at various stages of the systems development life cycle. One senior programmer on her project team quit, and she knew it would take extra time for a new person to get up to speed. It was still early in the project, but Sue knew they were falling behind. What could she do to meet the operational date of May 1st?

THE IMPORTANCE OF PROJECT SCHEDULES

Many information technology projects are failures in terms of meeting scope, time, and cost projections. Managers often cite delivering projects on time as one of their biggest challenges. In fact, even though the 2003 CHAOS studies showed improved success rates for information technology projects, average time overruns increased. Fifty-one percent of information technology projects were "challenged" according to the 2003 study, and their average time overrun increased to 82 percent from a low of 63 percent in 2000.[1]

Managers also cite schedule issues as causing the most conflict on projects throughout a project's timeline. Figure 6-1 shows the results of research on causes of conflict on projects. This figure shows that, overall, schedule issues cause the most conflict over the life of a project. During project formation, or when a project is beginning, priorities and procedures cause more conflict than schedules. During the early phases of a project, only priorities cause more conflict than schedules. During the middle and end phases, schedule issues are the predominant source of conflict.

Perhaps part of the reason schedule problems are so common is that time is easily and simply measured. You can debate scope and cost overruns and make actual numbers appear closer to estimates, but once a project schedule

[1] The Standish Group, "Latest Standish Group CHAOS Report Shows Project Success Rates Have Improved by 50%," (*www.standishgroup.com*) (March 25, 2003).

is set, anyone can quickly estimate schedule performance by subtracting the original time estimate from how long it really took to complete the project. People often compare planned and actual project completion times without taking into account approved changes in the project. Time is also the one variable that has the least amount of flexibility. Time passes no matter what happens on a project.

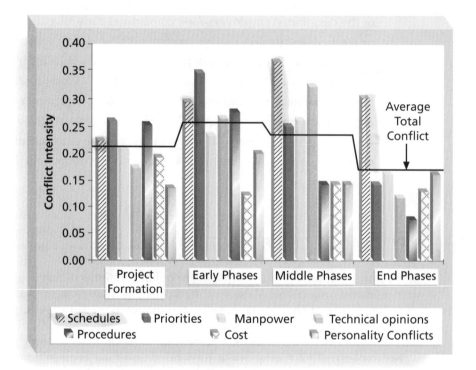

Figure 6-1. Conflict Intensity over the Life of a Project [2]

Individual work styles and cultural differences may also cause schedule conflicts. You will learn in Chapter 9, Project Human Resource Management, about the Myers Briggs Type Indicator. One dimension of this team-building tool deals with peoples' attitudes toward structure and deadlines. Some people prefer detailed schedules and emphasize task completion. Others prefer to keep things open and flexible. Different cultures and even entire countries have different attitudes about schedules. For example, some countries close businesses for several hours every afternoon to have siestas. Others countries may have different religious or secular holidays at certain times of the year when not

[2] Thamhain, H. J., and Wilemon, D. L., "Conflict Management in Project Life Cycles," *Sloan Management Review* (Summer 1975).

much work will be done. Cultures may also have different perceptions of work ethic—some may value hard work and strict schedules while others may value the ability to remain relaxed and flexible.

🎥 Media Snapshot

In contrast to the 2002 Salt Lake City Winter Olympic Games (see Chapter 4's Media Snapshot), planning and scheduling was very different for the 2004 Summer Olympic Games held in Athens, Greece. Many articles were written before the opening ceremonies predicting that the facilities would not be ready in time. "With just 162 days to go to the opening of the Athens Olympics, the Greek capital is still not ready for the expected onslaught…By now 22 of the 30 Olympic projects were supposed to be finished. This week the Athens Olympic Committee proudly announced 19 venues would be finished by the end of next month. That's a long way off target."[3]

However, many people were pleasantly surprised by the amazing opening ceremonies, beautiful new buildings, and state-of-the-art security and transportation systems in Athens. For example, traffic flow, considered a major pre-Games hurdle, was superb. One spectator at the games commented on the prediction that the facilities would not be ready in time, "Athens proved them all wrong…It has never looked better."[4] The Greeks even made fun of critics by having construction workers pretend to still be working as the ceremonies began. Athens Olympics chief Gianna Angelopoulos-Daskalaki deserves a gold medal for her performance in leading the many people involved in organizing the 2004 Olympic Games.

With all these possibilities for schedule conflicts, it's important to use good project time management so that project managers can help improve performance in this area. **Project time management**, simply defined, involves the processes required to ensure timely completion of a project. Achieving timely completion of a project, however, is by no means simple. There are six main processes involved in project time management:

- **Activity definition** involves identifying the specific activities that the project team members and stakeholders must perform to produce the project deliverables. An **activity** or **task** is an element of work normally found on the work breakdown structure (WBS) that has an expected duration, a cost, and resource requirements. The main outputs of this process are an activity list, activity attributes, milestone list, and requested changes.

- **Activity sequencing** involves identifying and documenting the relationships between project activities. The main outputs of this process include a project schedule network diagram, requested changes, and updates to the activity list and attributes.

---

[3] Kelly, Fran, "The World Today – Olympic planning schedule behind time," *ABC Online* (March 4, 2004).

[4] Weiner, Jay and Rachel Blount, "Olympics are safe but crowds are sparse," *Minneapolis Star Tribune* (22 August, 2004), A9.

- **Activity resource estimating** involves estimating how many **resources**—people, equipment, and materials—a project team should use to perform project activities. The main outputs of this process are activity resource requirements, a resource breakdown structure, requested changes, and updates to activity attributes and resource calendars.

- **Activity duration estimating** involves estimating the number of work periods that are needed to complete individual activities. Outputs include activity duration estimates and updates to activity attributes.

- **Schedule development** involves analyzing activity sequences, activity resource estimates, and activity duration estimates to create the project schedule. Outputs include a project schedule, schedule model data, a schedule baseline, requested changes, and updates to resources requirements, activity attributes, the project calendar, and the project management plan.

- **Schedule control** involves controlling and managing changes to the project schedule. Outputs include performance measurements, requested changes, recommended corrective actions, and updates to the schedule model data, the schedule baseline, organizational process assets, the activity list and attributes, and the project management plan.

You can improve project time management by performing these processes and by using some basic project management tools and techniques. Every manager is familiar with some form of scheduling, but most managers have not used several of the tools and techniques unique to project time management, such as Gantt charts, network diagrams, and critical path analysis.

ACTIVITY DEFINITION

Project schedules grow out of the basic documents that initiate a project. The project charter often mentions planned project start and end dates, which serve as the starting points for a more detailed schedule. The project manager starts with the project charter and develops a project scope statement and WBS, as discussed in Chapter 5, Project Scope Management. The project charter should also include some estimate of how much money will be allocated to the project. Using this information with the scope statement, WBS, WBS dictionary, project management plan, and information on organizational process assets, the project manager and project team begin developing a detailed list of activities and their attributes, a milestone list, and requested changes, if applicable.

The **activity list** is a tabulation of activities to be included on a project schedule. The list should include the activity name, an activity identifier or number, and a brief description of the activity. The **activity attributes** provide more schedule-related information about each activity, such as predecessors, successors, logical relationships, leads and lags, resource requirements, constraints, imposed dates, and assumptions related to the activity. The activity list and activity attributes should be in agreement with the WBS and WBS dictionary. Many project teams use an automated system to keep track of all of this activity-related information.

A **milestone** on a project is a significant event that normally has no duration. It often takes several activities and a lot of work to complete a milestone, but the milestone itself is like a marker to help in identifying necessary activities. Milestones are also useful tools for setting schedule goals and monitoring progress. For example, milestones on a project like the one in the opening case might include completion and customer sign-off of documents, such as design documents and test plans; completion of specific products, such as software modules or installation of new hardware; and completion of important process-related work, such as project review meetings, tests, and so on. You will learn more about milestones later in this chapter.

Activity information is a required input to the other time management processes. You cannot determine activity sequencing, resources, or durations, develop the schedule, or control the schedule until you have a good understanding of project activities.

Recall the triple constraint of project management—balancing scope, time, and cost goals—and note the order of these items. Ideally, the project team and key stakeholders first define the project scope, then the time or schedule for the project, and then the project's cost. The order of these three items reflects the basic order of the first four processes in project time management: activity definition (further defining the scope), activity sequencing (further defining the time), and activity resource and activity duration estimating (further defining the cost). These four project time management processes are the basis for creating a project schedule.

The goal of the activity definition process is to ensure that the project team has complete understanding of all the work they must do as part of the project scope so they can start scheduling the work. For example, a WBS item might be "Produce study report." The project team would have to understand what that means before they can make schedule-related decisions. How long should the report be? Does it require a survey or extensive research to produce it? What skill level does the report writer need to have? Further defining that task will help the project team determine how long it will take to do and who should do it.

The WBS is often dissected further during the activity definition process as the project team members further define the activities required for performing the work. For example, the task "Produce study report" might be broken down into several subtasks describing the steps involved in producing the report, such as developing a survey, administering the survey, analyzing the survey results, performing research, writing a draft report, editing the report, and finally producing the report.

As stated earlier, activities or tasks are elements of work performed during the course of a project; they have expected durations, costs, and resource requirements. Activity definition also results in supporting detail to document important product information as well as assumptions and constraints related to specific activities. The project team should review the activity list and activity attributes with project stakeholders before moving on to the next step in project time management. If they do not review these items, they could produce an unrealistic schedule and deliver unacceptable results. For example, if a project manager simply estimated that it would take one day for the "Produce study report" task and had an intern or trainee write a 10-page report to complete that task, the result could be a furious customer who expected extensive research, surveys, and a 100-page report. Clearly defining the work is crucial to all projects. If there are misunderstandings about activities, then requested changes may be required.

In the opening case, Sue Johnson and her project team had a contract and detailed specifications for the college's new online registration system. They also had to focus on meeting the May 1st date for an operational system so the college could start using the new system for the new semester's registration. To develop a project schedule, Sue and her team had to review the contract, detailed specifications, and desired operational date, create an activity list, activity attributes, and milestone list. After developing more detailed definitions of project activities, Sue and her team would review them with their customers to ensure that they were on the right track.

ACTIVITY SEQUENCING

After defining project activities, the next step in project time management is activity sequencing. Activity sequencing involves reviewing the activity list and attributes, project scope statement, milestone list, and approved change requests to determine the relationships between activities. It also involves evaluating the reasons for dependencies and the different types of dependencies.

Dependencies

A **dependency** or **relationship** relates to the sequencing of project activities or tasks. For example, does a certain activity have to be finished before another one can start? Can the project team do several activities in parallel? Can some overlap? Determining these relationships or dependencies between activities has a significant impact on developing and managing a project schedule.

There are three basic reasons for creating dependencies among project activities:

- **Mandatory dependencies** are inherent in the nature of the work being performed on a project. They are sometimes referred to as hard logic. For example, you cannot test code until after the code is written.
- **Discretionary dependencies** are defined by the project team. For example, a project team might follow good practice and not start the detailed design of a new information system until the users sign off on all of the analysis work. Discretionary dependencies are sometimes referred to as soft logic and should be used with care since they may limit later scheduling options.
- **External dependencies** involve relationships between project and non-project activities. The installation of a new operating system and other software may depend on delivery of new hardware from an external supplier. Even though the delivery of the new hardware may not be in the scope of the project, you should add an external dependency to it because late delivery will affect the project schedule.

As with activity definition, it is important that project stakeholders work together to define the activity dependencies that exist on their project. Some organizations have guidelines based on the activity dependencies of similar projects. Some organizations rely on the expertise of the people working on the project and their contacts with other employees and colleagues in the profession. Some people like to write each activity letter or name on a sticky note or some other moveable paper to determine dependencies or sequencing. Still others work directly in project management software to establish relationships. Just as it is easier to write a research paper by first having some thoughts on paper before typing it on a computer, it is usually easier to do some manual form of activity sequencing before entering the information into project management software.

Many organizations do not understand the importance of defining activity dependencies and do not use them *at all* in project time management. If you do not define the sequence of activities, you cannot use some of the most powerful schedule tools available to project managers: network diagrams and critical path analysis. The main output of activity sequencing is a project schedule network diagram. Other outputs include updates to the activity list, the project management plan, and the project scope management plan that might result after performing activity sequencing.

Network Diagrams

Network diagrams are the preferred technique for showing activity sequencing. A **network diagram** is a schematic display of the logical relationships among, or sequencing of, project activities. Some people refer to network diagrams as project schedule network diagrams or PERT charts. PERT is described later in this chapter. Figure 6-2 shows a sample network diagram for Project X, which uses the arrow diagramming method (ADM) or activity-on-arrow (AOA) approach.

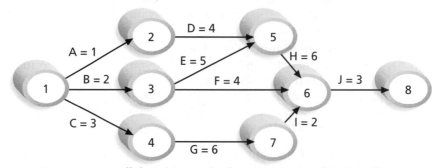

Note: Assume all durations are in days; A=1 means Activity A has a duration of 1 day.

Figure 6-2. Activity-on-Arrow (AOA) Network Diagram for Project X

Note the main elements on this network diagram. The letters A through J represent activities with dependencies that are required to complete the project. These activities come from the WBS and activity definition process described earlier. The arrows represent the activity sequencing or relationships between tasks. For example, Activity A must be done before Activity D; Activity D must be done before Activity H, and so on.

The format of this network diagram uses the **activity-on-arrow (AOA) approach** or the **arrow diagramming method (ADM)**—a network diagramming technique in which activities are represented by arrows and connected at points called nodes to illustrate the sequence of activities. A **node** is simply the starting and ending point of an activity. The first node signifies the start of a project, and the last node represents the end of a project.

Keep in mind that the network diagram represents activities that must be done to complete the project. It is not a race to get from the first node to the last node. *Every* activity on the network diagram must be completed in order for the project to finish.

It is also important to note that not every single item on the WBS needs to be on the network diagram, especially on large projects. Sometimes it is enough to put summary tasks on a network diagram or to break the project down into several smaller network diagrams. Some items that the team knows must be done and will be done regardless of other activities do not need to be included in the network diagram.

Assuming you have a list of the project activities and their start and finish nodes, follow these steps to create an AOA network diagram:

1. Find all of the activities that start at Node 1. Draw their finish nodes, and draw arrows between Node 1 and each of those finish nodes. Put the activity letter or name on the associated arrow. If you have a duration estimate, write that next to the activity letter or name, as shown in Figure 6-2. For example, A = 1 means that the duration of Activity A is one day, week, or other standard unit of time. Also be sure to put arrowheads on all arrows to signify the direction of the relationships.

2. Continue drawing the network diagram, working from left to right. Look for bursts and merges. **Bursts** occur when two or more activities follow a single node. A **merge** occurs when two or more nodes precede a single node. For example, in Figure 6-2, Node 1 is a burst since it goes into Nodes 2, 3, and 4. Node 5 is a merge preceded by Nodes 2 and 3.

3. Continue drawing the AOA network diagram until all activities are included on the diagram.

4. As a rule of thumb, all arrowheads should face toward the right, and no arrows should cross on an AOA network diagram. You may need to redraw the diagram to make it look presentable.

Even though AOA or ADM network diagrams are generally easy to understand and create, a different method is more commonly used: the precedence diagramming method.

The **precedence diagramming method (PDM)** is a network diagramming technique in which boxes represent activities. It is particularly useful for visualizing certain types of time relationships.

Figure 6-3 illustrates the types of dependencies that can occur among project activities. After you determine the reason for a dependency between activities (mandatory, discretionary, or external), you must determine the type of dependency. Note that the terms activity and task are used interchangeably, as are relationship and dependency. The four types of dependencies or relationships between activities include:

- **Finish-to-start:** a relationship where the "from" activity must finish before the "to" activity can start. For example, you cannot provide user training until after software, or a new system, has been installed. Finish-to-start is the most common type of relationship, or dependency, and AOA network diagrams use only finish-to-start dependencies.

- **Start-to-start:** a relationship in which the "from" activity cannot start until the "to" activity is started. For example, on several information technology projects, a group of activities all start simultaneously, such as the many tasks that occur when a new system goes live.

- **Finish-to-finish:** a relationship where the "from" activity must be finished before the "to" activity can be finished. One task cannot finish before another finishes. For example, quality control efforts cannot finish before production finishes, although the two activities can be performed at the same time.

- **Start-to-finish:** a relationship where the "from" activity must start before the "to" activity can be finished. This type of relationship is rarely used, but it is appropriate in some cases. For example, an organization might strive to stock raw materials just in time for the manufacturing process to begin. A delay in the manufacturing process starting should delay completion of stocking the raw materials.

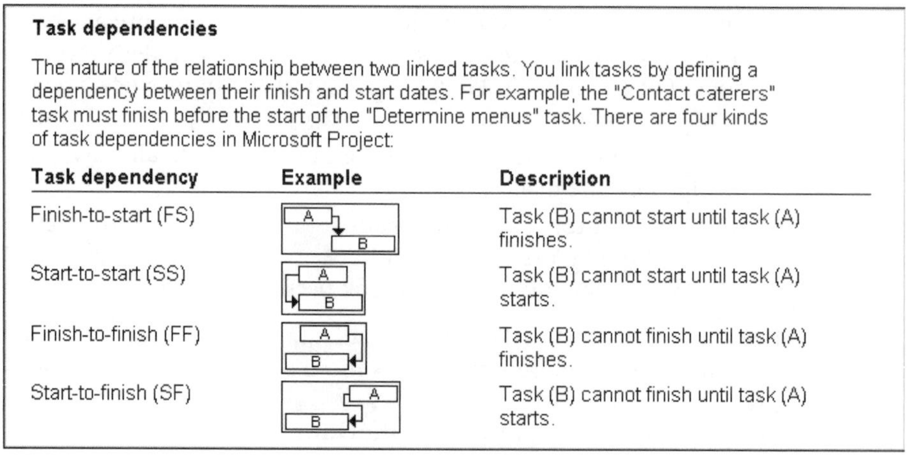

Task dependencies

The nature of the relationship between two linked tasks. You link tasks by defining a dependency between their finish and start dates. For example, the "Contact caterers" task must finish before the start of the "Determine menus" task. There are four kinds of task dependencies in Microsoft Project:

Task dependency	Example	Description
Finish-to-start (FS)	A / B	Task (B) cannot start until task (A) finishes.
Start-to-start (SS)	A / B	Task (B) cannot start until task (A) starts.
Finish-to-finish (FF)	A / B	Task (B) cannot finish until task (A) finishes.
Start-to-finish (SF)	A / B	Task (B) cannot finish until task (A) starts.

Figure 6-3. Task Dependency Types

Figure 6-4 illustrates Project X using the precedence diagramming method. Notice that the activities are placed inside boxes, which represent the nodes on this diagram. Arrows show the relationships between activities. This figure was created using Microsoft Project 2003, which automatically places additional information inside each node. Each task box includes the start and finish date, labeled Start and Finish, the task ID number, labeled ID, the task's duration, labeled Dur, and the names of resources, if any, assigned to the task, labeled Res. The border of the boxes for tasks on the critical path appears automatically in red in the Project 2003 network diagram view. In Figure 6-4, the boxes for critical tasks have a thicker border.

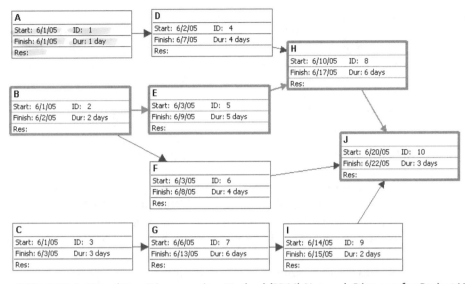

Figure 6-4. Precedence Diagramming Method (PDM) Network Diagram for Project X

The precedence diagramming method is used more often than AOA network diagrams and offers a number of advantages over the AOA technique. First, most project management software uses the precedence diagramming method. Second, the precedence diagramming method avoids the need to use dummy activities. **Dummy activities** have no duration and no resources but are occasionally needed on AOA network diagrams to show logical relationships between activities. They are represented with dashed arrow lines, and have zero for the duration estimate. For example, looking back at Figure 6-2, Activity A precedes Activity D and Activity B precedes Activities E and F. If there was also a dependency between Activities B and D, you would draw a dashed arrow line between Node 3 and Node 2. Third, the precedence diagramming method shows different dependencies among tasks, whereas AOA network diagrams use only finish-to-start dependencies. You will learn more about activity sequencing using Project 2003 in Appendix A.

ACTIVITY RESOURCE ESTIMATING

Before you can estimate the duration for each activity, you must have a good idea of the quantity and type of resources (people, equipment, and materials) that will be assigned to each activity. The nature of the project and the organization will affect resource estimating. Expert judgment, the availability of alternatives, and estimating data and software (discussed in more detail in Chapter 7, Project Cost Management) are tools available to assist in resource estimating. It is important that the people who help determine what resources are necessary include people who have experience and expertise in similar projects and with the organization performing the project.

Important questions to answer in activity resource estimating include:

- How difficult will it be to do specific activities on this project?
- Is there anything unique in the project's scope statement that will affect resources?
- What is the organization's history in doing similar activities? Has the organization done similar tasks before? What level of personnel did the work?
- Does the organization have people, equipment, and materials that are capable and available for performing the work? Are there any organizational policies that might affect the availability of resources?
- Does the organization need to acquire more resources to accomplish the work? Would it make sense to outsource some of the work? Will outsourcing increase or decrease the amount of resources needed and when they'll be available?

A project's activity list, activity attributes, project management plan, enterprise environmental factors, organizational process assets (such as policies regarding staffing and outsourcing), and resource availability information are all important input to answering these questions. During the early phases of a project, the project team may not know which specific people, equipment, and materials will be available. For example, they might know from past projects that there will be a mix of experienced and inexperienced programmers working on a project. They might also have information available that approximates the number of people or hours it normally takes to perform specific activities.

It is important to thoroughly brainstorm and evaluate alternatives related to resources, especially on projects that involve people from multiple disciplines and companies. Since most projects involve many human resources and the majority of costs are for salaries and benefits, it is often effective to solicit ideas from different people to help develop alternatives and address resource-related issues early in a project. The resource estimates should also be updated as more detailed information becomes available.

The main outputs of the resource estimating process include a list of activity resource requirements, a resource breakdown structure, requested changes, and updates to the activity attributes and resource calendars, if needed. For example, if junior employees will be assigned to many activities, the project manager might request that additional activities, time, and resources be approved to help train and mentor those employees. In addition to providing the basis for activity duration estimating, activity resource estimating provides vital information for project cost estimating (Chapter 7), project human resource management (Chapter 9), project communications management (Chapter 10), project risk management (Chapter 11), and project procurement management (Chapter 12). For example, a **resource breakdown structure** is a hierarchical structure that identifies the project's resources by category and type. Resource categories might include analysts, programmers, and testers. Under programmers, there might be types of programmers, such as Java programmers or COBOL programmers. This information would be helpful in determining resource costs, acquiring resources, and so on.

ACTIVITY DURATION ESTIMATING

After working with key stakeholders to define activities, determine their dependencies, and estimate their resources, the next process in project time management is to estimate the duration of activities. It is important to note that **duration** includes the actual amount of time worked on an activity *plus* elapsed time. For example, even though it might take one workweek or five workdays to do the actual work, the duration estimate might be two weeks to allow extra time needed to obtain outside information. The resources assigned to a task will also affect the task duration estimate. Do not confuse duration with **effort**, which is the number of workdays or work hours required to complete a task. A duration estimate of one day could be based on eight hours of work or eighty hours of work. Duration relates to the time estimate, not the effort estimate. Of course, the two are related, so project team members must document their assumptions when creating duration estimates and update the estimates as the project progresses. The people who will actually do the work, in particular, should have a lot of say in these duration estimates, since they are the ones whose performance will be evaluated on meeting them. If scope changes occur on the project, the duration estimates should be updated to reflect those changes. It is also helpful to review similar projects and seek the advice of experts in estimating activity durations.

There are several inputs to activity duration estimating. Enterprise environmental factors, organizational process assets, the project scope statement, activity list, activity attributes, activity resource requirements, resource calendars, and the project management plan all include information that affect duration estimates. In addition to reviewing past project information, the team should also review the accuracy of the duration estimates thus far on the project. For example, if they find that all of their estimates have been much too long or short, the team should update the estimates to reflect what they have learned. One of the most important considerations in making activity duration estimates is the availability of resources, especially human resources. What specific skills do people need to do the work? What are the skill levels of the people assigned to the project? How many people are expected to be available to work on the project at any one time?

The outputs of activity duration estimating include updates to the activity attributes, if needed, and duration estimates for each activity. Duration estimates are often provided as a discrete number, such as four weeks, or as a range, such as three to five weeks, or as a three-point estimate. A **three-point estimate** is an estimate that includes an optimistic, most likely, and pessimistic estimate, such as three weeks for the optimistic, four weeks for the most likely, and five weeks for the pessimistic estimate. The optimistic estimate is based on a best-case scenario, while the pessimistic estimate is based on a worst-case scenario. The most likely estimate, as it sounds, is an estimate based on a most likely or expected scenario. A three-point estimate is required for performing PERT estimates, as described later in this chapter, and for performing Monte Carlo simulation, described in Chapter 11, Project Risk Management. Other duration estimating techniques include analogous and parametric estimating and reserve analysis, as described in Chapter 7, Project Cost Management. Expert judgment is also an important tool for developing good activity duration estimates.

SCHEDULE DEVELOPMENT

Schedule development uses the results of all the preceding project time management processes to determine the start and end dates of the project. There are often several iterations of all the project time management processes before a project schedule is finalized. The ultimate goal of schedule development is to create a realistic project schedule that provides a basis for monitoring project progress for the time dimension of the project. The main outputs of this process are the project schedule, schedule model data, a schedule baseline, requested changes, and updates to resource requirements, activity attributes, the project calendar, and the project management plan. Some project teams create a computerized model to create a network diagram, enter resource requirements and availability by time period, and adjust other information to quickly generate alternative schedules. See Appendix A for information on using Project 2003 to assist in schedule development.

Several tools and techniques assist in the schedule development process:

- A Gantt chart is a common tool for displaying project schedule information.
- Critical path analysis is a very important tool for developing and controlling project schedules.
- Critical chain scheduling is a technique that accounts for resource constraints and uses project and feeding buffers.
- PERT analysis is a means for evaluating schedule risk on projects.

The following section provides samples of each of these tools and techniques and discusses their advantages and disadvantages. (See the *PMBOK® Guide 2004* to read how these main techniques and others are broken into additional categories.)

Gantt Charts

Gantt charts provide a standard format for displaying project schedule information by listing project activities and their corresponding start and finish dates in a calendar format. Gantt charts are sometimes referred to as bar charts since the activities' start and end dates are shown as horizontal bars. Figure 6-5 shows a simple Gantt chart for Project X created with Project 2003. Figure 6-6 shows a Gantt chart that is more sophisticated based on a software launch project. Recall that the activities on the Gantt chart should coincide with the activities on the WBS, which should coincide with the activity list and milestone list. Notice that the software launch project's Gantt chart contains milestones, summary tasks, individual task durations, and arrows showing task dependencies.

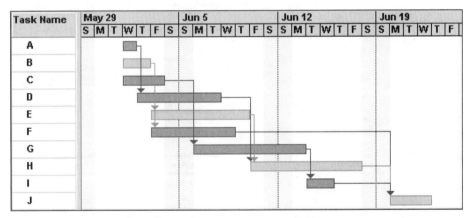

Figure 6-5. Gantt Chart for Project X

Notice the different symbols on the software launch project's Gantt chart (Figure 6-6):

- The black diamond symbol represents a milestone. In Figure 6-6, Task 1, "Marketing Plan distributed," is a milestone that occurs on March 17. Tasks 3, 4, 8, 9, 14, 25, 27, 43, and 45 are also milestones. For very large projects, top managers might want to see only milestones on a Gantt chart. Project 2003 allows you to filter information displayed on a Gantt chart so you can easily show specific tasks, such as milestones.

- The thick black bars with arrows at the beginning and end represent summary tasks. For example, Activities 12 through 15—"Develop creative briefs," "Develop concepts," "Creative concepts," and "Ad development"— are all subtasks of the summary task called "Advertising," Task 11. WBS activities are referred to as tasks and subtasks in most project management software.

- The light gray horizontal bars such as those found in Figure 6-6 for Tasks 5, 6, 7, 10, 12, 13, 15, 22, 24, 26, and 44, represent the duration of each individual task. For example, the light gray bar for Subtask 5, "Packaging," starts in mid-February and extends until early May.

- Arrows connecting these symbols show relationships or dependencies between tasks. Gantt charts often do not show dependencies, which is their major disadvantage. If dependencies have been established in Project 2003, they are automatically displayed on the Gantt chart.

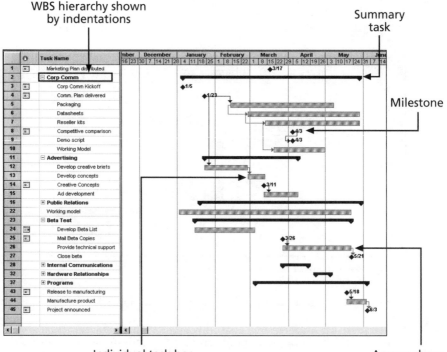

Figure 6-6. Gantt Chart for Software Launch Project

Adding Milestones to Gantt Charts

Milestones can be a particularly important part of schedules, especially for large projects. Many people like to focus on meeting milestones, so you can create milestones to emphasize important events or accomplishments on projects. Normally, you create milestones by entering tasks with zero duration. In Project 2003, you can also mark any task as a milestone by checking the appropriate box in the Advanced tab of the Task Information dialog box. The duration of the task will not change to zero, but the Gantt chart will show the milestone symbol to represent that task based on is start date. See Appendix A for more information.

To make milestones meaningful, some people use the SMART criteria to help define them. The **SMART criteria** are guidelines suggesting that milestones should be:

- **S**pecific
- **M**easurable
- **A**ssignable
- **R**ealistic
- **T**ime-framed

For example, distributing a marketing plan is specific, measurable, and assignable if everyone knows what should be in the marketing plan, how it should be distributed, how many copies should be distributed and to whom, and who is responsible for the actual delivery. Distributing the marketing plan is realistic and able to be time-framed if it is an achievable event and scheduled at an appropriate time.

Using Tracking Gantt Charts to Compare Planned and Actual Dates

You can use a special form of a Gantt chart to evaluate progress on a project by showing actual schedule information. Figure 6-7 shows a **Tracking Gantt chart**—a Gantt chart that compares planned and actual project schedule information. The planned schedule dates for activities are called the **baseline dates**, and the entire approved planned schedule is called the **schedule baseline**. The Tracking Gantt chart includes columns (hidden in Figure 6-7) labeled "Start" and "Finish" to represent actual start and finish dates for each task, as well as columns labeled "Baseline Start" and "Baseline Finish" to represent planned start and finish dates for each task. In this example, the project is completed, but several tasks missed their planned start and finish dates.

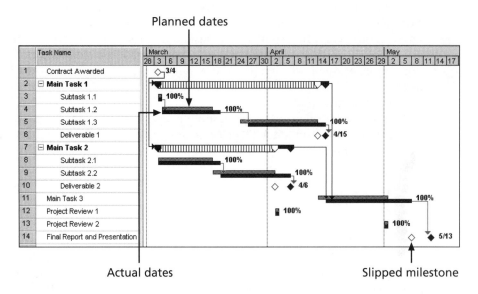

Figure 6-7. Sample Tracking Gantt Chart

To serve as a progress evaluation tool, a Tracking Gantt chart uses a few additional symbols:

■ Notice that the Gantt chart in Figure 6-7 often shows two horizontal bars for tasks. The top horizontal bar represents the planned or baseline duration for each task. The bar below it represents the actual duration. Subtasks 1.2 and 1.3 illustrate this type of display. If these two bars are the same length, and start and end on the same dates, then the actual schedule was the same as the planned schedule for that task. This scheduling occurred for Subtask 1.1, where the task started and ended as planned on March 4. If the bars do not start and end on the same dates, then the actual schedule differed from the planned or baseline schedule. If the top horizontal bar is shorter than the bottom one, the task took longer than planned, as you can see for Subtask 1.2. If the top horizontal bar is longer than the bottom one, the task took less time than planned. A striped horizontal bar, illustrated by Main Tasks 1 and 2, represents the planned duration for summary tasks. The black bar adjoining it shows progress for summary tasks. For example, Main Task 2 clearly shows the actual duration took longer than what was planned.

- A white diamond on the Tracking Gantt chart represents a **slipped milestone**. A slipped milestone means the milestone activity was actually completed later than originally planned. For example, the last task provides an example of a slipped milestone since the final report and presentation were completed later than planned.

- Percentages to the right of the horizontal bars display the percentage of work completed for each task. For example, 100 percent means the task is finished. 50 percent means the task is still in progress and is 50 percent completed.

A Tracking Gantt chart is based on the percentage of work completed for project tasks or the actual start and finish dates. It allows the project manager to monitor schedule progress on individual tasks and the whole project. For example, Figure 6-7 shows that this project is completed. It started on time, but it finished a little late, on May 13 (5/13) versus May 8.

The main advantage of using Gantt charts is that they provide a standard format for displaying planned and actual project schedule information. In addition, they are easy to create and understand. The main disadvantage of Gantt charts is that they do not usually show relationships or dependencies between tasks. If Gantt charts are created using project management software and tasks are linked, then the dependencies will be displayed, but not as clearly as they would be displayed on a network diagram.

Critical Path Method

Many projects fail to meet schedule expectations. **Critical path method (CPM)**—also called **critical path analysis**—is a network diagramming technique used to predict total project duration. This important tool will help you combat project schedule overruns. A **critical path** for a project is the series of activities that determine the *earliest* time by which the project can be completed. It is the *longest* path through the network diagram and has the least amount of slack or float. **Slack** or **float** is the amount of time an activity may be delayed without delaying a succeeding activity or the project finish date. There are normally several tasks done in parallel on projects, and most projects have multiple paths through a network diagram. The longest path or path containing the critical tasks is what is driving the completion date for the project. You are not finished with the project until you have finished *all* the tasks.

Calculating the Critical Path

To find the critical path for a project, you must first develop a good network diagram, which, in turn, requires a good activity list based on the WBS. Once you create a network diagram, you must also estimate the duration of each activity to determine the critical path. Calculating the critical path involves adding the durations for all activities on each path through the network diagram. The longest path is the critical path.

Figure 6-8 shows the AOA network diagram for Project X again. Note that you can use either the AOA or precedence diagramming method to determine the critical path on projects. Figure 6-8 shows all of the paths—a total of four—through the network diagram. Note that each path starts at the first node (1) and ends at the last node (8) on the AOA network diagram. This figure also shows the length or total duration of each path through the network diagram. These lengths are computed by adding the durations of each activity on the path. Since path B-E-H-J at 16 days has the longest duration, it is the critical path for the project.

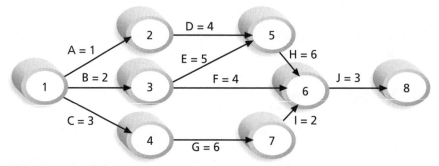

Note: Assume all durations are in days.

Path 1:	A-D-H-J	Length = 1+4+6+3 = 14 days
Path 2:	B-E-H-J	Length = 2+5+6+3 = 16 days
Path 3:	B-F-J	Length = 2+4+3 = 9 days
Path 4:	C-G-I-J	Length = 3+6+2+3 = 14 days

Since the critical path is the longest path through the network diagram, Path 2, B-E-H-J, is the critical path for Project X.

Figure 6-8. Determining the Critical Path for Project X

What does the critical path really mean? The critical path shows the shortest time in which a project can be completed. Even though the critical path is the *longest* path, it represents the *shortest* time it takes to complete a project. If one or more of the activities on the critical path takes longer than planned, the whole project schedule will slip *unless* the project manager takes corrective action.

Project teams can be creative in managing the critical path. For example, Joan Knutson, a well-known author and speaker in the project management field, often describes how a gorilla helped Apple computer complete a project on time. Team members worked in an area with cubicles, and whoever was in charge of a task currently on the critical path had a big, stuffed gorilla on top of his or her cubicle. Everyone knew that person was under the most time pressure, so they tried not to distract him or her. When a critical task was completed, the person in charge of the next critical task received the gorilla.

Growing Grass Can Be on the Critical Path

People are often confused about what the critical path is for a project or what it really means. Some people think the critical path includes the most critical activities. However, the critical path is concerned only with the time dimension of a project. The fact that its name includes the word *critical* does not mean that it includes all critical activities. For example, Frank Addeman, Executive Project Director at Walt Disney Imagineering, explained in a keynote address at the May 2000 PMI-ISSIG Professional Development Seminar that growing grass was on the critical path for building Disney's Animal Kingdom theme park! This 500-acre park required special grass for its animal inhabitants, and some of the grass took years to grow. Another misconception is that the critical path is the shortest path through the network diagram. In some areas, such as transportation modeling, similar network diagrams are drawn in which identifying the shortest path is the goal. For a project, however, each task or activity must be done in order to complete the project. It is not a matter of choosing the shortest path.

Other aspects of critical path analysis may cause confusion. Can there be more than one critical path on a project? Does the critical path ever change? In the Project X example, suppose that Activity A has a duration estimate of three days instead of one day. This new duration estimate would make the length of Path 1 equal to sixteen days. Now the project has two longest paths of equal duration, so there are two critical paths. Therefore, there *can* be more than one critical path on a project. Project managers should closely monitor performance of activities on the critical path to avoid late project completion. If there is more than one critical path, project managers must keep their eyes on all of them.

The critical path on a project can change as the project progresses. For example, suppose everything is going as planned at the beginning of the project. In this example, suppose Activities A, B, C, D, E, F, and G all start and finish as planned. Then suppose Activity I runs into problems. If Activity I takes more than four days, it will cause path C-G-I-J to be longer than the other paths, assuming they progress as planned. This change would cause path C-G-I-J to become the new critical path. Therefore, the critical path can change on a project.

Using Critical Path Analysis to Make Schedule Trade-Offs

It is important to know what the critical path is throughout the life of a project so the product manager *can* make trade-offs. If the project manager knows that one of the tasks on the critical path is behind schedule, he or she needs to decide what to do about it. Should the schedule be renegotiated with stakeholders? Should more resources be allocated to other items on the critical path to make up for that time? Is it okay if the project finishes behind schedule? By keeping track of the critical path, the project manager and his or her team take a proactive role in managing the project schedule.

A technique that can help project managers make schedule trade-offs is determining the free slack and total slack for each project activity. **Free slack** or **free float** is the amount of time an activity can be delayed without delaying the early start date of any immediately following activities. The **early start date** for an activity is the earliest possible time an activity can start based on the project network logic. **Total slack** or **total float** is the amount of time an activity can be delayed from its early start without delaying the planned project finish date.

Project managers calculate free slack and total slack by doing a forward and backward pass through a network diagram. A **forward pass** determines the early start and early finish dates for each activity. The **early finish date** for an activity is the earliest possible time an activity can finish based on the project network logic. The project start date is equal to the early start date for the first network diagram activity. Early start plus the duration of the first activity is equal to the early finish date of the first activity. It is also equal to the early start date of each subsequent activity unless an activity has multiple predecessors. When an activity has multiple predecessors, its early start date is the latest of the early finish dates of those predecessors. For example, Tasks D and E immediately precede Task H in Figure 6-8. The early start date for Task H, therefore, is the early finish date of Task E, since it occurs later than the early finish date of Task D. A **backward pass** through the network diagram determines the late start and late finish dates for each activity in a similar fashion. The **late start date** for an activity is the latest possible time an activity might begin without delaying the project finish date. The **late finish date** for an activity is the latest possible time an activity can be completed without delaying the project finish date.

Project managers can determine the early and late start and finish dates of each activity by hand. For example, Figure 6-9 shows a simple network diagram with three tasks, A, B, and C. Tasks A and B both precede Task C. Assume all duration estimates are in days. Task A has an estimated duration of 5 days, Task B has an estimated duration of 10 days, and Task C has an estimated duration of 7 days. There are only two paths through this small network diagram: path A-C has a duration of 12 days (5+7), and path B-C has a duration of 17 days (10+7). Since path B-C is longest, it is the critical path. There is no float or slack on this path, so the early and late start and finish dates are the same. However, Task A has 5 days of float or slack. Its early start date is day 0, and its late start date is day 5. Its early finish date is day 5, and its late finish date is day 10. Both the free and total float amounts for Task A are the same at 5 days.

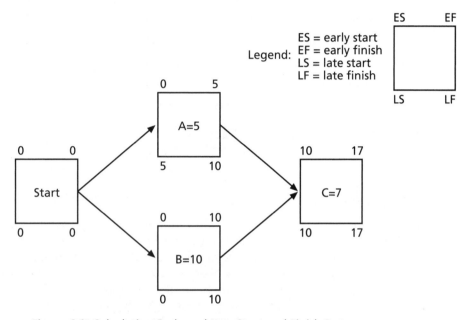

Figure 6-9. Calculating Early and Late Start and Finish Dates

A much faster and easier way to determine early and late start and finish dates and free and total slack amounts for activities is by using project management software. Table 6-1 shows the free and total slack for all activities on the network diagram for Project X after entering the data from Figure 6-8 and assuming Tasks A, B, and C started on 6/1/05. (The network diagram is shown in Figure 6-4.) The data in this table was created by selecting the Schedule Table view in Project 2003. Knowing the amount of float or slack allows project managers to know whether the schedule is flexible and how flexible it might be. For example, at 7 days (7d), Task F has the most free and total slack. The most slack on any other activity is only 2 days (2d). Understanding how to create and use slack information provides a basis for negotiating (or not negotiating) project schedules.

Table 6-1: Free and Total Float or Slack for Project X

Task Name	Start	Finish	Late Start	Late Finish	Free Slack	Total Slack
A	6/1/05	6/1/05	6/3/05	6/3/05	0d	2d
B	6/1/05	6/2/05	6/1/05	6/2/05	0d	0d
C	6/1/05	6/3/05	6/3/05	6/7/05	0d	2d
D	6/2/05	6/7/05	6/6/05	6/9/05	2d	2d
E	6/3/05	6/9/05	6/3/05	6/9/05	0d	0d
F	6/3/05	6/8/05	6/14/05	6/17/05	7d	7d
G	6/6/05	6/13/05	6/8/05	6/15/05	0d	2d
H	6/10/05	6/17/05	6/10/05	6/17/05	0d	0d
I	6/14/05	6/15/05	6/16/05	6/17/05	2d	2d
J	6/20/05	6/22/05	6/20/05	6/22/05	0d	0d

Using the Critical Path to Shorten a Project Schedule

It is common for stakeholders to want to shorten a project schedule estimate. A project team may have done their best to develop a project schedule by defining activities, determining sequencing, and estimating resources and durations for each activity. The results of this work may have shown that the project team needs ten months to complete the project. The sponsor might ask if the project can be done in eight or nine months. Rarely do people ask the project team to take longer than they suggested. By knowing the critical path, the project manager and his or her team can use several duration compression techniques to shorten the project schedule. One technique is to reduce the duration of activities on the critical path. The project manager can shorten the duration of critical path activities by allocating more resources to those activities or by changing their scope.

Recall that Sue Johnson in the opening case was having schedule problems with the online registration project because several users missed important project review meetings and one of the senior programmers quit. If Sue and her team created a realistic project schedule, produced accurate duration estimates, and established dependencies between tasks, they could analyze their status in terms of meeting the May 1 deadline. If some activities on the critical path had already slipped and they did not build in extra time at the end of the project, then they would have to take corrective actions to finish the project on time. Sue could request that her company or the college provide more people to work on the project in an effort to make up time. She could also request that the scope of activities be reduced to complete the project on time. Sue could also use project time management techniques, such as crashing or fast tracking, to shorten the project schedule.

Crashing is a technique for making cost and schedule trade-offs to obtain the greatest amount of schedule compression for the least incremental cost. For example, suppose one of the items on the critical path for the online registration project was entering course data for the new semester into the new system. If this task is yet to be done and was originally estimated to take two weeks based on the college providing one part-time data entry clerk, Sue could suggest that the college have the clerk work full time to finish the task in one week instead of two. This change would not cost Sue's company more money, and it could shorten the project end date by one week. If the college could not meet this request, Sue could consider hiring a temporary data entry person for one week to help get the task done faster. By focusing on tasks on the critical path that could be done more quickly for no extra cost or a small cost, the project schedule can be shortened.

The main advantage of crashing is shortening the time it takes to finish a project. The main disadvantage of crashing is that it often increases total project costs. You will learn more about project costs in Chapter 7, Project Cost Management.

Another technique for shortening a project schedule is fast tracking. **Fast tracking** involves doing activities in parallel that you would normally do in sequence. For example, Sue Johnson's project team may have planned not to start any of the coding for the online registration system until *all* of the analysis was done. Instead, they could consider starting some coding activity before the analysis is completed.

The main advantage of fast tracking, like crashing, is that it can shorten the time it takes to finish a project. The main disadvantage of fast tracking is that it can end up lengthening the project schedule since starting some tasks too soon often increases project risk and results in rework. In fact, as the What Went Wrong example shows next, project managers often need to push for less aggressive schedules instead of finding ways to speed things up. Remember that many information technology projects fail due to organizational issues, which often require more time to address than technical or business issues.

ⓧ *What Went Wrong?*

Even though many information technology projects seem to be rushed, some project managers have realized that a slower approach is required for some projects to be successful. For example, project managers continue to face considerable problems when they try to rush the implementation of Customer Relationship Management (CRM) software. Scott Nelson, a Gartner, Inc. analyst, said that based on Gartner's survey results last year, about 55 percent of the respondents characterized their CRM rollouts as failures. In fact, 42 percent of the CRM end-user licenses bought last year are not even being used.

Why are CRM rollouts so troublesome? Most people say that adoption of the tools by end users can be very slow. End users are interested in the new capabilities, but the technology "causes fear," said Shawn Kaplan, director of marketing and business development at New York-based financial data provider Reuters America, Inc. Reuters rolled out Siebel Systems, Inc.'s CRM applications internally and planned to provide the functionality to financial advisers who use Reuters data by using Web services integration technology. That way, financial advisers will not even know they are using the system, according to Kaplan. Fairchild Semiconductor International, Inc. plans to take up to three years to complete its CRM project. American Trans Air, Inc. (ATA) also started small with its CRM software. "We have been taking baby steps," noted Robert Ellison, director of e-business and network services. "In fact, it was probably more like a crawl."[5]

Importance of Updating Critical Path Data

In addition to finding the critical path at the beginning of a project, it is important to update the schedule with actual data. After the project team completes activities, the project manager should document the actual durations of those activities. He or she should also document revised estimates for activities in progress or yet to be started. These revisions often cause a project's critical path to change, resulting in a new estimated completion date for the project. Again, proactive project managers and their teams stay on top of changes so they can make informed decisions and keep stakeholders informed of, and involved in, major project decisions.

Critical Chain Scheduling

Another technique that addresses the challenge of meeting or beating project finish dates is an application of the Theory of Constraints called critical chain scheduling. The **Theory of Constraints (TOC)** is a management philosophy developed by Eliyahu M. Goldratt and introduced in his book *The Goal*. The Theory of Constraints is based on the fact that, like a chain with its weakest link, any complex system at any point in time often has only one aspect or constraint that limits its ability to achieve more of its goal. For the system to attain any significant improvements, that constraint must be identified, and the whole system must be managed with it in mind. **Critical chain scheduling** is a method of scheduling that considers limited resources when creating a project schedule and includes buffers to protect the project completion date.

You can find the critical path for a project without considering the allocation of scarce resources. For example, task duration estimates and dependencies can be made without considering the availability of specific resources. Estimators might just assume that certain people, pieces of equipment, or other resources will be available. In contrast, an important concept in critical chain scheduling is the availability of scarce resources. If a particular resource is needed full time

[5] Songini, Marc L, "CRM Projects Continue to Inspire Caution, Users Say," *ComputerWorld* (March 10, 2003).

to complete two tasks that were originally planned to occur simultaneously, critical chain scheduling acknowledges that you must either delay one of those tasks until the resource is available or find another resource. In this case, accounting for limited resources often extends the project finish date, which is not most people's intent. Other important concepts related to critical chain scheduling include multitasking and time buffers.

Multitasking occurs when a resource works on more than one task at a time. This situation occurs frequently on projects. People are assigned to multiple tasks within the same project or different tasks on multiple projects. For example, suppose someone is working on three different tasks, Task 1, Task 2, and Task 3, for three different projects, and each task takes ten days to complete. If the person did not multitask, and instead completed each task sequentially, starting with Task 1, then Task 1 would be completed after day ten, Task 2 would be completed after day twenty, and Task 3 would be completed after day thirty, as shown in Figure 6-10a. However, because many people in this situation try to please all three people who need their tasks completed, they often work on the first task for some time, then the second, then the third, then go back to the first task, and so on, as shown in Figure 6-10b. In this example, the tasks were all half-done one at a time, then completed one at a time. Task 1 is now completed at the end of day twenty instead of day ten, Task 2 is completed at the end of day twenty-five instead of day twenty, and Task 3 is still completed on day thirty. This example illustrates how multitasking can delay task completions. Multitasking also often involves wasted setup time, which increases total duration.

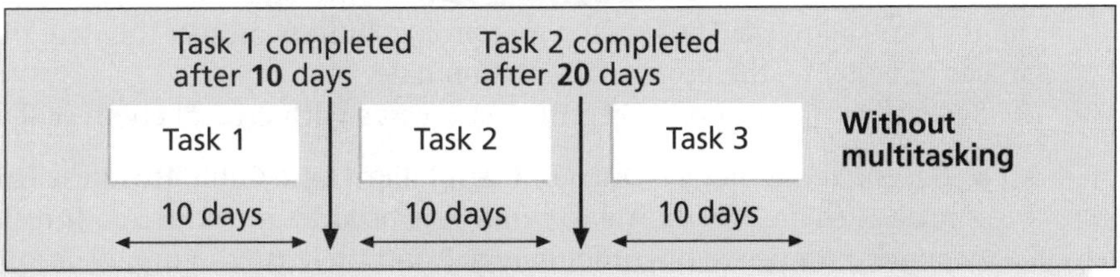

Figure 6-10a. Three Tasks Without Multitasking

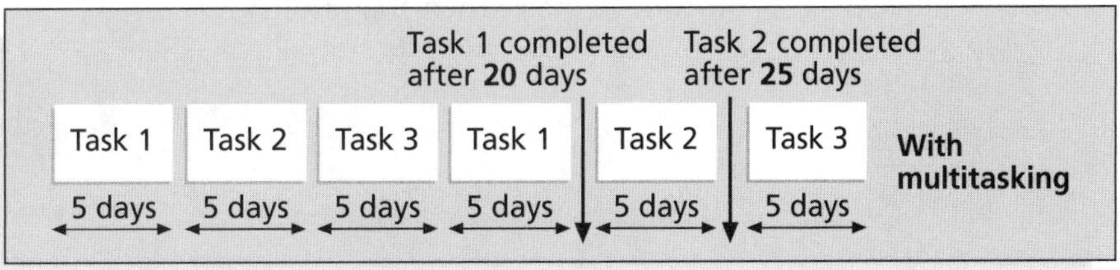

Figure 6-10b. Three Tasks With Multitasking

Critical chain scheduling assumes that resources do not multitask or at least minimize multitasking. Someone should not be assigned to two tasks simultaneously on the same project when critical chain scheduling is in effect. Likewise, critical chain theory suggests that projects be prioritized so people working on more than one project at a time know which tasks take priority. Preventing multitasking avoids resource conflicts and wasted setup time caused by shifting between multiple tasks over time.

An essential concept to improving project finish dates with critical chain scheduling is to change the way people make task estimates. Many people add a safety or **buffer**, which is additional time to complete a task, to an estimate to account for various factors. These factors include the negative effects of multitasking, distractions and interruptions, fear that estimates will be reduced, Murphy's Law, and so on. **Murphy's Law** states that if something can go wrong, it will. Critical chain scheduling removes buffers from individual tasks and instead creates a **project buffer**, which is additional time added before the project's due date. Critical chain scheduling also protects tasks on the critical chain from being delayed by using **feeding buffers**, which are additional time added before tasks on the critical chain that are preceded by non-critical-path tasks.

Figure 6-11 provides an example of a network diagram constructed using critical chain scheduling. Note that the critical chain accounts for a limited resource, X, and the schedule includes use of feeding buffers and a project buffer in the network diagram. The tasks marked with an X are part of the critical chain, which can be interpreted as being the critical path using this technique. The task estimates in critical chain scheduling should be shorter than traditional estimates because they do not include their own buffers. Not having task buffers should mean less occurrence of **Parkinson's Law**, which states that work expands to fill the time allowed. The feeding and project buffers protect the date that really needs to be met—the project completion date.

Several companies have reported successes with critical chain scheduling:

■ Lucent Technology's Outside Plant Fiber Optic Cable Business Unit used critical chain scheduling to reduce its product introduction interval by 50 percent, improve on-time delivery, and increase the organization's capacity to develop products.

■ Synergis Technologies Group successfully implemented critical chain scheduling to manage more than 200 concurrent projects in nine locations, making on-time delivery their top priority.

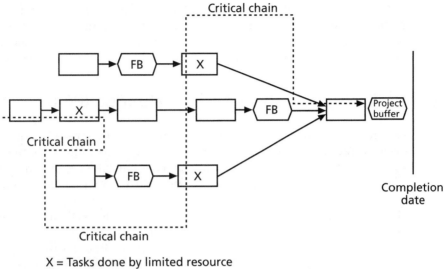

X = Tasks done by limited resource
FB = Feeding buffer

Figure 6-11. Example of Critical Chain Scheduling[6]

■ The Antarctic Support Associates project team switched to critical chain scheduling to enable the Laurence M. Gould research vessel to pass sea-trial tests and be ready to embark on its voyage to Antarctica on schedule in January 1998, rather than four months late as had been anticipated.[7]

■ The U.S. health care industry is learning to think differently by using the Theory of Constraints. "Consider a relatively simple system of a physician's office or clinic. The steps in the process could be patients checking in, filling out forms, having vital signs taken by a nurse, seeing the physician, seeing the nurse for a prescribed procedure such as vaccination, and so forth. These steps could take place in a simple linear sequence or chain…This process or chain only can produce an average of eight per hour. The chain is only as strong as its weakest link and the rate of the slowest resource in this example, the weakest link, is eight. This is true regardless of how fast each of the other resources can process individually, how much work is stuffed into the pipeline, or how complex the process or set of interconnected processes is to complete. Moreover, improving the performance of any link besides the constraint does nothing to improve the system as a whole."[8]

---

[6] Goldratt, Eliyahu, *Critical Chain*, (Great Barrington, MA: The North River Press), 1997, p. 218.

[7] Avraham Y. Goldratt Institute Web site, *www.goldratt.com* (January 2001).

[8] Breen, Anne M., Tracey Burton-Houle, and David C. Aron, "Applying the Theory of Constraints in Health Care: Part 1 - The Philosophy," *Quality Management in Health Care*, Volume 10, Number 3 (Spring 2002) (*www.goldratt.com/for-cause/applyingtocinhcpt1fco.htm*).

As you can see, critical chain scheduling is a fairly complicated yet powerful tool that involves critical path analysis, resource constraints, and changes in how task estimates are made in terms of buffers. Some consider critical chain scheduling one of the most important new concepts in the field of project management.

Program Evaluation and Review Technique (PERT)

Another project time management technique is the **Program Evaluation and Review Technique (PERT)**—a network analysis technique used to estimate project duration when there is a high degree of uncertainty about the individual activity duration estimates. PERT applies the critical path method to a weighted average duration estimate.

PERT uses **probabilistic time estimates**—duration estimates based on using optimistic, most likely, and pessimistic estimates of activity durations—instead of one specific or discrete duration estimate. In other words, PERT uses a three-point estimate, as described earlier. Like the critical path method, PERT is based on a network diagram, normally the precedence diagramming method. To use PERT, you calculate a weighted average for the duration estimate of each project activity using the following formula:

$$\text{PERT weighted average} = \frac{\text{optimistic time} + 4 \times \text{most likely time} + \text{pessimistic time}}{6}$$

By using the PERT weighted average for each activity duration estimate, the total project duration estimate accounts for the risk or uncertainty in the individual activity estimates.

Suppose Sue Johnson's project team in the opening case used PERT to determine the schedule for the online registration system project. They would have to collect numbers for the optimistic, most likely, and pessimistic duration estimates for each project activity. Suppose one of the activities was to design an input screen for the system. Someone might estimate that it would take about two weeks or ten workdays to do this activity. Without using PERT, the duration estimate for that activity would be ten workdays. Using PERT, the project team would also need to estimate the pessimistic and optimistic times for completing this activity. Suppose an optimistic estimate is that the input screen can be designed in eight workdays, and a pessimistic time estimate is 24 workdays. Applying the PERT formula, you get the following:

$$\text{PERT weighted average} = \frac{8 \text{ workdays} + 4 \times 10 \text{ workdays} + 24 \text{ workdays}}{6} = 12 \text{ workdays}$$

Instead of using the most likely duration estimate of ten workdays, the project team would use 12 workdays when doing critical path analysis. These additional two days could really help the project team in getting the work completed on time.

The main advantage of PERT is that it attempts to address the risk associated with duration estimates. Since many projects exceed schedule estimates, PERT may help in developing schedules that are more realistic. PERT has three disadvantages: it involves more work since it requires several duration estimates, there are better probabilistic methods for assessing risk (see information on Monte Carlo simulations in Chapter 11, Project Risk Management), and it is rarely used in practice. In fact, many people confuse PERT with network diagrams since the latter are often referred to as PERT charts. PERT uses network diagrams, but the activity durations are estimated using the PERT weighted average formula.

SCHEDULE CONTROL

The final process in project time management is schedule control. Like scope control, schedule control is a portion of the integrated change control process under project integration management. The goal of schedule control is to know the status of the schedule, influence the factors that cause schedule changes, determine that the schedule has changed, and manage changes when they occur.

The main inputs to schedule control are the schedule baseline, performance reports, approved change requests, and the schedule management plan. Tools and techniques include:

- Progress reports
- A schedule change control system, operated as part of the integrated change control system described in Chapter 4, Project Integration Management
- Project management software, such as Project 2003 or similar software
- Schedule comparison charts, such as the tracking Gantt chart
- Variance analysis, such as analyzing float or slack
- Performance management, such as earned value, described in Chapter 7, Project Cost Management

The main outputs of schedule control include performance measurements, requested changes, recommended corrective actions, and updates to the schedule baseline, schedule model data, activity list, activity attributes, project management plan, and organizational process assets, such as lessons-learned reports related to schedule control.

There are many issues involved in controlling changes to project schedules. It is important first to ensure that the project schedule is realistic. Many projects, especially in information technology, have very unrealistic schedule expectations. It is also important to use discipline and leadership to emphasize the importance of following and meeting project schedules. Although the various tools and techniques assist in developing and managing project schedules, project managers must handle several people-related issues to keep projects on track. "Most projects fail because of people issues, not from failure to draw a good PERT chart."[9] Project managers can perform a number of reality checks that will help them manage changes to project schedules. Several soft skills can help project managers to control schedule changes.

Reality Checks on Scheduling

One of the first reality checks a project manager should make is to review the draft schedule usually included in the project charter. Although this draft schedule might include only a project start and end date, the project charter sets some initial schedule expectations for the project. Next, the project manager and his or her team should prepare a more detailed schedule and get stakeholders' approval. To establish the schedule, it is critical to get involvement and commitment from all project team members, top management, the customer, and other key stakeholders.

Several experts have written about the lack of realistic schedule estimates for information technology projects. Ed Yourdon, a well-known and respected software development expert, describes "death march" projects as projects doomed for failure from the start, due to unrealistic expectations, especially for meeting time constraints.[10] Many schedules for embedded software systems are "propped up by prayer and hopeful wishes."[11] Yielding to pressure from top management or other groups, such as marketing, to determine project schedules is nothing short of lying.[12] Experts in chaos theory suggest that project managers must account for the complex behavior of the organization by arranging for additional resources to be committed to the project. Using an analogy of how to avoid traffic jams on highways, they suggest that project managers schedule resources on the project so that no single resource is utilized more than 75 percent.[13] It is very important, therefore, to set realistic project schedules and allow for contingencies throughout the life of a project.

Another type of reality check comes from progress meetings with stakeholders. The project manager is responsible for keeping the project on track, and key stakeholders like to stay informed, often through high-level periodic reviews. Managers like to see progress made on projects approximately each

---

[9] Bolton, Bart, "IS Leadership," *ComputerWorld* (May 19, 1997).

[10] Yourdon, Ed, "Surviving a Death March Project," *Software Development* (July 1997).

[11] Ganssle, Jack G., "Lies, Damn Lies, and Schedule," *Embedded Systems Programming* (December 1997).

[12] Ibid.

[13] Monson, Robert, "The Role of Complexity and Chaos in Project Management," working paper (February 1999).

month. Project managers often illustrate progress with a tracking Gantt chart showing key deliverables and activities. The project manager needs to understand the schedule, including why activities are or are not on track, and take a proactive approach to meeting stakeholder expectations. It is also important to verify schedule progress, as Sue Johnson from the opening case discovered (see the Case Wrap-Up). Just because a team member says a task was completed on time does not always mean that it was. Project managers must review the actual work and develop a good relationship with team members to ensure work is completed as planned or changes are reported as needed.

Top management hates surprises, so the project manager must be clear and honest in communicating project status. By no means should project managers create the illusion that the project is going fine when, in fact, it is having serious problems. When serious conflicts arise that could affect the project schedule, the project manager must alert top management and work with them to resolve the conflicts.

Working with People Issues

As in any endeavor involving many people, it is often not the technical aspects that are most difficult but rather the people. Creating a good network diagram and detailed project schedule are significant skills for project managers, but good project managers realize that their main job is to lead the people involved in the project. Depending on the size of the project, many project managers have one or more staff members responsible for coordinating input from many other people to create and update the project schedule. Delegating the details of a project schedule allows the project manager to focus on the big picture and lead the entire project, ensuring it remains on schedule. Several leadership skills that help project managers control schedule changes include:

- Empowerment
- Incentives
- Discipline
- Negotiation

It is important for the project manager to empower project team members to take responsibility for their activities. Having team members help create a detailed schedule and provide timely status information empowers them to take responsibility for their actions. As a result, they should feel more committed to the project.

The project manager can also use financial or other incentives to encourage people to meet schedule expectations. Alternatively, it is sometimes effective to use coercive power or negative incentives to stop people from missing deadlines. For example, one project manager started charging her team members 25 cents every time they were late in submitting their weekly time sheets for a project. She was amazed at how few people were late after implementing that simple policy.

Project managers must also use discipline to control project schedules. Several information technology project managers have discovered that setting firm dates for key project milestones helps minimize schedule changes. It is very easy for scope creep to raise its ugly head on information technology projects. Insisting that important schedule dates be met and that proper planning and analysis be completed up front helps everyone focus on doing what is most important for the project. This discipline results in meeting project schedules.

☑ *What Went Right?*

Chris Higgins used the discipline he learned in the U.S. Army to transform project management into a cultural force at Bank of America. Higgins learned that taking time on the front end of a project could save significant time and money on the back end. As a quartermaster in the Army, when Higgins's people had to pack tents, he devised a contest to find the best way to fold a tent and determine the precise spots to place the pegs and equipment for the quickest possible assembly. Higgins used the same approach when he led an interstate banking initiative to integrate incompatible check processing, checking account, and savings account platforms in various states. Law mandated that the banks solve the problem in one year or less. Higgins's project team was pushing to get to the coding phase of the project quickly, but Higgins held them back. He made the team members analyze, plan, and document requirements for the system in such detail that it took six months just to complete that phase. However, the discipline up front enabled the software developers on the team to do all of the coding in only three months, and the project was completed on time.[14]

Canadian Imperial Bank of Commerce (CIBC) provides a more recent example of successfully controlling the schedule for another information technology project in the banking industry. CIBC transformed 20,000 workstations and 1,200 dispersed financial branches in just one year. They created a Web-based tool to enable large, geographically dispersed teams to access information simultaneously. Each of the 1,200 sites had 75 milestones to track, including the baseline, latest plan, and actual finish dates, resulting in 90,000 data points. According to Jack Newhouse, the company's director of application support, their Web-based tracking tool "was a critical component to success… Accurate, timely data was an invaluable management tool."[15]

Project managers and their team members also need to practice good negotiation skills. Customers and management often pressure people to shorten project schedules. Some people blame information technology schedule overruns on poor estimating, but others state that the real problem is that software developers and other information technology professionals are terrible at defending their estimates. Unlike the general population, most information technology professionals are introverts and have a difficult time holding their

[14] Melymuke, Kathleen, "Spit and Polish," *ComputerWorld* (February 16, 1998).

[15] Chauduri, Tom and David Schlotzhauer, "So many projects, so little time," *PM Network* (October 2003): p. 58.

own when extroverts in marketing, sales, or top management tell them to shorten their schedule estimates. Many information technology professionals are also younger than the stakeholders who are pushing for shorter schedules, making them less apt to question senior members' opinions. It is important, therefore, for the project manager and team members to defend their estimates and learn to negotiate with demanding stakeholders.

USING SOFTWARE TO ASSIST IN PROJECT TIME MANAGEMENT

Several types of software are available to assist in project time management. Software for facilitating communications helps project managers exchange schedule-related information with project stakeholders. Decision support models can help project managers analyze various trade-offs that can be made related to schedule issues. However, project management software, such as Microsoft Project 2003, was designed specifically for performing project management tasks. You can use project management software to draw network diagrams, determine the critical path for a project, create Gantt charts, and report, view, and filter specific project time management information.

Many projects involve hundreds of tasks with complicated dependencies. After you enter the necessary information, project management software automatically generates a network diagram and calculates the critical path(s) for the project. It also highlights the critical path in red on the network diagram. Project management software also calculates the free and total float or slack for all activities. For example, the data in Table 6-1 was created using Project 2003 by selecting the schedule table view from the menu bar. Using project management software eliminates the need to perform cumbersome calculations manually and allows for "what if" analysis as activity duration estimates or dependencies change. Recall that knowing which activities have the most slack gives the project manager an opportunity to reallocate resources or make other changes to compress the schedule or help keep it on track.

Project 2003 easily creates Gantt charts and Tracking Gantt charts, which make tracking actual schedule performance versus the planned or baseline schedule much easier to do. It is important, however, to enter actual schedule information in a timely manner in order to benefit from using the Tracking Gantt chart feature. Some organizations use e-mail or other communications software to send up-to-date task and schedule information to the person responsible for updating the schedule. He or she can then quickly authorize these updates and enter them directly into the project management software. This process provides an accurate and up-to-date project schedule in Gantt chart form.

Project 2003 also includes many built-in reports, views, and filters to assist in project time management. Table 6-2 lists some of these features. For example, a project manager can quickly run a report to list all tasks that are to start soon. He or she could then send out a reminder to the people responsible for these tasks. If the project manager were presenting project schedule information to top management, he or she could create a Gantt chart showing only summary tasks or milestones. You can also create custom reports, views, tables, and filters. You will learn more about the project time management features of Project 2003 in Appendix A.

Table 6-2: Project 2003 Features Related to Project Time Management

REPORTS	VIEWS AND TABLE VIEWS	FILTERS
Overview reports: critical tasks and milestones	Gantt chart, network diagram, Tracking Gantt chart, schedule, tracking, variance, constraint dates, and delay	All tasks, completed critical tasks, incomplete tasks, and milestone tasks
Current activities reports: unstarted tasks, tasks starting soon, tasks in progress, completed tasks, should have started tasks, and slipping tasks		
Assignment reports: who does what and when		

Words of Caution on Using Project Management Software

Many people misuse project management software because they do not understand the concepts behind creating a network diagram, determining the critical path, or setting a schedule baseline. They might also rely too heavily on sample files or templates in developing their own project schedules. Understanding the underlying concepts (even being able to work with the tools manually) is critical to successful use of project management software, as is understanding the specific needs of your project.

Many top managers, including software professionals, have made blatant errors using various versions of Microsoft Project and similar tools. For example, one top manager did not know about establishing dependencies among project activities and entered every single start and end date for hundreds of activities. When asked what would happen if the project started a week or two late, she responded that she would have to reenter all of the dates.

This manager did not understand the importance of establishing relationships among the tasks. Establishing relationships among tasks allows the software to update formulas automatically when the inputs change. If the project start date slips by one week, the project management software will update all the other dates automatically, as long as they are not hard-coded into the software. (Hard-coding involves entering all activity dates manually instead of letting the software calculate them based on durations and relationships.) If one activity cannot start before another ends, and the first activity's actual start date is two days late, the start date for the succeeding activity will automatically be moved back two days. To achieve this type of functionality, tasks that have relationships must be linked in project management software.

Another top manager on a large information technology project did not know about setting a baseline in Microsoft Project. He spent almost one day every week copying and pasting information from Microsoft Project into a spreadsheet and using complicated "IF" statements to figure out what activities were behind schedule. He had never received any training on Microsoft Project and did not know about many of its capabilities. To use any software effectively, users must have adequate training in the software and an understanding of its underlying concepts.

Many project management software programs also come with templates or sample files. It is very easy to use these files without considering unique project needs. For example, a project manager for a software development project can use the Project 2003 Software Development template file, the files containing information from similar projects done in the past, or sample files purchased from other companies. All of these files include suggested tasks, durations, and relationships. There are benefits to using templates or sample files, such as less setup time and a reality check if the project manager has never managed this type of project before. However, there are also drawbacks to this approach. There are many assumptions in these template files that may not apply to the project, such as a design phase that takes three months to complete or the performance of certain types of testing. Project managers and their teams may over-rely on templates or sample files and ignore unique concerns for their particular projects.

CASE WRAP-UP

It was now March 15, just a month and a half before the new online registration system was supposed to go live. The project was in total chaos. Sue Johnson thought she could handle all of the conflicts that kept appearing on the project, and she was too proud to admit to her top management or the college president that things were not going well. She spent a lot of time preparing a detailed schedule for the project, and she thought she was using their project management software well enough to keep up with project status. However, the five main programmers on the project all figured out a way to generate automatic updates for their tasks every week, saying that everything was completed as planned. They paid very little attention to the actual plan and hated filling out status information. Sue did not verify most of their work to check that it was actually completed. In addition, the head of the Registrar's Office was uninterested in the project and delegated sign-off responsibility to one of his clerks who really did not understand the entire registration process. When Sue and her team started testing the new system, she learned they were using last year's course data. Using last year's course data caused additional problems because the college was moving from quarters to semesters in the new semester. How could they have missed that requirement? Sue hung her head in shame as she walked into a meeting with her manager to ask for help. She learned the hard way how difficult it was to keep a project on track. She wished she had spent more time talking face-to-face with key project stakeholders, especially her programmers and the Registrar's Office representatives, to verify that the project was on schedule and that the schedule was updated accurately.

CHAPTER SUMMARY

Project time management is often cited as the main source of conflict on projects. Most information technology projects exceed time estimates. The main processes involved in project time management include activity definition, activity sequencing, activity resource estimating, activity duration estimating, schedule development, and schedule control.

Activity definition involves identifying the specific activities that must be done to produce the project deliverables. It usually results in a more detailed WBS.

Activity sequencing determines the relationships or dependencies between activities. Three reasons for creating relationships are that they are mandatory based on the nature of the work, discretionary based on the project team's experience, or external based on non-project activities. Activity sequencing must be done in order to use critical path analysis.

Network diagrams are the preferred technique for showing activity sequencing. The two methods used for creating these diagrams are the arrow diagramming method and the precedence diagramming method. There are four types of relationships between tasks: finish-to-start, finish-to-finish, start-to-start, and start-to-finish.

Activity resource estimating involves determining the quantity and type of resources (people, equipment, and materials) that will be assigned to each activity. The nature of the project and the organization will affect resource estimating.

Activity duration estimating creates estimates for the amount of time it will take to complete each activity. These time estimates include the actual amount of time worked plus elapsed time.

Schedule development uses results from all of the other project time management processes to determine the start and end dates for the project. Project managers often use Gantt charts to display the project schedule. Tracking Gantt charts show planned and actual schedule information.

The critical path method predicts total project duration. The critical path for a project is the series of activities that determines the earliest completion date for the project. It is the longest path through a network diagram. If any activity on the critical path slips, the whole project will slip unless the project manager takes corrective action.

Crashing and fast tracking are two techniques for shortening project schedules. Project managers and their team members must be careful about accepting unreasonable schedules, especially for information technology projects.

Critical chain scheduling is an application of the Theory of Constraints (TOC) that uses critical path analysis, resource constraints, and buffers to help meet project completion dates.

The Program Evaluation and Review Technique (PERT) is a network analysis technique used to estimate project duration when there is a high degree of uncertainty about the individual activity duration estimates. It uses optimistic, most likely, and pessimistic estimates of activity durations. PERT is seldom used today.

Even though scheduling techniques are very important, most projects fail because of people issues, not from a poor network diagram. Project managers must involve all stakeholders in the schedule development process. It is critical to set realistic project schedules and allow for contingencies throughout the life of a project. Several leadership skills that help project managers control schedule changes include empowerment, discipline, using incentives, and negotiation.

Project management software can assist in project scheduling if used properly. With project management software, you can avoid the need to perform cumbersome calculations manually and perform "what if" analysis as activity duration estimates or dependencies change. Many people misuse project management software because they do not understand the concepts behind creating a network diagram, determining the critical path, or setting a schedule baseline. People must also avoid over-relying on sample files or templates when creating their unique project schedules.

DISCUSSION QUESTIONS

1. Why do you think schedule issues often cause the most conflicts on projects?

2. Why is activity definition the first process involved in project time management?

3. Why is it important to determine activity sequencing on projects? Discuss diagrams you have seen that are similar to network diagrams. Describe their similarities and differences.

4. How does activity resource estimating affect activity duration estimating?

5. Explain the difference between activity duration estimating and estimating the effort required to perform an activity.

6. Explain the following schedule development tools and concepts: Gantt charts, critical path method, PERT, and critical chain scheduling.

7. How can you minimize or control changes to project schedules?

8. List some of the reports you can generate with Project 2003 to assist in project time management.

9. Why is it difficult to use project management software well?

EXERCISES

1. Using Figure 6-2, enter the activities, their durations, and their relationships in Project 2003. Use a project start date of June 1, 2005. View the network diagram. Does it look like Figure 6-4? Print the network diagram on one page. Return to Gantt Chart view. Click View on the menu bar, select Table: Entry, and then click Schedule to recreate Table 6-1. You may need to move the split bar to the right to reveal all of the table columns. (See Appendix A for detailed information on using Project 2003.) Write a few paragraphs explaining what the network diagram and schedule table show concerning Project X's schedule.

2. Consider Table 6-3, Network Diagram Data for a Small Project. All duration estimates or estimated times are in days; and the network proceeds from Node 1 to Node 9.

Table 6-3: Network Diagram Data for a Small Project

ACTIVITY	INITIAL NODE	FINAL NODE	ESTIMATED DURATION
A	1	2	2
B	2	3	2
C	2	4	3
D	2	5	4
E	3	6	2
F	4	6	3
G	5	7	6
H	6	8	2
I	6	7	5
J	7	8	1
K	8	9	2

a. Draw an AOA network diagram representing the project. Put the node numbers in circles and draw arrows from node to node, labeling each arrow with the activity letter and estimated time.

b. Identify all of the paths on the network diagram and note how long they are, using Figure 6-8 as a guide for how to represent each path.

c. What is the critical path for this project and how long is it?

d. What is the shortest possible time it will take to complete this project?

e. Review the online tutorials for Gantt and PERT charts in the third Suggested Reading by Mark Kelly in this chapter. Write a one-page paper with your answers to the questions in the tutorials. Also include any questions you had in doing the tutorials.

3. Enter the information from Exercise 2 into Project 2003. View the network diagram and task schedule table to see the critical path and float or slack for each activity. Print the Gantt chart and network diagram views and the task schedule table. Write a short paper that interprets this information for someone unfamiliar with project time management.

4. You have been asked to determine a rough schedule for a nine-month Billing System Conversion project, as part of your job as a consultant to a Fortune 500 firm. The firm's old system was written in COBOL on a mainframe computer, and the maintenance costs are prohibitive. The new system will run on an off-the-shelf application. You have identified several high-level activities that must be done in order to initiate, plan, execute, control, and close the project. Table 6-4 shows your analysis of the project's tasks and schedule so far.

Table 6-4: COBOL Conversion Project Schedule

Tasks	Mar	Apr	May	Jun	Jul	Aug	Sep	Oct	Nov
Initiating									
Develop project charter									
Meet with stakeholders									
Planning									
Create detailed WBS and schedule									
Estimate project costs									
Create project team									
Create communication plan									
Organize a comprehensive project plan									
Executing									
Award and manage contract for software conversion									
Install new software on servers									
Install new hardware and software on client's machines									
Test new billing system									
Train users on new system									
Controlling									
Closing									

     a. Using the information in Table 6-4, draw horizontal bars to illustrate when you think each task would logically start and end. Then use Project 2003 to create a Gantt chart and network diagram based on this information.

     b. Identify at least two milestones that could be included under each of the process groups in Table 6-4. Then write a detailed description of each of these milestones that meets the SMART criteria.

5. Interview someone who uses some of the techniques discussed in this chapter. How does he or she feel about network diagrams, critical path analysis, Gantt charts, critical chain scheduling, using project management software, and managing the people issues involved in project time management? Write a paper describing the responses.

6. Review two different articles about critical chain scheduling. Write a two- to three-page paper describing how this technique can help improve project schedule management.

RUNNING CASE

Tony Prince is the project manager for the Recreation and Wellness Intranet Project, and team members include you, a programmer/analyst and aspiring project manager; Patrick, a network specialist; Nancy, a business analyst; and

Bonnie, another programmer/analyst. Other people supporting the project from other departments are Yusaff from human resources, and Cassandra from finance. Assume these are the only people who can be assigned and charged to work on project activities. Recall that your schedule and cost goals are to complete the project in six months for under $200,000.

TASKS

1. Review the WBS and Gantt chart you created for Chapter 5, Task 3. Propose three to five additional activities you think should be added to help you estimate resources and durations. Write a one-page paper describing these new activities.

2. Identify at least eight milestones for this project. Write a one-page paper describing each milestone using the SMART criteria. Discuss how determining these milestones might add additional activities or tasks to the Gantt chart. Remember that milestones normally have no duration, so you must have tasks that will lead to completing the milestone.

3. Using the Gantt chart you created for Chapter 5, Task 3, and the new activities and milestones you proposed in Tasks 1 and 2 above, create a new Gantt chart using Project 2003. Estimate the task durations and enter dependencies, as appropriate. Remember that your schedule goal for the project is six months. Print the Gantt chart and network diagram, each on one page.

4. Write a one-page paper summarizing how you would assign people to each activity. Include a table or matrix listing how many hours each person would work on each task. These resource assignments should make sense given the duration estimates made in Task 3 above. Remember that duration estimates are not the same as effort estimates since they include elapsed time.

ADDITIONAL RUNNING CASES AND OTHER APPENDICES

Appendix C provides additional case studies and questions you can use to practice applying the concepts, tools, and techniques you are learning throughout this and subsequent chapters. Review the running cases provided in Appendix C and on the companion Web site (*www.course.com/mis/schwalbe4e*). Appendix D includes templates for various project management documents. For additional sample documents based on real projects, visit the author's Web site at *www.kathyschwalbe.com*. Appendix E and the CD-ROM included with this text include a computer simulation where you can also practice applying the project management process groups and knowledge areas.

SUGGESTED READINGS

1. Breen, Anne M., Tracey Burton-Houle, and David C. Aron. "Applying the Theory of Constraints in Health Care: Part 1 - The Philosophy," *Quality Management in Health Care* (Volume 10, Number 3) (Spring 2002) (*www.goldratt.com/for-cause/applyingtocinhcpt1fco.htm*).

    This article summarizes some of the basic concepts behind critical chain scheduling and shows how it can be used to improve various health care processes. It explains how the existence of constraints represents an excellent opportunity for improvement by allowing organizations to focus their efforts in the most productive areas.

2. Goldratt, Eliyahu. *Critical Chain*. Great Barrington, MA: The North River Press (1997).

    This text is written as a business novel about a college professor and his project management class who together develop the concept of critical chain scheduling. Many other books and articles have since been written about critical chain scheduling. Goldratt now runs the Avraham Y. Goldratt Institute (www.goldratt.com), *the world's largest consulting firm specializing in the application of the Theory of Constraints (TOC) to achieve significant bottom-line results in organizations.*

3. Kelly, Mark. Project Management Tools: Gantt Charts and PERT Charts, McKinnon Secondary College, Australia (*www.mckinnonsc.vic.edu.au/la/it/ipmnotes/ganttpert*).

    These online tutorials are a great way to reinforce your understanding of Gantt charts and network diagrams/PERT charts.

4. McConnell, Steve. "How to Defend an Unpopular Schedule." *IEEE Software* (May 1996) (*www.stevemcconnell.com/bp03.htm*).

    This article emphasizes the fact that it is just as important, if not more so, to defend project schedule estimates as it is to create them. Most software developers must learn to defend their schedule estimates by practicing good negotiation techniques. Steve McConnell has written several books and articles related to software development, including his latest book, Code Complete, *Second Edition, published in 2004.*

5. Patrick, Francis S. "Critical Chain Scheduling and Buffer Management...Getting Out From Between Parkinson's Rock and Murphy's Hard Place." PM Network (April 1999) (*www.focusedperformance.com/articles/CCPM.htm*).

    This article discusses critical chain scheduling using many graphics and examples to clearly explain this important technique. The Focused Performance Web site also provides additional articles from this page related to critical chain scheduling.

KEY TERMS

- **activity** — an element of work, normally found on the WBS, that has an expected duration and cost, and expected resource requirements; also called **task**

- **activity attributes** — information about each activity, such as predecessors, successors, logical relationships, leads and lags, resource requirements, constraints, imposed dates, and assumptions related to the activity

- **activity definition** — identifying the specific activities that the project team members and stakeholders must perform to produce the project deliverables

- **activity duration estimating** — estimating the number of work periods that are needed to complete individual activities

- **activity list** — a tabulation of activities to be included on a project schedule

- **activity-on-arrow (AOA)** or **arrow diagramming method (ADM)** — a network diagramming technique in which activities are represented by arrows and connected at points called nodes to illustrate the sequence of activities

- **activity sequencing** — identifying and documenting the relationships between project activities

- **backward pass** — a project network diagramming technique that determines the late start and late finish dates for each activity in a similar fashion

- **baseline dates** — the planned schedule dates for activities in a Tracking Gantt chart

- **buffer** — additional time to complete a task, added to an estimate to account for various factors

- **burst** — when a single node is followed by two or more activities on a network diagram

- **crashing** — a technique for making cost and schedule trade-offs to obtain the greatest amount of schedule compression for the least incremental cost

- **critical chain scheduling** — a method of scheduling that takes limited resources into account when creating a project schedule and includes buffers to protect the project completion date

- **critical path** — the series of activities in a network diagram that determines the earliest completion of the project; it is the longest path through the network diagram and has the least amount of slack or float

- **critical path method (CPM)** or **critical path analysis** — a project network analysis technique used to predict total project duration

- **dependency** — the sequencing of project activities or tasks; also called a **relationship**

- **discretionary dependencies** — sequencing of project activities or tasks defined by the project team and used with care since they may limit later scheduling options

- **dummy activities** — activities with no duration and no resources used to show a logical relationship between two activities in the arrow diagramming method of project network diagrams
- **duration** — the actual amount of time worked on an activity *plus* elapsed time
- **early finish date** — the earliest possible time an activity can finish based on the project network logic
- **early start date** — the earliest possible time an activity can start based on the project network logic
- **effort** — the number of workdays or work hours required to complete a task
- **external dependencies** — sequencing of project activities or tasks that involve relationships between project and non-project activities
- **fast tracking** — a schedule compression technique in which you do activities in parallel that you would normally do in sequence
- **feeding buffers** — additional time added before tasks on the critical path that are preceded by non-critical-path tasks
- **finish-to-finish dependency** — a relationship on a project network diagram where the "from" activity must be finished before the "to" activity can be finished
- **finish-to-start dependency** — a relationship on a project network diagram where the "from" activity must be finished before the "to" activity can be started
- **forward pass** — a network diagramming technique that determines the early start and early finish dates for each activity
- **free slack (free float)** — the amount of time an activity can be delayed without delaying the early start of any immediately following activities
- **Gantt chart** — a standard format for displaying project schedule information by listing project activities and their corresponding start and finish dates in a calendar format; sometimes referred to as bar charts
- **late finish date** — the latest possible time an activity can be completed without delaying the project finish date
- **late start date** — the latest possible time an activity may begin without delaying the project finish date
- **mandatory dependencies** — sequencing of project activities or tasks that are inherent in the nature of the work being done on the project
- **merge** — when two or more nodes precede a single node on a network diagram
- **milestone** — a significant event that normally has no duration on a project; serves as a marker to help in identifying necessary activities, setting schedule goals, and monitoring progress
- **multitasking** — when a resource works on more than one task at a time
- **Murphy's Law** — if something can go wrong, it will
- **network diagram** — a schematic display of the logical relationships or sequencing of project activities

- **node** — the starting and ending point of an activity on an activity-on-arrow diagram
- **Parkinson's Law** — work expands to fill the time allowed
- **PERT weighted average** — $\dfrac{\text{optimistic time} + 4 \times \text{most likely time} + \text{pessimistic time}}{6}$
- **Precedence Diagramming Method (PDM)** — a network diagramming technique in which boxes represent activities
- **probabilistic time estimates** — duration estimates based on using optimistic, most likely, and pessimistic estimates of activity durations instead of using one specific or discrete estimate
- **Program Evaluation and Review Technique (PERT)** — a project network analysis technique used to estimate project duration when there is a high degree of uncertainty with the individual activity duration estimates
- **project buffer** — additional time added before the project's due date
- **project time management** — the processes required to ensure timely completion of a project
- **resource breakdown structure** — a hierarchical structure that identifies the project's resources by category and type
- **resources** — people, equipment, and materials
- **schedule baseline** — the approved planned schedule for the project
- **schedule control** — controlling and managing changes to the project schedule
- **schedule development** — analyzing activity sequences, activity duration estimates, and resource requirements to create the project schedule
- **slack** — the amount of time a project activity may be delayed without delaying a succeeding activity or the project finish date; also called **float**
- **slipped milestone** — a milestone activity that is completed later than planned
- **SMART criteria** — guidelines to help define milestones that are specific, measurable, assignable, realistic, and time-framed
- **start-to-finish dependency** — a relationship on a project network diagram where the "from" activity cannot start before the "to" activity is finished
- **start-to-start dependency** — a relationship on a project network diagram in which the "from" activity cannot start until the "to" activity starts
- **Theory of Constraints (TOC)** — a management philosophy that states that any complex system at any point in time often has only one aspect or constraint that is limiting its ability to achieve more of its goal
- **three-point estimate** — an estimate that includes an optimistic, most likely, and pessimistic estimate
- **total slack (total float)** — the amount of time an activity may be delayed from its early start without delaying the planned project finish date
- **Tracking Gantt chart** — a Gantt chart that compares planned and actual project schedule information

7

Project Cost Management

OPENING CASE

*J*uan Gonzales was a systems analyst and network specialist for a major city's waterworks department in Mexico. He enjoyed helping the city develop its infrastructure. His next career objective was to become a project manager so he could have even more influence. One of his colleagues invited him to attend an important project review meeting for large government projects, including the Surveyor Pro project, in which Juan was most interested. The Surveyor Pro project was a concept for developing a sophisticated information system that included expert systems, object-oriented databases, and wireless communications. The system would provide instant, graphical information to government surveyors to help them do their jobs. For example, after a surveyor touched a map on the screen of a handheld device, the system would prompt him or her for the type of information needed for that area. This system would help in planning and implementing many projects, from laying fiber-optic cable to laying water lines.

Juan was very surprised, however, when the majority of the meeting was spent discussing cost-related issues. The government officials were reviewing many existing projects to evaluate their performance to date and the potential impact on their budgets before discussing the funding for any new projects. Juan did not understand many of the terms and charts the presenters were showing. What was this earned value term they kept referring to? How were they estimating what it would cost to complete projects or how long it would take? Juan thought he would learn more about the new technologies the Surveyor Pro project would use, but he discovered that the cost estimate and projected benefits were of most interest to the government officials at the meeting. It also seemed as if a lot of effort would go toward detailed financial studies before any technical work could even start. Juan wished he had taken some accounting and finance courses so he could understand the acronyms and concepts people were discussing. Although Juan had a degree in electrical engineering, he had no formal education, and little experience in finance. If Juan could understand information systems and networks, he was confident that he could understand financial issues on projects, too. He jotted down questions to discuss with his colleagues after the meeting.

THE IMPORTANCE OF PROJECT COST MANAGEMENT

Just as information technology projects have a poor track record in meeting project goals, they also have a poor track record in meeting budget goals. Even though the 2003 CHAOS studies showed improved success rates for information technology projects, the average cost **overrun**—the additional percentage or dollar amount by which actual costs exceed estimates—was still 43 percent. Fortunately, the amount of cost overruns continues to decrease. About 23 percent of projects had a cost overrun of less than 20 percent of the estimate, which was a 5 percent increase over the 2000 numbers and a three-fold increase since 1994. The lost dollar value for U.S. information technology projects in 2002 was estimated at $55 billion, including cancelled projects and cost overruns, compared to $140 billion in 1994.[1] Even with these improvements, most completed projects still exceeded their approved budgets, and even more were cancelled before completion. This chapter describes important concepts in project cost management, particularly, creating good estimates and using earned value management (EVM) to assist in cost control.

(X) What Went Wrong?

According to the *San Francisco Chronicle*'s front-page story, "Computer Bumbling Costs the State $1 Billion," the state of California had a series of expensive information technology project failures in the late 1990s, costing taxpayers nearly $1 billion. Some of the poorly managed projects included the Department of Motor Vehicles registration and driver's license databases, a statewide child support database, the State Automated Welfare System, and a Department of Corrections system for tracking inmates. Senator John Vasconcellos thought it was ironic that the state that was leading in the creation of computers was also the state most behind in using computer technology to improve its services.[2]

Consider also the U.S. Internal Revenue Service's expensive reengineering and information technology project failures. The Internal Revenue Service (IRS) managed a series of project failures in the 1990s that cost taxpayers over $50 billion a year—roughly as much money as the annual net profit of the entire computer industry that year.[3] The April 1, 2004 issue of *CIO Magazine* highlighted more recent IRS cost overruns in the article, "For the IRS There's No EZ Fix". The E-Services, Customer Account Data Engine, Integrated Financial System, Modernized E-File, and Custodial Accounting projects were all past due and had cost overruns totaling almost $250 million.

Project cost overruns also occur when companies try to purchase new systems from an outside source. Connecticut General Life Insurance Co. (CIGNA) sued PeopleSoft, Inc. over an aborted installation of a finance system. PeopleSoft increased the original project estimate from $5.16 million to $11.7 million and indicated that the final cost could go even higher. CIGNA wanted PeopleSoft to refund the $5.16 million since they had to develop a homegrown application.[4]

What Is Cost?

A popular cost accounting textbook states, "Accountants usually define cost as a resource sacrificed or foregone to achieve a specific objective."[5] Webster's dictionary defines cost as "something given up in exchange." Costs are often measured in monetary amounts, such as dollars, that must be paid to acquire goods

---

[1] The Standish Group, "Latest Standish Group CHAOS Report Shows Project Success Rates Have Improved by 50%," A Standish Group Research Note (3/25/03).

[2] Lucas, Greg, "Computer Bumbling Costs the State $1 Billion," *San Francisco Chronicle* (2/21/99).

[3] James, Geoffrey, "IT Fiascoes . . . and How to Avoid Them," *Datamation* (November 1997).

[4] Songini, Marc L., "PeopleSoft project ends in court," *ComputerWorld* (September 2001).

[5] Horngren, Charles T., George Foster, and Srikanti M. Datar, *Cost Accounting*, 8th ed. Englewood Cliffs, NJ: Prentice-Hall, 1994.

and services. Because projects cost money and consume resources that could be used elsewhere, it is very important for project managers to understand project cost management.

Many information technology professionals, however, often react to cost overrun information with a smirk. They know that many of the original cost estimates for information technology projects are low to begin with or based on very unclear project requirements, so naturally there will be cost overruns. Not emphasizing the importance of realistic project cost estimates from the outset is only one part of the problem. In addition, many information technology professionals think preparing cost estimates is a job for accountants. On the contrary, preparing good cost estimates is a very demanding, important skill that all information technology professionals need to acquire.

Another perceived reason for cost overruns is that many information technology projects involve new technology or business processes. Any new technology or business process is untested and has inherent risks. Thus, costs grow and failures are to be expected, right? Wrong. Using good project cost management can change this false perception. In fact, the 1995 Standish Group study mentions that many of the information technology projects in the study were as mundane as a new accounting package or an order-entry system. The state of California, the IRS, and the CIGNA examples in the What Went Wrong feature also demonstrate that many information technology projects cannot blame untested technology for cost problems. What was needed was better project cost management.

What Is Project Cost Management?

Recall from Chapter 1 that the triple constraint of project management involves balancing scope, time, and cost goals. Chapters 5 and 6 discuss project scope and time management, and this chapter describes project cost management. **Project cost management** includes the processes required to ensure that a project team completes a project within an approved budget. Notice two crucial phrases in this definition: "a project" and "approved budget." Project managers must make sure *their* projects are well defined, have accurate time and cost estimates, and have a realistic budget that *they* were involved in approving. It is the project manager's job to satisfy project stakeholders while continuously striving to reduce and control costs. There are three project cost management processes:

- **Cost estimating** involves developing an approximation or estimate of the costs of the resources needed to complete a project. The main outputs of the cost estimating process are activity cost estimates and supporting detail, requested changes, and updates to the cost management plan. According to the *PMBOK® Guide 2004*, the cost management plan should be created as part of the project management plan under project integration management.

- **Cost budgeting** involves allocating the overall cost estimate to individual work items to establish a baseline for measuring performance. The main outputs of the cost budgeting process are a cost baseline, project funding requirements, requested changes, and updates to the cost management plan.
- **Cost control** involves controlling changes to the project budget. The main outputs of the cost control process are performance measurements, forecasted completion information, requested changes, recommended corrective action, and updates to the project management plan (which includes the cost management plan), cost estimate, a cost baseline, and organizational process assets.

To understand each of the project cost management processes, you must first understand the basic principles of cost management. Many of these principles are not unique to project management; however, project managers need to understand how these principles relate to their specific projects.

BASIC PRINCIPLES OF COST MANAGEMENT

Many information technology projects are never initiated because information technology professionals do not understand the importance of basic accounting and finance principles. Important concepts, such as net present value analysis, return on investment, and payback analysis were discussed in Chapter 4, Project Integration Management. Likewise, many projects that are started never finish because of cost management problems. Most members of an executive board have a better understanding of and are more interested in financial terms than information technology terms. Therefore, information technology project managers need to be able to present and discuss project information in financial terms as well as in technical terms. In addition to net present value analysis, return on investment, and payback analysis, project managers must understand several other cost management principles, concepts, and terms. This section describes general topics such as profits, life cycle costing, cash flow analysis, tangible and intangible costs and benefits, direct costs, sunk costs, learning curve theory, and reserves. Another important topic and one of the key tools and techniques for controlling project costs—earned value management—is described in detail in the section on cost control.

Profits are revenues minus expenditures. To increase profits, a company can increase revenues, decrease expenses, or try to do both. Most executives are more concerned with profits than with other issues. When justifying investments in new information systems and technology, it is important to focus on the impact on profits, not just revenues or expenses. Consider an e-commerce application that you estimate will increase revenues for a $100 million company by 10 percent. You cannot measure the potential benefits of the e-commerce application without knowing the profit margin. **Profit margin** is the ratio of revenues to profits. If revenues of $100 generate $2 in profits, there is a 2 percent profit margin. If the company loses $2 for every $100 in revenue, there is a –2 percent profit margin.

Life cycle costing allows you to see a big-picture view of the cost of a project throughout its life cycle. This helps you develop an accurate projection of a project's financial benefits. Life cycle costing considers the total cost of ownership, or development plus support costs, for a project. For example, a company might complete a project to develop and implement a new customer service system in one or two years, but the new system could be in place for ten years. Project managers should create estimates of the costs and benefits of the project for its entire life cycle, or ten years in the preceding example. Recall that the net present value analysis for the project would include the entire ten-year period of costs and benefits (see Chapter 4). Top management and project managers need to consider the life cycle costs of projects when they make financial decisions.

Organizations have a history of not spending enough money in the early phases of information technology projects, which impacts total cost of ownership. For example, it is much more cost-effective to spend money on defining user requirements and doing early testing on information technology projects than to wait for problems to appear after implementation. Table 7-1 summarizes the typical costs of correcting software defects during different stages of the systems development life cycle. If you identified and corrected a software defect during the user requirements phase of a project, you would need to add $100 to $1,000 to the project's total cost of ownership. In contrast, if you waited until after implementation to correct the same software defect, you would need to add possibly millions of dollars to the project's total cost of ownership. For example, many automobiles include software components. If there is a defect after delivery of the automobiles, it would cost a lot of money to recall the vehicles to correct the defect.

Table 7-1: Costs of Software Defects [6]

When Defect Is Detected	Typical Costs of Correction
User Requirements	$100 – $1,000
Coding/Unit Testing	$1,000 or more
System Testing	$7,000 – $8,000
Acceptance Testing	$1,000 – $100,000
After Implementation	Up to millions of dollars

[6] Collard, Ross, *Software Testing and Quality Assurance*, working paper (1997).

What Went Right?

A good example of how understanding costs can improve the bottom line comes from a leading telecommunications equipment company. It estimated the cost of a software bug or defect at three stages: after coding, after manual inspection, and after beta release. The costs to correct the defect increased with each stage from $2,000 to $10,000 to $100,000, similar to the findings in Table 7-1. The company also estimated that when it released one million lines of new code, it had an average of 440 defects in the early stage, 250 in the middle stage, and 125 in the late stage. These bugs cost the company more than $15 million. The company decided it had to reduce these costs, so it implemented an automated inspection process. The results were reduced costs for fixing bugs of more than $11 million.[7]

Organizations often purchase information technology products and services to help reduce costs and improve operations. For example, Drug Trading Company, Limited, a Canadian retail pharmacy services company, reduced information technology costs and achieved tighter customer and partner relationships by using J.D. Edwards software (an enterprise/supply chain solution now owned by PeopleSoft). The company can now respond faster to customer requests, with a lower total cost of ownership (TCO). "We're targeting a very quick ROI and a significant reduction in TCO—by as much as 40 percent—once we're in full production with J.D. Edwards. Our strategic objectives are to improve the supply pipeline and provide convenient 24X7 communications channels for customers and vendors," says Grant Schwartz, Corporate Director of IT for Drug Trading Company.[8]

Cash flow analysis is a method for determining the estimated annual costs and benefits for a project and the resulting *annual* cash flow. Project managers must conduct cash flow analysis to determine net present value. Most consumers understand the basic concept of cash flow. If they do not have enough money in their wallets or checking accounts, they cannot purchase something. Top management must consider cash flow concerns when selecting projects in which to invest. If top management selects too many projects that have high cash flow needs in the same year, the company will not be able to support all of its projects and maintain its profitability. It is also important to define clearly the year on which the company bases the dollar amounts. For example, if a company bases all costs on 2005 estimates, it would need to account for inflation and other factors when projecting costs and benefits in future-year dollars.

Tangible and intangible costs and benefits are categories for determining how definable the estimated costs and benefits are for a project. **Tangible costs** or **benefits** are those costs or benefits that an organization can easily measure in dollars. For example, suppose the Surveyor Pro project described in the opening case included a preliminary feasibility study. If a company completed this study for $100,000, the tangible cost of the study is $100,000. If

[7] The Standish Group, "Show Me the Money: Return on Inspection," A Standish Group Research Note (2002).

[8] Customer Testimonial, "Drug Trading Company Targets Reduced IT Costs and Tighter Customer and Partner Relationships," Bitpipe, Inc. (2002).

Juan's government estimated that it would have cost $150,000 to do the study itself, the tangible benefits of the study would be $50,000 if it assigned the people who would have done the study to other projects. Conversely, **intangible costs** or **benefits** are costs or benefits that are difficult to measure in monetary terms. Suppose Juan and a few other people, out of personal interest, spent some time using government-owned computers, books, and other resources to research areas related to the study. Although their hours and the government-owned materials were not billed to the project, they could be considered intangible costs. Intangible benefits for projects often include items like goodwill, prestige, and general statements of improved productivity that an organization cannot easily translate into dollar amounts. Because intangible costs and benefits are difficult to quantify, they are often harder to justify.

Direct costs are costs that can be directly related to producing the products and services of the project. You can attribute direct costs directly to a certain project. For example, the salaries of people working full time on the project and the cost of hardware and software purchased specifically for the project are direct costs. Project managers should focus on direct costs, since they can control them.

Indirect costs are costs that are not directly related to the products or services of the project, but are indirectly related to performing the project. For example, the cost of electricity, paper towels, and so on in a large building housing a thousand employees who work on many projects would be indirect costs. Indirect costs are allocated to projects, and project managers have very little control over them.

Sunk cost is money that has been spent in the past. Consider it gone, like a sunken ship that can never be returned. When deciding what projects to invest in or continue, you should not include sunk costs. For example, in the opening case, suppose Juan's office had spent $1 million on a project over the past three years to create a geographic information system, but they never produced anything valuable. If his government were evaluating what projects to fund next year and someone suggested that they keep funding the geographic information system project because they had already spent $1 million on it, he or she would be incorrectly making sunk cost a key factor in the project selection decision. Many people fall into the trap of considering how much money has been spent on a failing project and, therefore, hate to stop spending money on it. This trap is similar to gamblers not wanting to stop gambling because they have already lost money. Sunk costs should be forgotten.

Learning curve theory states that when many items are produced repetitively, the unit cost of those items decreases in a regular pattern as more units are produced. For example, suppose the Surveyor Pro project would potentially produce 1000 handheld devices that could run the new software and access information via satellite. The cost of the first handheld device or unit would be much higher than the cost of the 1000th unit. Learning curve theory should help estimate costs on projects involving the production of large quantities of items. Learning curve theory also applies to the amount of time it takes to complete some tasks. For example, the first time a new employee performs a specific task, it will probably take longer than the tenth time that employee performs a very similar task.

Reserves are dollars included in a cost estimate to mitigate cost risk by allowing for future situations that are difficult to predict. **Contingency reserves** allow for future situations that may be partially planned for (sometimes called **known unknowns**) and are included in the project cost baseline. For example, if an organization knows it has a 20 percent rate of turnover for information technology personnel, it should include contingency reserves to pay for recruiting and training costs for information technology personnel. **Management reserves** allow for future situations that are unpredictable (sometimes called **unknown unknowns**). For example, if a project manager gets sick for two weeks or an important supplier goes out of business, management reserve could be set aside to cover the resulting costs.

COST ESTIMATING

Project managers must take cost estimates seriously if they want to complete projects within budget constraints. After developing a good resource requirements list, project managers and their project teams must develop several estimates of the costs for these resources. This section describes various types of cost estimates, tools and techniques for cost estimation, typical problems associated with information technology cost estimates, and a detailed example of a cost estimate for an information technology project.

Types of Cost Estimates

One of the main outputs of project cost management is a cost estimate. Project managers normally prepare several types of cost estimates for most projects. Three basic types of estimates include the following:

- A **rough order of magnitude (ROM) estimate** provides an estimate of what a project will cost. ROM estimates can also be referred to as a ballpark estimate, a guesstimate, a swag, or a broad gauge. This type of estimate is done very early in a project or even before a project is officially started. Project managers and top management use this estimate to help make project selection decisions. The timeframe for this type of estimate is often three or more years prior to project completion. A ROM estimate's accuracy is typically –25 percent to +75 percent, meaning the project's actual costs could be 25 percent below the ROM estimate or 75 percent above. For information technology project estimates, this accuracy range is often much wider. Many information technology professionals automatically double estimates for software development because of the history of cost overruns on information technology projects.

■ A **budgetary estimate** is used to allocate money into an organization's budget. Many organizations develop budgets at least two years into the future. Budgetary estimates are made one to two years prior to project completion. The accuracy of budgetary estimates is typically –10 percent to +25 percent, meaning the actual costs could be 10 percent less or 25 percent more than the budgetary estimate.

■ A **definitive estimate** provides an accurate estimate of project costs. Definitive estimates are used for making many purchasing decisions for which accurate estimates are required and for estimating final project costs. For example, if a project involves purchasing 1000 personal computers from an outside supplier in the next three months, a definitive estimate would be required to aid in evaluating supplier proposals and allocating the funds to pay the chosen supplier. Definitive estimates are made one year or less prior to project completion. A definitive estimate should be the most accurate of the three types of estimates. The accuracy of this type of estimate is normally –5 percent to +10 percent, meaning the actual costs could be 5 percent less or 10 percent more than the definitive estimate. Table 7-2 summarizes the three basic types of cost estimates.

Table 7-2: Types of Cost Estimates

TYPE OF ESTIMATE	WHEN DONE	WHY DONE	HOW ACCURATE
Rough Order of Magnitude (ROM)	Very early in the project life cycle, often 3–5 years before project completion	Provides estimate of cost for selection decisions	–25% to +75%
Budgetary	Early, 1–2 years out	Puts dollars in the budget plans	–10% to +25%
Definitive	Later in the project, less than 1 year out	Provides details for purchases, estimates actual costs	–5% to +10%

The number and type of cost estimates vary by application area. For example, the Association for the Advancement of Cost Engineering International identifies five types of cost estimates for construction projects: order of magnitude, conceptual, preliminary, definitive, and control. The main point is that estimates are usually done at various stages of a project and should become more accurate as time progresses.

In addition to creating cost estimates, it is also important to provide supporting details for the estimates. The supporting details include the ground rules and assumptions used in creating the estimate, a description of the project (scope statement, WBS, and so on) used as a basis for the estimate, and details on the cost estimation tools and techniques used to create the estimate. These supporting details should make it easier to prepare an updated estimate or similar estimate as needed.

A **cost management plan** is a document that describes how the organization will manage cost variances on the project. For example, if a definitive cost estimate provides the basis for evaluating supplier cost proposals for all or part of a project, the cost management plan describes how to respond to proposals that are higher or lower than the estimates. Some organizations assume that a cost proposal within 10 percent of the estimate is acceptable and only negotiate items that are more than 10 percent higher or 20 percent lower than the estimated costs. The cost management plan is part of the overall project management plan described in Chapter 4, Project Integration Management.

A large percentage of total project costs are often labor costs. Many organizations estimate the number of people or hours they need by department or skill over the life cycle of a project. For example, when Northwest Airlines developed cost estimates for its reservation system project, ResNet, it determined the maximum number of people it could assign to the project each year by department. Table 7-3 shows this information. (Figure 9-6 in Chapter 9, Project Human Resource Management, provides similar resource information in graphical form, where the number of resources are provided by job category, such as business analyst, programmer, and so on.) Note the small number of contractors Northwest Airlines planned to use. Labor costs are often much higher for contractors, so it is important to distinguish between internal and external resources. (See the companion Web site for this text to read the detailed case study on ResNet, including cost estimates.)

Table 7-3: Maximum Departmental Headcounts by Year

DEPARTMENT	YEAR 1	YEAR 2	YEAR 3	YEAR 4	YEAR 5	TOTALS
Information Systems	24	31	35	13	13	116
Marketing Systems	3	3	3	3	3	15
Reservations	12	29	33	9	7	90
Contractors	2	3	1	0	0	6
Totals	41	66	72	25	23	227

Cost Estimation Tools and Techniques

As you can imagine, developing a good cost estimate is difficult. Fortunately, there are several tools and techniques available to assist in creating them. Four commonly used tools and techniques are analogous cost estimating, bottom-up estimating, parametric modeling, and using computerized tools.

Analogous estimates, also called **top-down estimates**, use the actual cost of a previous, similar project as the basis for estimating the cost of the current project. This technique requires a good deal of expert judgment and is generally less costly than others are, but it is also less accurate. Analogous estimates are most reliable when the previous projects are similar in fact, not just in appearance. In addition, the groups preparing cost estimates must have the needed expertise to determine whether certain parts of the project will be more or less expensive than analogous projects. However, if the project to be estimated involves a new programming language or working with a new type of hardware or network, the analogous estimate technique could easily result in too low an estimate.

Bottom-up estimates involve estimating individual work items or activities and summing them to get a project total. It is sometimes referred to as Activity Based Costing. The size of the individual work items and the experience of the estimators drive the accuracy of the estimates. If a detailed WBS is available for a project, the project manager could have each person responsible for a work package develop his or her own cost estimate for that work package. The project manager would then add all of the cost estimates to create cost estimates for each higher-level WBS item and finally for the entire project. See Appendix A's section on project cost management for detailed information on entering resource costs and assigning resources to tasks to create a bottom-up estimate. Using smaller work items increases the accuracy of the cost estimate because the people assigned to do the work develop the cost estimate instead of someone unfamiliar with the work. The drawback with bottom-up estimates is that they are usually time-intensive and therefore expensive to develop.

Parametric modeling uses project characteristics (parameters) in a mathematical model to estimate project costs. A parametric model might provide an estimate of $50 per line of code for a software development project based on the programming language the project is using, the level of expertise of the programmers, the size and complexity of the data involved, and so on. Parametric models are most reliable when the historical information that was used to create the model is accurate, the parameters are readily quantifiable, and the model is flexible in terms of the size of the project. For example, in the 1980s, engineers at McDonnell Douglas Corporation (now part of Boeing) developed a parametric model for estimating aircraft costs based on a large historical database. The model included the following parameters: the type of aircraft (fighter aircraft, cargo aircraft, or passenger aircraft), how fast the plane would fly, the thrust-to-weight ratio of the engine, the estimated weights of various parts of the aircraft, the number of aircraft produced, the amount of time available to produce them, and so on. In contrast to this sophisticated model, some parametric models involve very simple heuristics or rules of thumb. For example, a large office automation project might use a ballpark figure of $10,000 per workstation based on a history of similar office automation projects developed during the same time period. Parametric models that are more complicated are usually computerized.

One popular parametric model is the **Constructive Cost Model (COCOMO)**, which is used for estimating software development costs based on parameters such as the source lines of code or function points. Barry Boehm, a well-known expert in the field of software development and cost estimating, developed COCOMO. **Function points** are technology-independent assessments of the functions involved in developing a system. For example, the number of inputs and outputs, the number of files maintained, and the number of updates are examples of function points. The number of function points can then be used to determine the **Source Lines of Code (SLOC)**, which is a human-written line of code that is not a blank line or comment. The acronym KSLOC stands for thousands of SLOC. See the sample cost estimate later in this chapter for a brief example of using function points. **COCOMO II** is a newer, computerized version of Boehm's model that allows you to estimate the cost, effort, and schedule when planning a new software development activity. Boehm suggests that only algorithmic or parametric models do not suffer from the limits of human decision-making capability. Therefore, Boehm and many other software experts favor using algorithmic models when estimating software project costs.

Computerized tools, such as spreadsheets and project management software, can make working with different cost estimates and cost estimation tools easier. Computerized tools, when used properly, can also help improve the accuracy of estimates. In addition to spreadsheets and project management software, more sophisticated tools are available for estimating software project costs. Sample products include Galorath's SEER™ Suite of Tools, Cost Xpert's Cost Xpert 3.3, Owens & Minors CostTrackSM, PRICE Estimating Suite, and many more. See the fourth Suggested Reading in this chapter for information on finding computerized tools to assist in cost estimating.

Another common technique for cost estimating is using resource cost rates. For example, you can estimate the number of people or hours required for a project or activity and multiply that quantity by a cost rate. You can also use cost rates for materials. For example, you might estimate the quantity of printers you might need for a project and multiply that number by the cost rate per printer. Other considerations to make when preparing cost estimates are reserves, as described earlier, the cost of quality, described in Chapter 8, Project Quality Management, and other cost estimating methods such as vendor bid analysis, as described in Chapter 12, Project Procurement Management.

Typical Problems with Information Technology Cost Estimates

Although there are many tools and techniques to assist in creating project cost estimates, many information technology project cost estimates are still very inaccurate, especially those involving new technologies or software development. Tom DeMarco, a well-known author on software development, suggests four reasons for these inaccuracies and some ways to overcome them.[9]

1. Developing an estimate for a large software project is a complex task requiring a significant amount of effort. Many estimates must be done quickly and before clear system requirements have been produced. For example, the Surveyor Pro project described in the opening case involves a lot of complex software development. Before fully understanding what information surveyors really need in the system, someone would have to create a ROM estimate and budgetary estimates for this project. Rarely are the more precise, later estimates less than the earlier estimates for information technology projects. It is important to remember that estimates are done at various stages of the project, and project managers need to explain the rationale for each estimate.

2. The people who develop software cost estimates often do not have much experience with cost estimation, especially for large projects. There is also not enough accurate, reliable project data available on which to base estimates. If an organization uses good project management techniques and develops a history of keeping reliable project information, including estimates, it should help improve the organization's estimates. Enabling information technology people to receive training and mentoring on cost estimating will also improve cost estimates.

3. Human beings are biased toward underestimation. For example, senior information technology professionals or project managers might make estimates based on their own abilities and forget that many junior people will be working on a project. Estimators might also forget to allow for extra costs needed for integration and testing on large information technology projects. It is important for project managers and top management to review estimates and ask important questions to make sure the estimates are not biased.

4. Management might ask for an estimate, but really desire a more accurate number to help them create a bid to win a major contract or get internal funding. This problem is similar to the situation discussed in Chapter 6, Project Time Management, in which top managers or other stakeholders want project schedules to be shorter than the estimates. It is important for project managers to help develop good cost and schedule estimates and to use their leadership and negotiation skills to stand by those estimates.

[9] DeMarco, Tom, *Controlling Software Projects*. New York: Yourdon Press, 1982.

Sample Cost Estimate

One of the best ways to learn how the cost estimating process works is by studying sample cost estimates. Every cost estimate is unique, just as every project is unique. You can see a short sample cost estimate in Chapter 3 for JWD Consulting's Project Management Intranet Site Project. You can also view the ResNet cost estimate on the companion Web site for this text.

This section includes a step-by-step approach for developing a cost estimate for the Surveyor Pro project described in the opening case. Of course, it is much shorter and simpler than a real cost estimate would be, but it illustrates a process to follow and uses several of the tools and techniques described earlier. For more detailed information on creating a cost estimate, see the NASA Cost Estimating Handbook and other references provided in the Suggested Readings.

Before beginning any cost estimate, you must first gather as much information as possible about the project and ask how the organization plans to use the cost estimate. If the cost estimate will be the basis for contract awards and performance reporting, it should be a definitive estimate and as accurate as possible, as described earlier.

It is also important to clarify the ground rules and assumptions for the estimate. For the Surveyor Pro project cost estimate, these include the following:

- This project was preceded by a detailed study and proof of concept to show that it was possible to develop the hardware and software needed by surveyors and link the new devices to existing information systems. The proof of concept project produced a prototype handheld device and much of the software to provide basic functionality and link to the Global Positioning Systems (GPS) and other government databases used by surveyors. There is some data available to help estimate future labor costs, especially for the software development. There is also some data to help estimate the cost of the handheld devices.

- The main goal of this project is to produce 100 handheld devices, continue developing the software (especially the user interface), test the new system in the field, and train 100 surveyors in selected cities on how to use the new system. A follow-up contract is expected for a much larger number of devices based on the success of this project.

- There is a WBS for the project, as shown below:

  1. Project Management
  2. Hardware
       2.1 Handheld devices
       2.2 Servers

      3. Software

          3.1 Licensed software

          3.2 Software development

      4. Testing

      5. Training and Support

      6. Reserves

- Costs must be estimated by WBS and by month. The project manager will report progress on the project using earned value analysis, which requires this type of estimate.

- Costs will be provided in U.S. dollars. Since the project length is one year, inflation will not be included.

- The project will be managed by the government's project office. There will be a half-time project manager and three team members assigned to the project full time. The three team members will help manage various parts of the project and provide their expertise in the areas of software development, training, and support.

- The project involves purchasing the handheld devices from the same company that developed the prototype device. Based on producing 100 devices, the cost rate is estimated to be $600 per unit. The project will require four additional servers to run the software required for the devices and for managing the project.

- The project requires purchased software licenses for accessing the GPS and three other external systems. Software development includes developing a graphical user interface for the devices, an online help system, and a new module for tracking surveyor performance using the device.

- The devices and software will require extensive testing.

- Training will include instructor-led classes in five different locations. The project team believes it will be best to outsource most of the training, including developing course materials, holding the sessions, and providing help desk support for three months as the surveyors start using their devices in the field. Two project team members will assist with the training.

- Because there are several risks related to this project, include 20 percent of the total estimate as reserves.

- You must develop a computer model for the estimate, making it easy to change several inputs, such as the number of labor hours for various activities or labor rates.

Fortunately, the project team can easily access cost estimates and actual information from similar projects. There is a great deal of information available from the proof of concept project, and the team can also talk to contractor personnel from the past project to help them develop the estimate. There are also some computer models available, such as a software-estimating tool based on function points.

Since the estimate must be provided by WBS and by month, the team first reviews a draft of the project schedule and makes further assumptions, as needed. They decide first to estimate the cost of each WBS item and then determine when the work will be performed, even though costs may be incurred at different times than the work is performed. Their budget expert has approved this approach for the estimate. Below are further assumptions and information for estimating the costs for each WBS category:

1. Project Management: Estimate based on compensation for the project manager and 25 percent of team members' time. The remaining team member hours will be divided equally between the software development, testing, and training activities. The budget expert for this project suggested using a labor rate of $100/hour for the project manager and $75/hour for each team member, based on working an average of 160 hours per month, full time. Therefore, the total hours for the project manager under this category are 960 ($160/2 \times 12 = 960$). Costs are also included for the project team members, assuming a total of 160 hours per month for all project personnel ($160 \times 12 = 1920$). An additional amount will be added for all contracted labor, estimated by multiplying 10 percent of their total estimates for software development and testing costs ($10\% \times (\$594,000 + \$69,000)$).

2. Hardware

    2.1 Handheld devices: 100 devices estimated by contractor at $600 per unit.

    2.2 Servers: Four servers estimated at $4,000 each, based on recent server purchases.

3. Software

    3.1 Licensed software: License costs will be negotiated with each supplier. Since there is a strong probability of large future contracts and great publicity if the system works well, costs are expected to be lower than usual. A cost of $200/handheld device will be used.

    3.2 Software development: This estimate will use two approaches: a labor estimate and a function point estimate. The higher estimate will be used. If the estimates are more than 20 percent apart, a third approach to providing the estimate will be required. The supplier who did the development on the proof of concept project will provide the labor estimate input, and local technical experts will make the function point estimates.

4. Testing: Based on similar projects, testing will be estimated as 10 percent of the total hardware and software cost.

5. Training and Support: Based on similar projects, training will be estimated on a per-trainee basis, plus travel costs. The cost per trainee (100 total) will be $1,000, and travel will be $700/day/person for the instructors and project team members. It is estimated that there will be a total of 12 travel days. Labor costs for the project team members will be added to this estimate to assist in training and providing support after the training.

6. Reserves: As directed, reserves will be estimated at 20 percent of the total estimate.

The project team then develops a cost model using the above information. Figure 7-1 shows a spreadsheet that summarizes the costs by WBS item based on the above information. Notice that the WBS items are listed in the first column, and sometimes the items are broken down into more detail based on how the costs are estimated. For example, the project management category includes three items to calculate costs for the project manager, the project team members, and the contractors, since all of these people will perform some project management activities. Also notice that there are columns for entering the number of units or hours and the cost per unit or hour. Several items are estimated using this approach. There are also some short comments within the estimate, such as reserves being 20 percent of the total estimate. Also notice that you can easily change several input variables, such as number of hours or cost per hour, to quickly change the estimate.

Surveyor Pro Project Cost Estimate Created October 5, 2006

	# Units/Hrs.	Cost/Unit/Hr.	Subtotals	WBS Level 1 Totals	% of Total
WBS Items					
1. Project Management				**$306,300**	**20%**
Project manager	960	$100	$96,000		
Project team members	1920	$75	$144,000		
Contractors (10% of software development and testing)			$66,300		
2. Hardware				**$76,000**	**5%**
2.1 Handheld devices	100	$600	$60,000		
2.2 Servers	4	$4,000	$16,000		
3. Software				**$614,000**	**40%**
3.1 Licensed software	100	$200	$20,000		
3.2 Software development*			$594,000		
4. Testing (10% of total hardware and software costs)			$69,000	**$69,000**	**5%**
5. Training and Support				**$202,400**	**13%**
Trainee cost	100	$500	$50,000		
Travel cost	12	$700	$8,400		
Project team members	1920	$75	$144,000		
6. Reserves (20% of total estimate)			$253,540	**$253,540**	**17%**
Total project cost estimate				**$1,521,240**	

* See software development estimate

Figure 7-1. Surveyor Pro Project Cost Estimate

There is also an asterisk by the software development item in Figure 7-1, referring to another reference for more detailed information on how this more complicated estimate was made. Recall that one of the assumptions was that software development must be estimated using two approaches, and that the higher estimate be used as long as both estimates were no more than 20 percent apart. The labor estimate in this case was slightly higher than the function point estimate, so that number was used ($594,000 versus $562,158). Details are provided in Figure 7-2 on how the function point estimate was made. As you can see, there are many assumption made in producing the function point estimate. Again, by putting the information into a cost model, you can easily change several inputs to adjust the estimate. See the referenced article and other sources for more information on function point estimates.

Surveyor Pro Software Development Estimate Created October 5, 2006

1. Labor Estimate	# Units/Hrs.	Cost/Unit/Hr.	Subtotals	Calculations
Contractor labor estimate	3000	$150	$450,000	3000*150
Project team member estimate	1920	$75	$144,000	1920*75
Total labor estimate			$594,000	Sum above two values
2. Function point estimate**	Quantity	Conversion Factor	Function Points	Calculations
External inputs	10	4	40	10*4
External interface files	3	7	21	3*7
External outputs	4	5	20	4*5
External queries	6	4	24	6*4
Logical internal tables	7	10	70	7*10
Total function points			175	Sum above function point values
Java 2 languange equivalency value			46	Assumed value from reference
Source lines of code (SLOC) estimate			8,050	175*46
Productivity*KSLOC^Penalty (in months)			29.28	3.13*8.05^1.072 (see reference)
Total labor hours (160 hours/month)			4,684.65	29.28*160
Cost/labor hour ($120/hour)			$120	Assumed value from budget expert
Total function point estimate			$562,158	4684.65*120

**Approach based on paper by William Roetzheim, "Estimating Software Costs," Cost Xpert Group, Inc. (2003) using the COCOMO II default linear productivity factor (3.13) and penalty factor (1.072).

Figure 7-2. Surveyor Pro Software Development Estimate

It is very important to have several people review the project cost estimate. It is also helpful to analyze the total dollar value as well as the percentage of the total amount for each major WBS category. For example, a senior executive could quickly look at the Surveyor Pro project cost estimate and decide if the numbers are reasonable and the assumptions are well documented. In this case, the government had budgeted $1.5 million for the project, so the estimate was right in line with that amount. The WBS Level 1 items also seemed to be at appropriate percentages of the total cost based on similar past projects. In some cases, a project team might also be asked to provide a range estimate for each item instead of one discrete amount. For example, they might estimate that the testing costs will be between $60,000 and $80,000 and document their assumptions in determining those values.

After the total cost estimate is approved, the team can then allocate costs for each month based on the project schedule and when costs will be incurred. Many organizations also require that the estimated costs be allocated into certain budget categories, as described in the next section.

COST BUDGETING

Project cost budgeting involves allocating the project cost estimate to individual work items over time. These work items are based on the work breakdown structure for the project. The WBS, therefore, is a required input to the cost budgeting process. Likewise, the project scope statement, WBS dictionary, activity cost estimates and supporting detail, project schedule, resource calendars, contracts, and cost management plan also provide useful information for cost budgeting. The main goal of the cost budgeting process is to produce a cost baseline for measuring project performance and project funding requirements. It may also result in requested changes to the project and updates to the cost management plan to help meet project cost constraints.

For example, the Surveyor Pro project team could use the cost estimate from Figure 7-1 along with the project schedule and other information to allocate costs for each month. Figure 7-3 provides an example of a cost baseline for this project. Again, it's important for the team to document assumptions they made when developing the cost baseline and have several experts review it.

Surveyor Pro Project Cost Baseline Created October 10, 2006*

WBS Items	1	2	3	4	5	6	7	8	9	10	11	12	Totals
1. Project Management													
Project manager	8,000	8,000	8,000	8,000	8,000	8,000	8,000	8,000	8,000	8,000	8,000	8,000	96,000
Project team members	12,000	12,000	12,000	12,000	12,000	12,000	12,000	12,000	12,000	12,000	12,000	12,000	144,000
Contractors		6,027	6,027	6,027	6,027	6,027	6,027	6,027	6,027	6,027	6,027	6,027	66,300
2. Hardware													
2.1 Handheld devices				30,000	30,000								60,000
2.2 Servers				8,000	8,000								16,000
3. Software													
3.1 Licensed software				10,000	10,000								20,000
3.2 Software development		60,000	60,000	80,000	127,000	127,000	90,000	50,000		594,000			594,000
4. Testing			6,000	8,000	12,000	15,000	15,000	13,000		69,000			69,000
5. Training and Support													
Trainee cost									50,000				50,000
Travel cost									8,400				8,400
Project team members							24,000	24,000	24,000	24,000	24,000	24,000	144,000
6. Reserves				10,000	10,000	30,000	30,000	60,000	40,000	40,000	30,000	3,540	253,540
Totals	20,000	86,027	92,027	172,027	223,027	198,027	185,027	173,027	148,427	753,027	80,027	53,567	1,521,240

*See the lecture slides for this chapter on the companion Web site for a larger view of this and other figures in this chapter.

Figure 7-3. Surveyor Pro Project Cost Baseline

Most organizations have a well-established process for preparing budgets. For example, many organizations require budget estimates to include the number of full-time equivalent (FTE) staff, often referred to as headcount, for each month of the project. This number provides the basis for estimating total compensation costs each year. Many organizations also want to know the amount of money projected to be paid to suppliers for their labor costs or other purchased goods and services. Other common budget categories include travel, depreciation, rents/leases, and other supplies and expenses. It is important to understand these budget categories before developing an estimate to make sure data is collected accordingly. Organizations use this information to track costs across projects and non-project work and look for ways to reduce costs. They also use the information for legal and tax purposes.

In addition to providing input for budgetary estimates, cost budgeting provides a cost baseline. A **cost baseline** is a time-phased budget that project managers use to measure and monitor cost performance. Estimating costs for each major project activity over time provides project managers and top management with a foundation for project cost control, as described in the next section. See Appendix A for information on using Project 2003 for cost control.

Cost budgeting, as well as requested changes or clarifications, may result in updates to the cost management plan, a subsidiary part of the project management plan. Cost budgeting also provides information for project funding requirements. For example, some projects have all funds available when the project begins, but others must rely on periodic funding to avoid cash flow problems. If the cost baseline shows that more funds are required in certain months than are expected to be available, the organization must make adjustments to avoid financial problems.

COST CONTROL

Project cost control includes monitoring cost performance, ensuring that only appropriate project changes are included in a revised cost baseline, and informing project stakeholders of authorized changes to the project that will affect costs. The cost baseline, performance reports, change requests, and project funding requirements are inputs to the cost control process. Outputs of this process are project management plan updates, corrective action, revised estimates for project completion, requested changes, and updates to organizational process assets, such as lessons-learned documents.

🎥 Media Snapshot

Several 2004 newspaper headlines around the globe describe massive cost overruns. Several also describe potential lawsuits or other legal actions due to poor management and cost control of various types of projects. Few articles mention technical problems as the main causes of these cost overruns.

■ **Australia**: Problems with the installation of an ERP system at Australia-based Crane Group Ltd. led to an estimated cost overrun of $11.5 million. The difficulties may also lead to financial claims against PeopleSoft Inc., according to Crane officials. [10]

[10] Songini, Marc L., "Australian Firm Wrestles With ERP Delays," *ComputerWorld* (July 12, 2004).

■ **India**: As many as 274 projects currently under implementation in the Central sector are suffering serious cost and time overruns. These problems are also delaying the potential benefits from these projects, resulting in employment and income generation implications. According to a report prepared for the Prime Minister's Office, *only 65 of these projects are being monitored on a regular basis.* The biggest cost overrun is accounted for by projects in the power sector where the original cost estimates for 37 power projects estimated at Rs 55,254.39 crore (more than $12 billion U.S. dollars) is anticipated to rise up to Rs 70,679.03 crore (more than $15 billion U.S. dollars).[11]

■ **Pakistan**: Pakistan has sustained a cost overrun of Rs 1.798 billion (more than $30 million U.S. dollars) in the execution of the 66.5 megawatt Jagran Hydropower Project in the Neelum Valley. Official sources told Daily Times that the overrun was caused by *massive wrongdoings*, which caused the cost of the project to skyrocket to Rs 4.401 billion (more than 74 million U.S. dollars). The sources said that Pakistan suffered the huge overrun due to massive mismanagement, embezzlement of funds, and changes made in the project without the approval of a competent authority.[12]

■ **United States**: Northern California lawmakers were outraged over Governor Arnold Schwarzenegger's announcement that commuters should have to pay construction costs on Bay Area bridges. Under one scenario, commuters who travel over any of the seven state toll bridges in Northern California could see their bridge usage bills rise by more than $40 a month. Seven years ago, the cost for the new Bay Bridge, which will span San Francisco Bay between Oakland and Treasure Island, was estimated at $1.1 billion. It has now been revised to $5.1 billion. *Maybe it takes the Terminator to help control costs!*[13]

Several tools and techniques assist in project cost control. As shown in Appendix A, Project 2003 has many cost management features to help you enter budgeted costs, set a baseline, enter actuals, calculate variances, and run various cost reports. In addition to using software, however, there must be some change control system to define procedures for changing the cost baseline. This cost control change system is part of the integrated change control system described in Chapter 4, Project Integration Management. Since many projects do not progress exactly as planned, new or revised cost estimates are often required, as are estimates to evaluate alternate courses of action. Performance review meetings can be a powerful tool for helping to control project costs. People often perform better when they know they must report on their progress. Another very important tool for cost control is performance measurement. Although many general accounting approaches are available for measuring cost performance, earned value management (EVM) is a very powerful cost control technique that is unique to the field of project management.

[11] Srinivasan, G., "274 Central sector projects suffer cost, time overruns," *The Hindu Business Line* (May 4, 2004).

[12] Mustafa, Khalid, "Rs 1.8 billion cost overrun in Jagran hydropower project," *Daily Times* (November 19, 2002).

[13] Gannett Company, "Governor Refuses to Pay for Bay Bridge Cost Overruns," News10 (August 17, 2004).

Earned Value Management

Earned value management (EVM) is a project performance measurement technique that integrates scope, time, and cost data. Given a cost performance baseline, project managers and their teams can determine how well the project is meeting scope, time, and cost goals by entering actual information and then comparing it to the baseline. A **baseline** is the original project plan plus approved changes. Actual information includes whether or not a WBS item was completed or approximately how much of the work was completed, when the work actually started and ended, and how much it actually cost to do the completed work.

In the past, earned value management was used primarily on large government projects. Today, however, more and more companies are realizing the value of using this tool to help control costs. In fact, an e-mail discussion in late 2004 by several academic experts in earned value management and a real practitioner revealed the need to clarify how to actually calculate earned value. Brenda Taylor, a senior project manager for Management Solutions in Johannesburg, South Africa, questioned calculating earned value by simply multiplying the planned value to date by a percentage complete value. She suggested using the rate of performance instead, as described below.

Earned value management involves calculating three values for each activity or summary activity from a project's WBS.

1. The **planned value (PV)**, formerly called the budgeted cost of work scheduled (BCWS), also called the budget, is that portion of the approved total cost estimate planned to be spent on an activity during a given period. Table 7-3 shows an example of earned value calculations. Suppose a project included a summary activity of purchasing and installing a new Web server. Suppose further that, according to the plan, it would take one week and cost a total of $10,000 for the labor hours, hardware, and software involved. The planned value (PV) for that activity that week is, therefore, $10,000.

2. The **actual cost (AC)**, formerly called the actual cost of work performed (ACWP), is the total direct and indirect costs incurred in accomplishing work on an activity during a given period. For example, suppose it actually took two weeks and cost $20,000 to purchase and install the new Web server. Assume that $15,000 of these actual costs were incurred during Week 1 and $5,000 was incurred during Week 2. These amounts are the actual cost (AC) for the activity each week.

3. The **earned value (EV)**, formerly called the budgeted cost of work performed (BCWP), is an estimate of the value of the physical work actually completed. It is based on the original planned costs for the project or activity and the rate at which the team is completing work on the project or activity to date. The **rate of performance (RP)** is the ratio of actual work completed to the percentage of work planned to have been

completed at any given time during the life of the project or activity. For example, suppose the server installation was halfway completed by the end of week 1. The rate of performance would be 50 percent (50/100) because by the end of week 1, the planned schedule reflects that the task should be 100 percent complete and only 50 percent of that work has been completed. In Table 7-4, the earned value estimate after one week is therefore $5,000.[14]

Table 7-4: Earned Value Calculations for One Activity After Week One

ACTIVITY	WEEK 1	
Earned Value (EV)	5,000	
Planned Value (PV)	10,000	
Actual Cost (AC)	15,000	
Cost Variance (CV)	–7,500	*–10,000*
Schedule Variance (SV)	–2,500	*– 5,000*
Cost Performance Index (CPI)	50%	*33%*
Schedule Performance Index (SPI)	75%	*50%*

The earned value calculations in Table 7-4 are carried out as follows:

$$EV = 10{,}000 \times 50\% = 5{,}000$$

$$CV = 7{,}500 - 15{,}000 = -7{,}500$$

$$SV = 7{,}500 - 10{,}000 = -2{,}500$$

$$CPI = 7{,}500/15{,}000 = 50\%$$

$$SPI = 7{,}500/10{,}000 = 75\%$$

Table 7-5 summarizes the formulas used in earned value management. Note that the formulas for variances and indexes start with EV, the earned value. Variances are calculated by subtracting the actual cost or planned value from EV, and indexes are calculated by dividing EV by the actual cost or planned value. After you total the EV, AC, and PV data for all activities on a project, you can use the CPI and SPI to project how much it will cost and how long it will take to finish the project based on performance to date. Given the budget at completion and original time estimate, you can divide by the appropriate index to calculate the estimate at completion (EAC) and estimated time to complete, assuming performance remains the same. (See the *PMBOK® Guide 2004* or additional references for other ways to calculate EAC.) There are no standard acronyms for the term estimated time to complete or original time estimate.

[14] Taylor, Brenda, P2 Senior Project Manager, Management Solutions, Johannesburg, South Africa.

Table 7-5: Earned Value Formulas

TERM	FORMULA
Earned Value	EV = PV to date × RP
Cost Variance	CV = EV − AC
Schedule Variance	SV = EV − PV
Cost Performance Index	CPI = EV/AC
Schedule Performance Index	SPI = EV/PV
Estimate at Completion (EAC)*	EAC = BAC/CPI
Estimated Time to Complete	Original Time Estimate/SPI

*Note: You can also calculate the Estimate to Complete (EAC) by subtracting the actual cost to date from the EAC. See the *The PMBOK® Guide 2004* for additional earned value calculations.

Cost variance (CV) is the earned value minus the actual cost. If cost variance is a negative number, it means that performing the work cost more than planned. If cost variance is a positive number, it means that performing the work cost less than planned.

Schedule variance (SV) is the earned value minus the planned value. A negative schedule variance means that it took longer than planned to perform the work, and a positive schedule variance means that it took less time than planned to perform the work.

The **cost performance index (CPI)** is the ratio of earned value to actual cost and can be used to estimate the projected cost of completing the project. If the cost performance index is equal to one, or 100 percent, then the planned and actual costs are equal—the costs are exactly as budgeted. If the cost performance index is less than one or less than 100 percent, the project is over budget. If the cost performance index is greater than one or more than 100 percent, the project is under budget.

The **schedule performance index (SPI)** is the ratio of earned value to planned value and can be used to estimate the projected time to complete the project. Similar to the cost performance index, a schedule performance index of one, or 100 percent, means the project is on schedule. If the schedule performance index is greater than one or 100 percent, then the project is ahead of schedule. If the schedule performance index is less than one or 100 percent, the project is behind schedule.

Note that in general, *negative numbers for cost and schedule variance indicate problems in those areas.* Negative numbers mean the project is costing more than planned or taking longer than planned. Likewise, *CPI and SPI less than one or less than 100 percent also indicate problems.*

Earned value calculations for all project activities (or summary level activities) are required to estimate the earned value for the entire project. Some activities may be over budget or behind schedule, but others may be under budget and ahead of schedule. By adding all of the earned values for all project activities, you can determine how the project as a whole is performing.

Figure 7-4 provides sample earned value information for a one-year project. This project had a planned total cost of $100,000. The spreadsheet shows actual cost and performance information for the first five months, or through the end of May. Notice the "RV" column at the upper-right side, or in column Q, of the spreadsheet. The RV is the ratio of the planned percent complete and the actual percent complete. The earned value (EV) for each activity in this example is calculated by multiplying this rate of performance (RP) value by the planned value. For example, the task in row 7 has a PV of 8,000 as of the end of May. The planned percent complete was 75 percent, and the actual percent complete was 50 percent. The RP is then 50 percent divided by 75 percent, or 67 percent. The earned value for that task is then $8,000 times 67 percent, or $5,333. The "Monthly Actual Cost" row (row 16 of the spreadsheet) shows the actual cost each month for the project's activities through May. By calculating the total planned costs, the total actual costs, and the earned value costs, you can determine the cost variance, schedule variance, cost performance index, and schedule performance index for the entire project (see cells A20 through B26 in Figure 7-4). See the lecture slides for this chapter on the companion Web site for a larger view of this and other figures in this chapter

	A	B	C	D	E	F	G	H	I	J	K	L	M	N	O	P
1														To Date	Planned	Actual
2	Activity	Jan	Feb	Mar	Apr	May	Jun	Jul	Aug	Sep	Oct	Nov	Dec	PV	% Complete	% Complete
3	Plan and staff project	4,000	4,000											8,000	100	100
4	Analyze requirements		6,000	6,000										12,000	100	100
5	Develop ERDs			4,000	4,000									8,000	100	100
6	Design database tables				6,000	4,000								10,000	100	100
7	Design forms, reports, and queries					8,000	4,000							8,000	75	50
8	Construct working prototype						10,000									-
9	Test/evaluate prototype						2,000	6,000								-
10	Incorporate user feedback							4,000	6,000	4,000						-
11	Test system									4,000	4,000	2,000				-
12	Document system											3,000	1,000			-
13	Train users												4,000			-
14	Monthly Planned Value (PV)	4,000	10,000	10,000	10,000	12,000	16,000	10,000	6,000	8,000	4,000	5,000	5,000			
15	Cumulative Planned Value (PV)	4,000	14,000	24,000	34,000	46,000	62,000	72,000	78,000	86,000	90,000	95,000	100,000			
16	Monthly Actual Cost (AC)	4,000	11,000	11,000	12,000	15,000										
17	Cumulative Actual Cost (AC)	4,000	15,000	26,000	38,000	53,000										
18	Monthly Earned Value (EV)	4,000	10,000	10,000	10,000	9,333										
19	Cumulative Earned Value (EV)	4,000	14,000	24,000	34,000	43,333										
20	Project EV as of May 31	43,333														
22	Project PV as of May 31	46,000														
21	Project AC as of May 31	$ 53,000														
23	CV=EV-AC	$ (9,667)														
24	SV=EV-PV	$ (2,667)														
25	CPI=EV/AC	81.761%														
26	SPI=EV/PV	94.203%														
27	Estimate at Completion (EAC)	$ 122,308 (original plan of $100,000 divided by CPI)														
28	Estimated time to complete	12.74 (original plan of 12 months divided by SPI)														

Figure 7-4. Earned Value Calculations for a One-Year Project after Five Months

In this example, the cost variance is –$9,667, and the schedule variance is –$2,667. These values mean that the project is both over budget and behind schedule after five months. The cost performance and schedule performance indexes are 81.761 percent and 94.203 percent, respectively. The cost performance index can be used to calculate the **estimate at completion (EAC)**—an estimate of what it will cost to complete the project based on performance to date. Similarly, the schedule performance index can be used to calculate an estimated time to complete the project. For example, in Figure 7-4, the estimate at completion is $122,308, or $100,000/81.761 percent. The estimated time to complete the project is 12.74 months, or twelve months/94.203 percent (see cells A25 through B28).

You can graph earned value information to track project performance. Figure 7-5 shows an earned value chart for the one-year sample project from Figure 7-4. The chart includes three lines and two points, as follows:

- Planned value (PV), the cumulative planned amounts for all activities by month. Note that the planned value line extends for the estimated length of the project and ends at the BAC point.
- Actual cost (AC), the cumulative actual amounts for all activities by month.
- Earned value (EV), the cumulative earned value amounts for all activities by month.
- **Budget at completion (BAC)**, the original total budget for the project, or $100,000 in this example. The BAC point is plotted on the chart at the original time estimate of twelve months.
- Estimate at completion (EAC), estimated to be $122,308, as described earlier. This EAC point is plotted on the chart at the estimated time to complete of 12.74 months.

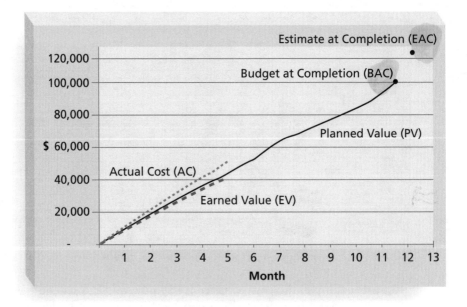

Figure 7-5. Earned Value Chart for Project after Five Months

Viewing earned value information in chart form helps you visualize how the project is performing. For example, you can see the planned performance by looking at the planned value line. If the project goes as planned, it will finish in 12 months and cost $100,000. Notice in the example that the actual cost line is always right on or above the earned value line. When the actual cost line is right on or above the earned value line, costs are equal to or more than planned. The planned value line is pretty close to the earned value line, just slightly higher in the last month. This relationship means that the project has been right on schedule until the last month, when the project fell behind schedule.

Top managers overseeing multiple projects often like to see performance information in a graphical form such as the earned value chart in Figure 7-5. For example, in the opening case, the government officials were reviewing earned value charts and estimates at completion for several different projects. Earned value charts allow you to see how projects are performing quickly. If there are serious cost and schedule performance problems, top management may decide to terminate projects or take other corrective action. The estimates at completion (EAC) are important inputs to budget decisions, especially if total funds are limited. Earned value management is an important technique because, when used effectively, it helps top management and project managers evaluate progress and make sound management decisions.

If earned value management is such a powerful cost control tool, then why doesn't every organization use it? Why do many government projects require it, but many commercial projects don't? Two reasons why organizations do not widely use earned value management are its focus on tracking actual performance versus planned performance and the importance of percentage completion data in making calculations. Many projects, particularly information technology projects, do not have good planning information, so tracking performance against a plan might produce misleading information. Several cost estimates are usually made on information technology projects, and keeping track of the most recent cost estimate and the actual costs associated with it could be cumbersome. In addition, estimating percentage completion of tasks might produce misleading information. What does it really mean to say that a task is actually 75 percent complete after three months? Such a statement is often not synonymous with saying the task will be finished in one more month or after spending an additional 25 percent of the planned budget.

To make earned value management simpler to use, organizations can modify the level of detail and still reap the benefits of the technique. For example, you can use percentage completion data such as 0 percent for items not yet started, 50 percent for items in progress, and 100 percent for completed tasks. As long as the project is defined in enough detail, this simplified percentage completion data should provide enough summary information to allow managers to see how well a project is doing overall. You can get very accurate total project performance information using these simple percentage complete amounts. For example, using simplified percentage complete amounts for a one-year project with weekly reporting and an average task or work packet size of one week, you can expect about a 1 percent error rate.[15]

You can also only enter and collect earned value data at summary levels of the WBS. Quentin Fleming, author of the book *Earned Value Project Management, Second Edition,* often gives presentations about earned value management. Many people express their frustration in trying to collect such detailed information. Quentin explains that you do not have to collect information at the work package level to use earned value management. It is most important to have a deliverable-oriented WBS, and many WBS items can summarize several sub-deliverables. For example, you might have a WBS for a house that includes items for each room in the house. Just collecting earned value data for each room would provide meaningful information instead of trying to collect the detailed information on each component of the room, such as flooring, furniture, lighting, and so on.

Earned value management is the primary method available for integrating performance, cost, and schedule data. It can be a powerful tool for project managers and top management to use in evaluating project performance. Project management software, such as Project 2003, includes tables for collecting earned value data and reports that calculate variance information. See the project cost management section of Appendix A for an example of using earned value management. Another approach to evaluating the performance of multiple projects is project portfolio management.

Project Portfolio Management

As mentioned in Chapter 1, many organizations now collect and control an entire suite of projects or investments as one set of interrelated activities in one place—a portfolio. Project managers need to understand how their projects fit into the bigger picture, and they need to help their organizations make wise investment decisions. Many project managers also want to move on to manage larger projects, become program managers, then vice presidents, and eventually CEOs. Understanding project portfolio management, therefore, is important for project and organizational success.

There can be a portfolio for information technology projects, for example, and portfolios for other types of projects. An organization can view project portfolio management as having five levels, from simplest to most complex, as follows:

1. Put all your projects in one database.
2. Prioritize the projects in your database.

---

[15] Brandon, Daniel M., Jr., "Implementing Earned Value Easily and Effectively," *Project Management Journal* (June 1998) 29 (2), p. 11–18.

3. Divide your projects into two or three budgets based on type of investment, such as utilities or required systems to keep things running, incremental upgrades, and strategic investments.

4. Automate the repository.

5. Apply modern portfolio theory, including risk-return tools that map project risk on a curve.

For example, Jane Walton, the project portfolio manager for information technology projects at Schlumberger, saved the company $3 million in one year by organizing the organization's 120 information technology projects into a portfolio. Manufacturing companies used project portfolio management in the 1960s, and Walton anticipated the need to justify investments in information technology projects just as managers have to justify capital investment projects. She found that 80 percent of the organization's projects overlapped, and 14 separate projects were trying to accomplish the same thing. Other managers, such as Douglas Hubbard, president of a consulting firm, see the need to use project portfolio management, especially for information technology projects. Hubbard suggests, "IT investments are huge, risky investments. It's time we do this."[16]

META Group research shows that organizations that evaluate information technology projects by what their business impacts are and what their potential business values will be implement projects that result in 25 percent more improvement to the bottom line. META Group also estimates that by 2005–2006, more than 50 percent of the CIOs for Global 2000 companies (the top 2,000 companies worldwide, based on revenue) will adopt project portfolio management tools and techniques for information technology projects, asset management, and budget planning and monitoring. Business executives state that using project portfolio management allows managers to make decisions faster and with more confidence.[17]

Project portfolio managers can start by using spreadsheet software to develop and manage project portfolios, or they can use sophisticated software designed to help manage project portfolios. Several software tools available today help project portfolio managers summarize earned value and project portfolio information, as described in the following section.

USING PROJECT MANAGEMENT SOFTWARE TO ASSIST IN PROJECT COST MANAGEMENT

Most organizations use software to assist in various activities related to project cost management. Spreadsheets are a common tool for cost estimating, cost

---

[16] Berinato, Scott, "Do the Math," *CIO Magazine* (October 1, 2001).

[17] META Group, "IT Investment Management: Portfolio Management Lessons Learned," A META Group White Paper (*www.metagroup.com*) (2002).

budgeting, and cost control. Many companies also use more sophisticated and centralized financial applications software to provide important cost-related information to accounting and finance personnel. This section will focus specifically on how you can use project management software in cost management. Appendix A includes a section on using the cost management features in Project 2003.

Project management software can be a very helpful tool during each project cost management process. It can help you study overall project information or focus on tasks that are over a specified cost limit. You can use the software to assign costs to resources and tasks, prepare cost estimates, develop cost budgets, and monitor cost performance. Project 2003 has several standard cost reports: cash flow, budget, over budget tasks, over budget resources, and earned value reports. For several of these reports, you must enter percentage completion information and actual costs, just as you need this information when manually calculating earned value or other analyses.

Although Microsoft Project 2003 has a fair amount of cost management features, many information technology project managers use other tools to manage cost information; they do not know that they can use Project 2003 for cost management or they simply do not track costs based on a WBS, as most project management software does. As with most software packages, users need training to use the software effectively and understand the features available. Instead of using dedicated project management software for cost management, some information technology project managers use company accounting systems; others use spreadsheet software to achieve more flexibility. Project managers who use other software often do so because these other systems are more generally accepted in their organizations and more people know how to use them. On the other hand, some companies have developed methods to link data between their project management software and their main accounting software systems.

Many organizations are starting to use enterprise project management software to organize and analyze all types of project data into project portfolios and across the entire enterprise. Enterprise project management software, such as Microsoft Solution for Enterprise Project Management (see Figures A-1 through A-3 in Appendix A), Pacific Edge's The Edge and The Edge for IT, and PlanView's Web Software, integrate information from multiple projects to show the project's status, or health. For example, Figure 7-6 provides a sample screen from PlanView. Note that the charts and text in the upper half of the screen show the number and percentage of projects in this project portfolio that are on target and in trouble in terms of schedule and cost variance. The bottom half of the screen lists the names of individual projects, percent complete, schedule variance, cost variance, budget variance, and risk percentage.

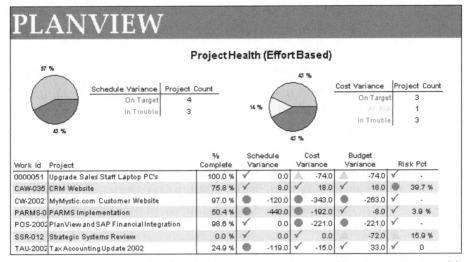

Figure 7-6. Sample Project Portfolio Management Screen Showing Project Health

Many organizations, even very large ones, are hesitant to invest in enterprise project management software. Fortunately, there are now several inexpensive, Web-based products on the market that help organizations manage portfolios of projects. For example, VPMi Enterprise Online (*www.vcsonline.com*) is an Application Service Provider (ASP) solution available for only $12 per user per month. It includes industry standard metrics defined in either traditional or graphical reports. The dynamic graphical charts allow users to see how well the organization is managing its projects, how much time is spent on low-value projects, and how clients view the products and services being provided. Users can also share data with other applications, such as Microsoft Project or Excel. As mentioned in Chapter 1, you can find links to hundreds of project management software packages from the Project Management Center Web site (*www.infogoal.com/pmc*) or similar sites.

CASE WRAP-UP

After talking to his colleagues about the meeting, Juan had a better idea of the importance of project cost management. He understood the value of doing detailed studies before making major expenditures on new projects, especially after learning about the high cost of correcting defects late in a project. He also learned how important it is to develop good cost estimates and keep costs on track. He really enjoyed seeing how the cost estimate for the Surveyor Pro project was developed and was eager to learn more about various estimating tools and techniques.

At the meeting, government officials cancelled several projects when the project managers showed how poorly the projects were performing and admitted that they did not do much planning and analysis early in the projects. Juan knew that he could not focus on just the technical aspects of projects if he wanted to move ahead in his career. He began to wonder whether several projects the city was considering were really worth the taxpayers' money. Issues of cost management added a new dimension to Juan's job.

CHAPTER SUMMARY

Project cost management is a traditionally weak area of information technology projects. Information technology project managers must acknowledge the importance of cost management and take responsibility for understanding basic cost concepts, cost estimating, budgeting, and cost control.

Project managers must understand several basic principles of cost management in order to be effective in managing project costs. Important concepts include profits and profit margins, life cycle costing, cash flow analysis, sunk costs, and learning curve theory.

Cost estimating is a very important part of project cost management. There are several types of cost estimates, including rough order of magnitude (ROM), budgetary, and definitive. Each type of estimate is done during different stages of the project life cycle, and each has a different level of accuracy. There are several tools and techniques for developing cost estimates, including analogous estimating, bottom-up estimating, parametric modeling, and computerized tools. A detailed example of a project cost estimate is provided, illustrating how to apply several of these concepts.

Cost budgeting involves allocating costs to individual work items over time. It is important to understand how particular organizations prepare budgets so estimates are made accordingly.

Project cost control includes monitoring cost performance, reviewing changes, and notifying project stakeholders of changes related to costs. Many basic accounting and finance principles relate to project cost management. Earned value management is an important method used for measuring project performance. Earned value management integrates scope, cost, and schedule information. Project portfolio management allows organizations to collect and control an entire suite of projects or investments as one set of interrelated activities.

Several software products assist with project cost management. Project 2003 has many cost management features, including earned value management. Enterprise project management software helps managers evaluate integrated project data in a graphical format.

DISCUSSION QUESTIONS

1. Discuss why many information technology professionals may overlook project cost management and how this might affect completing projects within budget.

2. Explain some of the basic principles of cost management, such as profits, life cycle costs, tangible and intangible costs and benefits, direct and indirect costs, reserves, and so on.

3. Give examples of when you would prepare rough order of magnitude, budgetary, and definitive cost estimates for an information technology project. Give an example of using each of the following techniques for creating a cost estimate: analogous, parametric, bottom-up, and computerized tools.

4. Explain what happens during the cost budgeting process.

5. Explain how earned value management can be used to measure project performance and speculate as to why it is not used more often. What are some general rules of thumb for deciding if cost variance, schedule variance, cost performance index, and schedule performance index numbers are good or bad?

6. What is project portfolio management? Can project managers use it with earned value management?

7. Describe several types of software that project managers can use to support project cost management.

EXERCISES

1. Given the following information for a one-year project, answer the following questions. Recall that PV is the planned value, EV is the earned value, AC is the actual cost, and BAC is the budget at completion.

PV = $23,000

EV = $20,000

AC = $25,000

BAC = $120,000

a. What is the cost variance, schedule variance, cost performance index (CPI), and schedule performance index (SPI) for the project?

b. How is the project doing? Is it ahead of schedule or behind schedule? Is it under budget or over budget?

c. Use the CPI to calculate the estimate at completion (EAC) for this project. Is the project performing better or worse than planned?

d. Use the schedule performance index (SPI) to estimate how long it will take to finish this project.

e. Sketch the earned value chart based for this project, using Figure 7-5 as a guide.

2. Create a cost estimate/model for building a new, state-of-the-art multimedia classroom for your organization within the next six months. The classroom should include 20 high-end personal computers with appropriate software for your organization, a network server, Internet access for all machines, an instructor station, and a projection system. Be sure to include personnel costs associated with the project management for this project. Document the assumptions you made in preparing the estimate and provide explanations for key numbers.

3. Research online information, textbooks, and local classes for using the cost management features of Project 2003. Also review the project cost management section of Appendix A. Ask three people in different information technology organizations that use Project 2003 if they use the cost management features and in what ways they use them. Write a brief report of what you learned from your research.

4. Read the article by William Roetzheim from the Suggested Readings in this chapter on estimating software costs. Find one or two other references on using SLOC and function points. In your own words, explain how to estimate software development costs using both of these approaches in a two- to three-page paper.

5. Create a spreadsheet to calculate your projected total costs, total revenues, and total profits for giving a seminar on cost estimating. Below are some of your assumptions:

- You will charge $600 per person for a two-day class.
- You estimate that 30 people will register for and attend the class, but you want to change this input.
- Your fixed costs include $500 total to rent a room for both days, setup fees of $400 for registration, and $300 for designing a postcard for advertising.
- You will not include any of your labor costs for this estimate, but you estimate that you will spend at least 150 hours developing materials, managing the project, and giving the actual class. You would like to know what your time is worth given different scenarios.

- You will order 5,000 postcards, mail 4,000, and distribute the rest to friends and colleagues.

- Your variable costs include the following:
  a. $5 per person for registration plus four percent of the class fee per person to handle credit card processing; assume everyone pays by credit card
  b. $.40 per postcard if you order 5,000 or more
  c. $.25 per postcard for mailing and postage
  d. $25 per person for beverages and lunch
  e. $30 per person for class handouts

Be sure to have input cells for any variables that might change, such as the cost of postage, handouts, and so on. Calculate your profits based on the following number of people who attend: 10, 20, 30, 40, 50, and 60. In addition, calculate what your time would be worth per hour based on the number of students. Try to use the Excel data table feature showing the profits based on the number of students. If you are unfamiliar with data tables, just repeat the calculations for each possibility of 10, 20, 30, 40, 50, and 60 students. Print your results on one page, highlighting the profits for each scenario and what your time is worth.

RUNNING CASE

Tony Prince and his team working on the Recreation and Wellness Intranet Project have been asked to refine the existing cost estimate for the project so they can evaluate supplier bids and have a solid cost baseline for evaluating project performance. Recall that your schedule and cost goals are to complete the project in six months for under $200,000.

Tasks

1. Prepare and print a one-page cost model for the project, similar to the one provided in Figure 7-1. Use the following WBS, and be sure to document assumptions you make in preparing the cost model. Assume a labor rate of $100/hour for the project manager and $60/hour for other project team members. Assume that none of the work is outsourced, labor costs for users are not included, and there are no additional hardware costs. The total estimate should be $200,000.
   1. Project Management
   2. Requirements Definition
   3. Web Site Design
      3.1 Registration for recreational programs
      3.2 Registration for classes and programs

      3.3 Tracking system

      3.4 Incentive system

    4. Web Site Development

      4.1 Registration for recreational programs

      4.2 Registration for classes and programs

      4.3 Tracking system

      4.4 Incentive system

    5. Testing

    6. Training, Roll Out, and Support

2. Using the cost model you created above, prepare a cost baseline by allocating the costs by WBS for each month of the project.

3. Assume you have completed three months of the project. The BAC was $200,000 for this six-month project. Also assume the following:

    PV = $120,000

    EV = $100,000

    AC = $90,000

  a. What is the cost variance, schedule variance, cost performance index (CPI), and schedule performance index (SPI) for the project?

  b. How is the project doing? Is it ahead of schedule or behind schedule? Is it under budget or over budget?

  c. Use the CPI to calculate the estimate at completion (EAC) for this project. Is the project performing better or worse than planned?

  d. Use the schedule performance index (SPI) to estimate how long it will take to finish this project.

  e. Sketch an earned value chart using the above information. See Figure 7-5 as a guide.

ADDITIONAL RUNNING CASES AND OTHER APPENDICES

Appendix C provides additional case studies and questions you can use to practice applying the concepts, tools, and techniques you are learning throughout this and subsequent chapters. Review the running cases provided in Appendix C and on the companion Web site (*www.course.com/mis/schwalbe4e*). Appendix D includes templates for various project management documents. For additional sample documents based on real projects, visit the author's Web site at *www.kathyschwalbe.com*. Appendix E and the CD-ROM included with this text include a computer simulation where you can also practice applying the project management process groups and knowledge areas.

SUGGESTED READINGS

1. Abbas, Wayne. "How Earned Value Got to Primetime: A Short Look Back and Glance Ahead," Project Management Forum (*www.pmforum.org/ library/papers/evprimetime.htm*) (2003).

   This article provides a brief history of earned value management and discusses why it has reached "primetime" as an important project management concept. Abbas also references several good Web sites with earned value information, including the Department of Defense site at www.acq.osd.mil/pm.

2. Boehm, Barry, et al. *Software Estimation with COCOMO II*. Prentice Hall (July 2000) and the COCOMO II Web site (*http://sunset.usc.edu/research/ COCOMO II*).

   COCOMO II is a model that assists in estimating the cost, effort, and schedule for a new software development activity. It consists of three sub-models, called the Applications Composition, Early Design, and Post-architecture models. Each one offers increased fidelity the further along one is in the project planning and design process. Barry Boehm's book and the COCOMO II Web site are good references for learning more about this estimating tool.

3. Fleming, Quentin W. and Joel M. Koppelman. *Earned Value Project Management, Second Edition*. Project Management Institute (2000).

   This book provides a detailed description of the basic concepts involved in earned value management and a brief history of its use. The authors show how earned value management can be used to assess the status of any kind of project as well as forecast project outcomes. You can also find several presentations and discussions of applying earned value to information technology projects at PMFORUM's Web site (www.pmforum.org/pmwt03/papers03-023more.htm).

4. Hamaker, Joseph W. "But What Will It Cost? The History of NASA Cost Estimating," NASA Web site (*http://www.jsc.nasa.gov/bu2/hamaker.html*) (2000).

   This article provides a wealth of information on how NASA develops cost estimates. Sections of the Web site include cost estimating models, handbooks to assist in cost estimating, salaries, and many links to great references and software. The "NASA Cost Estimating Handbook" (2002) and other resources are available at www.jsc.nasa.gov/bu2/. You can find links to many cost-estimating tools at www.jsc.nasa.gov/bu2/resources.html.

5. META Group. "IT Investment Management: Portfolio Management Lessons Learned," A META Group White Paper (*www.metagroup.com*) (2002).

   This white paper discusses the benefits of adopting portfolio management through customer testimonials and case-study examples. It includes an overview of portfolio management, a review of the current market, and an analysis of how to maximize the return on investment of information technology by leveraging portfolio management tools and techniques. Pacific Edge, a software firm that provides project portfolio management software, commissioned this paper.

6. Roetzheim, William. "Estimating Software Costs," Cost Xpert Group, Inc. (*http://www.costxpert.com/resource_center/SDarticles/SDArticle1.pdf*) (2003).

This article is the first in a series of four articles written by the CEO of the Cost Xpert Group, Inc. It summarizes the traditional measures of estimating the size of a program, such as Lines of Code and Function Points, and introduces other approaches. This article was used in preparing the software development function point estimate in this chapter.

KEY TERMS

- **actual cost (AC)** — the total of direct and indirect costs incurred in accomplishing work on an activity during a given period, formerly called the actual cost of work performed (ACWP)

- **analogous estimates** — a cost estimating technique that uses the actual cost of a previous, similar project as the basis for estimating the cost of the current project, also called **top-down estimates**

- **baseline** — the original project plan plus approved changes

- **bottom-up estimates** — a cost estimating technique based on estimating individual work items and summing them to get a project total

- **budget at completion (BAC)** — the original total budget for a project

- **budgetary estimate** — a cost estimate used to allocate money into an organization's budget

- **cash flow analysis** — a method for determining the estimated *annual* costs and benefits for a project

- **COCOMO II** — a newer, computerized cost-estimating tool based on Boehm's original model that allows one to estimate the cost, effort, and schedule when planning a new software development activity

- **computerized tools** — cost-estimating tools that use computer software, such as spreadsheets and project management software

- **Constructive Cost Model (COCOMO)** — a parametric model developed by Barry Boehm for estimating software development costs

- **contingency reserves** — dollars included in a cost estimate to allow for future situations that may be partially planned for (sometimes called **known unknowns**) and are included in the project cost baseline

- **cost baseline** — a time-phased budget that project managers use to measure and monitor cost performance

- **cost budgeting** — allocating the overall cost estimate to individual work items to establish a baseline for measuring performance

- **cost control** — controlling changes to the project budget
- **cost estimating** — developing an approximation or estimate of the costs of the resources needed to complete the project
- **cost management plan** — a document that describes how cost variances will be managed on the project
- **cost performance index (CPI)** — the ratio of earned value to actual cost; can be used to estimate the projected cost to complete the project
- **cost variance (CV)** — the earned value minus the actual cost
- **definitive estimate** — a cost estimate that provides an accurate estimate of project costs
- **direct costs** — costs that can be directly related to producing the products and services of the project
- **earned value (EV)** — the percentage of work actually completed multiplied by the planned cost, formerly called the **budgeted cost of work performed (BCWP)**
- **earned value management (EVM)** — a project performance measurement technique that integrates scope, time, and cost data
- **estimate at completion (EAC)** — an estimate of what it will cost to complete the project based on performance to date
- **function points** — technology-independent assessments of the functions involved in developing a system
- **indirect costs** — costs that are not directly related to the products or services of the project, but are indirectly related to performing the project.
- **intangible costs** or **benefits** — costs or benefits that are difficult to measure in monetary terms
- **learning curve theory** — a theory that states that when many items are produced repetitively, the unit cost of those items normally decreases in a regular pattern as more units are produced
- **life cycle costing** — considers the total cost of ownership, or development plus support costs, for a project
- **management reserves** — dollars included in a cost estimate to allow for future situations that are unpredictable (sometimes called **unknown unknowns**)
- **overrun** — the additional percentage or dollar amount by which actual costs exceed estimates
- **parametric modeling** — a cost-estimating technique that uses project characteristics (parameters) in a mathematical model to estimate project costs
- **planned value (PV)** — that portion of the approved total cost estimate planned to be spent on an activity during a given period, formerly called the budgeted cost of work scheduled (BCWS)
- **profit margin** — the ratio between revenues and profits
- **profits** — revenues minus expenses

- **project cost management** — the processes required to ensure that the project is completed within the approved budget
- **rate of performance (RP)** — the ratio of actual work completed to the percentage of work planned to have been completed at any given time during the life of the project or activity
- **reserves** — dollars included in a cost estimate to mitigate cost risk by allowing for future situations that are difficult to predict
- **rough order of magnitude (ROM) estimate** — a cost estimate prepared very early in the life of a project to provide a rough idea of what a project will cost
- **schedule performance index (SPI)** — the ratio of earned value to planned value; can be used to estimate the projected time to complete a project
- **schedule variance (SV)** — the earned value minus the planned value
- **Source Lines of Code (SLOC)** — a human written line of code that is not a blank line or comment
- **sunk cost** — money that has been spent in the past
- **tangible costs** or **benefits** — costs or benefits that can be easily measured in dollars

8

Project Quality Management

Objectives

After reading this chapter, you will be able to:

1. *Understand the importance of project quality management for information technology products and services*
2. *Define project quality management and understand how quality relates to various aspects of information technology projects*
3. *Describe quality planning and its relationship to project scope management*
4. *Discuss the importance of quality assurance*
5. *Explain the main outputs of the quality control process*
6. *Understand the tools and techniques for quality control, such as Pareto analysis, statistical sampling, Six Sigma, quality control charts, and testing*
7. *Summarize the contributions of noteworthy quality experts to modern quality management*
8. *Describe how leadership, the cost of quality, organizational influences, expectations, cultural differences, and maturity models relate to improving quality in information technology projects*
9. *Discuss how software can assist in project quality management*

OPENING CASE

A large medical instruments company just hired Scott Daniels, a senior consultant from a large consulting firm, to lead a project to resolve the quality problems with the company's new Executive Information System (EIS). A team of internal programmers and analysts worked with several company executives to develop this new system. Many executives who had never used computers before were hooked on the new, user-friendly EIS. They loved the way the system allowed them to track sales of various medical instruments quickly and easily by product, country, hospital, and sales representative. After successfully testing the new EIS with several executives, the company decided to make the system available to all levels of management.

Unfortunately, several quality problems developed with the new EIS after a few months of operation. People were complaining that they could not get into the system. The system started going down a couple of times a month, and the response time was reportedly getting slower. Users complained when they could not access information within a few seconds. Several people kept forgetting how to log in to the system, thus increasing the number of calls to the company's Help Desk. There were complaints that some of the reports in the system gave inconsistent information. How could a summary report show totals that were not consistent with a detailed report on the same information? The executive sponsor of the EIS wanted the problems fixed quickly and accurately, so he decided to hire an expert in quality from outside the company whom he knew from past projects. Scott Daniels' job was to lead a team of people from both the medical instruments company and his own firm to identify and resolve quality-related issues with the EIS and to develop a plan to help prevent quality problems from happening on future projects.

THE IMPORTANCE OF PROJECT QUALITY MANAGEMENT

Most people have heard jokes about what cars would be like if they followed a development history similar to that of computers. A well-known joke that has been traveling around the Internet goes as follows:

"At a recent computer exposition (COMDEX), Bill Gates, the founder and CEO of Microsoft Corporation, stated: "If General Motors had kept up with technology like the computer industry has, we would all be driving $25 cars that got 1,000 miles to the gallon." In response to Gates' comments, General Motors issued a press release stating: "If GM had developed technology like Microsoft, we would all be driving cars with the following characteristics:

1. For no reason whatsoever your car would crash twice a day.

2. Every time they repainted the lines on the road, you would have to buy a new car.

3. Occasionally, your car would die on the freeway for no reason, and you would just accept this, restart, and drive on.

4. Occasionally, executing a maneuver such as a left turn would cause your car to shut down and refuse to restart, in which case you would have to reinstall the engine.

5. Only one person at a time could use the car, unless you bought 'Car95' or 'CarNT.' But then you would have to buy more seats.

6. Macintosh would make a car that was powered by the sun, reliable, five times as fast, and twice as easy to drive, but would run on only five percent of the roads.

7. The oil, water temperature, and alternator warning lights would be replaced by a single 'general car default' warning light.

8. New seats would force everyone to have the same size hips.

9. The airbag system would say 'Are you sure?' before going off.

10. Occasionally, for no reason whatsoever, your car would lock you out and refuse to let you in until you simultaneously lifted the door handle, turned the key, and grabbed hold of the radio antenna.

11. GM would require all car buyers to also purchase a deluxe set of Rand McNally road maps (now a GM subsidiary), even though they neither need them nor want them. Attempting to delete this option would immediately cause the car's performance to diminish by 50 percent or more. Moreover, GM would become a target for investigation by the Justice Department.

12. Every time GM introduced a new model car, buyers would have to learn how to drive all over again because none of the controls would operate in the same manner as the old car.

13. You would press the Start button to shut off the engine." [1]

Most people simply accept poor quality from many information technology products. So what if your computer crashes a couple of times a month? Just make sure you back up your data. So what if you cannot log in to the corporate intranet or the Internet right now? Just try a little later when it is less busy. So what if the latest version of your word-processing software was shipped with several known bugs? You like the software's new features, and all new software has bugs. Is quality a real problem with information technology projects?

Yes, it is! Information technology is not just a luxury available in some homes, schools, or offices. Many companies throughout the world provide all employees with access to computers. The majority of people in the United States use the Internet, and usage in other countries continues to grow rapidly. It took only five years for 50 million people to use the Internet compared to 25 years for 50 million people to use telephones. Many aspects of our daily lives depend on high-quality information technology products. Food is produced and distributed with the aid of computers; cars have computer chips to track performance;

---

[1] This joke was found on hundreds of Web sites and printed in the *Consultants in Minnesota Newsletter*, Independent Computer Consultants Association, December 1998.

children use computers to help them learn in school; corporations depend on technology for many business functions; and millions of people rely on technology for entertainment and personal communications. Many information technology projects develop mission-critical systems that are used in life-and-death situations, such as navigation systems on board airplanes and computer components built into medical equipment. When one of these systems does not function correctly, it is much more than a slight inconvenience.

⊗ *What Went Wrong?*

- In 1981, a small timing difference caused by a computer program change created a 1 in 67 chance that the space shuttle's five on-board computers would not synchronize. The error caused a launch abort.[2]

- In 1986, two hospital patients died after receiving fatal doses of radiation from a Therac 25 machine. A software problem caused the machine to ignore calibration data.[3]

- In one of the biggest software errors in banking history, Chemical Bank mistakenly deducted about $15 million from more than 100,000 customer accounts one evening. The problem resulted from a single line of code in an updated computer program that caused the bank to process every withdrawal and transfer at its automated teller machines (ATMs) twice. For example, a person who withdrew $100 from an ATM had $200 deducted from his or her account, though the receipt indicated only a withdrawal of $100. The mistake affected 150,000 transactions from Tuesday night through Wednesday afternoon.[4]

- Britain's Coast Guard was unable to use its computers for several hours in May 2004 after being hit by the Sasser virus. The virus knocked out the electronic mapping systems, e-mail, and other computer functions, forcing workers to revert to pen, paper, and radios.[5]

Before you can improve the quality of information technology projects, it is important to understand the basic concepts of project quality management.

---

[2] *Design News* (February 1988).

[3] *Datamation* (May 1987).

[4] *The New York Times* (February 18, 1994).

[5] Fleming, Nic, "Virus sends coastguard computers off course" (*http://news.telegraph.co.uk/news/main.jhtml?xml=/news/2004/05/05/ncoast05.xml*) (May 15, 2004).

WHAT IS PROJECT QUALITY MANAGEMENT?

Project quality management is a difficult knowledge area to define. The International Organization for Standardization (ISO) defines **quality** as "the totality of characteristics of an entity that bear on its ability to satisfy stated or implied needs" (ISO8042:1994) or "the degree to which a set of inherent characteristics fulfils requirements." (ISO9000:2000). Many people spent many hours coming up with these definitions, yet they are still very vague. Other experts define quality based on conformance to requirements and fitness for use. **Conformance to requirements** means the project's processes and products meet written specifications. For example, if the project scope statement requires delivery of 100 Pentium 4 computers, you could easily check whether the correct computers had been delivered. **Fitness for use** means a product can be used as it was intended. If these Pentium 4 computers were delivered without monitors or keyboards and were left in boxes on the customer's shipping dock, the customer might not be satisfied because the computers would not be fit for use. The customer may have assumed that the delivery included monitors and keyboards, unpacking the computers, and installation so they would be ready to use.

The purpose of **project quality management** is to ensure that the project will satisfy the needs for which it was undertaken. Recall that project management involves meeting or exceeding stakeholder needs and expectations. The project team must develop good relationships with key stakeholders, especially the main customer for the project, to understand what quality means to them. *After all, the customer ultimately decides if quality is acceptable.* Many technical projects fail because the project team focuses only on meeting the written requirements for the main products being produced and ignores other stakeholder needs and expectations for the project. For example, the project team should know what successfully delivering 100 Pentium 4 computers means to the customer.

Quality, therefore, must be on an equal level with project scope, time, and cost. If a project's stakeholders are not satisfied with the quality of the project management or the resulting products of the project, the project team will need to adjust scope, time, and cost to satisfy stakeholder needs and expectations. Meeting only written requirements for scope, time, and cost is not sufficient. To achieve stakeholder satisfaction, the project team must develop a good working relationship with all stakeholders and understand their stated or implied needs.

Project quality management involves three main processes:

- **Quality planning** includes identifying which quality standards are relevant to the project and how to satisfy those standards. Incorporating quality standards into project design is a key part of quality planning. For an information technology project, quality standards might include allowing for system growth, planning a reasonable response time for a system, or ensuring that the system produces consistent and accurate information.

Quality standards can also apply to information technology services. For example, you can set standards for how long it should take to get a reply from a Help Desk or how long it should take to ship a replacement part for a hardware item under warranty. The main outputs of quality planning are a quality management plan, quality metrics, quality checklists, a process improvement plan, a quality baseline, and updates to the project management plan.

- **Quality assurance** involves periodically evaluating overall project performance to ensure that the project will satisfy the relevant quality standards. The quality assurance process involves taking responsibility for quality throughout the project's life cycle. Top management must take the lead in emphasizing the roles all employees play in quality assurance, especially senior managers' roles. The main outputs of this process are requested changes, recommended corrective actions, and updates to organization process assets and the project management plan.

- **Quality control** involves monitoring specific project results to ensure that they comply with the relevant quality standards while identifying ways to improve overall quality. This process is often associated with the technical tools and techniques of quality management, such as Pareto charts, quality control charts, and statistical sampling. You will learn more about these tools and techniques later in this chapter. The main outputs of quality control include quality control measurements, validated and recommended defect repair, recommended corrective and preventive actions, requested changes, validated deliverables, and updates to the quality baseline, organization process assets, and the project management plan.

QUALITY PLANNING

Project managers today have a vast knowledge base of information related to quality, and the first step to ensuring project quality management is planning. Quality planning implies the ability to anticipate situations and prepare actions that bring about the desired outcome. The current thrust in modern quality management is the prevention of defects through a program of selecting the proper materials, training and indoctrinating people in quality, and planning a process that ensures the appropriate outcome. In project quality planning, it is important to identify relevant quality standards for each unique project and to design quality into the products of the project and the processes involved in managing the project.

Design of experiments is a quality planning technique that helps identify which variables have the most influence on the overall outcome of a process. Understanding which variables affect outcome is a very important part of quality planning. For example, computer chip designers might want to determine which combination of materials and equipment will produce the most reliable chips at a reasonable cost. You can also apply design of experiments to project management issues such as cost and schedule trade-offs. For example, junior programmers or consultants cost less than senior programmers or consultants, but you cannot expect them to complete the same level of work in the same amount of time. An appropriately designed experiment to compute project costs and durations for various combinations of junior and senior programmers or consultants can allow you to determine an optimal mix of personnel, given limited resources.

Quality planning also involves communicating the correct actions for ensuring quality in a format that is understandable and complete. In quality planning for projects, it is important to describe important factors that directly contribute to meeting the customer's requirements. Organizational policies related to quality, the particular project's scope statement and product descriptions, and related standards and regulations are all important input to the quality planning process. The main outputs of quality planning are a quality management plan and checklists for ensuring quality throughout the project life cycle.

As mentioned in the discussion of project scope management (see Chapter 5), it is often difficult to completely understand the performance dimension of information technology projects. Even if hardware, software, and networking technology would stand still for a while, it is often difficult for customers to explain exactly what they want in an information technology project. Important scope aspects of information technology projects that affect quality include functionality and features, system outputs, performance, and reliability and maintainability.

- **Functionality** is the degree to which a system performs its intended function. **Features** are the system's special characteristics that appeal to users. It is important to clarify what functions and features the system *must* perform, and what functions and features are *optional*. In the EIS example in the opening case, the mandatory functionality of the system might be that it allows users to track sales of specific medical instruments by predetermined categories such as the product group, country, hospital, and sales representative. Mandatory features might be a graphical user interface with icons, menus, online Help, and so on.

- **System outputs** are the screens and reports the system generates. It is important to define clearly what the screens and reports look like for a system. Can the users easily interpret these outputs? Can users get all of the reports they need in a suitable format?

- **Performance** addresses how well a product or service performs the customer's intended use. To design a system with high quality performance, project stakeholders must address many issues. What volumes of data and transactions should the system be capable of handling? How many simultaneous users should the system be designed to handle? What is the projected growth rate in the number of users? What type of equipment must the system run on? How fast must the response time be for different aspects of the system under different circumstances? For the EIS in the opening case, several of the quality problems appear to relate to performance issues. The system is failing a couple of times a month, and users are unsatisfied with the response time. The project team may not have had specific performance requirements or tested the system under the right conditions to deliver the expected performance. Buying faster hardware might address these performance issues. Another performance problem that might be more difficult to fix is the fact that some of the reports are generating inconsistent results. This could be a software quality problem that may be difficult and costly to correct since the system is already in operation.

■ **Reliability** is the ability of a product or service to perform as expected under normal conditions. **Maintainability** addresses the ease of performing maintenance on a product. Most information technology products cannot reach 100 percent reliability, but stakeholders must define what their expectations are. For the EIS, what are the normal conditions for operating the system? Should reliability tests be based on 100 people accessing the system at once and running simple queries? Maintenance for the EIS might include uploading new data into the system or performing maintenance procedures on the system hardware and software. Are the users willing to have the system be unavailable several hours a week for system maintenance? Providing Help Desk support could also be a maintenance function. How fast a response do users expect for Help Desk support? How often can users tolerate system failure? Are the stakeholders willing to pay more for higher reliability and fewer failures?

These aspects of project scope are just a few of the requirement issues related to quality planning. Project managers and their teams need to consider all of these project scope issues in determining quality goals for the project. The main customers for the project must also realize their role in defining the most critical quality needs for the project and constantly communicate these needs and expectations to the project team. Since most information technology projects involve requirements that are not set in stone, it is important for all project stakeholders to work together to balance the quality, scope, time, and cost dimensions of the project. *Project managers, however, are ultimately responsible for quality management on their projects.*

Project managers should be familiar with basic quality terms, standards, and resources. For example, the International Organization for Standardization (ISO) provides information based on inputs from 146 different countries. They have an extensive Web site (*www.iso.org*), which is the source of ISO 9000 and more than 14,000 international standards for business, government and society. IEEE also provides many standards related to quality and has detailed information on their Web site (*www.ieee.org*). You will learn more about some of these standards later in this chapter and in the Suggested Readings.

QUALITY ASSURANCE

It is one thing to develop a plan for ensuring quality on a project; it is another to ensure delivery of quality products and services. **Quality assurance** includes all of the activities related to satisfying the relevant quality standards for a project. Another goal of quality assurance is continuous quality improvement.

Many companies understand the importance of quality assurance and have entire departments dedicated to this area. They have detailed processes in place to make sure their products and services conform to various quality requirements. They also know they must produce those products and services at competitive prices. To be successful in today's competitive business environment, successful companies develop their own best practices and evaluate other organizations' best practices to continuously improve the way they do business.

Top management and project managers can have the greatest impact on the quality of projects by doing a good job of quality assurance. The importance of leadership in improving information technology project quality is discussed in more detail later in this chapter.

Several tools used in quality planning can also be used in quality assurance. Design of experiments, as described under quality planning, can also help ensure and improve product quality. **Benchmarking** generates ideas for quality improvements by comparing specific project practices or product characteristics to those of other projects or products within or outside the performing organization. For example, if a competitor has an EIS with an average down time of only one hour a week, that might be a benchmark for which to strive. Fishbone or Ishikawa diagrams, as described later in this chapter, can assist in ensuring and improving quality by finding the root causes of quality problems.

Several organizations use templates for developing various plans related to quality assurance. Table 8-1 provides the table of contents for a Quality Assurance Plan for the Desktop, Telecommunications, and LAN environment developed by the U.S. Department of Energy (DOE). Note the main categories with a description of the plan itself, the management structure, required documentation, quality assurance procedures, problem-reporting procedures, quality assurance metrics, and quality assurance checklist forms. You can visit the referenced Web site for more information on creating a quality assurance plan.

Table 8-1: Table of Contents for a Quality Assurance Plan

1.0 Draft Quality Assurance Plan
    1.1 Introduction
    1.2 Purpose
    1.3 Policy Statement
    1.4 Scope

Table 8-1: Table of Contents for a Quality Assurance Plan (continued)

2.0 Management
    2.1 Organizational Structure
    2.2 Roles and Responsibilities
        2.2.1 Technical Monitor/Senior Management
        2.2.2 Task Leader
        2.2.3 Quality Assurance Team
        2.2.4 Technical Staff

3.0 Required Documentation

4.0 Quality Assurance Procedures
    4.1 Walkthrough Procedure
    4.2 Review Process
        4.2.1 Review Procedures
    4.3 Audit Process
        4.3.1 Audit Procedures
    4.4 Evaluation Process
    4.5 Process Improvement

5.0 Problem Reporting Procedures
    5.1 Noncompliance Reporting Procedures

6.0 Quality Assurance Metrics

Appendix

Quality Assurance Checklist Forms

U.S. Department of Energy Office of the Chief Information Officer, Software Quality & Systems Engineering, Templates, "XXX YYYY Quality Assurance Plan For the Desktop, Telecommunications, and LAN Environment" (*http://cio.doe.gov*) (March 15, 2002).

An important tool for quality assurance is a quality audit. A **quality audit** is a structured review of specific quality management activities that help identify lessons learned that could improve performance on current or future projects. In-house auditors or third parties with expertise in specific areas can perform quality audits, and quality audits can be scheduled or random. Industrial engineers often perform quality audits by helping to design specific quality metrics for a project and then applying and analyzing the metrics throughout the project. For example, the Northwest Airlines ResNet project (available on the companion Web site for this text) provides an excellent example of using quality audits to emphasize the main goals of a project and then track progress in reaching those goals. The main objective of the ResNet project was to develop a new reservation system to increase direct airline ticket sales and reduce the time it took for sales agents to handle customer calls. The measurement techniques for monitoring these goals helped ResNet's project manager and project team supervise various aspects of the project by focusing on meeting those goals. Measuring progress toward increasing direct sales and reducing call times also helped the project manager justify continued investments in ResNet.

QUALITY CONTROL

Many people only think of quality control when they think of quality management. Perhaps it is because there are many popular tools and techniques in this area. Before describing some of these tools and techniques, it is important to distinguish quality control from quality planning and quality assurance.

Although one of the main goals of quality control is to improve quality, the main outcomes of this process are acceptance decisions, rework, and process adjustments.

- **Acceptance decisions** determine if the products or services produced as part of the project will be accepted or rejected. If they are accepted, they are considered to be validated deliverables. If project stakeholders reject some of the products or services produced as part of the project, there must be rework. For example, the executive who sponsored development of the EIS in the opening case was obviously not satisfied with the system and hired an outside consultant, Scott Daniels, to lead a team to address and correct the quality problems.
- **Rework** is action taken to bring rejected items into compliance with product requirements or specifications or other stakeholder expectations. Rework often results in requested changes and validated defect repair, resulting from recommended defect repair or corrective or preventive actions. Rework can be very expensive, so the project manager must strive to do a good job of quality planning and quality assurance to avoid this need. Since the EIS did not meet all of the stakeholders' expectations for quality, the medical instruments company was spending additional money for rework.
- **Process adjustments** correct or prevent further quality problems based on quality control measurements. Process adjustments are often found by using quality control measurements, and they often result in updates to the quality baseline, organization process assets, and the project management plan. For example, Scott Daniels, the consultant in the opening case, might recommend that the medical instruments company purchase a faster server for the EIS to correct the response-time problems. This change would require changes to the project management plan since it would require more work to be done related to the project. The company

also hired Scott to develop a plan to help prevent future information technology project quality problems.

TOOLS AND TECHNIQUES FOR QUALITY CONTROL

Quality control includes many general tools and techniques. This section focuses on just a few of them—Pareto analysis, statistical sampling, Six Sigma, and quality control charts—and discusses how they can be applied to information technology projects. The section concludes with a discussion on testing, since information technology projects use testing extensively to ensure quality.

Pareto Analysis

Pareto analysis involves identifying the vital few contributors that account for most quality problems in a system. It is sometimes referred to as the 80-20 rule, meaning that 80 percent of problems are often due to 20 percent of the causes. **Pareto diagrams** are histograms, or column charts representing a frequency distribution, that help identify and prioritize problem areas. The variables described by the histogram are ordered by frequency of occurrence.

For example, suppose there was a history of user complaints about the EIS described in the opening case. Many organizations log user complaints, but many do not know how to analyze them. A Pareto diagram is an effective tool for analyzing user complaints and helping project managers decide which ones to address first. Figure 8-1 shows a Pareto diagram of user complaints by category that could apply to the EIS project. The bars represent the number of complaints for each category, and the connected line segments show the cumulative percent of the complaints. For example, Figure 8-1 shows that the most frequent user complaint was log-in problems, followed by the system locking up, the system being too slow, the system being hard to use, and the reports being inaccurate. The first complaint accounts for 55 percent of the total complaints. The first and second complaints have a cumulative percentage of almost 80 percent, meaning these two areas account for 80 percent of the complaints. Therefore, if the company wants to reduce the number of complaints, it should first focus on making it easier to log in to the system, since the majority of complaints fall under that category.

However, good project managers know it is also important to address the potential severity of all types of complaints. If the system was generating inaccurate reports—the smallest bar on the Pareto diagram—it could be a severe problem that would cost a lot of money to correct. The company should review all of the complaints before deciding what to do next.

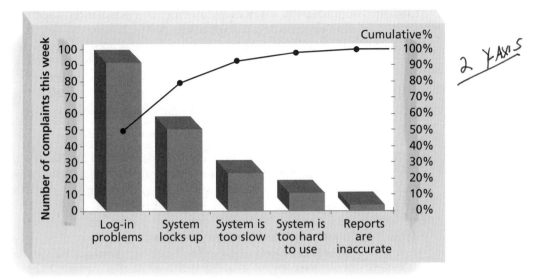

Figure 8-1. Sample Pareto Diagram

Statistical Sampling

Statistical sampling is a key concept in project quality management. Members of a project team who focus on quality control must have a strong understanding of statistics, but other project team members need to understand only the basic concepts. These concepts include statistical sampling, certainty factor, standard deviation, and variability. Standard deviation and variability are fundamental concepts for understanding quality control charts. This section briefly describes these concepts and describes how a project manager might apply them to information technology projects. Refer to statistics texts for additional details.

Statistical sampling involves choosing part of a population of interest for inspection. For example, suppose a company wants to develop an electronic data interchange (EDI) system for handling data on invoices from all of its suppliers. Assume also that in the past year, the total number of invoices was 50,000 from 200 different suppliers. It would be very time consuming and expensive to review every single invoice to determine data requirements for the new system. Even if the system developers did review all 200 invoice forms from the different suppliers, the data might be entered differently on every form. It is impractical to study every member of a population, such as all 50,000 invoices, so statisticians have developed techniques to help determine an appropriate sample size. If the system developers used statistical techniques, they might find that by studying only 100 invoices, they would have a good sample of the type of data they would need in designing the system.

The size of the sample depends on how representative you want the sample to be. A simple formula for determining sample size is:

Sample size = .25 X (certainty factor/acceptable error)^2

The certainty factor denotes how certain you want to be that the data sampled will not include variations that do not naturally exist in the population. You calculate the certainty factor from tables available in statistics books. Table 8-2 shows some commonly used certainty factors.

Table 8-2: Commonly Used Certainty Factors

DESIRED CERTAINTY	CERTAINTY FACTOR
95%	1.960
90%	1.645
80%	1.281

For example, suppose the developers of the EDI system described earlier would accept a 95 percent certainty that a sample of invoices would contain no variation unless it was present in the population of total invoices. They would then calculate the sample size as:

Sample size = 0.25 X (1.960/.05)^2 = 384

If the developers wanted 90 percent certainty, they would calculate the sample size as:

Sample size = 0.25 X (1.645/.10)^2 = 68

If the developers wanted 80 percent certainty, they would calculate the sample size as:

Sample size = 0.25 X (1.281/.20)^2 = 10

Assume the developers decide on 90 percent for the certainty factor. Then they would need to examine 68 invoices to determine the type of data the EDI system would need to capture.

Six Sigma

The work of many project quality experts contributed to the development of today's Six Sigma principles. There has been some confusion in the past few years about the term Six Sigma. This section summarizes recent information about this important concept and explains how organizations worldwide use Six Sigma principles to improve quality, decrease costs, and better meet customer needs.

In their book, *The Six Sigma Way*, authors Peter Pande, Robert Neuman, and Roland Cavanagh define **Six Sigma** as "a comprehensive and flexible *system* for achieving, sustaining and maximizing business success. Six Sigma is uniquely driven by close understanding of customer needs, disciplined use of facts, data, and statistical analysis, and diligent attention to managing, improving, and reinventing business processes."[6]

Six Sigma's target for perfection is the achievement of no more than 3.4 defects, errors, or mistakes per million opportunities. This target number is explained in more detail later in this section. An organization can apply the Six Sigma principles to the design and production of a product, a Help Desk, or other customer-service process.

Projects that use Six Sigma principles for quality control normally follow a five-phase improvement process called **DMAIC** (pronounced de-MAY-ick), which stands for Define, Measure, Analyze, Improve, and Control. DMAIC is a systematic, closed-loop process for continued improvement that is scientific and fact based. The following are brief descriptions of each phase of the DMAIC improvement process:

1. **Define**: Define the problem/opportunity, process, and customer requirements. Important tools used in this phase include a project charter, a description of customer requirements, process maps, and Voice of the Customer (VOC) data. Examples of VOC data include complaints, surveys, comments, and market research that represent the views and needs of the organization's customers.

2. **Measure**: Define measures, then collect, compile, and display data. Measures are defined in terms of defects per opportunity, as defined later in this section.

3. **Analyze**: Scrutinize process details to find improvement opportunities. A project team working on a Six Sigma project, normally referred to as a Six Sigma team, investigates and verifies data to prove the suspected root causes of quality problems and substantiates the problem statement. An important tool in this phase is the fishbone or Ishikawa diagram, as described later in this chapter.

4. **Improve**: Generate solutions and ideas for improving the problem. A final solution is verified with the project sponsor, and the Six Sigma team develops a plan to pilot test the solution. The Six Sigma team reviews the results of the pilot test to refine the solution, if needed, and then implements the solution where appropriate.

---

[6] Pande, Peter S., Robert P. Neuman, and Roland R. Cavanagh, *The Six Sigma Way*. New York: McGraw-Hill, 2000, p. xi.

5. **Control**: Track and verify the stability of the improvements and the predictability of the solution. Control charts are one tool used in the control phase, as described later in this chapter.

How Is Six Sigma Quality Control Unique?

How does using Six Sigma principles differ from using previous quality control initiatives? Many people remember other quality initiatives from the past few decades such as Total Quality Management (TQM) and Business Process Reengineering (BPR), to name a few. The origins of many Six Sigma principles and tools are found in these previous initiatives; however, there are several new ideas included in Six Sigma principles that help organizations improve their competitiveness and bottom-line results. Below are a few of these principles:

- Using Six Sigma principles is an organization-wide commitment. CEOs, top managers, and all levels of employees in an organization that embraces Six Sigma principles (often referred to as a Six Sigma organization) have seen remarkable improvements due to its use. There are often huge training investments, but these investments pay off as employees practice Six Sigma principles and produce higher quality goods and services at lower costs.

- Six Sigma training normally follows the "Belt" system, similar to a karate class in which students receive different color belts for each training level. In Six Sigma training, those in the Yellow Belt category receive the minimum level of training, which is normally two to three full days of Six Sigma training for project team members who work on Six Sigma projects on a part-time basis. Those in the Green Belt category usually participate in two to three full weeks of training. Those in the Black Belt category normally work on Six Sigma projects full-time and attend four to five full weeks of training. Project managers are often Black Belts. The Master Black Belt category describes experienced Black Belts who act as technical resources and mentors to people with lower-level belts.

- Organizations that successfully implement Six Sigma principles have the ability and willingness to adopt two seemingly contrary objectives at the same time. Authors James Collins and Jerry Porras describe this as the "We can do it all" or "Genius of the And" approach in their book, *Built to Last*. For example, Six Sigma organizations believe that they can be creative *and* rational, focus on the big picture *and* minute details, reduce errors *and* get things done faster, and make customers happy *and* make a lot of money.

- Six Sigma is not just a program or a discipline to organizations that have benefited from it. Six Sigma is an operating philosophy that is customer-focused and strives to drive out waste, raise levels of quality, and improve financial performance at *breakthrough* levels. A Six Sigma organization sets high goals and uses the DMAIC improvement process to achieve extraordinary quality improvements.

Many organizations do some of what now fits under the definition of Six Sigma, and many Six Sigma principles are not brand-new. What *is* new is its ability to bring together many different themes, concepts, and tools into a coherent management process that can be used on an organization-wide basis.

☑ *What Went Right?*

Many organizations have reported dramatic improvements as a result of implementing Six Sigma principles.

■ Motorola, Inc. pioneered the adoption of Six Sigma in the 1980s. Its reason for developing and implementing Six Sigma was clear—to stay in business. Japanese competitors were putting several U.S. and European companies out of business. Using Six Sigma provided Motorola with a simple, consistent way to track and compare performance to customer requirements and meet ambitious quality goals in terms of defect reduction. Chairman Bob Galvin set a goal of ten times improvement in defect reduction every two years, or 100 times improvement in four years. Motorola did stay in business, achieving excellent growth and profitability in the 1980s and 1990s. Motorola estimates the cumulative savings based on Six Sigma efforts to be about $14 billion.[7]

■ Allied Signal/Honeywell, a technology and manufacturing company, began several quality improvement activities in the early 1990s. By 1999, the company reported saving more than $600 million a year, thanks to training in and application of Six Sigma principles. Six Sigma teams reduced the costs of reworking defects and applied the same principles to designing new products, like aircraft engines. They reduced the time it took from design to certification of engines from 42 to 33 months. Allied Signal/Honeywell credits Six Sigma with a 6 percent productivity increase in 1998 and record profit margins of 13 percent. As one of the company's directors said, Six Sigma "changed the way we think and the way we communicate. We never used to talk about the process or the customer; now they're part of our everyday conversation."[8]

■ General Electric Company (GE) uses Six Sigma to focus on achieving customer satisfaction. Chairman Jack Welch urged his top managers to become "passionate lunatics" about Six Sigma. Welch said, "Six Sigma has forever changed GE. Everyone from the Six Sigma zealots emerging from their Black Belt teams, to the engineers, the auditors, and the scientists, to the senior leadership that will take this Company into its new millennium—is a true believer in Six Sigma, the way this Company now works."[9]

---

[7] Ibid. p. 7.
[8] Ibid. p. 9.
[9] Address to General Electric Company Annual Meetings, Cleveland, Ohio, April 21, 1999.

Six Sigma and Project Management

Joseph M. Juran stated, "All improvement takes place project by project, and in no other way."[10] Organizations implement Six Sigma by selecting and managing projects. An important part of project management is good project selection.

This statement is especially true for Six Sigma projects. Pande, Neuman, and Cavanagh conducted an informal poll to find out what the most critical and most commonly mishandled activity was in launching Six Sigma, and the unanimous answer was project selection. "It's a pretty simple equation, really: Well-selected and -defined improvement projects equal better, faster results. The converse equation is also simple: Poorly selected and defined projects equal delayed results and frustration."[11]

Organizations must also be careful to apply higher quality where it makes sense. A recent article in *Fortune* states that companies that have implemented Six Sigma have not necessarily boosted their stock values. Although GE boasted savings of more than $2 billion in 1999 due to its use of Six Sigma, other companies, such as Whirlpool, could not clearly demonstrate the value of their investments. Why can't all companies benefit from Six Sigma? Because minimizing defects does not matter if an organization is making a product that no one wants to buy. As one of Six Sigma's biggest supporters, Mikel Harry, puts it, "I could genetically engineer a Six Sigma goat, but if a rodeo is the marketplace, people are still going to buy a Four Sigma horse."[12]

As described in Chapter 4, Project Integration Management, there are several methods for selecting projects. However, what makes a project a potential Six Sigma project? First, there must be a quality problem or gap between the current and desired performance. Many projects do not meet this first criterion, such as building a house, merging two corporations, or providing an information technology infrastructure for a new organization. Second, the project should not have a clearly understood problem. Third, the solution should not be predetermined, and an optimal solution should not be apparent.

Once a project is selected as a good candidate for Six Sigma, many project management concepts, tools, and techniques described in this text come into play. For example, Six Sigma projects usually have a business case, a project charter, requirements documents, a schedule, a budget, and so on. Six Sigma projects are done in teams and have sponsors called champions. There are also, of course, project managers, often called team leaders in Six Sigma organizations. In other words, Six Sigma projects are simply types of projects that focus on supporting the Six Sigma philosophy by being customer-focused and striving to drive out waste, raise levels of quality, and improve financial performance at breakthrough levels.

[10] "What You Need to Know About Six Sigma," *Productivity Digest* (December 2001), p. 38.

[11] Pande, Peter S., Robert P. Neuman, and Roland R. Cavanagh, *The Six Sigma Way*. New York: McGraw-Hill, 2000, p. 137.

[12] Clifford, Lee, "Why You Can Safely Ignore Six Sigma," *Fortune* (January 22, 2001), p. 140.

Six Sigma and Statistics

An important concept in Six Sigma is improving quality by reducing variation. The term sigma means standard deviation. **Standard deviation** measures how much variation exists in a distribution of data. A small standard deviation means that data clusters closely around the middle of a distribution and there is little variability among the data. A large standard deviation means that data is spread out around the middle of the distribution and there is relatively greater variability. Statisticians use the Greek symbol σ (sigma) to represent the standard deviation.

Figure 8-2 provides an example of a **normal distribution**—a bell-shaped curve that is symmetrical regarding the **mean** or average value of the population (the data being analyzed). In any normal distribution, 68.3 percent of the population is within one standard deviation (1σ) of the mean, 95.5 percent of the population is within two standard deviations (2σ), and 99.7 percent of the population is within three standard deviations (3σ) of the mean.

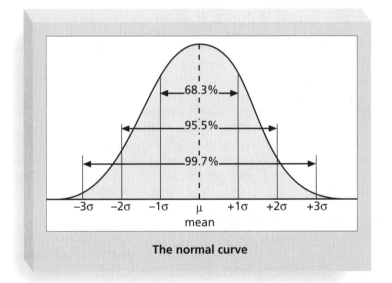

The normal curve

Figure 8-2. Normal Distribution and Standard Deviation

Standard deviation is a key factor in determining the acceptable number of defective units found in a population. Table 8-3 illustrates the relationship between sigma, the percentage of the population within that sigma range, and the number of defective units per billion. Note that this table shows that being plus or minus six sigma in pure statistical terms means only *two* defective units per *billion*. Why, then, is the target for Six Sigma programs 3.4 defects per million opportunities, as stated earlier in this chapter?

Table 8-3: Sigma and Defective Units

SPECIFICATION RANGE (IN +/- SIGMAS)	PERCENT OF POPULATION WITHIN RANGE	DEFECTIVE UNITS PER BILLION
1	68.27	317,300,000
2	95.45	45,400,000
3	99.73	2,700,000
4	99.9937	63,000
5	99.999943	57
6	99.9999998	2

Based on Motorola's original work on Six Sigma in the 1980s, the convention used for Six Sigma is a scoring system that accounts for more variation in a process than you would typically find in a few weeks or months of data gathering. In other words, time is an important factor in determining process variations. Table 8-4 shows the Six Sigma conversion table applied to Six Sigma projects. (See the Suggested Reading by Pande, Neuman, and Cavanagh for a detailed description of Six Sigma scoring.) The **yield** represents the number of units handled correctly through the process steps. A **defect** is any instance where the product or service fails to meet customer requirements. Because most products or services have multiple customer requirements, there can be several opportunities to have a defect. For example, suppose a company is trying to reduce the number of errors on customer billing statements. There could be several errors on a billing statement due to a misspelled name, incorrect address, wrong date of service, calculation error, and so on. There might be 100 opportunities for a defect to occur on one billing statement. Instead of measuring the number of defects per unit or billing statement, Six Sigma measures the number of defects based on the number of opportunities.

Table 8-4: Sigma Conversion Table

SIGMA	YIELD	DEFECTS PER MILLION OPPORTUNITIES (DPMO)
1	31.0%	690,000
2	69.2%	308,000
3	93.3%	66,800
4	99.4%	6,210
5	99.97%	230
6	99.99966%	3.4

As you can see, the Six Sigma conversion table shows that a process operating at six sigma means there are no more than 3.4 defects per million opportunities. However, most organizations today use the term *six sigma project* in a broad sense to describe projects that will help them in achieving, sustaining, and maximizing business success through better business processes.

Another term you may hear being used in the telecommunications industry is **six 9s of quality**. Six 9s of quality is a measure of quality control equal to 1 fault in 1 million opportunities. In the telecommunications industry, it means 99.9999 percent service availability or *30 seconds of down time a year*. This level of quality has also been stated as the target goal for the number of errors in a communications circuit, system failures, or errors in lines of code. To achieve six 9s of quality requires continual testing to find and eliminate errors or enough redundancy and back-up equipment in systems to reduce the overall system failure rate to that low a level.

Quality Control Charts and the Seven Run Rule

A **control chart** is a graphic display of data that illustrates the results of a process over time. The main use of control charts is to prevent defects, rather than to detect or reject them. Quality control charts allow you to determine whether a process is in control or out of control. When a process is in control, any variations in the results of the process are created by random events. Processes that are in control do not need to be adjusted. When a process is out of control, variations in the results of the process are caused by nonrandom events. When a process is out of control, you need to identify the causes of those nonrandom events and adjust the process to correct or eliminate them. Control charts are often used to monitor manufactured lots, but they can also be used to monitor the volume and frequency of change requests, errors in documents, cost and schedule variances, and other items related to project quality management.

Figure 8-3 illustrates an example of a control chart for a process that manufactures 12-inch rulers. Assume that these are wooden rulers created by machines on an assembly line. Each point on the chart represents a length measurement for a ruler that comes off the assembly line. The scale on the vertical axis goes from 11.90 to 12.10. These numbers represent the lower and upper specification limits for the ruler. In this case, this would mean that the customer for the rulers has specified that all rulers purchased must be between 11.90 and 12.10 inches long, or 12 inches plus or minus 0.10 inches. The lower and upper control limits on the quality control chart are 11.91 and 12.09 inches, respectively. This means the manufacturing process is designed to produce rulers between 11.91 and 12.09 inches long.

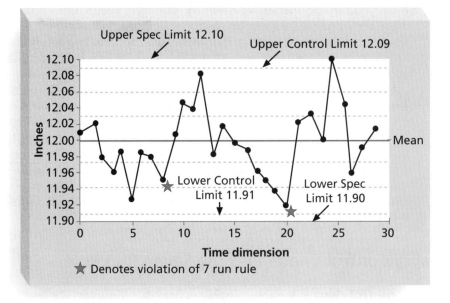

Figure 8-3. Sample Quality Control Chart

Note the dotted lines on the chart at 12.03 inches, 12.06 inches, and 12.09 inches. These dotted lines represent the points at one, two, and three standard deviations above the mean. The dotted lines at 11.97 inches, 11.94 inches, and 11.91 inches represent the points at one, two, and three standard deviations below the mean. Based on the statistical definition of ±3σ described earlier, 99.73 percent of the manufactured rulers should come off the assembly line with a measurement between 11.91 and 12.09 inches, if the manufacturing process is operating in control.

Looking for and analyzing patterns in process data is an important part of quality control. You can use quality control charts and the seven run rule to look for patterns in data. The **seven run rule** states that if seven data points in a row are all below the mean, above the mean, or are all increasing or decreasing, then the process needs to be examined for nonrandom problems. In Figure 8-3, data points that violate the seven run rule are starred. Note that you include the first point in a series of points that are all increasing or decreasing. In the ruler manufacturing process, these data points may indicate that a calibration device may need adjustment. For example, the machine that cuts the wood for the rulers might need to be adjusted or the blade on the machine might need to be replaced.

Testing

Many information technology professionals think of testing as a stage that comes near the end of information technology product development. Instead of putting serious effort into proper planning, analysis, and design of information technology projects, some organizations rely on testing just before a product ships to ensure some degree of quality. In fact, testing needs to be done during almost every phase of the systems development life cycle, not just before the organization ships or hands over a product to the customer.

Figure 8-4 shows one way of portraying the systems development life cycle. This example includes 17 main tasks involved in a software development project and shows their relationship to each other. For example, every project should start by initiating the project, conducting a feasibility study, and then performing project planning. The figure then shows that the work involved in preparing detailed requirements and the detailed architecture for the system can be performed simultaneously. The oval-shaped phases represent actual tests or tasks, which will include test plans to help ensure quality on software development projects.[13]

Several of the phases in Figure 8-4 include specific work related to testing. A **unit test** is done to test each individual component (often a program) to ensure that it is as defect-free as possible. Unit tests are performed before moving onto the integration test. **Integration testing** occurs between unit and system testing to test functionally grouped components. It ensures a subset(s) of the entire system works together. **System testing** tests the entire system as one entity. It focuses on the big picture to ensure that the entire system is working properly. **User acceptance testing** is an independent test performed by end users prior to accepting the delivered system. It focuses on the business fit of the system to the organization, rather than technical issues.

[13] Hollstadt & Associates, Inc., *Software Development Project Life Cycle Testing Methodology User's Manual*. Burnsville: MN, August 1998, p. 13.

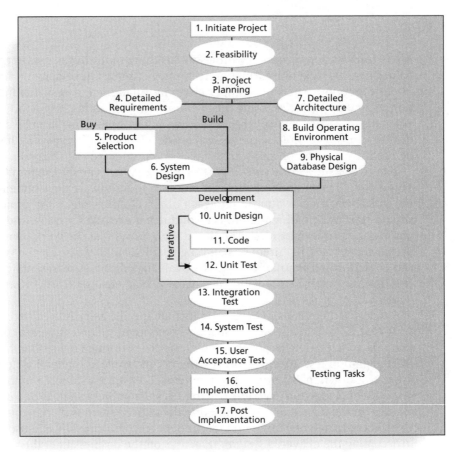

Figure 8-4. Testing Tasks in the Software Development Life Cycle

Figure 8-5 presents a Gantt chart that shows testing tasks that are appropriate for different phases of the systems development life cycle.[14] This is a simplified version of a detailed testing management plan used by a consulting firm. To help improve the quality of software development projects, it is important for organizations to follow a thorough and disciplined testing methodology. System developers and testers must also establish a partnership with all project stakeholders to make sure the system meets their needs and expectations and the tests are done properly. As described in the next section, there are tremendous costs involved in failure to perform proper testing.

[14] Ibid. p. 54–57.

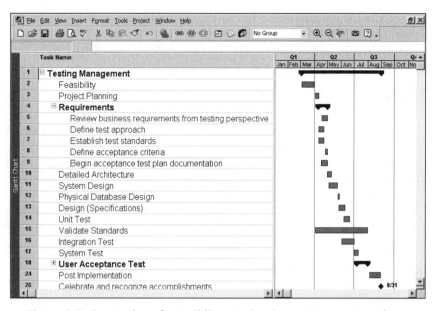

Figure 8-5. Gantt Chart for Building Testing into a Systems Development Project Plan

Testing alone, however, cannot always solve software defect problems, according to Watts S. Humphrey, a renowned expert on software quality and Fellow at Carnegie Mellon's Software Engineering Institute. He believes that the traditional code/test/fix cycle for software development is not enough. As code gets more complex, the number of defects missed by testing increases and becomes the problem not just of testers, but also of paying customers. Humphrey says that, on average, programmers introduce a defect for every nine or ten lines of code, and the finished software, after all testing, contains about five to six defects per thousand lines of code. Although there are many different definitions, Humphrey defines a **software defect** as anything that must be changed before delivery of the program. Testing does not sufficiently prevent software defects because the number of ways to test a complex system is huge. In addition, users will continue to invent new ways to use a system that its developers never considered, so certain functionalities may never have been tested or even included in the system requirements. Humphrey suggests that people rethink the software development process to provide *no* potential defects when you enter system testing. This means that developers must be responsible for providing error-free code at each stage of testing. Humphrey teaches a development process where programmers measure and track the kinds of errors they commit so they can use the data to improve their performance. He also acknowledges that top management must support developers by letting them self-direct their work. Programmers need to be motivated and excited to do high-quality work and have some control over how they do it.[15]

MODERN QUALITY MANAGEMENT

Modern quality management requires customer satisfaction, prefers prevention to inspection, and recognizes management responsibility for quality. Several noteworthy people helped develop the following theories, tools, and techniques that define modern quality management.[16] The suggestions from these quality experts led to many projects to improve quality and provided the foundation for today's Six Sigma projects. This section summarizes major contributions made by Deming, Juran, Crosby, Ishikawa, Taguchi, and Feigenbaum.

Deming and His Fourteen Points for Management

Dr. W. Edwards Deming is known primarily for his work on quality control in Japan. Dr. Deming went to Japan after World War II, at the request of the Japanese government, to assist them in improving productivity and quality. Deming, a statistician and former professor at New York University, taught Japanese manufacturers that higher quality meant greater productivity and lower cost. American industry did not recognize Deming's theories until Japanese manufacturers started producing products that seriously challenged American products, particularly in the auto industry. Ford Motor Company then adopted Deming's quality methods and experienced dramatic improvement in quality and sales thereafter. By the 1980s, after seeing the excellent work coming out of Japan, several U.S. corporations vied for Deming's expertise to help them establish quality improvement programs in their own factories. Many people are familiar with the Deming Prize, an award given to recognize high-quality organizations, and Deming's Cycle for Improvement: plan, do, check, and act. Most Six Sigma principles described earlier are based on the plan-do-check-act model created by Deming. Many are also familiar with Deming's 14 Points for Management, summarized below from Deming's text *Out of the Crisis*:

1. Create constancy of purpose for improvement of product and service.
2. Adopt the new philosophy.
3. Cease dependence on inspection to achieve quality.

---

[15] Eisenberg, Bart, "Achieving Zero-Defects Software," *Pacific Connection* (January 2004).
[16] Kerzner, Harold, *Project Management*, 6th ed. New York: Van Nostrand Reinhold, 1998, p. 1048.

4. End the practice of awarding business based on price tag alone. Instead, minimize total cost by working with a single supplier.

5. Improve constantly and forever every process for planning, production, and service.

6. Institute training on the job.

7. Adopt and institute leadership.

8. Drive out fear.

9. Break down barriers between staff areas.

10. Eliminate slogans, exhortations, and targets for the workforce.

11. Eliminate numerical quotas for the workforce and numerical goals for management.

12. Remove barriers that rob people of workmanship. Eliminate the annual rating or merit system.

13. Institute a vigorous program of education and self-improvement for everyone.

14. Put everyone in the company to work to accomplish the transformation.

Juran and the Importance of Top Management Commitment to Quality

Joseph M. Juran, like Deming, taught Japanese manufacturers how to improve their productivity. U.S. companies later discovered him as well. He wrote the first edition of the *Quality Control Handbook* in 1974, stressing the importance of top management commitment to continuous product quality improvement. In 1999, at the age of 94, Juran published the fifth edition of this famous handbook. He also developed the Juran Trilogy: quality improvement, quality planning, and quality control. Juran stressed the difference between the manufacturer's view of quality and the customer's view. Manufacturers often focus on conformance to requirements, but customers focus on fitness for use. Most definitions of quality now use fitness for use to stress the importance of satisfying stated or implied needs and not just meeting stated requirements or specifications. Juran developed ten steps to quality improvement:

1. Build awareness of the need and opportunity for improvement.

2. Set goals for improvement.

3. Organize to reach the goals (establish a quality council, identify problems, select projects, appoint teams, designate facilitators).

4. Provide training.

5. Carry out projects to solve problems.

6. Report progress.

7. Give recognition.

8. Communicate results.

9. Keep score.

10. Maintain momentum by making annual improvement part of the regular systems and processes of the company.

Crosby and Striving for Zero Defects

Philip B. Crosby wrote *Quality Is Free* in 1979 and is best known for suggesting that organizations strive for zero defects. He stressed that the costs of poor quality should include all the costs of not doing the job right the first time, such as scrap, rework, lost labor hours and machine hours, customer ill will and lost sales, and warranty costs. Crosby suggested that the cost of poor quality is so understated that companies can profitably spend unlimited amounts of money on improving quality. Crosby developed the following 14 steps for quality improvement:

1. Make it clear that management is committed to quality.

2. Form quality improvement teams with representatives from each department.

3. Determine where current and potential quality problems lie.

4. Evaluate the cost of quality and explain its use as a management tool.

5. Raise the quality awareness and personal concern of all employees.

6. Take actions to correct problems identified through previous steps.

7. Establish a committee for the zero-defects program.

8. Train supervisors to actively carry out their part of the quality improvement program.

9. Hold a "zero-defects day" to let all employees realize that there has been a change.

10. Encourage individuals to establish improvement goals for themselves and their groups.

11. Encourage employees to communicate to management the obstacles they face in attaining their improvement goals.

12. Recognize and appreciate those who participate.

13. Establish quality councils to communicate on a regular basis.

14. Do it all over again to emphasize that the quality improvement program never ends.

Crosby also developed the Quality Management Process Maturity Grid in 1978. This grid can be applied to an organization's attitude toward product usability. For example, the first stage in the grid is ignorance, where people might think they don't have any problems with usability. The final stage is wisdom, where people have changed their attitude so that usability defect prevention is a routine part of their operation.

Ishikawa and the Fishbone Diagram

Kaoru Ishikawa is best known for his 1972 book *Guide to Quality Control*. He developed the concept of quality circles and pioneered the use of Fishbone diagrams. **Fishbone diagrams**, sometimes called **Ishikawa diagrams**, trace complaints about quality problems back to the responsible production operations. In other words, they help find the root cause of quality problems. **Quality circles** are groups of nonsupervisors and work leaders in a single company department who volunteer to conduct group studies on how to improve the effectiveness of work in their department. Ishikawa suggested that Japanese managers and workers were totally committed to quality, but that most U.S. companies delegated the responsibility for quality to a few staff members.

Figure 8-6 provides an example of a fishbone or Ishikawa diagram that Scott Daniels, the consultant in the opening case, might create to trace the real cause of the problem of users not being able to log in to the EIS. Notice that it resembles the skeleton of a fish, hence its name. This fishbone diagram lists the main areas that could be the cause of the problem: the EIS system's hardware, the individual user's hardware, software, or training. This figure describes two of these areas, the individual user's hardware and training, in more detail. The root cause of the problem would have a significant impact on the actions taken to solve the problem. If many users could not get into the system because their computers did not have enough memory, the solution might be to upgrade memory for those computers. If many users could not get into the system because they forgot their passwords, there might be a much quicker, less expensive solution.

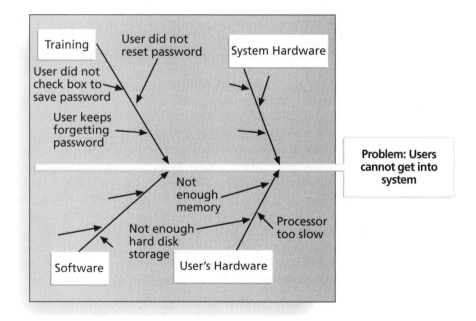

Figure 8-6. Sample Fishbone or Ishikawa Diagram

Taguchi and Robust Design Methods

Genichi Taguchi is best known for developing the Taguchi methods for optimizing the process of engineering experimentation. Key concepts in the Taguchi methods are that quality should be designed into the product and not inspected into it and that quality is best achieved by minimizing deviation from the target value. For example, if the target response time for accessing the EIS described in the opening case is half a second, there should be little deviation from this time. A 1998 article in *Fortune* magazine states, "Japan's Taguchi is America's new quality hero."[17] Many companies, including Xerox, Ford, Hewlett-Packard, and Goodyear, have recently used Taguchi's Robust Design methods to design high-quality products. **Robust Design methods** focus on eliminating defects by substituting scientific inquiry for trial-and-error methods.

Feigenbaum and Workers Responsibility for Quality

Armand V. Feigenbaum developed the concept of total quality control (TQC) in his 1983 book *Total Quality Control: Engineering and Management*. He proposed that the responsibility for quality should rest with the people who do the work. In TQC, product quality is more important than production rates, and workers are allowed to stop production whenever a quality problem occurs.

Malcolm Baldrige National Quality Award

The **Malcolm Baldrige National Quality Award** originated in 1987 to recognize companies that have achieved a level of world-class competition through quality management. The award was started in honor of Malcolm Baldrige, who was the U.S. secretary of commerce from 1981 until his death in a rodeo accident in July 1987. Baldrige was a proponent of quality management as a key element in improving the prosperity and long-term strength of U.S. organizations. The Malcolm Baldrige National Quality Award is given by the president of the United States to U.S. businesses—manufacturing and service, small and large—and to education and health care organizations. Organizations must apply for the award, and they must be judged outstanding in seven areas: leadership, strategic planning, customer and market focus, information and analysis, human resource focus, process management, and business results. Three awards may be given annually in each of these categories: manufacturing, service, small business and, from 1999, education and health care. The awards recognize achievements in quality and performance and raise awareness about the importance of quality as a competitive edge; the award is not given for specific products or services.

[17] Bylinsky, Gene, "How to Bring Out Better Products Faster," *Fortune* (November 23, 1998) p. 238[B].

ISO Standards

The International Organization for Standardization (ISO) is a network of national standards institutes from 145 countries that work in partnership with international organizations, governments, industries, businesses, and consumer representatives. **ISO 9000**, a quality system standard developed by the ISO, is a three-part, continuous cycle of planning, controlling, and documenting quality in an organization. ISO 9000 provides minimum requirements needed for an organization to meet its quality certification standards. The ISO 9000 family of international quality management standards and guidelines has earned a global reputation as the basis for establishing quality management systems. The ISO's Web site includes testimonials from many business leaders throughout the world explaining the benefits of following these standards.

- A government official in Malaysia believes that the universally accepted ISO 9000 standard can contribute significantly to improving quality and enhancing development of an excellent work culture. ISO 9000 leads to a more systematic management of quality and provides a means of consolidating quality management systems in public service.

- Managers in the United Kingdom say that ISO 9000 has become a firmly established cornerstone of their organizations' drive toward quality improvement. It leads to substantial cost savings and contributes strongly to customer satisfaction.

- A Brazilian newspaper reported that following ISO 9000 helped its home delivery service improve so much that the complaints dropped to a mere 0.06 percent of the total copies distributed.[18]

ISO has also developed ISO 15504, sometimes known as SPICE (Software Process Improvement and Capability dEtermination). **ISO 15504** is a framework for the assessment of software processes. The overall goals of the standard are to encourage organizations interested in improving quality of software products to employ proven, consistent, and reliable methods for assessing the state of their software development processes. They can also use their assessment results as part of coherent improvement programs. One of the outcomes of assessment and consequent improvement programs is reliable, predictable, and continuously improving software processes.

The contributions of several quality experts, quality awards, and quality standards are important parts of project quality management. The Project Management Institute was proud to announce in 1999 that their certification department had become the first certification department in the world to earn ISO 9000 certification, and that the *PMBOK® Guide 1996* had been recognized as an international standard. Emphasizing quality in project management helps

[18] International Organization for Standardization (Market feedback) (*www.iso.org*) (March 2003).

ensure that projects produce products or services that meet customer needs and expectations.

IMPROVING INFORMATION TECHNOLOGY PROJECT QUALITY

In addition to some of the suggestions provided for using good quality planning, quality assurance, and quality control, there are several other important issues involved in improving the quality of information technology projects. Strong leadership, understanding the cost of quality, providing a good workplace to enhance quality, and working toward improving the organization's overall maturity level in software development and project management can all assist in improving quality.

Leadership

As Joseph M. Juran said in 1945, "It is most important that top management be quality-minded. In the absence of sincere manifestation of interest at the top, little will happen below."[19] Juran and many other quality experts argue that the main cause of quality problems is a lack of leadership.

As globalization continues to increase and customers become more and more demanding, creating quality products quickly at a reasonable price is essential for staying in business. Having good quality programs in place helps organizations remain competitive. To establish and implement effective quality programs, top management must lead the way. A large percentage of quality problems are associated with management, not technical issues. Therefore, top management must take responsibility for creating, supporting, and promoting quality programs.

Motorola provides an excellent example of a high-technology company that truly emphasizes quality. Leadership is one of the factors that helped Motorola achieve its great success in quality management and Six Sigma. Top management emphasized the need to improve quality and helped all employees take responsibility for customer satisfaction. Strategic objectives in Motorola's long-range plans included managing quality improvement in the same way that new products or technologies were managed. Top management stressed the need to develop and use quality standards and provided resources such as staff, training, and customer inputs to help improve quality.

Leadership provides an environment conducive to producing quality. Management must publicly declare the company's philosophy and commitment to quality, implement company-wide training programs in quality concepts and principles, implement measurement programs to establish and track

---

[19] American Society for Quality (ASQ), *(www.asqc.org/about/history/juran.html)*.

quality levels, and actively demonstrate the importance of quality. When every employee understands and insists on producing high-quality products, then top management has done a good job of promoting the importance of quality.

The Cost of Quality

The **cost of quality** is the cost of conformance plus the cost of nonconformance. **Conformance** means delivering products that meet requirements and fitness for use. Examples of these costs include the costs associated with developing a quality plan, costs for analyzing and managing product requirements, and costs for testing. The **cost of nonconformance** means taking responsibility for failures or not meeting quality expectations.

A 2002 study by RTI International reported that software bugs cost the U.S. economy $59.6 billion each year, about .6 percent of gross domestic product. Most of the costs are borne by software users, and the rest by developers and vendors. RTI International also suggested that more than one third of these costs could be eliminated by an improved testing infrastructure to enable earlier and more effective identification and removal of software defects.[20]

Other studies estimate the costs per hour of down time for systems. Table 8-5 summarizes the net bottom-line costs per hour of down time caused by software defects for different businesses.[21] Both American Airlines and United Airlines have estimated that reservation system down time costs them more than $20,000 per minute in lost income. Table 8-5 illustrates more examples of the cost of nonconformance.

Table 8-5: Costs per Hour of Down Time Caused by Software Defects

BUSINESS	COST PER HOUR OF DOWN TIME
Automated teller machines (medium-sized bank)	$14,500
Package shipping service	$28,250
Telephone ticket sales	$69,000
Catalog sales center	$90,000
Airline reservation center (small airline)	$89,500

The five major cost categories related to quality include:

1. **Prevention cost**: The cost of planning and executing a project so that it is error-free or within an acceptable error range. Preventive actions such as training, detailed studies related to quality, and quality surveys of suppliers and subcontractors fall under this category. Recall from the discussion of

[20] RTI International, "Software Bugs Cost U.S. Economy $59.6 Billion Annually, RTI Study Finds," July 1, 2002.
[21] Contingency Planning Research, White Plains, NY, 1995.

cost management (see Chapter 7) that detecting defects in information systems during the early phases of the systems development life cycle is much less expensive than during the later phases. One hundred dollars spent refining user requirements could save millions by finding a defect before implementing a large system. The Year 2000 (Y2K) issue provides a good example of these costs. If organizations had decided during the 1960s, 1970s, and 1980s that all dates would need four characters to represent the year instead of two characters, they would have saved billions of dollars.

2. **Appraisal cost**: The cost of evaluating processes and their outputs to ensure that a project is error-free or within an acceptable error range. Activities such as inspection and testing of products, maintenance of inspection and test equipment, and processing and reporting inspection data all contribute to appraisal costs of quality.

3. **Internal failure cost**: A cost incurred to correct an identified defect before the customer receives the product. Items such as scrap and rework, charges related to late payment of bills, inventory costs that are a direct result of defects, costs of engineering changes related to correcting a design error, premature failure of products, and correcting documentation all contribute to internal failure cost.

4. **External failure cost**: A cost that relates to all errors not detected and not corrected before delivery to the customer. Items such as warranty cost, field service personnel training cost, product liability suits, complaint handling, and future business losses are examples of external failure costs.

5. **Measurement and test equipment costs**: The capital cost of equipment used to perform prevention and appraisal activities.

Many industries tolerate a very low cost of nonconformance, but not the information technology industry. Tom DeMarco is famous for several studies he conducted on the cost of nonconformance in the information technology industry. DeMarco found that the average large company devoted more than 60 percent of its software development efforts to maintenance. Around 50 percent of development costs are typically spent on testing and debugging software.[22]

Top management is primarily responsible for the high cost of nonconformance in information technology. Top managers often rush their organizations to develop new systems and do not give project teams enough time or resources to do a project right the first time. To correct these quality problems, top management must create a culture that embraces quality.

[22] DeMarco, Tom, *Controlling Software Projects: Management, Measurement and Estimation*. New York: Yourdon Press, 1982.

Media Snapshot

What do Melissa, Anna Kournikova, Code Red, and Sobig have to do with quality and information technology? They are all names of recent computer viruses that have cost companies millions of dollars. A quality issue faced by computer users around the world is lost productivity due to computer viruses and spam—unsolicited e-mail sent to multiple mailing lists, individuals, or newsgroups. A 2004 study by Nucleus Research Inc. estimates that spam will cost large companies nearly $2,000 per employee in lost productivity in 2004 alone, despite investments in software to block spam. Spam currently accounts for more than *70 percent* of total e-mail volume worldwide.

Computer viruses are also getting more technically advanced and infectious as they flood systems across the globe. In just one month (August 2003), at least 50 new Internet viruses surfaced, and losses related to computer viruses cost North American companies about $3.5 billion. Businesses have suffered at least *$65 billion in lost productivity* because of computer viruses since 1997.[23]

Organizational Influences, Workplace Factors, and Quality

A study done by Tom DeMarco and Timothy Lister produced interesting results related to organizations and relative productivity. Starting in 1984, DeMarco and Lister conducted "Coding War Games" over several years, in which time more than 600 software developers from 92 organizations participated. The games were designed to examine programming quality and productivity over a wide range of organizations, technical environments, and programming languages. The study demonstrated that organizational issues had a much greater influence on productivity than the technical environment or programming languages.

For example, DeMarco and Lister found that productivity varied by a factor of about one to ten across all participants. That is, one team may have finished a coding project in one day while another team took ten days to finish the same project. In contrast, productivity varied by an average of only 21 percent between pairs of software developers from the same organization. If one team from a specific organization finished the coding project in one day, the longest time it took for another team from the same organization to finish the project was 1.21 days.

DeMarco and Lister also found *no correlation* between productivity and programming language, years of experience, or salary. Furthermore, the study showed that providing a dedicated workspace and a quiet work environment were key factors in improving productivity. The results of the study suggest that top managers must focus on workplace factors to improve productivity and quality.[24]

---

[23] McGuire, David, "Report: Spam Costs Are Rising at Work" (*http://www.washingtonpost.com/wp-dyn/articles/A21657-2004Jun7.html*) (June 7, 2004).

[24] DeMarco, Tom and Timothy Lister, *Peopleware: Productive Projects and Teams*. New York: Dorset House, 1987.

DeMarco and Lister wrote a book called *Peopleware* in 1987, and the second edition was published in 1999. The underlying thesis of their book is that major problems with work performance and project failures are not technological in nature; they are sociological in nature. They suggest minimizing office politics and giving smart people physical space, intellectual responsibility, and strategic direction—and then just *letting* them work. The manager's function is not to *make* people work but to make it possible for people to work by removing political roadblocks.

Expectations and Cultural Differences in Quality

Many experienced project managers know that a crucial aspect of project quality management is managing expectations. Although many aspects of quality can be clearly defined and measured, many cannot. Different project sponsors, customers, users, and other stakeholders have different expectations about various aspects of projects. It's very important to understand these expectations and manage any conflicts that might occur due to differences in expectations. For example, in the opening case, several users were upset when they could not access information within a few seconds. In the past, it may have been acceptable to have to wait two or three seconds for a system to load, but many of today's computer users expect systems to run much faster. Project managers and their teams must consider quality-related expectations as they define the project scope.

Expectations can also vary based on an organization's culture or geographic region. Anyone who has traveled to different parts of an organization, a country, or the world understands that expectations are not the same everywhere. For example, one department in a company might expect workers to be in their work areas most of the work day and to dress a certain way. Another department in the same company might focus on whether or not workers produce expected results, no matter where they work or how they dress. People working in smaller towns expect little traffic driving to work, while people working in large cities expect traffic to be a problem or rely on mass transit systems.

People working in other countries for the first time are often amazed at different quality expectations. Visitors to other countries may complain about things they take for granted, such as easily making cell phone calls, buying train tickets, and getting up-to-date maps. It's important to realize that different countries are at different stages of development in terms of quality, which requires patience when working with them.

Maturity Models

Another approach to improving quality in software development projects and project management in general is the use of **maturity models**, which are frameworks for helping organizations improve their processes and systems. Maturity models describe an evolutionary path of increasingly organized and systematically more mature processes. Many maturity models have five levels, with the first level describing characteristics of the least organized or mature organizations, and level five describing the characteristics of the most organized and mature organizations. Three popular maturity models include the Software Quality Function Deployment (SQFD) model, the Capability Maturity Model (CMM) and Capability Maturity Model Integration (CMMI), and the project management maturity model.

Software Quality Function Deployment Model

The **Software Quality Function Deployment (SQFD) model** is an adaptation of the quality function deployment model suggested in 1986 as an implementation vehicle for Total Quality Management (TQM). SQFD focuses on defining user requirements and planning software projects. The result of SQFD is a set of measurable technical product specifications and their priorities. Having clearer requirements can lead to fewer design changes, increased productivity, and, ultimately, software products that are more likely to satisfy stakeholder requirements. The idea of introducing quality early in the design stage was based on Taguchi's emphasis on robust design methods.[25]

Capability Maturity Model and Capability Maturity Model Integration

Another popular maturity model is in continuous development at the Software Engineering Institute at Carnegie Mellon University. The Software Engineering Institute (SEI) is a federally funded research and development center established in 1984 by the U.S. Department of Defense with a broad mandate to address the transition of software engineering technology. The **Capability Maturity Model (CMM)** is a five-level model laying out a generic path to process improvement for software development in organizations. Watts Humphrey's book *Managing the Software Process* (1989) describes an earlier version of the CMM.[26]

The five levels of the CMM are:

1. *Initial*: The software development processes for organizations at this maturity level are ad hoc and occasionally even chaotic. Few processes are defined, and success often depends on individual effort.

---

[25] Yilmaz, R. R. and Sangit Chatterjee, "Deming and the Quality of Software Development," *Business Horizons*, Foundation for the School of Business at Indiana University, 40(6) (November-December 1997) 51(8).

[26] Humphrey, W. S, *Managing the Software Process*. Reading, MA: Addison-Wesley, 1989.

2. *Repeatable*: Organizations at this maturity level have established basic project management processes to track cost, schedule, and functionality for software projects. Process discipline is in place to repeat earlier successes on similar projects.

3. *Defined*: At this level, the software processes for both management and software engineering activities are documented, standardized, and integrated into a standard software process for the organization. All projects use an approved, tailored version of the standard process in the organization.

4. *Managed*: At this maturity level, organizations collect detailed measures of the software process and product quality. Both the software processes and products are quantitatively understood and controlled.

5. *Optimizing*: Operating at the highest level of the maturity model, organizations can enable continuous process improvement by using quantitative feedback from the processes and from piloting innovative ideas and technologies.[27]

The **Capability Maturity Model Integration (CMMI)** is replacing the older CMM ratings. CMMI addresses software engineering, system engineering, and program management. Several organizations no longer use CMM and focus on CMMI because it includes a broader set of measures, including project management skills that are crucial to system integrators. CMMI also has five levels, with Level 5 being the highest. Companies can earn CMMI ratings that apply to the whole organization, individual divisions, or specific engagements. Many companies that want to work in the government market have realized that they will not get many opportunities even to bid on projects unless they have a CMMI Level 3. According to one manager, "CMMI is really the future. People who aren't on the bandwagon now are going to find themselves falling behind."[28]

Project Management Maturity Models

In the late 1990s, several organizations began developing project management maturity models based on the Capability Maturity Model. Just as organizations realized the need to improve their software development processes and systems, they also realized the need to enhance their project management processes and systems for all types of projects.

The PMI Standards Development Program published the Organizational Project Management Maturity Model (OPM3) in December 2003. More than

---

[27] Paulk, Mark C., Bill Curtis, Mary Beth Chrissis, and Charles V. Weber, *Capability Maturity Model for Software*, Version 1.1, Technical Report, CMU/SEI-93-TR-024, ESC-TR-93-177 (February 1993).

[28] Hardy, Michael, "Strength in numbers: New maturity ratings scheme wins support among systems integrators" (*http://www.fcw.com/fcw/articles/2004/0329/feat-cmmi-03-29-04.asp*) (March 29, 2004).

200 volunteers from around the world were part of the OPM3 team. The model is based on market research surveys that had been sent to more than 30,000 project management professionals and incorporates 180 best practices and more than 2,400 capabilities, outcomes, and key performance indicators.[29] According to John Schlichter, the OPM3 program director, "The standard would help organizations to assess and improve their project management capabilities as well as the capabilities necessary to achieve organizational strategies through projects. The standard would be a project management maturity model, setting the standard for excellence in project, program, and portfolio management best practices, and explaining the capabilities necessary to achieve those best practices."[30] The OPM3 Web site (see the sixth Suggested Reading for this chapter) provides information to assist organizations in implementing OPM3.

Several other companies provide similar project management maturity models. The International Institute for Learning, Inc. calls the five levels in its model common language, common processes, singular methodology, benchmarking, and continuous improvement. ESI International Inc.'s model's five levels are called ad hoc, consistent, integrated, comprehensive, and optimizing. Regardless of the names of each level, the goal is clear: organizations want to improve their ability to manage projects.

William Ibbs and Young H. Kwak have been researching project management maturity levels in a series of studies, partially funded by the PMI Educational Foundation. They developed an assessment methodology to measure project management maturity, consisting of 148 multiple-choice questions based on the project management knowledge areas and process groups. Developing this systematic assessment methodology was a major contribution to the study of project management maturity. Organizations can benchmark their project management maturity using factual, impartial techniques such as those contained in this questionnaire, giving them a good reference point from which to begin making process improvements. Preliminary findings of Ibbs's and Kwak's research show that certain industries, such as engineering and construction, have a higher maturity level than other industries, such as information systems/software development.[31]

Many organizations are assessing where they stand in terms of project management maturity, just as they did for software development maturity with the SQFD and CMMI maturity models. Organizations are recognizing that they must make a commitment to the discipline of project management to improve project quality.

[29] Schlichter, John, The Project Management Institute's Organizational Project Management Maturity Model: An Update on the PMI's OPM3 Program (September 2002) (*www.pmforum.org/library/opm3update1.htm*).

[30] Schlichter, John, "The History of OPM3," *Project Management World Today* (June 2003).

[31] Ibbs, C. William and Young Hoon Kwak, "Assessing Project Management Maturity," *Project Management Journal* (March 2000).

USING SOFTWARE TO ASSIST IN PROJECT QUALITY MANAGEMENT

This chapter provides examples of several tools and techniques used in project quality management. Software can be used to assist with several of these tools and techniques. For example, you can create Pareto diagrams using spreadsheet and charting software. You can use statistical software packages to help you determine standard deviations and perform many types of statistical analyses. You can create Gantt charts using project management software to help you plan and track work related to project quality management. There are also several specialized software products to assist people with managing Six Sigma projects, creating quality control charts, and assessing maturity levels. Project teams need to decide what types of software will help them manage their particular projects.

As you can see, quality itself is a very broad topic, and it is only one of the nine project management knowledge areas. Project managers must focus on defining how quality relates to their specific projects and ensure that those projects will satisfy the needs for which they were undertaken.

CASE WRAP-UP

Scott Daniels assembled a team to identify and resolve quality-related issues with the EIS and to develop a plan to help the medical instruments company prevent future quality problems. The first thing Scott's team did was to research the problems with the EIS. They created a fishbone diagram similar to the one in Figure 8-6. They also created a Pareto diagram (Figure 8-1) to help analyze the many complaints the Help Desk received and documented about the EIS. After further investigation, Scott and his team found out that many of the managers using the system were very inexperienced in using computer systems beyond basic office automation systems. They also found out that most users received no training at all on how to properly access or use the new EIS. There did not appear to be any major problems with the hardware for the EIS or the users' individual computers. The complaints about reports not giving consistent information all came from one manager, who had actually misread the reports, so there were no problems with the way the software was designed. Scott was very impressed with the quality of the entire project, except for the training. Scott reported his team's findings to the project sponsor of the EIS, who was relieved to find out that the quality problems were not as serious as many people feared they were.

CHAPTER SUMMARY

Many news headlines regarding the poor quality of information technology projects demonstrate that quality is a serious issue. Several mission-critical information technology systems have caused deaths, and quality problems in many business systems have resulted in major financial losses.

Customers are ultimately responsible for defining quality. Important quality concepts include satisfying stated or implied stakeholder needs, conforming to requirements, and delivering items that are fit for use.

Project quality management includes quality planning, quality assurance, and quality control. Quality planning identifies which quality standards are relevant to the project and how to satisfy them. Quality assurance involves evaluating overall project performance to ensure that the project will satisfy the relevant quality standards. Quality control includes monitoring specific project results to ensure that they comply with quality standards and identifying ways to improve overall quality.

There are many tools and techniques related to project quality management. Fishbone diagrams help find the root cause of problems. Pareto charts help identify the vital few contributors that account for the most quality problems. Statistical sampling helps define a realistic number of items to include in analyzing a population. Six Sigma helps companies improve quality by reducing defects. Standard deviation measures the variation in data. Control charts display data to help keep processes in control. Testing is very important in developing and delivering high-quality information technology products.

Many people contributed to the development of modern quality management. Deming, Juran, Crosby, Ishikawa, Taguchi, and Feigenbaum all made significant contributions to the field. Many organizations today use their ideas, which also influenced Six Sigma principles. The Malcolm Baldrige National Quality Award, ISO 9000, and ISO 15504 have also helped organizations emphasize the importance of improving quality.

There is much room for improvement in information technology project quality. Strong leadership helps emphasize the importance of quality. Understanding the cost of quality provides an incentive for its improvement. Providing a good workplace can improve quality and productivity. Understanding stakeholders' expectations and cultural differences are also related to project quality management. Developing and following maturity models can help organizations systematically improve their project management processes to increase the quality and success rate of projects.

There are several types of software available to assist in project quality management. It is important for project teams to decide which software will be most helpful for their particular projects.

DISCUSSION QUESTIONS

1. Discuss some of the examples of poor quality in information technology projects presented in the "What Went Wrong?" section. Could most of these problems have been avoided? Why do you think there are so many examples of poor quality in information technology projects?

2. What are the main processes included in project quality management?

3. How do functionality, system outputs, performance, reliability, and maintainability requirements affect quality planning?

4. What type of information goes in a quality assurance plan?

5. What are the three main categories of outputs for quality control?

6. Provide examples of when you would use a Pareto diagram, statistical sampling, quality control charts, and testing on an information technology project.

7. Discuss the history of modern quality management. How have experts such as Deming, Juran, Crosby, and Taguchi affected the quality movement and today's use of Six Sigma?

8. Discuss three suggestions for improving information technology project quality that were not made in this chapter.

9. Describe three types of software that can assist in project quality management.

EXERCISES

1. Assume your organization wants to hire new instructors for your project management course. Develop a list of quality standards that you could use in making this hiring decision.

2. Create a Pareto diagram based on the information in the table below. First, create a spreadsheet in Excel, using the data in the table below. List the most frequent customer problems first. Add a column called "% of Total" and another one called "Cumulative %." Then enter formulas to calculate those items. Next, use the Excel Chart Wizard to create a Pareto diagram based on this data. Use the Line – Column on 2 Axis custom type chart so your resulting chart looks similar to the one in Figure 8-1.

Customer Complaints	Frequency/Week
Customer is on hold too long	90
Customer gets transferred to wrong area or cut off	20
Service rep cannot answer customer's questions	120
Service rep does not follow through as promised	40

3. To illustrate a normal distribution, shake and roll a pair of dice 30 times and graph the results. It is more likely for someone to roll a six, seven, or eight than a two or twelve, so these numbers should come up more often. To create the graph, use graph paper or draw a grid. Label the x-axis with the numbers 2 through 12. Label the y-axis with the numbers 1 through 10. Fill in the appropriate grid for each roll of the dice. Do your results resemble a normal distribution? Why or why not?

4. Research the criteria for the Malcolm Baldrige National Quality Award or a similar quality award provided by another organization. Investigate a company that has received this award. What steps did the company take to earn this quality award? What are the benefits of earning a quality award?

5. Review the information in this chapter about Six Sigma principles and Six Sigma organizations. Brainstorm ideas for a potential Six Sigma project that could improve quality on your campus, at your workplace, or in your community. Write a two-page paper describing one project idea and explain why it would be a Six Sigma project. Review and discuss how you could use the DMAIC process on this project.

6. Review the concepts in this chapter related to improving the quality of software. Write a one-page paper describing how you could apply these concepts to software development projects.

Running Case

The Recreation and Wellness Intranet Project team is working hard to ensure that the new system they develop meets expectations. The team has a detailed scope statement, but the project manager, Tony Prince, wants to make sure they're not forgetting any requirements that might affect how different people view the quality of the project. He knows that the project's sponsor and other senior managers are most concerned with getting people to use the system, improve their health, and reduce healthcare costs. System users will want the system to be very user-friendly, informative, fun to use, and fast.

Tasks

1. Develop a list of quality standards or requirements related to meeting the stakeholder expectations described above. Also provide a brief description of each requirement. For example, a requirement might be that 90 percent of employees have logged into the system within two weeks after the system rolls out.

2. Based on the list created for Task 1, determine how you will measure progress on meeting the requirements. For example, you might have employees log into the system as part of the training program and track who attends the training. You could also build a feature into the system to track usage by user name, department, and so on.

3. After analyzing survey information, you decide to create a Pareto diagram to easily see which types of recreational programs and company-sponsored classes most people were interested in. First, create a spreadsheet in Excel, using the data in the table below. List the most frequently requested programs or classes first. Add a column called "% of Total" and another one called "Cumulative %." Then enter formulas to calculate those items. Next, use the Excel Chart Wizard to create a Pareto diagram based on this data. Use the Line – Column on 2 Axis custom type chart so your resulting chart looks similar to the one in Figure 8-1.

REQUESTED PROGRAMS/CLASSES	# OF TIMES REQUESTED
Walking program	7,115
Volleyball program	2,054
Weight reduction class	8,875
Stop smoking class	4,889
Stress reduction class	1,894
Soccer program	3,297
Table tennis program	120
Softball program	976

ADDITIONAL RUNNING CASES AND OTHER APPENDICES

Appendix C provides additional case studies and questions you can use to practice applying the concepts, tools, and techniques you are learning throughout this and subsequent chapters. Review the running cases provided in Appendix C and on the companion Web site (*www.course.com/mis/shwalbe4e*). Appendix D includes templates for various project management documents. For additional sample documents based on real projects, visit the author's Web site at *www.kathyschwalbe.com*. Appendix E and the CD-ROM included with this text include a computer simulation where you can also practice applying the project management process groups and knowledge areas.

SUGGESTED READINGS

1. Chrissis, Mary Beth, Konrad, Mike, and Shrum, Sally. *CMMI : Guidelines for Process Integration and Product Improvement.* Addison-Wesley, Boston (2003).

   Capability maturity model integration (CMMI) is a framework of best practices that address the development and maintenance of products and services. It covers the product life cycle from conception through delivery and maintenance. This book discusses how to develop a CMMI model for a specific organization and choose from the 25 different process areas.

2. Deming, W. Edwards. *Out of the Crisis.* Cambridge, MA: Massachusetts Institute of Technology (1988).

   Despite Deming's final book, The New Economics for Industry, Government, Education *(1995),* Out of the Crisis *is Deming's best-known classic and is the source of constant reference. Deming uses real-life case studies to share with the reader the "simplicity" of his profound knowledge, his 14 points, and the Deadly Diseases of quality. Deming has authored and coauthored several other books on quality.*

3. Futrell, Robert, Shafer, Donald F., and Shafer, Linda I. Quality Software Project Management, Prentice Hall (2002).

   This book describes the product development techniques, project management skills, and people management skills necessary to manage a software project. Topics include estimating duration and cost, defining the software requirement specification, and selecting the project team.

4. Juran, Joseph. *Juran's Quality Handbook*, 5th ed. New York: McGraw Hill (1999).

   Juran published the fifth edition of this famous handbook at the age of 94. This major revision of the classic reference on quality provides a comprehensive body of knowledge needed to engineer and manage for quality into the twenty-first century. There is also an excellent Web site with information about quality and Juran at www.juran.com.

5. Pande, Peter S., Neuman, Robert P., and Cavanagh, Roland R. *The Six Sigma Way.* New York: McGraw-Hill (2000).

   Many books and articles have been written in the past few years about Six Sigma. The Six Sigma Way, *published in 2000, continues to be a top-seller. This comprehensive book is a self-help guide to adapting and using Six Sigma under various conditions. Part one of the book provides an executive summary of Six Sigma. Part two describes how to gear up and adapt Six Sigma to various organizations. Part three provides detailed information on implementing Six Sigma, including a detailed roadmap and description of tools. The appendix includes several checklists and a detailed Six Sigma conversion table.*

6. Project Management Institute, Organizational Project Management Maturity Model (OPM3) (*www.pmi.org/opm3*) (2004).

> *This Web site provides information about PMI's Organizational Project Management Maturity Model (OPM3) program, created to assist organizations in implementing organization strategy through the successful, consistent, and predictable delivery of projects. The OPM3 includes a method for assessing organizations' project management maturity levels as well as a step-by-step method for increasing and maintaining an organization's ability to deliver projects as promised.*

7. Rout, Terrance P. "ISO/IEC 15504 - Evolution to an International Standard." Software Process: Improvement and Practice 8(1): 27-40 (2003).

> *Terrance Rout has written many publications on improving software quality. In this article, he describes the work currently being undertaken to help ISO/IEC TR 15504 progress to the status of a full international standard.*

KEY TERMS

- **acceptance decisions** — decisions that determine if the products or services produced as part of the project will be accepted or rejected

- **appraisal cost** — the cost of evaluating processes and their outputs to ensure that a project is error-free or within an acceptable error range

- **benchmarking** — a technique used to generate ideas for quality improvements by comparing specific project practices or product characteristics to those of other projects or products within or outside the performing organization

- **Capability Maturity Model (CMM)** — a five-level model laying out a generic path to process improvement for software development in organizations

- **Capability Maturity Model Integration (CMMI)** — a process improvement model that addresses software engineering, system engineering, and program management and is replacing the older CMM ratings

- **conformance** — delivering products that meet requirements and fitness for use

- **conformance to requirements** — the project processes and products meet written specifications

- **control chart** — a graphic display of data that illustrates the results of a process over time

- **cost of nonconformance** — taking responsibility for failures or not meeting quality expectations

- **cost of quality** — the cost of conformance plus the cost of nonconformance

- **defect** — any instance where the product or service fails to meet customer requirements

- **Define, Measure, Analyze, Improve, Control (DMAIC)** — a systematic, closed-loop process for continued improvement that is scientific and fact based
- **design of experiments** — a quality technique that helps identify which variables have the most influence on the overall outcome of a process
- **external failure cost** — a cost related to all errors not detected and corrected before delivery to the customer
- **features** — the special characteristics that appeal to users
- **fishbone diagrams** — diagrams that trace complaints about quality problems back to the responsible production operations or root cause; sometimes called **Ishikawa diagrams**
- **fitness for use** — a product can be used as it was intended
- **functionality** — the degree to which a system performs its intended function
- **integration testing** — testing that occurs between unit and system testing to test functionally grouped components to ensure a subset(s) of the entire system works together
- **internal failure cost** — a cost incurred to correct an identified defect before the customer receives the product
- **ISO 9000** — a quality system standard developed by the International Organization for Standardization (ISO) that includes a three-part, continuous cycle of planning, controlling, and documenting quality in an organization
- **ISO 15504** — a framework for the assessment of software processes developed by the International Organization for Standardization (ISO)
- **maintainability** — the ease of performing maintenance on a product
- **Malcolm Baldrige National Quality Award** — an award started in 1987 to recognize companies that have achieved a level of world-class competition through quality management
- **maturity model** — a framework for helping organizations improve their processes and systems
- **mean** — the average value of a population
- **measurement and test equipment costs** — the capital cost of equipment used to perform prevention and appraisal activities
- **normal distribution** — a bell-shaped curve that is symmetrical about the mean of the population
- **Pareto analysis** — identifying the vital few contributors that account for most quality problems in a system
- **Pareto diagrams** — histograms that help identify and prioritize problem areas
- **performance** — how well a product or service performs the customer's intended use
- **prevention cost** — the cost of planning and executing a project so that it is error-free or within an acceptable error range

- **process adjustments** — adjustments made to correct or prevent further quality problems based on quality control measurements
- **project quality management** — ensuring that a project will satisfy the needs for which it was undertaken
- **quality** — the totality of characteristics of an entity that bear on its ability to satisfy stated or implied needs or the degree to which a set of inherent characteristics fulfill requirements
- **quality assurance** — periodically evaluating overall project performance to ensure that the project will satisfy the relevant quality standards
- **quality audit** — structured review of specific quality management activities that helps identify lessons learned and can improve performance on current or future projects
- **quality circles** — groups of nonsupervisors and work leaders in a single company department who volunteer to conduct group studies on how to improve the effectiveness of work in their department
- **quality control** — monitoring specific project results to ensure that they comply with the relevant quality standards and identifying ways to improve overall quality
- **quality planning** — identifying which quality standards are relevant to the project and how to satisfy them
- **reliability** — the ability of a product or service to perform as expected under normal conditions
- **rework** — action taken to bring rejected items into compliance with product requirements or specifications or other stakeholder expectations
- **Robust Design methods** — methods that focus on eliminating defects by substituting scientific inquiry for trial-and-error methods
- **seven run rule** — if seven data points in a row on a quality control chart are all below the mean, above the mean, or are all increasing or decreasing, then the process needs to be examined for nonrandom problems
- **six 9s of quality** — a measure of quality control equal to 1 fault in 1 million opportunities
- **Six Sigma** — a comprehensive and flexible system for achieving, sustaining, and maximizing business success that is uniquely driven by close understanding of customer needs, disciplined use of facts, data, statistical analysis, and diligent attention to managing, improving, and reinventing business processes
- **software defect** — anything that must be changed before delivery of the program
- **Software Quality Function Deployment (SQFD) model** — a maturity model that focuses on defining user requirements and planning software projects
- **standard deviation** — a measure of how much variation exists in a distribution of data

- **statistical sampling** — choosing part of a population of interest for inspection
- **system outputs** — the screens and reports the system generates
- **system testing** — testing the entire system as one entity to ensure that it is working properly
- **unit test** — a test of each individual component (often a program) to ensure that it is as defect-free as possible
- **user acceptance testing** — an independent test performed by end users prior to accepting the delivered system
- **yield** — the number of units handled correctly through the development process

9

Project Human Resource Management

Objectives

After reading this chapter, you will be able to:

1. *Explain the importance of good human resource management on projects, including the current state and future implications of human resource management, especially on information technology projects*
2. *Define project human resource management and understand its processes*
3. *Summarize key concepts for managing people by understanding the theories of Abraham Maslow, Frederick Herzberg, David McClelland, and Douglas McGregor on motivation, H. J. Thamhain and D. L. Wilemon on influencing workers, and Stephen Covey on how people and teams can become more effective*
4. *Discuss human resource planning and be able to create a project organizational chart, responsibility assignment matrix, and resource histogram*
5. *Understand important issues involved in project staff acquisition and explain the concepts of resource assignments, resource loading, and resource leveling*
6. *Assist in team development with training, team-building activities, and reward systems*
7. *Explain and apply several tools and techniques to help manage a project team and summarize general advice on managing teams*
8. *Describe how project management software can assist in project human resource management*

OPENING CASE

*T*his was the third time someone from the Information Technology department tried to work with Ben, the head of the F-44 aircraft program. Ben, who had been with the company for almost 30 years, was known for being rough around the edges and very demanding.

The company was losing money on the F-44 upgrade project because the upgrade kits were not being delivered on time. The Canadian government had written severe late-penalty fees into its contract, and other customers were threatening to take their business elsewhere. Ben blamed it all on the Information Technology department for not letting his staff access the F-44 upgrade project's information system directly so they could work with their customers and suppliers more effectively. The information system was based on very old technology that only a couple of people in the company knew how to use. It often took days or even weeks for Ben's group to get the information they needed.

Ed Davidson, a senior programmer, attended a meeting with Sarah Ellis, an internal information technology business consultant. Sarah was in her early thirties and had advanced quickly in her company, primarily due to her keen ability to work well with all types of people. Sarah's job was to uncover the real problems with the F-44 aircraft program's information technology support, and then develop a solution with Ben and his team. If she found it necessary to invest in more information technology hardware, software, or staff, Sarah would write a business case to justify these investments and then work with Ed, Ben, and his group to implement the suggested solution as quickly as possible. Ben and three of his staff entered the conference room. Ben threw his books on the table and started yelling at Ed and Sarah. Ed could not believe his eyes or ears when Sarah stood nose to nose with Ben and started yelling right back at him.

THE IMPORTANCE OF HUMAN RESOURCE MANAGEMENT

Many corporate executives have said, "People are our most important asset." People determine the success and failure of organizations and projects. Most project managers agree that managing human resources effectively is one of the toughest challenges they face. Project human resource management is a vital component of project management, especially in the information technology field—in which qualified people are often hard to find and keep. It is important to understand the current state of human resource management in the information technology industry and its implications for the future.

Current State of Human Resource Management

In the 1990s, there was a growing shortage of personnel in information technology. The Information Technology Association of America (ITAA) calculated that in 1997 there were more than 345,000 job openings for programmers, systems analysts, computer scientists, and computer engineers in information technology and non–information technology companies with 100 or more employees. These estimated job openings represented about 10 percent of current employment in these four occupations.[1] In April 2000, ITAA published its "Bridging the Gap" workforce study, which clearly documented the continued staffing crisis facing the information technology industry. This study revealed that one in 14 American workers was involved in information technology, and that 844,000 information technology jobs were unfilled in 2000, with even more vacancies anticipated in future years.

Unfortunately, ITAA and other groups projecting employment in the first decade of the new millennium were wrong about these estimated vacancies. No one could foresee how changes in the global economy, the effects of terrorism, or the declining stock market would affect job openings. A more recent ITAA report showed that the U.S. information technology workforce was at least stabilizing and showed a slight recovery in 2004. Below are some findings from a September 2004 press release:

- The total number of people working in information technology in the United States in early 2004 was more than 10.5 million, compared to 10.3 million in early 2003 and 9.9 million in early 2002.

- Hiring managers planned to fill approximately 270,000 fewer jobs in 2004 than they did during 2003.

- Approximately 89 percent of new jobs came from non–information technology companies. The banking, finance, manufacturing, food service, and transportation industries employ about 79 percent of the information technology workforce.

- Employment increased 5 percent for technical support and network system designers.

- Employers like to hire people with a solid record of accomplishment. It helps to have previous experience in a related field and a four-year college degree. Hiring managers say interpersonal skills are the most important soft skill for information technology workers.[2]

In 2000, the World Information Technology and Services Alliance (WITSA) released its study, "Digital Planet 2000: The Global Information Economy," which

---

[1] Information Technology Association of America (ITAA), "Help Wanted: The IT Workforce Gap at the Dawn of a New Century," (1997) *www.itaa.org*.

[2] Information Technology Association of America (ITAA), "Recovery Slight for IT Job Market in 2004," (September 8, 2004) *www.itaa.org*.

revealed that the global high-tech industry generated more than $2.1 trillion in 1999, and was expected to surpass $3 trillion in 2003.[3] The 2002 study reported that the global high-tech industry did grow, but only from $2.3 trillion in 2000 to $2.4 trillion in 2001. The "Digital Planet 2002" study reports that the Internet and e-commerce remains a notable bright spot in the global economy, despite flattened spending in most of the developed world. Global e-commerce (business-to-business and business-to-consumer) reached $633 billion in 2001, an increase of 79 percent from the previous year. "Even an economic downturn cannot stifle the value and convenience of electronic commerce to businesses and consumers," said WITSA president Harris N. Miller at a press conference on February 28, 2002. The United States continues to lead in overall information and communication technology (ICT) spending, but its overall percentage of global spending continues to decline. "Another positive story Digital Plan 2002 tells is that as the benefits and efficiency of ICT spread across the globe, mature markets such as the U.S. will become less dominant. China, Poland, and other developing countries are already playing an increasing role in the global ICT marketplace, showing the upside opportunities for global ICT spending are still enormous," Miller said.[4]

These studies highlight the continued need for skilled information technology workers and project managers to lead their projects. As the job market changes, however, people should upgrade their skills to remain marketable and flexible. Many highly qualified information technology workers lost their jobs in the past few years. Workers need to develop a support system and financial reserves to make it through difficult economic times. They also need to consider different types of jobs and industries where they can use their skills and earn a living. Negotiation and presentation skills also become crucial in finding and keeping a good job. Workers need to know how they personally contribute to an organization's bottom line.

However, even as organizations demand increased profits, and available jobs become harder to find, it is still crucial that organizations do not take advantage of their workers' time. Jeannette Cabanis-Brewin, editor-in-chief of the Center for Business Practices publications and former editor for *PM Network* and the *Project Management Journal*, specializes in the organizational and human side of project management. Cabanis-Brewin, also a proud grandmother, wrote an article complaining that her grandchildren did not visit for Christmas in 2000 because their parents could not take a few days off work.[5] Cabanis-Brewin's editorial stressed that too many people in top management have not learned their lessons from watching Scrooge in the movie *A Christmas Carol*.

---

[3] Information Technology Association of America (ITAA), "Global IT Spending to Rocket from Current $2 Trillion to $3 Trillion, New Study Finds," *Update for IT Executives* (2001) p. 6 (15) *www.itaa.org*.

[4] World Information Technology and Services Alliance (WITSA), "WITSA Global Research Shows World's Largest Consumer of Technology Increased Spending Less Than 1% Last Year, Internet Spending Up, Despite Worldwide Economic Condition" (February 28, 2002), *www.witsa.org*.

[5] Cabanis-Brewin, Jeannette, "An Open Letter to Scrooge, Inc.," *Best Practices e-Advisor* (January 7, 2001) p.16.

Excessive working hours are not the only problem in the information technology industry. In addition, the field of information technology has become unappealing to certain groups, such as women. The number of women entering the information technology field peaked in 1984. Today, that number continues to decline. More than half of all students attending colleges today are women, yet less than 20 percent of the graduates awarded bachelor's degrees in computer science and related fields are female. Anita Borg, a computer engineer at Xerox Palo Alto Research Center and founder of the Institute for Women and Technology, states that if women had been going into computer fields at the same rate as men since 1984, the shortage of information technology workers in the late 1990s would not have existed.[6]

White House findings in 2000 showed that employment in the U.S. information technology industry grew 81 percent. Women, who represent 47 percent of the work force at large, made up only 29 percent of several information technology categories. Despite the demand for computing professionals outstripping the supply, more men than women are earning undergraduate, Master's, and doctoral degrees in computer science. The Association for Computing Machinery Committee on Women in Computing (ACM-W), which surveyed the shrinking pipeline of women in computer science in 1998, released a study in 2000 that found the pipeline was continuing to shrink. Females in high school, college, graduate school, and beyond are not choosing computer science or related degrees. Why not? Denise Gurer, 3Com Corporation technologist and ACM-W co-chair says, "Girls and women are not turned off by technology, but by how it is used in our society.... We can attract more girls and women to technology areas if we make classrooms more inclusive of technology uses that interest them."[7] Girls and women are interested in using technology to save time, to be more productive and creative, and to develop relationships. Girls are often turned off by technology when they see computers used to play unproductive games, which are often violent or sexist, to track sports statistics, to download unappealing music, or to avoid having real relationships with other people.

The ACM-W studies found that factors contributing to the gender gap in information technology include the following:

- The stereotype that girls are not good at math
- Computer games that target boys
- Online discussion groups that are largely male-oriented and sexist
- The fact that few women hold positions of power in the computer industry[8]

---

[6] Olsen, Florence, "Institute for Women and Technology Works to Bridge Computing's Gender Gap," *The Chronicle of Higher Education* (February 25, 2000).

[7] DeBlasi, Patrick J., "The Shrinking Pipeline Unlikely to Reverse: Nearly 30 Percent Decline of Women Choosing Undergraduate Degrees in Computer Science," *Association of Computing Machinery (ACM) Pressroom* (June 5, 2000).

[8] Ohlson, Kathleen, "ACM launches gender gap study," *Online News* (November 13, 1998).

Implications for the Future of Human Resource Management

It is crucial for organizations to practice what they preach about human resources. If people truly are their greatest asset, organizations must work to fulfill their human resource needs *and* the needs of individual people in their organizations, regardless of the job market. If organizations want to be successful at implementing information technology projects, they need to understand the importance of project human resource management and take actions to make effective use of people. Proactive organizations are addressing current and future human resource needs, by, for example, improving benefits, redefining work hours and incentives, and finding future workers. Many organizations have changed their benefits policies to meet worker needs. Most workers assume that their companies provide some perks, such as casual dress codes, flexible work hours, and tuition assistance. Other companies provide on-site day care, dry cleaning services, and staff to run errands such as walking your dog.

The January 20, 2003 issue of *Fortune* magazine included its annual list of the best companies to work for, stressing the importance of treating workers well. SAS, a software development company headquartered in Cary, North Carolina, continues to make the list. It has a very low turnover rate because it treats employees very well, with on-site child care for only $300 per month, a pianist in the cafeteria, a gym, and well-groomed walking paths, plus challenging work and good pay. Some organizations go even further and use special perks as a competitive advantage. For example, an organization may promote policies such as allowing workers to bring their pets to work, providing free vacation rentals, paying for children's college tuition, or providing workers with nice cars. Jeff Finney, a computer programmer in suburban Atlanta, accepted delivery of a BMW from his employer, Revenue Systems, Inc. All 45 employees at this firm get to lease BMWs at the bosses' expense. Finney, who normally drives a pickup, says, "It is probably one of the best perks I've seen any company give."[9] Although some of these perks disappear in hard economic times, successful organizations continue to reward their workers with bonuses or profit sharing when they can prove their direct contributions to the bottom line.

Other implications for the future of human resource management relate to the hours organizations expect many information technology professionals to work and how they reward performance. Today people brag about the fact that they can work *less* than 40 hours a week or work from home several days a week—not that they work a lot of overtime and haven't had a vacation in years. If companies plan their projects well, they can avoid the need for overtime, or they can make it clear that overtime is optional. Many companies also

---

[9] Irvine, Martha, "Executive-style perks spread to rank and file," *Minneapolis Star Tribune* (September 29, 1998) D-2.

outsource more and more of their project work, as described in Chapter 12, Project Procurement Management, to manage the fluctuating demand for workers.

Companies can also provide incentives that use performance, not hours worked, as the basis of rewards. If performance can be objectively measured, as it can in many aspects of information technology jobs, then it should not matter where employees do their work or how long it takes them to do it. For example, if a technical writer can produce a high-quality publication at home in one week, the company and the writer are better off than if the company insisted that the writer come to the office and take two weeks to produce the publication. Objective measures of work performance and incentives based on meeting those criteria are important considerations.

🎥 Media Snapshot

People often brag about the productivity of American workers. A September 2004 article in the *Minneapolis Star Tribune* by Joan Williams and Ariane Hegewisch, however, revealed some interesting facts:

> …Here's the dirty little secret: U.S. productivity is No. 1 in the world when productivity is measured as gross domestic product per worker, but our lead vanishes when productivity is measured as GDP per hour worked, according to the Organization for Economic Cooperation and Development, whose members are the world's 30 most developed nations. Productivity per hour is higher in France, with the United States at about the same level as other advanced European economies. As it turns out, the U.S. "productivity advantage" is just another way of saying that we work more hours than workers in any other industrialized country except South Korea. Is that something to brag about? Europeans take an average of six to seven weeks of paid annual leave, compared with just 12 days in the United States. Twice as many American as European workers put in more than 48 hours per week. Particularly sobering is the fact that in two out of three American families with small children in which both parents work, the couples work more than 80 total hours per week, also more than double the European rate.[10]

Economists say that Americans prefer the higher income gained from working extra hours, while Europeans prefer more family time and leisure. However, sociologists have shown that many Americans, especially men, would like to have more family or leisure time. Recent surveys show that many Americans are willing to sacrifice up to a quarter of their salaries in return for more time off!

So why don't Americans work less? Some do need the money, but many others cannot find jobs where they can work fewer hours. Even more Americans fear losing their jobs, healthcare benefits, or promotion opportunities if they work fewer hours or even request the opportunity to do so. As a result, many families who would prefer both parents work less than 40 hours a week end up having one parent working 50 or more hours a week while the

---

[10] Williams, Joan and Ariane Hegewisch, "Confusing productivity with long workweek," *Minneapolis Star Tribune* (September 6, 2004) (*www.startribune.com*).

other parent is unemployed altogether. Williams and Hegewisch describe this situation as "a squandering of human capital that is typically overlooked in discussions of productivity and GDP. A new and more promising way to fuel economic growth would be to offer good jobs with working hours that enable fathers as well as mothers to maintain an active involvement with family life as well as an active career."[11]

The need to develop future talent in information technology also has important implications for everyone. Who will maintain the systems we have today when the last of the "baby boomers" generation retires? Who will continue to develop new products and services using new technologies that have not yet been developed? Some schools are requiring all students to take computer literacy courses. Some individuals and organizations are offering incentives to minorities and women to enter technical fields. For example, some wealthy alumni offer to pay all college expenses for minority students from the alumni's high school or grade school. Several colleges and government agencies have programs to help recruit more women and minorities into technical fields. Some companies provide options for working parents to help them balance work and family. All of these efforts will help to develop the human resources needed for future information technology projects.

WHAT IS PROJECT HUMAN RESOURCE MANAGEMENT?

Project human resource management includes the processes required to make the most effective use of the people involved with a project. Human resource management includes all project stakeholders: sponsors, customers, project team members, support staff, suppliers supporting the project, and so on. Human resource management includes the following four processes:

- **Human resource planning** involves identifying and documenting project roles, responsibilities, and reporting relationships. Key outputs of this process include roles and responsibilities, an organizational chart for the project, and a staffing management plan.
- **Acquiring the project team** involves getting the needed personnel assigned to and working on the project. Key outputs of this process are project staff assignments, resource availability information, and updates to the staffing management plan.
- **Developing the project team** involves building individual and group skills to enhance project performance. Team-building skills are often a challenge for many project managers. The main output of this process is assessing team performance.

[11] Ibid.

- **Managing the project team** involves tracking team member performance, motivating team members, providing timely feedback, resolving issues and conflicts, and coordinating changes to help enhance project performance. Outputs of this process include requested changes, recommended corrective and preventive actions, updates to organizational process assets, and updates to the project management plan.

Some topics related to human resource management (understanding organizations, stakeholders, and different organizational structures) were introduced in Chapter 2, The Project Management and Information Technology Context. This chapter will expand on some of those topics and introduce other important concepts in project human resource management, including theories about managing people, resource loading, and resource leveling. You will also learn how to use software to assist in project human resource management.

KEYS TO MANAGING PEOPLE

Industrial-organizational psychologists and management theorists have devoted much research and thought to the field of managing people at work. Psychosocial issues that affect how people work and how well they work include motivation, influence and power, and effectiveness. This section will review Abraham Maslow's, Frederick Herzberg's, David McClelland's, and Douglas McGregor's contributions to an understanding of motivation; H. J. Thamhain's and D. L. Wilemon's work on influencing workers and reducing conflict; the effect of power on project teams; and Stephen Covey's work on how people and teams can become more effective. Finally, you will look at some implications and recommendations for project managers.

Motivation Theories

Psychologists, managers, coworkers, teachers, parents, and most people in general still struggle to understand what motivates people, or, why they do what they do. **Intrinsic motivation** causes people to participate in an activity for their own enjoyment. For example, some people love to read, write, or play an instrument because it makes them feel good. **Extrinsic motivation** causes people to do something for a reward or to avoid a penalty. For example, some young children would prefer *not* to play an instrument, but they do because they receive a reward or avoid a punishment for doing so. Why do some people require no external motivation whatsoever to produce high-quality work while others require significant external motivation to perform routine tasks? Why can't you get someone who is extremely productive at work to do simple tasks at home? Humankind will continue to try to answer these types of questions. A basic understanding of motivational theory will help anyone who has to work or live with other people to understand themselves and others.

Maslow's Hierarchy of Needs

Abraham Maslow, a highly respected psychologist who rejected the dehumanizing negativism of psychology in the 1950s, is best known for developing a hierarchy of needs. In the 1950s, proponents of Sigmund Freud's psychoanalytic theory were promoting the idea that human beings were not the masters of their destiny and that all their actions were governed by unconscious processes dominated by primitive sexual urges. During the same period, behavioral psychologists saw human beings as controlled by the environment. Maslow argued that both schools of thought failed to recognize unique qualities of human behavior: love, self-esteem, belonging, self-expression, and creativity. He argued that these unique qualities enable a person to make independent choices, which gives them full control of their destiny.

Figure 9-1 shows the basic pyramid structure of Maslow's **hierarchy of needs**, which states that people's behaviors are guided or motivated by a sequence of needs. At the bottom of the hierarchy are physiological needs. Once physiological needs are satisfied, safety needs guide behavior. Once safety needs are satisfied, social needs come to the forefront, and so on up the hierarchy. The order of these needs and their relative sizes in the pyramid are significant. Maslow suggests that each level of the hierarchy is a prerequisite for the levels above. For example, it is not possible for a person to consider self-actualization if he or she has not addressed basic needs concerning security and safety. People in an emergency situation, such as a flood or hurricane, are not going to worry about personal growth. Personal survival will be their main motivation. Once a particular need is satisfied, however, it no longer serves as a potent motivator of behavior.

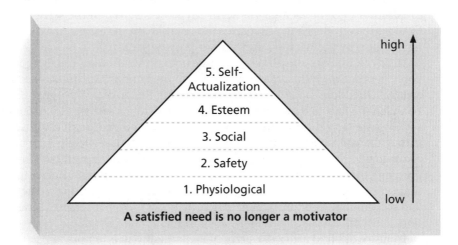

Figure 9-1. Maslow's Hierarchy of Needs

The bottom four needs in Maslow's hierarchy—physiological, safety, social, and esteem needs—are referred to as deficiency needs, and the highest level, self-actualization, is considered a growth need. Only after meeting deficiency needs can individuals act upon growth needs. Self-actualized people are problem-focused, have an appreciation for life, are concerned about personal growth, and have the ability to have peak experiences.

Most people working on an information technology project will probably have their basic physiological and safety needs met. If someone has a sudden medical emergency or is laid off from work, however, physiological and safety needs will move to the forefront. To motivate project team members, the project manager needs to understand each person's motivation, especially with regard to social, esteem, and self-actualization or growth needs. Team members new to a company and city might be motivated by social needs. To address social needs, some companies organize gatherings and social events for new workers. Other project members may find these events to be an invasion of personal time they would rather spend with their friends and family or working on an advanced degree.

Maslow's hierarchy conveys a message of hope and growth. People can work to control their own destinies and naturally strive to achieve higher and higher needs. Successful project managers know they must focus on meeting project goals, but they also know that they must understand team members' personal goals and needs to provide appropriate motivation and maximize team performance.

Herzberg's Motivation-Hygiene Theory

Frederick Herzberg is best known for distinguishing between motivational factors and hygiene factors when considering motivation in work settings. He called factors that cause job satisfaction motivators and factors that cause dissatisfaction hygiene factors. The term "hygiene" is used in the sense that these factors are considered maintenance factors that are necessary to avoid dissatisfaction but, by themselves, they do not provide satisfaction.

Head of Case Western University's psychology department, Herzberg wrote the book *Work and the Nature of Man* in 1966 and the famous *Harvard Business Review* article, "One More Time: How Do You Motivate Employees?" in 1968. Herzberg analyzed the factors that affected productivity among a sample of 1,685 employees. Popular beliefs at that time were that work output was most improved through larger salaries, more supervision, or a more attractive work environment. According to Herzberg, these hygiene factors would cause dissatisfaction if not present, but would not motivate workers to do more if present. Herzberg found that people were motivated to work mostly by feelings of personal achievement and recognition. Motivators, Herzberg concluded, included achievement, recognition, the work itself, responsibility, advancement, and growth.

In his books and articles, Herzberg explained why attempts to use positive factors such as reducing time spent at work, upward spiraling wages, offering fringe benefits, providing human relations and sensitivity training, and so on did not instill motivation. He argued that people want to actualize themselves. They need stimuli for their growth and advancement needs, in accordance with Maslow's hierarchy of needs. Factors such as achievement, recognition, responsibility, advancement, and growth produce job satisfaction and are work motivators.[12]

McClelland's Acquired-Needs Theory

David McClelland proposed that an individual's specific needs are acquired or learned over time and shaped by life experiences. The main categories of acquired needs include achievement, affiliation, and power. Normally one or two of these needs will be dominant in individuals.

- **Achievement:** People with a high need for achievement (nAch) seek to excel and tend to avoid both low-risk and high-risk situations to improve their chances for achieving something worthwhile. Achievers need regular feedback and often prefer to work alone or with other high achievers. Managers should give high achievers challenging projects with achievable goals. Achievers should receive frequent performance feedback, and although money is not an important motivator to them, it is an effective form of feedback.

- **Affiliation:** People with a high need for affiliation (nAff) desire harmonious relationships with other people and need to feel accepted by others. They tend to conform to the norms of their work group and prefer work that involves significant personal interaction. Managers should try to create a cooperative work environment to meet the needs of people with a high need for affiliation.

- **Power:** People with a need for power (nPow) desire either personal power or institutional power. People who need personal power want to direct others and can be seen as bossy. People who need institutional power or social power want to organize others to further the goals of the organization. Management should provide those seeking institutional or social power with the opportunity to manage others, emphasizing the importance of meeting organizational goals.

The Thematic Apperception Test (TAT) is a tool to measure the individual needs of different people using McClelland's categories. The TAT presents subjects with a series of ambiguous pictures and asks them to develop a spontaneous story for each picture, assuming they will project their own needs into the story.

---

[12] Herzberg, Frederick, "One More Time: How Do You Motivate Employees?" *Harvard Business Review* (February 1968) p.51–62.

McGregor's Theory X and Theory Y

Douglas McGregor was one of the great popularizers of a human relations approach to management, and he is best known for developing Theory X and Theory Y. In his research, documented in his 1960 book *The Human Side of Enterprise*, McGregor found that although many managers spouted the right ideas, they actually followed a set of assumptions about worker motivation that he called Theory X (sometimes referred to as classical systems theory). People who believe in Theory X assume that workers dislike and avoid work if possible, so managers must use coercion, threats, and various control schemes to get workers to make adequate efforts to meet objectives. They assume that the average worker wants to be directed and prefers to avoid responsibility, has little ambition, and wants security above all else. Research seemed to demonstrate clearly that these assumptions were not valid. McGregor suggested a different series of assumptions about human behavior that he called Theory Y (sometimes referred to as human relations theory). Managers who believe in Theory Y assume that individuals do not inherently dislike work, but consider it as natural as play or rest. The most significant rewards are the satisfaction of esteem and self-actualization needs, as described by Maslow. McGregor urged managers to motivate people based on these more valid Theory Y notions.

In 1981, William Ouchi introduced another approach to management in his book *Theory Z: How American Business Can Meet the Japanese Challenge*. Theory Z is based on the Japanese approach to motivating workers, which emphasizes trust, quality, collective decision making, and cultural values. Whereas Theory X and Theory Y emphasize how management views employees, Theory Z also describes how workers perceive management. Theory Z workers, it is assumed, can be trusted to do their jobs to their utmost ability, as long as management can be trusted to support them and look out for their well-being. Theory Z emphasizes things such as job rotation, broadening of skills, generalization versus specialization, and the need for continuous training of workers.

Thamhain and Wilemon's Influence and Power

Many people working on a project do not report directly to project managers, and project managers often do not have control over project staff who report to them. For example, people are free to change jobs. If they are given work assignments they do not like, many workers will simply quit or transfer to other departments or projects. H. J. Thamhain and D. L. Wilemon investigated the approaches project managers use to deal with workers and how those approaches relate to project success. They identified nine influence bases available to project managers:

1. Authority: the legitimate hierarchical right to issue orders
2. Assignment: the project manager's perceived ability to influence a worker's later work assignments

3. Budget: the project manager's perceived ability to authorize others' use of discretionary funds

4. Promotion: the ability to improve a worker's position

5. Money: the ability to increase a worker's pay and benefits

6. Penalty: the project manager's perceived ability to dispense or cause punishment

7. Work challenge: the ability to assign work that capitalizes on a worker's enjoyment of doing a particular task, which taps an intrinsic motivational factor

8. Expertise: the project manager's perceived special knowledge that others deem important

9. Friendship: the ability to establish friendly personal relationships between the project manager and others

Top management grants authority to the project manager. Assignment, budget, promotion, money, and penalty influence bases may or may not be inherent in a project manager's position. Unlike authority, they are not automatically available to project managers as part of their position. Others' perceptions are important in establishing the usefulness of these influence bases. For example, any manager can influence workers by providing challenging work; the ability to provide challenging work (or take it away) is not a special ability of project managers. In addition, project managers must earn the ability to influence by using expertise and friendship.

Thamhain and Wilemon found that projects were more likely to fail when project managers relied too heavily on using authority, money, or penalty to influence people. *When project managers used work challenge and expertise to influence people, projects were more likely to succeed.* The effectiveness of work challenge in influencing people is consistent with Maslow's and Herzberg's research on motivation. The importance of expertise as a means of influencing people makes sense on projects that involve special knowledge, as in most information technology projects.

Influence is related to the topic of power. **Power** is the potential ability to influence behavior to get people to do things they would not otherwise do. Power has a much stronger connotation than influence, especially since it is often used to force people to change their behavior. There are five main types of power:

■ **Coercive power** involves using punishment, threats, or other negative approaches to get people to do things they do not want to do. This type of power is similar to Thamhain's and Wilemon's influence category called penalty. For example, a project manager can threaten to fire workers or subcontractors to try to get them to change their behavior. If the project manager really has the power to fire people, he or she could follow through on the threat. Recall, however, that influencing using penalties is correlated with unsuccessful projects. Still, coercive power can be very effective in stopping negative behavior. For example, if students tend to hand in assignments late, an instructor can have a policy in his or her syllabus stating late penalties, such as a 20 percent grade reduction for each day an assignment is late.

- **Legitimate power** is getting people to do things based on a position of authority. This type of power is similar to the authority basis of influence. If top management gives project managers organizational authority, project managers can use legitimate power in several situations. They can make key decisions without involving the project team, for example. Overemphasis of legitimate power or authority also correlates with project failure.

- **Expert power** involves using personal knowledge and expertise to get people to change their behavior. If people perceive that project managers are experts in certain situations, they will follow their suggestions. For example, if a project manager has expertise in working with a particular information technology supplier and their products, the project team will be more likely to follow the project manager's suggestions on how to work with that vendor and its products.

- **Reward power** involves using incentives to induce people to do things. Rewards can include money, status, recognition, promotions, special work assignments, or other means of rewarding someone for desired behavior. Many motivation theorists suggest that only certain types of rewards, such as work challenge, achievement, and recognition, truly induce people to change their behavior or work hard.

- **Referent power** is based on an individual's personal charisma. People hold someone with referent power in very high regard and will do what they say based on their regard for the person. People such as Martin Luther King, Jr., John F. Kennedy, and Bill Clinton had referent power. Very few people possess the natural charisma that underlies referent power.

It is important for project managers to understand what types of influence and power they can use in different situations. New project managers often overemphasize their position—their legitimate power or authority influence—especially when dealing with project team members or support staff. They also neglect the importance of reward power or work challenge influence. People often respond much better to a project manager who motivates them with challenging work and provides positive reinforcement for doing a good job. It is important for project managers to understand the basic concepts of influence and power, and to practice using them to their own and their project team's advantage.

Covey and Improving Effectiveness

Stephen Covey, author of *The 7 Habits of Highly Effective People*, expanded on the work done by Maslow, Herzberg, and others to develop an approach for helping people and teams become more effective. Covey's first three habits of effective people—be proactive, begin with the end in mind, and put first things first—help people achieve a private victory by becoming independent. After achieving independence, people can then strive for interdependence by developing the next three habits—think win/win, seek first to understand then to be understood, and synergize. (**Synergy** is the concept that the whole is equal to more than the sum of its parts.) Finally, everyone can work on Covey's seventh habit—sharpen the saw—to develop and renew their physical, spiritual, mental, and social/emotional selves.

Project managers can apply Covey's seven habits to improve effectiveness on projects, as follows:

1. *Be proactive.* Covey, like Maslow, believes that people have the ability to be proactive and choose their responses to different situations. Project managers must be proactive, anticipate, and plan for problems and inevitable changes on projects. They can also encourage their team members to be proactive in working on their project activities.

2. *Begin with the end in mind.* Covey suggests that people focus on their values, what they really want to accomplish, and how they really want to be remembered in their lives. He suggests writing a mission statement to help achieve this habit. Many organizations and projects have mission statements that help them focus on their main purpose.

3. *Put first things first.* Covey developed a time management system and matrix to help people prioritize their time. He suggests that most people need to spend more time doing things that are important, but not urgent. Important but not urgent activities include planning, reading, and exercising. Project managers need to spend a lot of time working on important and not urgent activities, such as developing various project plans, building relationships with major project stakeholders, and mentoring project team members. They also need to avoid focusing only on important and urgent activities—putting out fires.

4. *Think win/win.* Covey presents several paradigms of interdependence, with think win/win being the best choice in most situations. When you use a win/win paradigm, parties in potential conflict work together to develop new solutions that make them all winners. Project managers should strive to use a win/win approach in making decisions, but sometimes, especially in competitive situations, they must use a win/lose paradigm.

5. *Seek first to understand, then to be understood.* **Empathic listening** is listening with the intent to understand. It is even more powerful than active listening because you forget your personal interests and focus on truly understanding the other person. To really understand other people, you must learn to focus on others first. When you practice empathic listening, you can begin two-way communication. This habit is critical for project managers so they can really understand their stakeholders' needs and expectations.

6. *Synergize.* In projects, a project team can synergize by creating collaborative products that are much better than a collection of individual efforts. Covey also emphasizes the importance of valuing differences in others to achieve synergy. Synergy is essential to many highly technical projects; in fact, several major breakthroughs in information technology occurred because of synergy. For example, in his Pulitzer Prize–winning book, *The Soul of a New Machine*, Tracy Kidder documented the 1970s synergistic efforts of a team of Data General researchers to create a new 32-bit super-minicomputer.[13]

7. *Sharpen the saw.* When you practice sharpening the saw, you take time to renew yourself physically, spiritually, mentally, and socially. The practice of self-renewal helps people avoid burnout. Project managers must make sure that they and their project team have time to retrain, reenergize, and occasionally even relax to avoid burnout.

Douglas Ross, author of *Applying Covey's Seven Habits to a Project Management Career* (see the sixth Suggested Reading for this chapter), related Covey's seven habits to project management. Ross suggests that Habit 5—Seek first to understand, then to be understood—differentiates good project managers from average or poor project managers. People have a tendency to focus on their own agendas instead of first trying to understand other people's points of view. Empathic listening can help project managers and team members find out what motivates different people. Understanding what motivates key stakeholders and customers can mean the difference between project success and project failure. Once project managers and team members begin to practice empathic listening, they can communicate and work together to tackle problems more effectively.

Before you can practice empathic listening, you first have to get people to talk to you. In many cases, you must work on developing a rapport with the other person before he or she will really talk to you. **Rapport** is a relation of harmony, conformity, accord, or affinity. Without rapport, people cannot begin to communicate. For example, in the opening case, Ben was not ready to talk to anyone from the Information Technology department, even if Ed and Sarah were ready to listen. Ben was angry about the lack of support he received from

---

[13] Kidder, Tracy, *The Soul of a New Machine*. New York: Modern Library (1997).

the Information Technology department and bullied anyone who reminded him of that group. Before Sarah could begin to communicate with Ben, she had to establish rapport.

One technique for establishing rapport is using a process called mirroring. **Mirroring** is the matching of certain behaviors of the other person. People tend to like people who are like themselves, and mirroring helps you take on some of the other person's characteristics. It also helps them realize if they are behaving unreasonably, as in this case. You can mirror someone's voice tone and/or tempo, breathing, movements, or body postures. After Ben started yelling at her, Sarah quickly decided to mirror his voice tone, tempo, and body posture. She stood nose to nose with Ben and started yelling right back at him. This action made Ben realize what he was doing, and it also made him notice and respect this colleague from the Information Technology department. Once Ben overcame his anger, he could start communicating his needs. In most cases, such extreme measures are not needed.

Recall the importance of getting users involved in information technology projects. For organizations to be truly effective in information technology project management, they must find ways to help users and developers of information systems work together. It is widely accepted that organizations make better decisions about information technology projects when business professionals and information technology staff collaborate. It is also widely accepted that this task is easier said than done. Many companies have been very successful in integrating technology and business departments, but many other companies continue to struggle with this issue.

☑ *What Went Right?*

Forrester Research, a well-known technology research company in Cambridge, Massachusetts, interviewed 25 Chief Information Officers (CIOs) at companies including L.L. Bean, Shell, GE, Taco Bell, Texas Instruments, and Coca-Cola to identify the best practices for ensuring partnerships between people in business and technology areas. Its research study identified three successful practices:

- Businesspeople, not information technology people, take the lead in determining and justifying investments in new computer systems. At several companies, business managers defined the expected benefits of new systems, with information technology support, and made the pitch for the investment to top management. At some companies, top business managers are running information technology departments. At GE, for example, the head of the corporate strategy unit assumed responsibility for the information technology division.

- CIOs continually push their staff to recognize that the needs of the business must drive all technology decisions. The CIOs at Textron and Texas Instruments, for example, go so far as to insist that everyone express all presentations, reports, and even discussions about information technology projects in business terms, not in the technical language used by information technology insiders.

■ Several companies that are successful at creating partnerships between people in business and technology areas are reshaping their information technology units to look and perform like consulting firms. For instance, Texas Instruments borrowed an idea from McKinsey & Company by assigning central information technology employees to one of 17 centers of excellence for specific technology disciplines. When a business challenge arises, managers pull together ad hoc teams of people drawn from those centers and the business itself to work on the problem. When the problem is solved, managers reassign team members to other problems. This approach allows managers to use resources more flexibly and keeps employees' skills fresh.

The Forrester Research study also found that successful CIOs downplayed the "T" (technology) of IT, and played up the "I" (information). The CIOs believed that choosing the right technology was important, but success rested more on having the right people involved in the right project, putting the right processes in place, and having a business-results focus at all times.[14]

It is important to understand and pay attention to concepts of motivation, influence, power, and improving effectiveness in all project processes. It is likewise important to remember that projects operate within an organizational environment. The challenge comes from applying these theories to the many unique individuals involved in particular projects in particular organizations.

As you can see, there are many important topics related to motivation, influence, power, and effectiveness that are relevant to project management. Projects are done by and for people, so it is important for project managers and team members to understand and practice key concepts related to these topics.

HUMAN RESOURCE PLANNING

Human resource planning for a project involves identifying and documenting project roles, responsibilities, and reporting relationships. This process generates an organizational chart for the project and a staffing management plan, and determines roles and responsibilities, which are often shown in a responsibility assignment matrix (RAM).

Before creating an organizational chart for a project, top management and the project manager must identify what types of people the project really needs to ensure project success. If the key to success lies in having the best Java programmers you can find, the organizational planning should reflect that need. If the real key to success is having a top-notch project manager and team leaders whom people respect in the company, that need should drive human resource planning.

[14] Kiely, Thomas, "Making Collaboration Work," *Harvard Business Review* (January-February 1997) p.10–11.

(X) What Went Wrong?

The organizational culture of a company has a tremendous influence on how effectively human resources are utilized. For example, one author describes a well-run software development organization where programmers received top pay and status. As a result, the ratio of programmers to managers was about ten to one. In that organization, the company developed and installed a customized publication system for less than $1 million. The author then described the next software development organization where he worked. Programming was considered a job of underlings. The typical software project had a manager, a project leader, a project manager, a marketing manager, an architect, a systems engineer, and one or two programmers. In that environment, the company spent more than $30 million in personnel costs to try to develop an almost identical publication system, without a single line of usable code ever being written.[15]

It is also crucial not to ignore training needs of users when developing project plans. At the new $2.2 billion Kuala Lumpur International Airport in Malaysia, flights were delayed for hours when the new total airport management system was unable to cope with incorrect commands entered by improperly trained employees. Many information technology project managers are pressured to meet unrealistic schedule deadlines. Since training often comes at the end of the schedule, it is often compressed or neglected altogether before new systems go live. In addition, it takes a special kind of person to train users on a new system, and project managers cannot expect the same people who wrote the code to be also the best people to provide the user training.[16]

Project Organizational Charts

Recall from Chapter 2 that the nature of information technology projects often means that project team members come from different backgrounds and possess a wide variety of skill sets. It can be very difficult to manage such a diverse group of people, so it is important to provide a clear organizational structure for a project. After identifying important skills and the types of people needed to staff a project, the project manager should work with top management and project team members to create an organizational chart for the project. Figure 9-2 provides a sample organizational chart for a large information technology project. Note that the project personnel include a deputy project manager, subproject managers, and teams. **Deputy project managers** fill in for project managers in their absence and assist them as needed, which is similar to the role of a vice president. **Subproject managers** are responsible for managing the subprojects into which a large project might be divided. There are often subproject managers that focus on managing various software (S/W) and hardware (H/W) components of large projects. This structure is typical for large projects. With many people working on a project, clearly defining and allocating project work is essential.

[15] James, Geoffrey, "IT Fiascoes . . . and How to Avoid Them," *Datamation* (November 1997).

[16] Moad, Jeff, "Grounding those high-flying IT projects," *PC Week* (July 17, 1998).

(Visit the companion Web site for this text to see the project organization chart that Northwest Airlines used for their large ResNet project.) Smaller information technology projects usually do not have deputy project managers or subproject managers. On smaller projects, the project managers might have just team leaders reporting directly to them.

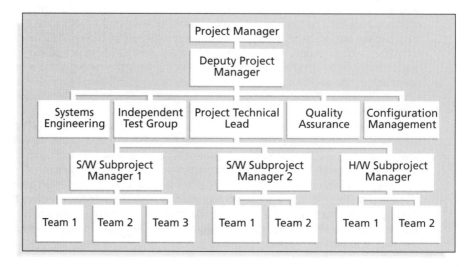

Figure 9-2. Sample Organizational Chart for a Large Information Technology Project

Figure 9-3 provides a framework for defining and assigning work. This process consists of four steps:

1. Finalizing the project requirements
2. Defining how the work will be accomplished
3. Breaking down the work into manageable elements
4. Assigning work responsibilities

The work definition and assignment process is carried out during the proposal and startup phases of a project. Note that the process is iterative, meaning it often takes more than one pass to refine it. A Request for Proposal (RFP) or draft contract often provides the basis for defining and finalizing work requirements, which are then documented in a final contract and technical baseline. If there were not an RFP, then the internal project charter and scope statement would provide the basis for defining and finalizing work requirements, as described in Chapter 5, Project Scope Management. The project team leaders then decide on a technical approach for how to do the work. Should work be broken down using a product-oriented approach or a phased approach? Will the project team outsource some of the work or subcontract to other companies? Once the project team has decided on a technical approach, they develop a work breakdown structure (WBS) to establish manageable elements of work (see Chapter 5, Project Scope Management). They then develop activity definitions to define the work involved in each activity on the WBS further (see Chapter 6, Project Time Management). The last step is assigning the work.

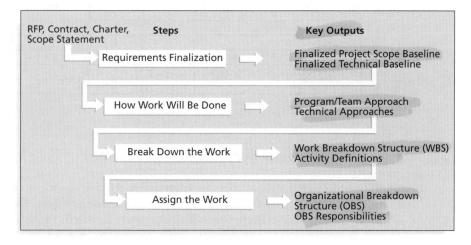

Figure 9-3. Work Definition and Assignment Process

Once the project manager and project team have broken down the work into manageable elements, the project manager assigns work to organizational units. The project manager often bases these work assignments on where the work fits in the organization and uses an organizational breakdown structure to conceptualize the process. An **organizational breakdown structure (OBS)** is a specific type of organizational chart that shows which organizational units are responsible for which work items. The OBS can be based on a general organizational chart and then broken down into more detail, based on specific units within departments in the company or units in any subcontracted companies.

Responsibility Assignment Matrices

After developing an OBS, the project manager is in a position to develop a responsibility assignment matrix (RAM). A **responsibility assignment matrix (RAM)** is a matrix that maps the work of the project as described in the WBS to the people responsible for performing the work as described in the OBS. Figure 9-4 shows an example of a RAM. The RAM allocates work to responsible and performing organizations, teams, or individuals, depending on the desired level of detail. For smaller projects, it would be best to assign individual people to WBS activities. For very large projects, it is more effective to assign the work to organizational units or teams.

WBS activities ⟶

OBS units	1.1.1	1.1.2	1.1.3	1.1.4	1.1.5	1.1.6	1.1.7	1.1.8
Systems Engineering	R	R P					R	
Software Development			R P					
Hardware Development				R P				
Test Engineering	P							
Quality Assurance					R P			
Configuration Management						R P		
Integrated Logistics Support							P	
Training								R P

R = Responsible organizational unit
P = Performing organizational unit

Figure 9-4. Sample Responsibility Assignment Matrix (RAM)

In addition to using a RAM to assign detailed work activities, you can also use a RAM to define general roles and responsibilities on projects. This type of RAM can include the stakeholders in the project. Figure 9-5 provides a RAM that shows whether stakeholders are accountable or just participants in part of a project, and whether they are required to provide input, review, or sign off on parts of a project. This simple tool can be a very effective way for the project manager to communicate roles and expectations of important stakeholders on projects.

	Stakeholders				
Items	A	B	C	D	E
Unit Test	S	A	I	I	R
Integration Test	S	P	A	I	R
System Test	S	P	A	I	R
User Acceptance Test	S	P	I	A	R

A = Accountable
P = Participant
R = Review Required
I = Input Required
S = Sign-off Required

Figure 9-5. RAM Showing Stakeholder Roles

Some organizations use **RACI charts** to show Responsibility, Accountability, Consultation, and Informed roles for project stakeholders. As shown in Table 9-1, a RACI chart lists tasks vertically, individuals or groups horizontally, and each intersecting cell contains an R, A, C, or I. Each task may have multiple A, C, or I entries, but there can be only one R entry to clarify which particular individual or group has responsibility for each task.

Table 9-1: Sample RACI Chart

	GROUP A	GROUP B	GROUP C	GROUP D	GROUP E
Test Plans	R	A	C	C	I
Unit Test	C	I	R	A	I
Integration Test	A	R	I	C	C
System Test	I	C	A	I	R
User Acceptance Test	A	I	C	R	A

Staffing Management Plans and Resource Histograms

Another output of human resource planning is a staffing management plan. A **staffing management plan** describes when and how people will be added to and taken off the project team. It is often part of the project management plan, as described in Chapter 4. The level of detail may vary based on the type of project. For example, if an information technology project is projected to need 100 people on average over a year, the staffing management plan would describe the types of people needed to work on the project, such as Java programmers, business analysts, technical writers, and so on, and the number of each type of person needed each month. It would also describe how these resources would be acquired, trained, rewarded, reassigned after the project, and so on. All of these issues are important to meeting the needs of the project, the employees, and the organization.

The staffing management plan often includes a **resource histogram**, which is a column chart that shows the number of resources assigned to a project over time. Figure 9-6 provides an example of a histogram that might be used for a six-month information technology project. Notice that the columns represent the number of people needed in each area—managers, business analysts, programmers, and technical writers. By stacking the columns, you can see the total number of people needed each month. After determining the project staffing needs, the next steps in project human resource management are to acquire the necessary staff and then develop the project team.

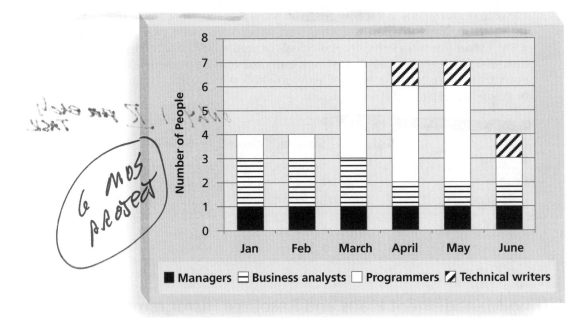

Figure 9-6. Sample Resource Histogram

ACQUIRING THE PROJECT TEAM

During the late 1990s, the information technology job market became extremely competitive. It was a seller's market with corporations competing fiercely for a shrinking pool of qualified, experienced information technology professionals. In the early 2000s, the market declined tremendously, so employers could be very selective in recruiting. Regardless of the current job market, acquiring qualified information technology professionals is critical. There's a saying that the project manager who is the smartest person on the team has done a poor job of recruiting! In addition to recruiting team members, it is also important to assign the appropriate type and number of people to work on projects at the appropriate times. This section addresses important topics related to acquiring the project team: resource assignment, resource loading, and resource leveling.

Resource Assignment

After developing a staffing management plan, project managers must work with other people in their organizations to assign particular personnel to their projects or to acquire additional human resources needed to staff the project.

Project managers with strong influencing and negotiating skills are often good at getting internal people to work on their projects. However, the organization must ensure that people are assigned to the projects that best fit their skills and the needs of the organization. The main outputs of this process are project staff assignments, resource availability information, and updates to the staffing management plan. Many project teams also find it useful to create a project team directory.

Organizations that do a good job of staff acquisition have good staffing plans. These plans describe the number and type of people who are currently in the organization and the number and type of people anticipated to be needed for the project based on current and upcoming activities. An important component of staffing plans is maintaining a complete and accurate inventory of employees' skills. If there is a mismatch between the current mix of people's skills and needs of the organization, it is the project manager's job to work with top management, human resource managers, and other people in the organization to address staffing and training needs.

It is also important to have good procedures in place for hiring subcontractors and recruiting new employees. Since the Human Resource department is normally responsible for hiring people, project managers must work with their human resource managers to address any problems in recruiting appropriate people. It is also a priority to address retention issues, especially for information technology professionals.

One innovative approach to hiring and retaining information technology staff is to offer existing employees incentives for helping recruit and retain personnel. For example, several consulting companies give their employees one dollar for every hour a new person they helped hire works. This provides an incentive for current employees to help attract new people and to keep both them and the person they recruited working at that company. Another approach that several companies are taking to attract and retain information technology professionals is to provide benefits based on their personal needs. For example, some people might want to work only four days a week or have the option of working a couple of days a week from home. As it gets more difficult to find good information technology professionals, organizations must become more innovative and proactive in addressing this issue.

Several organizations, publications, and Web sites address the need for good staff acquisition and retention. For example, ICEX, Inc. conducted several research studies to identify best practices. They have found that world-class information technology organizations have groups dedicated to managing information technology staff recruiting. They have also found that the most effective recruiting message emphasizes that "this is a great place to work and you'll have a chance to work on innovative information technology projects."[17] William C. Taylor, cofounder of *Fast Company* magazine and a public speaker, also believes

---

[17] Quillard, Judith, "Leading Practices for Recruiting IT Staff," ICEX Web site (1998) *www.icex.com*.

that people today are more demanding and have higher expectations of their jobs than just earning a paycheck. His company's research has found that people leave their jobs because:

- They feel they do not make a difference.
- They do not get proper recognition.
- They are not learning anything new or growing as a person.
- They do not like their coworkers.
- They want to earn more money.[18]

Key to it keeping

It is very important to consider the needs of individuals and the organization when making recruiting and retention decisions and to study the best practices of leading companies in these areas. It is also important to address a growing trend in project team members—many of them work in a virtual environment. See the section on managing the project team for suggestions on working with virtual team members.

Resource Loading

Chapter 6, Project Time Management, described using network diagrams to help manage a project's schedule. One of the problems or dangers inherent in scheduling processes is that they often do not address the issues of resource utilization and availability. (Hence the development of critical chain scheduling, as described in Chapter 6.) Schedules tend to focus primarily on time rather than on both time and resources, which includes people. An important measure of a project manager's success is how well he or she balances the trade-offs among performance, time, and cost. During a period of crisis, it is occasionally possible to add additional resources—such as additional staff—to a project at little or no cost. Most of the time, however, resolving performance, time, and cost trade-offs entails additional costs to the organization. The project manager's goal must be to achieve project success without increasing the costs or time required to complete the project. The key to accomplishing this goal is effectively managing human resources on the project.

Once people are assigned to projects, there are two techniques available to project managers that help them use project staff most effectively: resource loading and resource leveling. **Resource loading** refers to the amount of individual resources an existing schedule requires during specific time periods. Resource loading helps project managers develop a general understanding of

---

[18] Taylor, William C., Keynote address at the Project Management Institute's 2000 Annual Seminars & Symposium (September 2000).

the demands a project will make on the organization's resources, as well as on individual people's schedules. Project managers often use histograms, as described in Figure 9-6, to depict period-by-period variations in resource loading. A histogram can be very helpful in determining staffing needs or in identifying staffing problems.

A resource histogram can also show when work is being overallocated to a certain person or group. **Overallocation** means more resources than are available are assigned to perform work at a given time. For example, Figure 9-7 provides a sample resource histogram created in Microsoft Project 2003. This histogram illustrates how much one individual, Joe Franklin, is assigned to work on the project each week. The percentage numbers on the vertical axis represent the percentage of Joe's available time that is allocated for him to work on the project. The top horizontal axis represents time in weeks. Note that Joe Franklin is overallocated most of the time. For example, for most of March and April and part of May, Joe's work allocation is 300 percent of his available time. If Joe is normally available eight hours per day, this means he would have to work twenty-four hours a day to meet this staffing projection! Many people don't use the resource assignment features of project management software properly. See Appendix A for detailed information on using Project 2003.

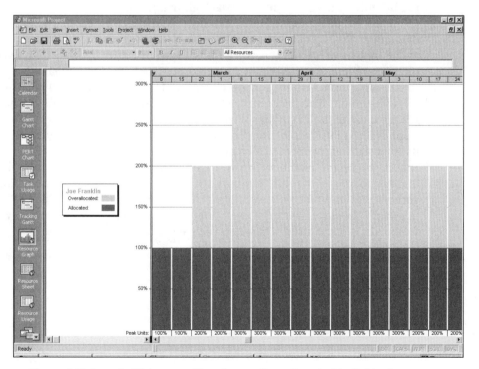

Figure 9-7. Sample Histogram Showing an Overallocated Individual

Resource Leveling

Resource leveling is a technique for resolving resource conflicts by delaying tasks. It is a form of network analysis in which resource management concerns drive scheduling decisions (start and finish dates). The main purpose of resource leveling is to create a smoother distribution of resource usage. Project managers examine the network diagram for areas of slack or float, and to identify resource conflicts. For example, you can sometimes remove overallocations by delaying noncritical tasks, which does not result in an overall schedule delay. Other times you will need to delay the project completion date to reduce or remove overallocations. See Appendix A for information on using Project 2003 to level resources using both of these approaches. You can also view resource leveling as addressing the resource constraints described in critical chain scheduling (see Chapter 6, Project Time Management).

Overallocation is one type of resource conflict. If a certain resource is overallocated, the project manager can change the schedule to remove resource overallocation. If a certain resource is underallocated, the project manager can change the schedule to try to improve the use of the resource. Resource leveling, therefore, aims to minimize period-by-period variations in resource loading by shifting tasks within their slack allowances.

Figure 9-8 illustrates a simple example of resource leveling. The network diagram at the top of this figure shows that Activities A, B, and C can all start at the same time. Activity A has a duration of two days and will take two people to complete; Activity B has a duration of five days and will take four people to complete; and Activity C has a duration of three days and will take two people to complete. The histogram on the lower-left of this figure shows the resource usage if all activities start on day one. The histogram on the lower-right of Figure 9-8 shows the resource usage if Activity C is delayed two days, its total slack allowance. Notice that the lower-right histogram is flat or leveled; that is, its pieces (activities) are arranged to take up the least space (saving days and numbers of workers). You may recognize this strategy from the computer game Tetris, in which you earn points for keeping the falling shapes as level as possible. The player with the most points (most level shape allocation) wins. Resources are also used best when they are leveled.

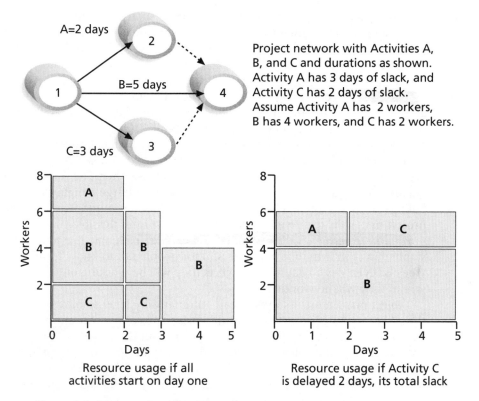

A=2 days

2

B=5 days

1   4

C=3 days   3

Project network with Activities A,
B, and C and durations as shown.
Activity A has 3 days of slack, and
Activity C has 2 days of slack.
Assume Activity A has 2 workers,
B has 4 workers, and C has 2 workers.

Workers (vertical axis left chart)
8, 6, 4, 2
A
B   B
B
C   C

Days
0 1 2 3 4 5

Resource usage if all
activities start on day one

Workers (vertical axis right chart)
8, 6, 4, 2
A   C
B

Days
0 1 2 3 4 5

Resource usage if Activity C
is delayed 2 days, its total slack

Figure 9-8. Resource Leveling Example

Resource leveling has several benefits. First, when resources are used on a more constant basis, they require less management. For example, it is much easier to manage a part-time project member who is scheduled to work 20 hours per week on a project for the next three months than it is to manage the same person who is scheduled to work 10 hours one week, 40 the next, 5 the next, and so on.

Second, resource leveling may enable project managers to use a just-in-time inventory type of policy for using subcontractors or other expensive resources. For example, a project manager might want to level resources related to work that must be done by particular subcontractors such as testing consultants. This leveling might allow the project to use four outside consultants full-time to do testing for four months instead of spreading the work out over more time or needing to use more than four people. The latter approach is usually more expensive.

Third, resource leveling results in fewer problems for project personnel and accounting departments. Increasing and decreasing labor levels and particular human resources often produce additional work and confusion. For example, if a person with expertise in a particular area is only assigned to a project two days a week and another person they need to work with is not assigned to the project those same days, they cannot work well together. The Accounting department might complain when subcontractors charge a higher rate for billing less than 20 hours a week on a project. The accountants will remind project managers to strive for getting the lowest rates possible.

Finally, resource leveling often improves morale. People like to have some stability in their jobs. It is very stressful for people not to know from week to week or even day to day what projects they will be working on and with whom they will be working.

Project management software can automatically level resources. However, the project manager must be careful in using the results without making adjustments. Automatic leveling often pushes out the project's completion date. Resources may also be reallocated to work at times that are inappropriate with other constraints. A wise project manager would have one of his or her team members who is proficient in using the project management software ensure that the leveling is done appropriately.

DEVELOPING THE PROJECT TEAM

Even if a project manager has successfully recruited enough skilled people to work on a project, he or she must ensure that people can work together as a team to achieve project goals. Many information technology projects have had very talented individuals working on them. However, it takes teamwork to complete most projects successfully. The main goal of **team development** is to help people work together more effectively to improve project performance.

Dr. Bruce Tuckman published his four-stage model of team development in 1965 and modified it to include an additional stage in the 1970s. The **Tuckman model** describes five stages of team development:

1. *Forming* involves the introduction of team members, either at the initiation of the team, or as new members are introduced. This stage is necessary, but little work is actually achieved.

2. *Storming* occurs as team members have different opinions as to how the team should operate. People test each other, and there is often conflict within the team.

3. *Norming* is achieved when team members have developed a common working method, and cooperation and collaboration replace the conflict and mistrust of the previous phase.

4. *Performing* occurs when the emphasis is on reaching the team goals, rather than working on team process. Relationships are settled, and team members are likely to build loyalty towards each other. At this stage, the team is able to manage tasks that are more complex and cope with greater change.

5. *Adjourning* involves the break-up of the team after they successfully reach their goals and complete the work.

There is an extensive body of literature on team development. This section will highlight a few important tools and techniques for team development, including training, team-building activities, and reward and recognition systems.

Training

Project managers often recommend that people take specific training courses to improve individual and team development. For example, Sarah from the opening case had gone through training in emotional intelligence and dealing with difficult people. She was familiar with the mirroring technique and felt comfortable using that approach with Ben. Many other people would not have reacted so quickly and effectively in the same situation. If Ben and Sarah did reach agreement on what actions they could take to resolve the F-44 aircraft program's information technology problems, it might result in a new project to develop and deliver a new system for Ben's group. If Sarah became the project manager for this new project, she would understand the need for special training in interpersonal skills for specific people in her and Ben's departments. Individuals could take special training classes to improve their personal skills. If Sarah thought the whole project team could benefit from taking training together to learn to work as a team, she could arrange for a special team-building session for the entire project team and key stakeholders.

It is very important to provide training in a just-in-time fashion. For example, if Sarah was preparing for a technical assignment where she would need to learn a new programming language, training to deal with difficult people would not help her much. However, the training was very timely for her new consulting position. Many organizations provide e-learning opportunities for their employees so they can learn specific skills at any time and any place. They have also found e-learning to sometimes be more cost-effective than traditional instructor-led training courses. It is important to make sure that the timing and delivery method for the training is appropriate for specific situations and individuals. Organizations have also found that it is often more economical to train current employees in particular areas than it is to hire new people who already possess those skills.

Several organizations that have successfully implemented Six Sigma principles have taken a unique and effective approach to training. They only let high-potential employees attend Six Sigma Black Belt training, which is a substantial investment in terms of time and money. In addition, they do not let employees into a particular Black Belt course until they have had a potential Six Sigma project approved that relates to their current job. Attendees can then apply the new concepts and techniques they learn in the classes to their work settings. High-potential employees feel rewarded by being picked to take this training, and the organization benefits by having these employees implement high-payoff projects because of the training.

Team-Building Activities

Many organizations provide in-house team-building training activities, and many also use specialized services provided by external companies that specialize in this area. Two common approaches to team-building activities include using physical challenges and psychological preference indicator tools.

Several organizations have teams of people go through certain physically challenging activities to help them develop as a team. Military basic training or boot camps provide one example. Men and women who wish to join the military must first make it through basic training, which often involves several strenuous physical activities such as rappelling off towers, running and marching in full military gear, going through obstacle courses, passing marksmanship training, and mastering survival training. Many organizations use a similar approach by sending teams of people to special locations where they work as a team to navigate white water rapids, climb mountains or rocks, participate in ropes courses, and so on.

Even more organizations have teams participate in mental team-building activities in which they learn about themselves, each other, and how to work as a group most effectively. It is important for people to understand and value each other's differences in order to work effectively as a team. Two common exercises used in mental team building include the Myers-Briggs Type Indicator and the Wilson Learning Social Styles Profile.

The **Myers-Briggs Type Indicator (MBTI)** is a popular tool for determining personality preferences. During World War II, Isabel B. Myers and

Katherine C. Briggs developed the first version of the MBTI based on psychologist Carl Jung's theory of psychological type. The four dimensions of psychological type in the MBTI include:

- Extrovert/Introvert (E/I): This first dimension determines if you are generally extroverted or introverted. The dimension also signifies whether people draw their energy from other people (extroverts) or from inside themselves (introverts).

- Sensation/Intuition (S/N): This second dimension relates to the manner in which you gather information. Sensation (or Sensing) type people take in facts, details, and reality and describe themselves as practical. Intuitive type people are imaginative, ingenious, and attentive to hunches or intuition. They describe themselves as innovative and conceptual.

- Thinking/Feeling (T/F): This third dimension represents thinking judgment and feeling judgment. Thinking judgment is objective and logical, and feeling judgment is subjective and personal.

- Judgment/Perception (J/P): This fourth dimension concerns people's attitude toward structure. Judgment type people like closure and task completion. They tend to establish deadlines and take them seriously, expecting others to do the same. Perceiving types prefer to keep things open and flexible. They regard deadlines more as a signal to start rather than complete a project and do not feel that work must be done before play or rest begins.[19]

There is much more involved in personality types, and many books are available on this topic. In 1998, David Keirsey and Ray Choiniere published *Please Understand Me II: Temperament Character Intelligence*. This book includes an easy-to-take and interpret test called The Keirsey Temperament Sorter, which is a personality type preference test based on the work done by Jung, Myers, and Briggs.

In 1985, an interesting study of the MBTI types of the general population in the United States and information systems (IS) developers revealed some significant contrasts.[20] Both groups of people were most similar in the judgment/perception dimension, with slightly more than half of each group preferring the judgment type (J). There were significant differences, however, in the other three dimensions. Most people would not be surprised to hear that most information systems developers are introverts. This study found that 75 percent of IS developers were introverts (I), and only 25 percent of the general population were introverts. This personality type difference might help explain some

[19] Briggs, Isabel Myers, with Peter Myers, *Gifts Differing: Understanding Personality Type*. Palo Alto, CA: Consulting Psychologists Press (1995).

[20] Lyons, Michael L., "The DP Psyche," *Datamation* (August 15, 1985).

of the problems users have communicating with developers. Another sharp contrast found in the study was that almost 80 percent of IS developers were thinking types (T) compared to 50 percent of the general population. IS developers were also much more likely to be intuitive (N) (about 55 percent) than the general population (about 25 percent). These results fit with Keirsey's classification of NT (Intuitive/Thinking types) people as rationals. Educationally, they tend to study the sciences, enjoy technology as a hobby, and pursue systems work. Keirsey also suggests that no more than 7 percent of the general population are NTs. Would you be surprised to know that Bill Gates is classified as a rational?[21]

Project managers can often benefit from knowing their team members' MBTI profiles by adjusting their management styles for each individual. For example, if the project manager is a strong N and one of the team members is a strong S, the project manager should take the time to provide more concrete, detailed explanations when discussing that person's task assignments. Project managers may also want to make sure they have a variety of personality types on their teams. For example, if all team members are strong introverts, it may be difficult for them to work well with users and other important stakeholders who are often extroverts.

Many organizations also use the Social Styles Profile in team-building activities. Psychologist David Merril, who helped develop the Wilson Learning Social Styles Profile, describes people as falling into four approximate behavioral profiles, or zones. People are perceived as behaving primarily in one of four zones, based on their assertiveness and responsiveness:

- "Drivers" are proactive and task-oriented. They are firmly rooted in the present, and they strive for action. Adjectives to describe drivers include pushy, severe, tough, dominating, harsh, strong-willed, independent, practical, decisive, and efficient.

- "Expressives" are proactive and people-oriented. They are future-oriented and use their intuition to look for fresh perspectives on the world around them. Adjectives to describe expressives include manipulating, excitable, undisciplined, reacting, egotistical, ambitious, stimulating, wacky, enthusiastic, dramatic, and friendly.

- "Analyticals" are reactive and task-oriented. They are past-oriented and strong thinkers. Adjectives to describe analyticals include critical, indecisive, stuffy, picky, moralistic, industrious, persistent, serious, expecting, and orderly.

[21] The Web Site for the Keirsey Temperament Sorter and Keirsey Temperament Theory *http://keirsey.com/personality/nt.html.*

■ "Amiables" are reactive and people-oriented. Their time orientation varies depending on whom they are with at the time, and they strongly value relationships. Adjectives to describe amiables include conforming, unsure, ingratiating, dependent, awkward, supportive, respectful, willing, dependable, and agreeable.[22]

Figure 9-9 shows these four social styles and how they relate to assertiveness and responsiveness. Note that the main determinants of the social style are your levels of assertiveness—if you are more likely to tell people what to do or ask what should be done—and how you respond to tasks—by focusing on the task itself or on the people involved in performing the task.

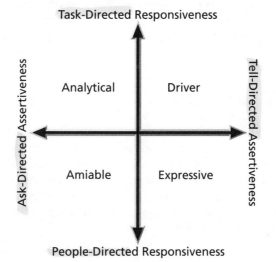

Figure 9-9. Social Styles

Knowing the social styles of project stakeholders can help project managers understand why certain people may have problems working together. For example, drivers are often very impatient working with amiables, and analyticals often have difficulties understanding expressives. Project managers can use their facilitation skills to help all types of people communicate better with each other and focus on meeting project goals.

[22] Robbins, Harvey A. and Michael Finley, *The New Why Teams Don't Work: What Goes Wrong and How to Make It Right.* San Francisco, CA: Berrett-Koehler Publishers (1999).

Reward and Recognition Systems

Another important tool for promoting team development is the use of team-based reward and recognition systems. If management rewards teamwork, they will promote or reinforce people to work more effectively in teams. Some organizations offer bonuses, trips, or other rewards to workers that meet or exceed company or project goals. In a project setting, project managers can recognize and reward people who willingly work overtime to meet an aggressive schedule objective or go out of their way to help a teammate. Project managers should not reward people who work overtime just to get extra pay or because of their own poor work or planning.

Project managers must continually assess their team's performance. When they find areas where individuals or the entire team can improve, it's their job to find the best way to develop their people and improve performance.

MANAGING THE PROJECT TEAM

In addition to developing the project team, the project manager must lead them in performing various project activities. After assessing team performance and related information, the project manager must decide if changes should be requested to the project, if corrective or preventive actions should be recommended, or if updates are needed to the project management plan or organizational process assets. Project managers must use their soft skills to find the best way to motivate and manage each team member.

Tools and Techniques for Managing Project Teams

There are several tools and techniques available to assist in managing project teams:

- Observation and conversation: It is hard to assess how your team members are performing or how they are feeling about their work if you never see or discuss these issues. Many project managers like to practice "management by walking around" to physically see and hear their team members at work. Informal or formal conversations about how a project is going can provide crucial information. For virtual workers, project managers can still observe and discuss work and personal issues via e-mail, telephone, or other communications media.

- Project performance appraisals: Just as general managers provide performance appraisals for their workers, so can project managers. The need for and type of project performance appraisals will vary depending on the length of the project, complexity of the project, organizational policies, contract requirements, and related communications. Even if a project manager does not provide official project performance appraisals for team members, it is still important to provide timely performance feedback. If a team member hands in sloppy or late work, the project manager should determine the reason for this behavior and take appropriate action. Perhaps the team member had a death in the family and could not concentrate. Perhaps the team member was planning to leave the project. The reasons for the behavior would have a strong impact on what action the project manager should take.

- **Conflict management:** Few projects are completed without any conflicts. Some types of conflict are actually desirable on projects, but many are not. As described in Chapter 10, Project Communications Management, there are several ways to handle conflicts. It's important for project managers to understand strategies for handling conflicts and to proactively manage conflict.

- **Issue logs:** Many project managers keep an **issue log** to document, monitor, and track issues that need to be resolved for the project team to work effectively. Issues could include items where people have differences of opinion, situations that need more clarification or investigation, or general concerns that need to be addressed. It is important to acknowledge issues that can hurt team performance and take action to resolve them. The project manager should assign someone to resolve each issue and assign a target date for resolution.

General Advice on Managing Teams

Effective project managers must be good team leaders. Suggestions for ensuring that teams are productive include the following:

- Be patient and kind with your team. Assume the best about people; do not assume that your team members are lazy and careless.
- Fix the problem instead of blaming people. Help people work out problems by focusing on behaviors.
- Establish regular, effective meetings. Focus on meeting project objectives and producing positive results.
- Allow time for teams to go through the basic team-building stages of forming, storming, norming, performing, and adjourning. Don't expect teams to work at the highest performance level right away.
- Limit the size of work teams to three to seven members.
- Plan some social activities to help project team members and other stakeholders get to know each other better. Make the social events fun and not mandatory.
- Stress team identity. Create traditions that team members enjoy.

- Nurture team members and encourage them to help each other. Identify and provide training that will help individuals and the team as a whole become more effective.
- Acknowledge individual and group accomplishments.
- Take additional actions to work with virtual team members. If possible, have a face-to-face or phone meeting at the start of a virtual project or when introducing a virtual team member. Screen people carefully to make sure they can work effectively in a virtual environment. Clarify how virtual team members will communicate.

As you can imagine, team development and management are critical concerns on many information technology projects. Many information technology project managers must break out of their rational/NT preference and focus on empathically listening to other people to address their concerns and create an environment in which individuals and teams can grow and prosper.

USING SOFTWARE TO ASSIST IN HUMAN RESOURCE MANAGEMENT

Earlier in this chapter, you read that a simple responsibility assignment matrix (Figures 9-4 and 9-5) or histograms (Figures 9-6 and 9-7) are useful tools that can help you effectively manage human resources on projects. You can use several different software packages, including spreadsheets or project management software such as Microsoft Project 2003, to create matrixes and histograms. Many people do not realize that Project 2003 provides a variety of human resource management tools, some of which include assigning and tracking resources, resource leveling, resource usage reports, overallocated resource reports, and to-do lists. You will learn how to use many of these functions and features in Appendix A. (Note that collaborative software to help people communicate is described in Chapter 10, Project Communications Management.)

You can use Project 2003 to assign resources—including equipment, materials, facilities, or people—to tasks. Project 2003 enables you to allocate individual resources to individual projects or to pool resources and share them across multiple projects. By defining and assigning resources in Project 2003, you can:

- Keep track of the whereabouts of resources through stored information and reports on resource assignments.
- Identify potential resource shortages that could force a project to miss scheduled deadlines and possibly extend the duration of a project.
- Identify underutilized resources and reassign them, which may enable you to shorten a project's schedule and possibly reduce costs.
- Use automated leveling to make level resources easier to manage.

Just as many project management professionals are not aware of the powerful cost-management features of Project 2003, many are also unaware of its powerful human resource management features. The Microsoft Enterprise Project Management Solution provides additional human resource management capabilities. You can also purchase add-in software for Microsoft's products or purchase software from other companies to perform a variety of project human resource management functions. With the aid of this type of software, project managers can have more information available in useful formats to help them decide how to manage human resources most effectively.

Project resource management involves much more than using software to assess and track resource loading, level resources, and so on. People are the most important asset on most projects, and human resources are very different from other resources. You cannot simply replace people the same way that you would replace a piece of equipment. People need far more than a tune-up now and then to keep them performing well. It is essential to treat people with consideration and respect, to understand what motivates them, and to communicate carefully with them. What makes good project managers great is not their use of tools, but rather their ability to enable project team members to deliver the best work they possibly can on a project.

CASE WRAP-UP

After Sarah yelled right back at Ben, he said, "You're the first person who's had the guts to stand up to me." After that brief introduction, Sarah, Ben, and the other meeting participants had a good discussion about what was really happening on the F-44 upgrade project. Sarah was able to write a justification to get Ben's group special software and support to download key information from the old system so they could manage their project better. When Sarah stood nose to nose with Ben and yelled at him, she used a technique for establishing rapport called mirroring. Although Sarah was by no means a loud and obnoxious person, she saw that Ben was and decided to mirror his behavior and attitude. She put herself in his shoes for a while, and doing so helped break the ice so Sarah, Ben, and the other people at the meeting could really start communicating and working together as a team to solve their problems.

CHAPTER SUMMARY

People are the most important assets in organizations and on projects. Therefore, it is essential for project managers to be good human resource managers.

The major processes involved in project human resource management include human resource planning, acquiring project team members, developing the project team, and managing the project team.

Psychosocial issues that affect how people work and how well they work include motivation, influence and power, and effectiveness.

Maslow developed a hierarchy of needs that suggests physiological, safety, social, esteem, and self-actualization needs motivate behavior. Once a need is satisfied, it no longer serves as a motivator.

Herzberg distinguished between motivators and hygiene factors. Hygiene factors such as larger salaries or a more attractive work environment will cause dissatisfaction if not present, but do not motivate workers to do more if present. Achievement, recognition, the work itself, responsibility, and growth are factors that contribute to work satisfaction and motivate workers.

McClelland proposed the acquired-needs theory, suggesting that an individual's needs are acquired or learned over time and shaped by their life experiences. The three types of acquired needs are a need for achievement, a need for affiliation, and a need for power.

McGregor developed Theory X and Theory Y to describe different approaches to managing workers, based on assumptions of worker motivation. Research supports the use of Theory Y, which assumes that people see work as natural and indicates that the most significant rewards are the satisfaction of esteem and self-actualization needs that work can provide. According to Ouchi's Theory Z, workers can be trusted to do their jobs to their utmost ability, as long as management can be trusted to support them and look out for their well-being. Theory Z emphasizes things such as job rotation, broadening of skills, generalization versus specialization, and the need for continuous training of workers.

Thamhain and Wilemon identified nine influence bases available to project managers: authority, assignment, budget, promotion, money, penalty, work challenge, expertise, and friendship. Their research found that project success is associated with project managers who use work challenge and expertise to influence workers. Project failure is associated with using too much influence by authority, money, or penalty.

Power is the potential ability to influence behavior to get people to do things they would not otherwise do. The five main types of power are coercive power, legitimate power, expert power, reward power, and referent power.

Project managers can use Steven Covey's seven habits of highly effective people to help themselves and project teams become more effective. The seven habits include being proactive; beginning with the end in mind; putting first things first; thinking win/win; seeking first to understand, then to be understood;

achieving synergy; and sharpening the saw. Using empathic listening is a key skill of good project managers.

Human resource planning involves identifying, assigning, and documenting project roles, responsibilities, and reporting relationships. A responsibility assignment matrix (RAM), staffing management plans, resource histograms, and RACI charts are key tools for defining roles and responsibilities on projects.

Project team acquisition is getting the appropriate staff assigned to and working on the project. This is an important issue in today's competitive environment. Companies must use innovative approaches to find and retain good information technology staff.

Resource loading shows the amount of individual resources an existing schedule requires during specific time frames. Histograms show resource loading and identify overallocation of resources.

Resource leveling is a technique for resolving resource conflicts, such as overallocated resources, by delaying tasks. Leveled resources require less management, lower costs, produce fewer personnel and accounting problems, and often improve morale.

Two crucial skills of a good project manager are team development and team management. Teamwork helps people work more effectively to achieve project goals. Project managers can recommend individual training to improve skills related to teamwork, organize team-building activities for the entire project team and key stakeholders, and provide reward and recognition systems that encourage teamwork. Project managers can use several tools and techniques, including observation and conversation, project performance appraisals, conflict management, and issue logs to help them effectively manage their teams.

Spreadsheets and project management software such as Microsoft Project 2003 can help project managers in project human resource management. Software makes it easy to produce responsibility assignment matrixes, create resource histograms, identify overallocated resources, level resources, and provide various views and reports related to project human resource management.

Project human resource management involves much more than using software to facilitate organizational planning and assign resources. What makes good project managers great is their ability to enable project team members to deliver the best work they possibly can on a project.

DISCUSSION QUESTIONS

1. Discuss the changes in the job market for information technology workers. How does the job market and current state of the economy affect human resource management?

2. Summarize the processes involved in project human resource management.

3. Briefly summarize the works of Maslow, Herzberg, McClelland, McGregor, Ouchi, Thamhain and Wilemon, and Covey. How do their theories relate to project management?

4. Describe situations where it would be appropriate to create a project organizational chart, a responsibility assignment matrix, a RACI chart, and a resource histogram. Describe what these charts or matrices look like.

5. Discuss the difference between resource loading and resource leveling, and provide an example of when you would use each technique.

6. Explain two types of team-building activities described in this chapter.

7. Summarize different tools and techniques project managers can use to help them manage project teams. What can they do to manage virtual team members?

8. How can you use Project 2003 to assist in project human resource management?

EXERCISES

1. Your company is planning to launch an important new project starting January 1, which will last one year. You estimate that you will need one full-time project manager, two full-time business analysts for the first six months, two full-time senior programmers for the whole year, four full-time junior programmers for the months of July, August, and September, and one full-time technical writer for the last three months. Use Microsoft Excel to create a stacked column chart showing a resource histogram for this project, similar to the one shown in Figure 9-6. Be sure to include a legend to label the types of resources needed. Use appropriate titles and axis labels.

2. Take the MBTI test and research information on this tool. There are several Web sites that have different versions of the test, such as *www.humanmetrics.com*, *www.personalitytype.com*, and *www.keirsey.com*. Write a two-page paper describing your MBTI type and what you think about this test as a team-building tool.

3. Read Douglas Ross's article "Applying Covey's Seven Habits to a Project Management Career" (see the sixth Suggested Reading in this chapter) or another article that describes the seven habits. Summarize three of Covey's habits in your own words and give examples of how these habits would apply to project management.

4. Research recruiting and retention strategies at three different companies. What distinguishes one company from another in this area? Are strategies such as signing bonuses, tuition reimbursement, and business casual dress codes standard for new information technology workers? What strategies appeal most to you?

5. Write a one-page paper summarizing the main features of Microsoft Project 2003 that can assist project managers in human resource management. In addition, interview someone who uses Project 2003. Ask him or her if their organization uses any project human resource management

features as described in this chapter and Appendix A, and document their reasons for using or not using certain features.

RUNNING CASE

Several people working on the Recreation and Wellness Intranet Project are confused about who needs to do what for the testing portion of the project. Recall that the team members include you, a programmer/analyst and aspiring project manager; Patrick, a network specialist; Nancy, a business analyst; and Bonnie, another programmer/analyst. Tony Prince is the project manager, and he has been working closely with managers in other departments to make sure everyone knows what's going on with the project.

1. Prepare a responsibility assignment matrix based on the following information: The main tasks that need to be done for testing include writing a test plan, unit testing, integration testing for each of the main system modules (registration, tracking, and incentives), system testing, and user acceptance testing. In addition to the project team members, there is a team of user representatives available to help with testing, and Tony has also hired an outside consulting firm to help as needed. Prepare a RACI chart to help clarify roles and responsibilities for these testing tasks. Document key assumptions you make in preparing the chart.

2. The people working for the outside consulting firm and user representatives have asked you to create a resource histogram to show how many people you think the project will need for the testing and when. Assume that the consulting firm has junior and senior testers and that the user group has workers and managers. You estimate that you'll need both groups involvement in testing over a period of six weeks. Assume you'll need one senior tester for all six weeks, two junior testers for the last four weeks, two user-group workers for the first week, four user-group workers for the last three weeks, and two user-group managers for the last two weeks. Create a resource histogram, similar to the one in Figure 9-6, based on this information.

3. One of the issues in Tony's issue log is working effectively with the user group during testing. Tony knows that several of his project team members are very introverted and strong thinking types, while several members of the user group are very extroverted and strong feeling types. Write a one-page paper describing options for resolving this issue.

ADDITIONAL RUNNING CASES AND OTHER APPENDICES

Appendix C provides additional case studies and questions you can use to practice applying the concepts, tools, and techniques you are learning throughout this and subsequent chapters. Review the running cases provided in Appendix C and on the companion Web site (*www.course.com/mis/schwalbe4e*). Appendix D includes templates for various project management documents. For additional sample documents based on real projects, visit the author's Web site at *www.kathyschwalbe.com.* Appendix E and the CD-ROM included with this text include a computer simulation where you can also practice applying the project management process groups and knowledge areas.

SUGGESTED READINGS

1. Brooks, Frederick P., Jr. *The Mythical Man-Month: Essays on Software Engineering.* Reading, MA: Addison-Wesley, 1995.

   The Mythical Man-Month *is one of the classics in the field of software program management. Brooks wrote the first version of this book in 1975. He draws on his experience as the head of operating systems development for IBM's famous 360 mainframe and some of his most recent experiences.*

2. Covey, Stephen. *The 7 Habits of Highly Effective People.* New York: Simon & Schuster, 1990.

   Stephen Covey is famous for his books, audiotapes, videotapes, and seminars on improving effectiveness. This book continues to be a business bestseller. Covey also has written related books based on these seven habits. Visit his Web site at www.franklincovey.com.

3. Goleman, Daniel. *Emotional Intelligence.* New York: Bantam Books, 1997.

   Another important topic related to human resource development is emotional intelligence. Daniel Goleman has written this book and several others to help people understand what emotional intelligence is and how to help people, especially young children, develop all aspects of intelligence.

4. Herzberg, Frederick. "One More Time: How Do You Motivate Employees?" *Harvard Business Review* (February 1968): 51–62.

   This classic article describes the results of Herzberg's studies on what motivates people. He distinguishes between hygiene factors and motivating factors in a work setting and offers suggestions on how to motivate workers effectively.

5. Keirsey, David, and Ray Choiniere. *Please Understand Me II: Temperament, Character, Intelligence.* Del Mar, CA: Prometheus Nemesis Book Co., May 1998.

   Keirsey and Bates's Please Understand Me, *first published in 1978, sold nearly two million copies in its first 20 years, becoming a bestseller all over the world. Keirsey has continued to investigate personality differences, and recently published this sequel to refine his theory of the four temperaments and to define the facets of character that distinguish one person from another. You can get more information on this book and take an online version of the Keirsey Temperament Sorter test at www.keirsey.com.*

6. Ross, Douglas. "Applying Covey's Seven Habits to a Project Management Career." *PM Network*, Project Management Institute (April 1996): 26–30.

   Ross summarizes key points from Covey's seven habits and relates them to improving effectiveness in project management. This article helps project managers translate Covey's principles to their own teams, projects, and personal lives.

7. Verma, Vijay K. *Organizing Projects for Success: The Human Aspects of Project Management*. Upper Darby, PA: Project Management Institute, 1995.

   This book focuses specifically on managing human resources in a project environment. There are separate chapters on interfacing with major stakeholders, designing organizational structures, addressing important issues in project organization design, designing a project organizational structure, and making matrix structures work. Verma has also written several other books related to project human resource management.

KEY TERMS

- **coercive power** — using punishment, threats, or other negative approaches to get people to do things they do not want to do
- **deputy project managers** — people who fill in for project managers in their absence and assist them as needed, similar to the role of a vice president
- **empathic listening** — listening with the intent to understand
- **expert power** — using one's personal knowledge and expertise to get people to change their behavior
- **extrinsic motivation** — causes people to do something for a reward or to avoid a penalty
- **hierarchy of needs** — a pyramid structure illustrating Maslow's theory that people's behaviors are guided or motivated by a sequence of needs
- **intrinsic motivation** — causes people to participate in an activity for their own enjoyment
- **issue log** — a tool for managing project teams where the project manager documents, monitors, and tracks issues that need to be resolved in order for the project to run smoothly
- **legitimate power** — getting people to do things based on a position of authority
- **mirroring** — the matching of certain behaviors of the other person
- **Myers-Briggs Type Indicator (MBTI)** — a popular tool for determining personality preferences
- **organizational breakdown structure (OBS)** — a specific type of organizational chart that shows which organizational units are responsible for which work items

- **overallocation** — when more resources than are available are assigned to perform work at a given time
- **power** — the potential ability to influence behavior to get people to do things they would not otherwise do
- **RACI charts** — charts that show Responsibility, Accountability, Consultation, and Informed roles for project stakeholders
- **rapport** — a relation of harmony, conformity, accord, or affinity
- **referent power** — getting people to do things based on an individual's personal charisma
- **resource histogram** — a column chart that shows the number of resources assigned to a project over time
- **resource leveling** — a technique for resolving resource conflicts by delaying tasks
- **resource loading** — the amount of individual resources an existing schedule requires during specific time periods
- **responsibility assignment matrix (RAM)** — a matrix that maps the work of the project as described in the WBS to the people responsible for performing the work as described in the organizational breakdown structure (OBS)
- **reward power** — using incentives to induce people to do things
- **staffing management plan** — a document that describes when and how people will be added to and taken off a project team
- **subproject managers** — people responsible for managing the subprojects that a large project might be broken into
- **synergy** — an approach where the whole is greater than the sum of the parts
- **team development** — building individual and group skills to enhance project performance
- **Tuckman model** — describes five stages of team development: forming, storming, norming, performing, and adjourning

10

Project Communications Management

Objectives

After reading this chapter, you will be able to:

1. *Understand the importance of good communications on projects*
2. *Explain the elements of project communications planning, including how to create a communications management plan and perform a stakeholder communications analysis*
3. *Describe various methods for distributing project information and the advantages and disadvantages of each, discuss the importance of addressing individual communication needs, and calculate the number of communications channels on a project*
4. *Understand how the main outputs of performance reporting help stakeholders stay informed about project resources*
5. *Recognize the importance of good communications management for stakeholder relationships and for resolving issues*
6. *List various methods for improving project communications, such as managing conflicts, running effective meetings, using e-mail and other technologies effectively, and using templates*
7. *Describe how software can enhance project communications management*

OPENING CASE

Peter Gumpert worked his way up the corporate ladder in a large telecommunications company. He was intelligent, competent, and a strong leader, but a new fiber-optic undersea telecommunications program was much larger and more complicated than anything he had worked on, let alone managed. The undersea telecommunications program consisted of several distinct projects, and Peter was the program manager in charge of overseeing them all. The changing marketplace for undersea telecommunications systems and the large number of projects involved made communications and flexibility critical concerns for Peter. For missing milestone and completion dates, his company would suffer huge financial penalties ranging from thousands of dollars per day for smaller projects to more than $250,000 per day for larger projects. Many projects depended on the success of other projects, so Peter had to understand and actively manage those critical interfaces.

Peter held several informal and formal discussions with the project managers reporting to him on this program. He worked with them and his project executive assistant, Christine Braun, to develop a communications plan for the program. He was still unsure, however, of the best way to distribute information and manage all of the inevitable changes that would occur. He also wanted to develop consistent ways for all of the project managers to develop their plans and track performance without stifling their creativity and autonomy. Christine suggested that they consider using some new communications technologies to keep important project information up to date and synchronized. Although Peter knew a lot about telecommunications and laying fiber-optic lines, he was not an expert in using information technology to improve the communication process. In fact, that was part of the reason he asked Christine to be his assistant. Could they really develop a process for communicating that would be flexible and easy to use? Time was of the essence as more projects were being added to the undersea telecommunications program every week.

THE IMPORTANCE OF PROJECT COMMUNICATIONS MANAGEMENT

Many experts agree that the greatest threat to the success of any project, especially information technology projects, is a failure to communicate. Recall that the 2001 Standish Group study found the top four factors related to information technology project success were executive support, user involvement, an experienced project manager, and clear business objectives.[1] All of these factors depend on the project manager and team having good communications skills, especially with non-information technology personnel.

The information technology field is constantly changing, and these changes bring with them a great deal of technical jargon. When computer professionals have to communicate with non–computer professionals, technical jargon can often complicate matters and confuse those who aren't technically savvy. Even though most people use computers today, the gap between users and developers increases as technology advances. Of course, not every computer professional is a poor communicator, but most people in any field can improve their communication skills.

In addition, most educational systems for information technology graduates promote strong technical skills over strong communication and social skills. Most information technology–related degree programs have many technical requirements, but few require courses in communications (speaking, writing, listening), psychology, sociology, and the humanities. People often assume it is easy to pick up these soft skills, but they *are* important skills and, as such, people must learn and develop them.

Many studies have shown that information technology professionals need these soft skills just as much or even more than other skills. You cannot totally separate technical skills and soft skills when working on information technology projects. For projects to succeed, every project team member needs both types of skills and needs to develop them continuously through formal education and on-the-job training.

An article in the *Journal of Information Systems Education* on the importance of communications skills for information technology professionals presented the following conclusion:

> Based on the results of this research we can draw some general conclusions. First, it is evident that IS professionals engage in numerous verbal communication activities that are informal in nature, brief in duration, and with a small number of people at a time. Second, we can infer that most of the communication is indeed verbal in nature but sometimes it is supported by notes or graphs on a board or a handout and also by computer output. Third, it is clear that people expect their peers to listen carefully during a conversation and respond correctly to the issues at hand. Fourth, all IS professionals must be aware of the fact that they will have to engage in some form of informal public speaking. Fifth, it is evident that IS professionals must be able to communicate effectively in order to be successful in their current position but they must also be able to do so in order to move to higher positions. Since our respondents, on average, seem to have moved

---

[1] The Standish Group, "Extreme CHAOS" (2001).

throughout their IS career, from lower to higher positions, and they ranked verbal skills more important for their advancement than for their current job, the ability to communicate verbally seems to be a key factor in career advancement.[2]

This chapter will highlight key aspects of project communications management, provide some suggestions for improving communications management, and describe how software can assist in project communications management.

The goal of project communications management is to ensure timely and appropriate generation, collection, dissemination, storage, and disposition of project information. There are four main processes in project communications management:

- **Communications planning** involves determining the information and communications needs of the stakeholders: who needs what information, when will they need it, and how will the information be given to them. The output of this process is a communications management plan.

- **Information distribution** involves making needed information available to project stakeholders in a timely manner. The main outputs of this process are updates to organizational process assets and requested changes.

- **Performance reporting** involves collecting and disseminating performance information, including status reports, progress measurement, and forecasting. The outputs of this process are performance reports, forecasts, requested changes, recommended corrective actions, and updates to organizational process assets.

- **Managing stakeholders** involves managing communications to satisfy the needs and expectations of project stakeholders and to resolve issues. The outputs of this process are resolved issues, approved change requests and corrective actions, and updates to organizational process assets and the project management plan.

COMMUNICATIONS PLANNING

Because communication is so important on projects, every project should include a **communications management plan**—a document that guides project communications. This plan should be part of the overall project management plan (described in Chapter 4, Project Integration Management). The communications management plan will vary with the needs of the project, but some type of written plan should always be prepared. For example, for small projects, such as the

---

[2] Sivitanides, Marcos P., James R. Cook, Roy B. Martin, and Beverly A. Chiodo, "Verbal Communication Skills Requirements for Information Systems Professionals," *Journal of Information Systems Education* (Spring 1995): p.7(1).

Project Management Intranet Site Project described in Chapter 3, the communications management plan can be part of the team contract. The communications management plan should address the following items:

1. Stakeholder communications requirements
2. Information to be communicated, including format, content, and level of detail
3. Who will receive the information and who will produce it
4. Suggested methods or technologies for conveying the information
5. Frequency of communication
6. Escalation procedures for resolving issues
7. Revision procedures for updating the communications management plan
8. A glossary of common terminology

It is important to know what kinds of information will be distributed to which stakeholders. By analyzing stakeholder communications, you can avoid wasting time or money on creating or disseminating unnecessary information. The project's organizational chart is a starting point for identifying internal stakeholders. You must also include key stakeholders outside of the project organization such as the customer, the customer's top management, and subcontractors.

Table 10-1 provides a sample stakeholder communications analysis that shows which stakeholders should get which written communications. Note that the stakeholder communications analysis includes information such as the contact person for the information, when the information is due, and the preferred format for the information. You can create a similar table to show which stakeholders should attend which project meetings. It is always a good idea to include comment sections with these types of tables to record special considerations or details related to each stakeholder, document, meeting, and so on. Having stakeholders review and approve all stakeholder communications analysis will ensure that the information is correct and useful.

Table 10-1: Sample Stakeholder Communications Analysis

STAKEHOLDERS	DOCUMENT NAME	DOCUMENT FORMAT	CONTACT PERSON	DUE
Customer management	Monthly status report	Hard copy	Tina Erndt, Tom Silva	First of month
Customer business staff	Monthly status report	Hard copy	Julie Grant, Sergey Cristobal	First of month
Customer technical staff	Monthly status report	E-mail	Li Chau , Nancy Michaels	First of month
Internal management	Monthly status report	Hard copy	Bob Thomson	First of month

Table 10-1: Sample Stakeholder Communications Analysis (continued)

STAKEHOLDERS	DOCUMENT NAME	DOCUMENT FORMAT	CONTACT PERSON	DUE
Internal business and technical staff	Monthly status report	Intranet	Angie Liu	First of month
Training subcontractor	Training plan	Hard copy	Jonathan Kraus	11/1/2006
Software subcontractor	Software implementation plan	E-mail	Najwa Gates	6/1/2006

Comments: Put the titles and dates of documents in e-mail headings and have recipients acknowledge receipt.

Many projects do not include enough initial information on communications. Project managers, top management, and project team members assume using existing communications channels to relay project information is sufficient. The problem with using existing communications channels is that each of these groups (as well as other stakeholders) has different communications needs. Creating some sort of communications management plan and reviewing it with project stakeholders early in a project helps prevent or reduce later communication problems. If organizations work on many projects, developing some consistency in handling project communications helps the organization run smoothly.

Consistent communication helps organizations improve project communications, especially for programs composed of multiple projects. For example, Peter Gumpert, the undersea telecommunications program manager in the opening case, would benefit greatly from having a communications management plan that all of the project managers who report to him help develop and follow. Since several of the projects probably have some of the same stakeholders, it is even more important to develop a coordinated communications management plan. For example, if customers receive status reports from Peter's company that have totally different formats and do not coordinate information from related projects within the same company, they will question the ability of Peter's company to manage large programs.

Information regarding the content of essential project communications comes from the work breakdown structure (WBS). In fact, many WBSs include a section for project communications to ensure that reporting key information is a project deliverable. If reporting essential information is an activity defined in the WBS, it becomes even more important to develop a clear understanding of what project information to report, when to report it, how to report it, who is responsible for generating the report, and so on.

INFORMATION DISTRIBUTION

Getting project information to the right people at the right time and in a useful format is just as important as developing the information in the first place. The stakeholder communications analysis serves as a good starting point for information distribution. Project managers and their teams must decide who receives what information, but they must also decide the best way to distribute the information. Is it sufficient to send written reports for project information? Are meetings alone effective in distributing project information? Are meetings and written communications both required for project information? What is the best way to distribute information to virtual team members?

After answering these questions, project managers and their teams must decide the best way to distribute the information. Important considerations for information distribution include the use of technology, formal and informal communications, and the complexity of communications.

Using Technology to Enhance Information Distribution

Technology can facilitate the process of distributing information, when used properly. Using an internal project management information system, you can organize project documents, meeting minutes, customer requests, requests to change status, and so on, and make them available in an electronic format. You can store this information in local software or make it available on an intranet, an extranet, or the Internet, if the information is not sensitive. Storing templates and samples of project documents electronically can make accessing standard forms easier, thus making the information distribution process easier. You will learn more about using software to assist in project communications management later in this chapter.

Formal and Informal Methods for Distributing Information

It is not enough for project team members to submit status reports to their project managers and other stakeholders and assume that everyone who needs to know that information will read the reports. Some technical professionals might assume that submitting the appropriate status reports is sufficient. Informal verbal communications are equally effective ways to distribute information, but technical professionals tend to neglect techniques that are more informal. Often, many non-technical professionals—from colleagues to managers—may prefer to have a two-way conversation about project information, rather than reading detailed reports, e-mails, or Web pages to try to find pertinent information.

Instead of focusing on getting information by reading technical documents, many colleagues and managers want to know the people working on their projects and develop a trusting relationship with them. They use informal discussions about the project to develop these relationships. Therefore, project managers must be good at nurturing relationships through good communication. Many experts believe that the difference between good project managers and excellent project managers is their ability to nurture relationships and use empathic listening skills, as described in Chapter 9, Project Human Resource Management.

Effective distribution of information depends on project managers and project team members having good communication skills. Communicating includes many different dimensions such as writing, speaking, and listening, and project personnel need to use all of these dimensions in their daily routines. In addition, different people respond positively to different levels or types of communication. For example, a project sponsor may prefer to stay informed through informal discussions held once a week over coffee. The project manager needs to be aware of and take advantage of this special communication need. The project sponsor will give better feedback about the project during these informal talks than he or she could give through some other form of communication. Informal conversations allow the project sponsor to exercise his or her role of leadership and provide insights and information that are critical to the success of the project and the organization as a whole. Short face-to-face meetings are often more effective than electronic communications, particularly for sensitive information.

⊗ What Went Wrong?

A well-publicized example of misuse of e-mail comes from the 1998 Justice Department's high-profile, antitrust suit against Microsoft. E-mail emerged as a star witness in the case. Many executives sent messages that they should have never put in writing. The court used e-mail as evidence, even though the senders of the notes said the information was being interpreted out of context.[3]

Several companies have strict policies about using e-mail, instant messaging, or other technologies at work. Employees have been fired for sending pornographic information via e-mail or for viewing inappropriate Web sites at work.

Another, more amusing, example of miscommunication comes from a director of communications at a large firm:

> I was asked to prepare a memo reviewing our company's training programs and materials. In the body of the memo in one of the sentences, I mentioned the "*pedagogical approach*" used by one of the training manuals. The day after I routed the memo to the executive committee, I was called into the HR director's office, and told that the executive vice president wanted me out of the building by lunch. When

[3] Harmon, Amy, "E-mail Comes Back to Haunt Companies," *Minneapolis Star Tribune* (from the *New York Times*) (November 29, 1998).

I asked why, I was told that she wouldn't stand for perverts (pedophiles?) working in her company. Finally, he showed me her copy of the memo, with her demand that I be fired—and the word "*pedagogical*" circled in red. The HR manager was fairly reasonable, and once he looked the word up in his dictionary and made a copy of the definition to send back to her, he told me not to worry. He would take care of it. Two days later, a memo to the entire staff came out directing us that no words which could not be found in the local Sunday newspaper could be used in company memos. A month later, I resigned. In accordance with company policy, I created my resignation memo by pasting words together from the Sunday paper.[4]

Distributing Important Information in an Effective and Timely Manner

Many written reports neglect to provide the important information that good managers and technical people have a knack for asking about. For example, it is important to include detailed technical information that will affect critical performance features of products or services the company is producing as part of a project. It is even more important to document any changes in technical specifications that might affect product performance. For example, if the undersea telecommunications program included a project to purchase and provide special diving gear, and the supplier who provided the oxygen tanks enhanced the tanks so divers could stay under water longer, it would be very important to let other people know about this new capability. The information should not be buried in an attachment with the supplier's new product brochure.

People also have a tendency to not want to report bad information. If the oxygen tank vendor was behind on production, the person in charge of the project to purchase the tanks might wait until the last minute to report this critical information. Oral communication via meetings and informal talks helps bring important information—positive or negative—out into the open.

Oral communication also helps build stronger relationships among project personnel and project stakeholders. People make or break projects, and people like to interact with each other to get a true feeling for how a project is going. Many people cite research that says in a face-to-face interaction, 58 percent of communication is through body language, 35 percent through how the words are said, and a mere 7 percent through the content or words that are spoken. The author of this information (see *Silent Messages* by Albert Mehrabian, 1980) was careful to note that these percentages were specific findings for a specific set of variables. Even if the actual percentages are different in verbal project communications today, it is safe to say that it is important to pay attention to more than just the actual words someone is saying. A person's tone of voice and body language say a lot about how they really feel.

4 Projectzone, "Humor" (*http://www.corporatedump.com/dilbertmanagers.html*) (2004).

Since information technology projects often require a lot of coordination, it is a good idea to have short, frequent meetings. For example, some information technology project managers require all project personnel to attend a "stand-up" meeting every week or even every morning, depending on the project needs. Stand-up meetings have no chairs, and the lack of chairs forces people to focus on what they really need to communicate. If people can't meet face to face, they are often in constant communications via cell phones, e-mail, instant messaging, or other technologies.

To encourage face-to-face, informal communications, some companies have instituted policies that workers cannot use e-mail between certain hours of the business day or even entire days of the week. For example, in the summer of 2004, Jeremy Burton, vice president of marketing at a large Silicon Valley company, decreed that in his department, Fridays would be e-mail-free. The 240 people in his department had to use the phone or meet face-to-face with people, and violators who did use e-mail were fined.[5]

Selecting the Appropriate Communications Medium

Table 10-2 provides guidelines from Practical Communications, Inc., a communications consulting firm, about how well different types of media, such as hard copy, phone calls, voice mail, e-mail, meetings, and Web sites, are suited to different communication needs. For example, if you were trying to assess commitment of project stakeholders, a meeting would be the most appropriate medium to use. A phone call would be adequate, but the other media would not be appropriate. Project managers must assess the needs of the organization, the project, and individuals in determining which communication medium to use, and when.

Table 10-2: Media Choice Table

KEY: 1 = EXCELLENT   2= ADEQUATE   3 = INAPPROPRIATE

How Well Medium is Suited to:	Hard Copy	Phone Call	Voice Mail	E-mail	Meeting	Web Site
Assessing commitment	3	2	3	3	1	3
Building consensus	3	2	3	3	1	3
Mediating a conflict	3	2	3	3	1	3
Resolving a misunderstanding	3	1	3	3	2	3
Addressing negative behavior	3	2	3	2	1	3
Expressing support/appreciation	1	2	2	1	2	3

[5] Walker, Marion, "E-mail is out at this office, at least on Fridays," *Star Tribune* (November 10, 2004).

Table 10-2: Media Choice Table (continued)

KEY: 1 = EXCELLENT   2= ADEQUATE   3 = INAPPROPRIATE

HOW WELL MEDIUM IS SUITED TO:	HARD COPY	PHONE CALL	VOICE MAIL	E-MAIL	MEETING	WEB SITE
Encouraging creative thinking	2	3	3	1	3	3
Making an ironic statement	3	2	2	3	1	3
Conveying a reference document	1	3	3	3	3	2
Reinforcing one's authority	1	2	3	3	1	1
Providing a permanent record	1	3	3	1	3	3
Maintaining confidentiality	2	1	2	3	1	3
Conveying simple information	3	1	1	1	2	3
Asking an informational question	3	1	1	1	3	3
Making a simple request	3	1	1	1	3	3
Giving complex instructions	3	3	2	2	1	2
Addressing many people	2	3 or 1*	2	2	3	1

*Depends on system functionality
Galati, Tess. *Email Composition and Communication (EmC2)*. Practical Communications, Inc. (*www.praccom.com*) (2001).

🎥 Media Snapshot

Although most projects do not use live video as a medium for sending project information, the technology is becoming more available and less expensive. You can reach many people at once using live video, and viewers can see and hear important information.

For example, Microsoft had been experimenting with its new conferencing product, Livemeeting. Anoop Gupta, a vice president of Microsoft's real-time collaboration group, says that one in every five face-to-face meetings can be replaced with Web conferencing tools, and Microsoft estimates that it will save $70 million in reduced travel in one year alone.[6]

However, any live communication broadcast can also backfire, especially if millions of people are watching. In fact, one event had a major impact on the entire broadcasting industry, causing television and radio stations to use several second delays to prevent offensive video or audio from reaching the airwaves. If you didn't watch the half-time show of the 2004 Super Bowl, you were probably one of the millions of people who searched the Internet for information on Janet Jackson's "wardrobe malfunction." In fact, Jackson's breast exposure was the most-searched event in the history of the Internet.[7] Reuters reported on September 22, 2004 that the Federal

[6] Lohr, Steve, "Ambitious Package to Raise Productivity (and Microsoft's Profit)," *The New York Times* (August 16, 2004).

[7] SearchEngineJournal, "Janet Jackson Tops Internet Search Terms" (*http://www.searchenginejournal.com/index.php?p=250&c=1*) (February 5, 2004).

Communication Commission had officially voted to fine each of the 20 stations owned by the CBS television network $27,500 for violating indecency rules. The fine was the maximum allowed by law at the time, and Congress is considering legislation to increase the fine to as much as $500,000 per incident.[8]

Understanding Group and Individual Communication Needs

Many top managers think they can just add more people to a project that is falling behind schedule. Unfortunately, this approach often causes more setbacks because of the increased complexity of communications. In his popular book, *The Mythical Man-Month*, Frederick Brooks illustrates this concept very clearly. People are not interchangeable parts. You cannot assume that a task originally scheduled to take two months of one person's time can be done in one month by two people. A popular analogy is that you cannot take nine women and produce a baby in one month!

In addition to understanding that people are not interchangeable, it is also important to understand individuals' personal preferences for communications. As described in Chapter 9, people have different personality traits, which often affect their communication preferences. For example, if you want to praise a project team member for doing a good job, an introvert would be more comfortable receiving that praise in private while an extrovert would like everyone to hear about his or her good work. An intuitive person would want to understand how something fits into the big picture, while a sensing person would prefer to have more focused, step-by-step details. A strong thinker would want to know the logic behind information, while a feeling person would want to know how the information affects him or her personally as well as other people. Someone who is a judging person would be very driven to meet deadlines with few reminders while a perceiving person would need more assistance in developing and following plans.

Rarely does the receiver interpret a message exactly as the sender intended. Therefore, it is important to provide many methods of communication and an environment that promotes open dialogue. It is important for project managers and their teams to be aware of their own communication styles and preferences and those of other project stakeholders. As described in the previous chapter, many information technology professionals have different personality traits than the general population, such as being more introverted, intuitive, and oriented to thinking (as opposed to feeling). These personality differences can lead to miscommunication with people who are extroverted, sensation-oriented, and feeling types. For example, a user guide written by an information technology

[8] Reuters, "TV Stations Fined for Janet Jackson Breast Flash," *www.reuters.com* (September 22, 2004).

professional might not provide the detailed steps most users need. Many users also prefer face-to-face meetings to learn how to use a new system instead of trying to follow a cryptic user guide.

Geographic location and cultural background also affect the complexity of project communications. If project stakeholders are in different countries, it is often difficult or impossible to schedule times for two-way communication during normal working hours. Language barriers can also cause communication problems. The same word may have very different meanings in different languages. Times, dates, and other units of measure are also interpreted differently. People from some cultures also prefer to communicate in ways that may be uncomfortable to others. For example, managers in some countries still do not allow workers of lower ranks or women to give formal presentations. Some cultures also reserve written documents for binding commitments.

Setting the Stage for Communicating Bad News

It is also important to put information in context, especially if it's bad news. An amusing example of putting bad news in perspective is in the following letter from a college student to her parents. Variations of this letter can be found on numerous Web sites.

> Dear Mom and Dad, or should I say Grandma & Grandpa,
>
> Yes, I am pregnant. No, I'm not married yet since Larry, my boyfriend, is out of a job. Larry's employers just don't seem to appreciate the skills he has learned since he quit high school. Larry looks much younger than you, Dad, even though he is three years older. I'm quitting college and getting a job so we can get an apartment before the baby is born. I found a beautiful apartment above a 24-hour auto repair garage with good insulation so the exhaust fumes and noise won't bother us.
>
> I'm very happy. I thought you would be too.
>
> Love, Ashley
>
> P.S.   There is no Larry. I'm not pregnant. I'm not getting married. I'm not quitting school, but I am getting a "D" in Chemistry. I just wanted you to have some perspective.

Determining the Number of Communications Channels

Another important aspect of information distribution is the number of people involved in a project. As the number of people involved increases, the complexity of communications increases because there are more communications channels or pathways through which people can communicate. There is a simple formula for determining the number of communications channels as the number of people involved in a project increases. You can calculate the number of communications channels as follows:

number of communications channels = $\dfrac{n(n-1)}{2}$

where *n* is the number of people involved.

For example, two people have one communications channel: $(2(2-1))/2 = 1$. Three people have three channels: $(3(3-1))/2 = 3$. Four people have six channels, five people have ten, and so on. Figure 10-1 illustrates this concept. You can see that as the number of people communicating increases above three, the number of communications channels increases rapidly. Project managers should try to limit the size of teams or sub teams to avoid making communications too complex. For example, if a team of three people are working together on one particular project task, they have three communications channels. If you add two more people to their team, you would have ten communications channels, an increase of seven. If you added three more people instead of two, you'd have twelve communication channels. You can see how quickly communications becomes more complex as you increase team size.

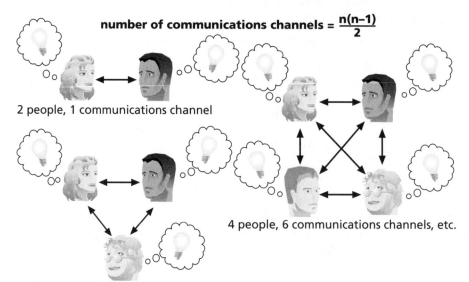

number of communications channels = $\dfrac{n(n-1)}{2}$

2 people, 1 communications channel

3 people, 3 communications channels

4 people, 6 communications channels, etc.

Figure 10-1. The Impact of the Number of People on Communications Channels

Good communicators consider many factors before deciding how to distribute information, including the size of the group, the type of information, and the appropriate communication medium to use. People tend to overuse e-mail because it is an easy, inexpensive way to send information to a lot of people. When asked why you cannot always send an e-mail to a team of 100 people, just as you would to a team of five, one CIO answered, "As a group increases in size, you have a whole slew of management challenges. Communicating badly exponentially increases the possibility of making fatal mistakes. A large-scale project has a lot of moving parts, which makes it that much easier to break down. Communication is the oil that keeps everything working properly. It's much easier to address an atmosphere of distrust among a group of five team members than it is with a team of 500 members."[9]

However, there are situations in which you cannot have face-to-face meetings and must e-mail a large group of people. Many information technology professionals work on virtual projects where they never meet their project sponsors, other team members, or other project stakeholders. In a virtual project environment, it is crucial for project managers to develop clear communication procedures. They can and must use e-mail, instant messaging, discussion threads, project Web sites, and other technologies to communicate most information. They might be able to use phone calls or other media occasionally, but in general, they must rely on good written communications.

As you can see, information distribution involves more than creating and sending status reports or holding periodic meetings. Many good project managers know their personal strengths and weaknesses in this area and surround themselves with people who complement their skills, just as Peter Gumpert in the opening case did in asking Christine to be his assistant. It is good practice to share the responsibility for project communications management with the entire project team.

PERFORMANCE REPORTING

Performance reporting keeps stakeholders informed about how resources are being used to achieve project objectives. Work performance information and measurements, forecasted completion dates, quality control measurements, the project management plan, approved change requests, and deliverables are all important inputs to performance reporting. Two key outputs of performance reporting are performance reports and forecasts. Performance reports are normally provided as status reports or progress reports. Many people use the two terms interchangeably, but some people distinguish between them as follows:

- **Status reports** describe where the project stands at a specific point in time. Recall the importance of the triple constraint. Status reports address where the project stands in terms of meeting scope, time, and cost goals. How much money has been spent to date? How long did it take to do certain tasks? Is work being accomplished as planned? Status reports can take various formats depending on the stakeholders' needs.

[9] Hildenbrand, Carol, "Loud and Clear," *CIO Magazine* (April 15, 1996), *http://www.cio.com/archive/041596_qa.html.*

- **Progress reports** describe what the project team has accomplished during a certain period. Many projects have each team member prepare a monthly or sometimes weekly progress report. Team leaders often create consolidated progress reports based on the information received from team members. A sample template for a monthly progress report is provided later in this chapter.

Forecasts predict future project status and progress based on past information and trends. How long will it take to finish the project based on how things are going? How much more money will be needed to complete the project? Project managers can also use earned value management (see Chapter 7, Project Cost Management) to answer these questions by estimating the budget at completion and projected completion date based on how the project is progressing.

Another important technique for performance reporting is the status review meeting. Status review meetings, as described in Chapter 4, Project Integration Management, are a good way to highlight information provided in important project documents, empower people to be accountable for their work, and have face-to-face discussions about important project issues. Many program and project managers hold periodic status review meetings to exchange important project information and motivate people to make progress on their parts of the project. Likewise, many top managers hold monthly or quarterly status review meetings where program and project managers must report overall status information.

Status review meetings sometimes become battlegrounds where conflicts between different parties come to a head. Project managers or higher-level top managers should set ground rules for status review meetings to control the amount of conflict and should work to resolve any potential problems. It is important to remember that project stakeholders should work together to address performance problems.

MANAGING STAKEHOLDERS

Project managers must understand and work with various stakeholders; therefore, they should specifically address how they can use communications to satisfy the needs and expectations of project stakeholders. In addition, project managers need to devise a way to identify and resolve issues. Two important tools to assist in these areas include using an expectations management matrix and an issue log.

Recall that project success is often measured in different ways. Many studies define project success as meeting project scope, time, *and* cost goals. Many practitioners, however, define project success as satisfying the customer/sponsor, knowing that it's rare to meet scope, time, and cost goals without modifying at least one goal. Project sponsors can usually rank scope, time, and cost goals in order of importance and provide guidelines on how to balance the triple constraint. This ranking is shown in an **expectations management matrix**, which can help clarify expectations. For example, Table 10-3 shows an expectations management matrix that Peter's project managers from the opening case could use to help manage their key stakeholders. The expectations management matrix includes a list of measures of success as well as priorities, expectations, and guidelines related to each measure. You could add additional measures of success, such as meeting quality expectations, achieving a certain customer satisfaction rating, meeting ROI projections after the project is completed, and so on to the matrix to meet individual project needs.

Table 10-3: Expectations Management Matrix

MEASURE OF SUCCESS	PRIORITY	EXPECTATIONS	GUIDELINES
Scope	2	The scope statement clearly defines mandatory requirements and optional requirements.	Focus on meeting mandatory requirements before considering optional ones.
Time	1	There is no give in the project completion date. Every major deadline must be met, and the schedule is very realistic.	The project sponsor and program manager must be alerted if there are any issues that might affect meeting schedule goals.
Cost	3	This project is crucial to the organization. If you can clearly justify the need for more funds, they can be made available.	There are strict rules for project expenditures and escalation procedures. Cost is very important, but it takes a back seat to meeting schedule and then scope goals.
Etc.			

Another tool to help manage stakeholders is an issue log. An **issue** is a matter under question or dispute that could impede project success. An **issue log** is a tool to document and monitor the resolution of project issues. Table 10-4 shows part of an issue log that one of Peter's project managers could use to help document and manage the resolution of issues. The issue log includes columns for the issue number, issue description, impact of the issue on the project, the date the issue was reported and by whom, who the issue resolution was assigned to, the priority of the issue (High, Medium, or Low), the due date to report back on the issue, and comments related to the issue. Project managers can tailor the format of issue logs as needed. It's important to resolve issues as soon as possible, so that the project can proceed through all activities. It's also important not to get too bogged down in issues. Some project managers might choose not to document low-priority issues or small issues that can be resolved without logging them.

Table 10-4: Issue Log

Issue #	Issue Description	Impact on Project	Date Reported	Reported By	Assigned To	Priority (M/H/L)	Due Date	Status	Comments
1	Servers cost 10% more than planned	Slight increase in project cost	5/15	Jean	Oded	M	6/15	Closed	The sponsor agreed to provide additional funds to meet the deadline.
2	Two people left the project	Need to reassign personnel	9/26	Gaurav	Karen	H	10/2	Open	If Karen cannot reassign people within a week, she should talk to Peter directly.
etc.									

*You can find a template for an issue log on the companion Web site under Appendix D.

Notice that understanding the stakeholders' expectations can help in managing issues. If the project manager knows that cost is not as high a priority as the schedule, he or she will know that it shouldn't be too difficult to ask the project sponsor for needed funds, as long as there is good logic behind the request. Unresolved issues can be a major source of conflict and result in not meeting stakeholder expectations.

SUGGESTIONS FOR IMPROVING PROJECT COMMUNICATIONS

You have seen that good communication is vital to the management and success of information technology projects; you have also learned that project communications management can ensure that essential information reaches the right people at the right time, that feedback and reports are appropriate and useful, and that there is a formalized process of stakeholder management. This section highlights a few areas that all project managers and project team The text below provides guidelines for managing conflict, developing better communication skills, running effective meetings, using e-mail effectively, using templates for project communications, and developing a communications infrastructure.

Using Communication Skills to Manage Conflict

Most large information technology projects are high-stake endeavors that are highly visible within organizations. They require tremendous effort from team members, are expensive, commandeer significant resources, and can have an extensive impact on the way work is done in an organization. When the stakes are high, conflict is never far away; when the potential for conflict is high, good communication is a necessity.

Chapter 6, Project Time Management, explained that schedule issues cause the most conflicts over the project life cycle and provided suggestions for improving project scheduling. Other conflicts occur over project priorities, staffing, technical issues, administrative procedures, personalities, and cost. It is crucial for project managers to develop and use their human resources and communication skills to help identify and manage conflict on projects. Project managers should lead their teams in developing norms for dealing with various types of conflicts that might arise on their projects. For example, team members should know that disrespectful behavior toward any project stakeholder is inappropriate, and that team members are expected to try to work out small conflicts themselves before elevating them to higher levels. Blake and Mouton (1964) delineated five basic modes for handling conflicts: confrontation, compromise, smoothing, forcing, and withdrawal.

1. *Confrontation.* When using the **confrontation mode**, project managers directly face a conflict using a problem-solving approach that allows affected parties to work through their disagreements. This approach is also called the problem-solving mode.
2. *Compromise.* With the **compromise mode**, project managers use a give-and-take approach to resolving conflicts. They bargain and search for solutions that bring some degree of satisfaction to all the parties in a dispute.
3. *Smoothing.* When using the **smoothing mode**, the project manager de-emphasizes or avoids areas of differences and emphasizes areas of agreement.
4. *Forcing.* The **forcing mode** can be viewed as the win-lose approach to conflict resolution. Project managers exert their viewpoint at the potential expense of another viewpoint. Managers who are very competitive or autocratic in their management style might favor this approach.
5. *Withdrawal.* When using the **withdrawal mode**, project managers retreat or withdraw from an actual or potential disagreement. This approach is the least desirable conflict-handling mode.

Research indicates that project managers favor using confrontation for conflict resolution over the other four modes. The term confrontation may be misleading. This mode really focuses on addressing conflicts using a problem-solving approach. Using Stephen Covey's paradigms of interdependence, this mode focuses on a win-win approach. All parties work together to find the best way to solve the conflict. The next most favored approach to conflict resolution is compromise. Successful project managers are less likely to use smoothing, forcing, or withdrawal than they are to use confrontation or compromise.

Project managers must also realize that not all conflict is bad. In fact, conflict can often be good. Conflict often produces important results, such as new ideas, better alternatives, and motivation to work harder and more collabora-tively. Project team members may become stagnant or develop **groupthink**—conformance to the values or ethical standards of a group—if there are no conflicting viewpoints on various aspects of a project. Research by Karen Jehn, Professor of Management at Wharton, suggests that task-related conflict, which is derived from differences over team objectives and how to achieve them, often improves team performance. Emotional conflict, however, which stems from personality clashes and misunderstandings, often depresses team perfor-mance.[10] Project managers should create an environment that encourages and maintains the positive and productive aspects of conflict.

Several organizations are emphasizing the importance of conflict manage-ment. For example, an innovative program at California State University, Monterey Bay, has incorporated conflict resolution as one of the program's 11 major learning objectives that students must know and understand in order to graduate. The academic program, based at the Institute for Community Collaborative Studies, focuses on preparing students for careers in the field of health and human services. The program's belief is that collaboration is an essential aspect of success for workers in the modern health and human ser-vices delivery field. Conflict resolution is a core competency for developing col-laboration skills and is important for project managers in any field.

Developing Better Communication Skills

Some people seem to be born with great communication skills. Others seem to have a knack for picking up technical skills. It is rare to find someone with a natural ability for both. Both communication and technical skills, however, can be developed. Most information technology professionals enter the field because of their technical skills. Most find, however, that communication skills are the key to advancing in their careers, especially if they want to become good project managers.

Most companies spend a lot of money on technical training for their employees, even when employees might benefit more from communications training. Individual employees are also more likely to enroll voluntarily in classes on the latest technology than those on developing their soft skills. Communication skills training usually includes role-playing activities in which participants learn concepts such as building rapport, as described in Chapter 9, Project Human Resource Management. Training sessions also give participants a chance to develop specific skills in small groups. Training sessions that focus on presentation skills usually use video to record the participants' presenta-tions. Most people are surprised to see some of their mannerisms on tape and enjoy the challenge of improving their skills. A minimal investment in com-munication and presentation training can have a tremendous payback to individuals, their projects, and their organizations. These skills also have a

---

[10] *Wharton Leadership Digest*, "Constructive Team Conflict" (March, 1997): p.1(6).

much longer shelf life than many of the skills learned in technical training courses.

As organizations become more global, they realize that they must also invest in ways to improve communication with people from different countries and cultures. For example, many Americans are raised to speak their minds, while in some other cultures people are offended by outspokenness. Not understanding how to communicate effectively with other cultures and people of diverse backgrounds hurts projects and businesses. Many training courses are available to educate people in cultural awareness, international business, and international team building.

It takes leadership to help improve communication. If top management lets employees give poor presentations, write sloppy reports, offend people from different cultures, or behave poorly at meetings, the employees will not want to improve their communication skills. Top management must set high expectations and lead by example. Some organizations send all information technology professionals to training that includes development of technical *and* communication skills. Successful organizations allocate time in project schedules for preparing drafts of important reports and presentations and incorporating feedback on the drafts. It is good practice to include time for informal meetings with customers to help develop relationships and provide staff to assist in relationship management. As with any other goal, improving communication can be achieved with proper planning, support, and leadership from top management.

Running Effective Meetings

A well-run meeting can be a vehicle for fostering team building and reinforcing expectations, roles, relationships, and commitment to the project. However, a poorly run meeting can have a detrimental effect on a project. For example, a terrible **kickoff meeting**—a meeting held at the beginning of a project or project phase where all major project stakeholders discuss project objectives, plans, and so on—may cause some important stakeholders to decide not to support the project further. Many people complain about the time they waste in unnecessary or poorly planned and poorly executed meetings. Following are some guidelines to help improve time spent at meetings:

- *Determine if a meeting can be avoided.* Do not have a meeting if there is a better way of achieving the objective at hand. For example, a project manager might know that he or she needs approval from a top manager to hire another person for the project team. It could take a week or longer to schedule even a ten-minute meeting on the top manager's calendar. Instead, an e-mail or phone call describing the situation and justifying the request is a faster, more effective approach than having a meeting. However, many times you do need a face-to-face meeting, and it would not be appropriate to try to use e-mail or a phone call. Consider which medium would be most effective, as described earlier.

- *Define the purpose and intended outcome of the meeting.* Be specific about what should happen as a result of the meeting. Is the purpose to brainstorm ideas, provide status information, or solve a problem? Make the purpose of a meeting very clear to all meeting planners and participants. For example, if a project manager calls a meeting of all project team members without knowing the true purpose of the meeting, everyone will start focusing on their own agendas and very little will be accomplished. All meetings should have a purpose and intended outcome.

- *Determine who should attend the meeting.* Do certain stakeholders have to be at a meeting to make it effective? Should only the project team leaders attend a meeting, or should the entire project team be involved? Many meetings are most effective with the minimum number of participants possible, especially if decisions must be made. Other meetings require many attendees. It is important to determine who should attend a meeting based on the purpose and intended outcome of the meeting.

- *Provide an agenda to participants before the meeting.* Meetings are most effective when the participants come prepared. Did they read reports before the meeting? Did they collect necessary information? Some professionals refuse to attend meetings if they do not have an agenda ahead of time. Insisting on an agenda forces meeting organizers to plan the meeting and gives potential attendees the chance to decide whether they really need to attend the meeting.

- *Prepare handouts and visual aids, and make logistical arrangements ahead of time.* By creating handouts and visual aids, the meeting organizers must organize their thoughts and ideas. This usually helps the entire meeting run more effectively. It is also important to make logistical arrangements by booking an appropriate room, having necessary equipment available, and providing refreshments or entire meals, if appropriate. It takes time to plan for effective meetings. Project managers and their team members must take time to prepare for meetings, especially important ones with key stakeholders.

- *Run the meeting professionally.* Introduce people, restate the purpose of the meeting, and state any ground rules that attendees should follow. Have someone facilitate the meeting to make sure important items are discussed, watch the time, encourage participation, summarize key issues, and clarify decisions and action items. Designate someone to take minutes and send the minutes out soon after the meeting. Minutes should be short and focus on the crucial decisions and action items from the meeting.

- *Build relationships.* Depending on the culture of the organization and project, it may help to build relationships by making meetings fun experiences. For example, it may be appropriate to use humor, refreshments, or prizes for good ideas to keep meeting participants actively involved. If used effectively, meetings are a good way to build relationships.

Using E-Mail Effectively

Since most people use e-mail now, communications should improve, right? Not necessarily. In fact, few people have received any training or guidelines on when or how to use e-mail. As discussed earlier in this chapter, e-mail is not an appropriate medium for several types of communications. The media choice table (Table 10-2) suggests that e-mail is not appropriate for assessing commitment, building consensus, mediating a conflict, resolving a misunderstanding, making an ironic statement, conveying a reference document, reinforcing one's authority, or maintaining confidentiality.

Even if people do know when to use e-mail for project communications, they also need to know how to use it. New features are added to e-mail programs with every new release, but often users are unaware of these features and do not receive any training on how to use them. Do you know how to file your e-mail messages, or do you have hundreds of e-mail messages sitting in your Inbox? Do you know how to use your address book or how to create distribution lists? Have you ever used sorting features to find e-mail messages by date or author or key words? Do you use filtering software to prevent spam? Do you automatically delete messages you think are spam or forwarded jokes or stories from friends you just don't care to read?

Even if you know how to use all the features of an e-mail system, you will likely need to learn how to put ideas into words clearly. For example, the subject line for any e-mail messages you write should clearly state the intention of the e-mail. A business professional that is not a very good writer may prefer to talk to people than to send an e-mail message. Poor writing often leads to misunderstandings and confusion.

Unwanted e-mail also increases as advertisers use e-mail as an inexpensive way to reach potential customers. Most e-mail systems can block unsolicited e-mail or spam, but few users know how to implement or customize this feature. As the Media Snapshot in Chapter 8 described, people are less productive if they don't know how to block or quickly delete spam. Project managers should do whatever they can to help their project stakeholders use e-mail effectively and not waste time with unwanted e-mails.

The following guidelines will help you use e-mail more effectively:

- Information sent via e-mail should be appropriate for that medium, versus other media. If you can communicate the information better with a phone call or meeting, for example, then do so.
- Be sure to send the e-mail to the right people. Do not automatically "reply to all" if you do not need to.
- Use meaningful subject lines so readers can quickly see what information the message will contain. If the entire message can be put in the subject line, put it there. For example, if a meeting is cancelled, just type that in

the subject line. Also, do not continue replying to e-mail messages without changing the subject. The subject should always relate to the latest correspondence.

- Limit the content of the e-mail to one main subject. Send a second or third e-mail if it relates to a different subject.
- The body of the e-mail should be as clear and concise as possible and you should always reread your e-mail before you send it. Also, be sure to check your spelling using the spell check function. If you have three questions you need answered, number them as question 1, 2, and 3.
- Limit the number and size of e-mail attachments. If you can include a link to an online version of a document instead of attaching a file, do so.
- Delete e-mail that you do not need to save or respond to. Do not even open e-mail that you know is not important, such as spam. Use the e-mail blocking feature of the software, if available, to block unwanted junk mail.
- Make sure your virus protection software is up to date. Never open e-mail attachments if you do not trust the source.
- Respond to e-mail quickly, if possible. It will take you longer to open and read it again later. In addition, if you send an e-mail that does not require a response, make that clear as well.
- If you need to keep e-mail, file each message appropriately. Create folders with meaningful names to file the e-mail messages you want to keep. File them as soon as possible.
- Learn how to use important features of your e-mail software such as mailing lists, sorting, and so on.

Using Templates for Project Communications

Many intelligent people have a hard time writing a performance report or preparing a ten-minute technical presentation for a customer review. Some people in these situations are too embarrassed to ask for help. To make preparing project communications easier, project managers need to provide examples and templates for common project communications items such as project descriptions, project charters, monthly performance reports, oral status reports, and so on. Good documentation from past projects can be an ample source of examples. Samples and templates of both written and oral reports are particularly helpful for people who have never before had to write project documents or give project presentations. Finding, developing, and sharing relevant templates and sample documents are important tasks for many project managers. Several examples of project documentation such as a business case, project charter, scope statement, stakeholder analysis, WBS, Gantt chart, cost estimate, and so on are provided throughout this text. Appendix D includes information on these and other templates, as well as resource information for finding additional templates. The companion Web site for this text includes the actual files used in creating the templates. A few of these templates and guidelines for preparing them are provided in this section.

Figure 10-2 shows a sample template for a one-page project description. This form could be used to show a "snapshot" of an entire project on one page. For example, top managers might require that all project managers provide a brief project description as part of a quarterly management review meeting. Peter Gumpert, the program manager in the opening case, might request this type of document from all of the project managers working for him to get an overall picture of what each project involves. According to Figure 10-2, a project description should include the project objective, scope, assumptions, cost information, and schedule information. This template suggests including information from the project's Gantt chart to highlight key deliverables and other milestones.

Project X Descripton

Objective: Describe the objective of the project in one or two sentences. Focus on the business benefits of doing the project.

Scope: Briefly describe the scope of the project. What business functions are involved, and what are the main products the project will produce?

Assumptions: Summarize the most critical assumptions for the project.

Cost: Provide the total estimated cost of the project. If desired, list the total cost each year.

Schedule: Provide summary information from the project's Gantt chart, as shown. Focus on summary tasks and milestones.

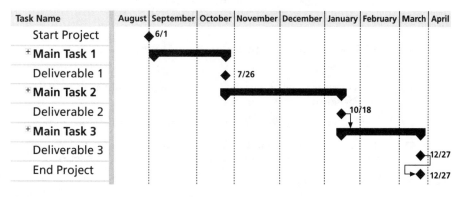

Figure 10-2. Sample Template for a Project Description

Table 10-5 shows a template for a monthly progress report. Sections of the progress report include accomplishments from the current period, plans for the next period, issues, and project changes. Table 3-13 in Chapter 3 provided a sample of a weekly status report that includes similar information. Recall that progress reports focus on accomplishments during a specific time period while status reports focus on where the project stands at a certain point in time. Because progress and status reports are important ways to communicate project information, it is important for project teams to tailor them to meet their specific needs. Some organizations, such as JWD Consulting from Chapter 3, combine both progress and status information on the same template.

Table 10-5: Sample Template for a Monthly Progress Report

I. Accomplishments for January (or appropriate month):

- Describe most important accomplishments. Relate them to project's Gantt chart.

- Describe other important accomplishments, one bullet for each. If any issues were resolved from the previous month, list them as accomplishments.

II. Plans for February (or following month):

- Describe most important items to accomplish in the next month. Again, relate them to project's Gantt chart.

- Describe other important items to accomplish, one bullet for each.

III. Issues: Briefly list important issues that surfaced or are still important. Managers hate surprises and want to help the project succeed, so be sure to list issues.

IV. Project Changes (Dates and Description): List any approved or requested changes to the project. Include the date of the change and a brief description.

Table 10-6 provides an exhaustive list of all of the documentation that should be organized and filed at the end of a major project. From this list, you can see that a large project can generate a lot of documentation. In fact, some project professionals have observed that documentation for designing an airplane usually weighs more than the airplane itself. (Smaller projects usually generate much less documentation!)

Table 10-6: Final Project Documentation Items

I. Project description

II. Project proposal and backup data (request for proposal, statement of work, proposal correspondence, and so on)

III. Original and revised contract information and client acceptance documents

IV. Original and revised project plans and schedules (WBS, Gantt charts and network diagrams, cost estimates, communications management plan, etc.)

V. Design documents

VI. Final project report

VII. Deliverables, as appropriate

Table 10-6: Final Project Documentation Items (continued)

VIII.	Audit reports
IX.	Lessons-learned reports
X.	Copies of all status reports, meeting minutes, change notices, and other written and electronic communications

The project manager and project team members should all prepare a **lessons-learned report**—a reflective statement documenting important things they have learned from working on the project. The project manager often combines information from all of the lessons-learned reports into a project summary report. See Chapter 3 for an example of this type of lessons-learned report. Some items discussed in lessons-learned reports include reflections on whether project goals were met, whether the project was successful or not, the causes of variances on the project, the reasoning behind corrective actions chosen, the use of different project management tools and techniques, and personal words of wisdom based on team members' experiences. On some projects, all project members are required to write a brief lessons-learned report; on other projects, just the team leads or project manager writes the report. These reports provide valuable reflections by people who know what really worked or did not work on the project. Everyone learns in different ways and has different insights into a project, so it is helpful to have more than one person provide inputs on the lessons-learned reports. These reports can be an excellent resource and help future projects run more smoothly.

To reinforce the benefits of lessons-learned reports, some companies require new project managers to read several past project managers' lessons-learned reports and discuss how they will incorporate some of their ideas into their own projects. Other companies actually charge a small fee for purchasing lessons-learned reports, so the authors can make a little money for documenting useful lessons-learned reports and the buyers are more motivated to read them.

In the past few years, more and more project teams have started putting all or part of their project information, including various templates and lessons-learned reports, on project Web sites. Project Web sites provide a centralized way of delivering project documents and other communications. Project teams can develop project Web sites using Web-authoring tools, such as Microsoft FrontPage or Macromedia Dreamweaver. Figure 10-3 provides a sample project Web site students created in doing a class project to find good templates to use for project management. The home page for the project Web site should include summary information about the project, such as the background and objectives of the project. The home page should also include contact information, such as names and e-mail addresses for the project manager, other team members, or the Webmaster. Links should be provided to items such as project documents, a team member list, meeting minutes, a discussions area, if applicable, and other materials related to the project. If the project involves creating research reports, software, design documents, or other items that can also be accessed via the Web site, links can be provided to those files as well. The project team should also address other issues in creating and using a project Web site, such as security, access, and type of content that should be included on the site.

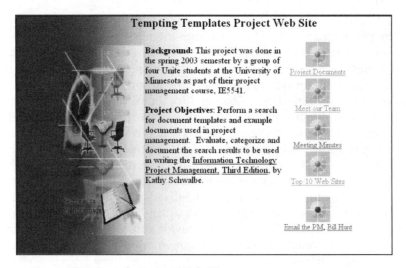

Figure 10-3. Sample Project Web Site

For more sophisticated Web sites, project teams can also use one of the many software products created specifically to assist in project communications through the Web. These products vary considerably in price and functionality, as described in Chapter 1.

See Appendix D and the companion Web site for many templates that can be used in creating various project documents. You can also access the actual site shown in Figure 10-3 from the companion Web site for this text.

When the project team develops their project communications management plan, they should also determine what templates to use for key documentation. To make it even more convenient to use templates, the organization should make project templates readily available online for all projects. The project team should also understand top management's and customers' documentation expectations for each particular project. For example, if a project sponsor or customer wants a one-page monthly progress report for a specific project, but the project team delivers a 20-page report, there are communication problems. In addition, if particular customers or top managers want specific items in all final project reports, they should make sure the project team is aware of those expectations and modifies any templates for those reports to take these requirements into account.

Developing a Communications Infrastructure

A **communications infrastructure** is a set of tools, techniques, and principles that provides a foundation for the effective transfer of information among people. Tools include e-mail, project management software, groupware, fax machines, phones, videoconferencing systems, document management systems, and word processing software. Techniques include reporting guidelines and templates, meeting ground rules and procedures, decision-making processes, problem-solving approaches, conflict resolution and negotiation techniques, and the like. Principles include providing an environment for open dialogue using "straight talk" and following an agreed-upon work ethic.

Since the defense industry has been involved in project management for a long time, many government and defense contractors have formal communications infrastructures already in place. For example, in the early 1980s—before most people ever used personal computers—the U.S. Air Force had standard forms for reporting project progress information, outlines for developing project final reports, regulations describing the progression of major projects, inspections of project archives, and forms and procedures for creating project Gantt charts. **Project archives** are a complete set of organized project records that provide an accurate history of the project. The U.S. Air Force even provided special stickers you could place on Gantt charts to show changes, such as a white diamond for a slipped milestone. (No project manager wanted to ask for that particular sticker!)

Today, several organizations have designed customized systems for collecting and reporting project information, and most use various forms of information technology as part of their project communications infrastructure. Some organizations have intranets to keep track of all project information. Some use enterprise project management software while others use Lotus Notes or other groupware to maintain consistent and complete project information. Still other industries are developing communications infrastructures specifically to help in managing projects.

In his 1999 book, *Business @ the Speed of Thought: Using a Digital Nervous System*, Microsoft CEO Bill Gates suggests that organizations must develop a communications infrastructure or "digital nervous system" that allows for rapid movement of information inside a company as well as with customers, suppliers, and other business partners. Gates suggests that how organizations gather, manage, and use information to empower people will determine whether they win or lose in a competitive business environment.[11] In an interview, Gates

[11] Gates, Bill, *Business @ the Speed of Thought: Using a Digital Nervous System.* New York: Warner Books, 1999.

mentioned systems to aid project management as an example of a great use of technology that really makes a difference in the companies that implement such systems.[12] In 2002, Microsoft released its enterprise version of project management software to address the need for this type of software. See Appendix A for more information describing the differences between the enterprise and standard versions of Microsoft Project 2003.

USING SOFTWARE TO ASSIST IN PROJECT COMMUNICATIONS

Many organizations are discovering how valuable project management software can be in communicating information about individual projects and multiple projects across the organization. Project management software can provide different views of information to help meet various communication needs. For example, senior managers might only need to see summary screens with colors indicating the overall health of all projects. Middle managers often want to see the status of milestones for all of the projects in their area. Project team members often need to see all project documentation. Often, one of the biggest communication problems on projects is providing the most recent project plans, Gantt charts, specifications, meeting information, change requests, and so on to all or selective stakeholders in a timely fashion. Most project management software allows users to insert hyperlinks to other project-related files. In Project 2003, you can insert a hyperlink from a task or milestone listed in a Gantt chart to another file that contains relevant information. For example, there might be a milestone that the project charter was signed. You can insert a hyperlink to the Word file containing the project charter from within the Gantt chart. You could also link appropriate tasks or milestones to Excel files that contain a staffing management plan or cost estimate or to PowerPoint files with important presentations or other information. The Project 2003 file and all associated hyperlinked files could then be placed on a local area network server or Web server, allowing all project stakeholders easy access to important project information. See Appendix A for more information on using Project 2003 to assist in project communications management and Appendix D for information on using project templates.

Even though organizations routinely use many types of hardware and software to enhance communications, they need to take advantage of new technologies and adjust existing systems to serve the special communications needs of customers and project teams. In addition to diverse customer and project needs, they also have to address the changing expectations of consumers and the workforce. For example, several television shows have harnessed communications technology to engage their audiences by letting them vote online or

[12] Schlender, Brent, "E-Business According to Gates," *Fortune* (April 12, 1999): p.73.

via telephone for their favorite singers, e-mail celebrities, or access information on their Web sites. Many people, especially younger people, use instant messaging every day to communicate with friends. Some business and technical professionals also find instant messaging to be a useful tool for quickly communicating with colleagues, customers, suppliers, and others. Web logs, or **blogs**, are easy-to-use journals on the Web that allow users to write entries, respond to another poster's comments, create links, upload pictures, and post comments to journal entries. Blogs have also become popular as a communication technology in the past year. If television shows and non-technical people can use advanced communications technologies, why can't project stakeholders?

Employers have made changes to meet changing expectations and needs in communications. Today, more than 37 percent of people telecommute or work remotely at least part time, according to Gartner, Inc.[13] On several information technology projects, project managers have found that their team members can be more productive when they are allowed to work from home. Other project managers have no choice in the matter when some or all of their project team members work remotely. As described in Chapter 9, Project Human Resource Management, studies show that providing a quiet work environment and a dedicated workspace increases programmer productivity. Most people who work from home have well-equipped, comfortable home offices with fewer distractions and more space than corporate offices. Workers also appreciate the added bonus of avoiding traffic and having more flexible work schedules. However, it is important to make sure work is well defined and communications are in place to allow for remote workers to work effectively.

☑ *What Went Right?*

Frito-Lay, Inc. has taken advantage of various types of communication technology for decades. Its sales force used handheld computers in the mid-1980s. Today, Frito-Lay is taking advantage of new technologies to manage its software development projects. "We do all our own development, and many of our developers work from home full or part time—some out of state," said Anthony King, director of software and engineering systems for Frito-Lay in Tucson, Arizona. Frito-Lay's applications were so large that using its virtual private network (VPN) was very slow, almost to the point of not being functional. In 2001, the company installed GoToMyPC from Expertcity, Inc. in Santa Barbara, California. This Web-based technology allows developers to work much faster from a remote environment and tracks users' online time. This feature came in handy when auditors wanted to know how King monitored remote workers.[14]

Everyday citizens use new communications technologies to bring people together from around the world to support projects, especially when it helps their loved ones. As service men and women were called to duty in the Persian Gulf in 2003, people used technology in innovative ways to show their support. NBC's *Today Show* interviewed the wife of a

---

[13] Melymuka, Kathleen, "Far from the mother ship: Managing remote workers," *ComputerWorld* (December 9, 2002).

serviceman who set up a Web site called Hugs to Kuwait to facilitate communications with her husband's unit and help their family members at home. She got so many people interested in what she was doing that groups from around the world decided to show their support. Girl Scout troops adopted units and sent soldiers letters of support as well as much-appreciated Girl Scout cookies.

Several products are now available to assist individual consumers and organizations with communications. Many products were developed or enhanced in the early 2000s to address the issue of providing fast, convenient, consistent, and up-to-date project information. Webcasts are now a common tool for presenting video, graphics, sound, voice, and participant feedback live over the Web. Most working adults or college students have cell phones, and today it is not uncommon to see someone take and send a picture or send and receive text messages with a cell phone. Many high school or college students are instant messaging their friends to plan social activities or occasionally discuss academic topics.

These same technologies can enhance project communications. For even more powerful and integrated communications, enterprise project management software provides many workgroup functions that allow a team of people at different locations to work together on projects and share project information. Workgroup functions allow the exchange of messages through e-mail, an intranet, wireless devices, or the Web. For example, you can use Project 2003 to alert members about new or changed task assignments, and members can return status information and notify other workgroup members about changes in the schedule or other project parameters.

Microsoft's Enterprise Project Management (EPM) Solution and similar products also provides the following tools to enhance communications:

- Portfolio management: By providing a centralized and consolidated view of programs and projects, the user can evaluate and prioritize activities across the organization. This feature makes it possible to maximize productivity, minimize costs, and keep activities aligned with strategic objectives.

- Resource management: Maximizing human resources is often the key to minimizing project costs. This feature enables the user to maximize resource use across the organization to help effectively plan and manage the workforce.

- Project collaboration: Sharing project information is often a haphazard endeavor. Project collaboration enables an organization to share knowledge immediately and consistently to improve communications and decision making, eliminate redundancies, and take advantage of best practices for project management.

Communication is among the more important factors for success in project management. While technology can aid in the communications process and be the easiest aspect of the process to address, it is not the most important. Far more important is improving an organization's ability to communicate. Improving the ability to communicate often requires a cultural change in an organization that takes a lot of time, hard work, and patience. Information technology personnel, in particular, often need special coaching to improve their communications skills. The project manager's chief role in the communications process is that of facilitator. Project managers must educate all stakeholders—management, team members, and customers—on the importance of good project communications and ensure that the project has a communications management plan to help make good communication happen.

CASE WRAP-UP

Christine Braun worked closely with Peter Gumpert and his project managers to develop a communications management plan for all of the fiber-optic undersea telecommunications systems projects. Peter was very skilled at running effective meetings, so everyone focused on meeting specific objectives. Peter stressed the importance of keeping himself, the project managers, and other major stakeholders informed about the status of all projects. He emphasized that the project managers were in charge of their projects, and that he did not intend to tell them how to do their jobs. He just wanted to have accurate and consistent information to help coordinate all of the projects and make everyone's jobs easier. When some of the project managers balked at the additional work of providing more project information in different formats, Peter openly discussed the issues with them in more detail. He then authorized each project manager to use additional staff to help develop and follow standards for all project communications.

Christine used her strong technical and communications skills to create a Web site that included samples of important project documents, presentations, and templates for other people to download and use on their own projects. After determining the need for more remote communications and collaboration between projects, Christine and other staff members researched the latest hardware and software products available. Peter authorized funds for a new project led by Christine to evaluate and then purchase several wireless handheld devices and enterprise project management software that could be accessed via the Web. Any project stakeholder could check out one of these handheld devices and get one-on-one training on how to use it with the new Web-based software. Even Peter learned how to use one and took it along on all of his business trips.

CHAPTER SUMMARY

A failure to communicate is often the greatest threat to the success of any project, especially information technology projects. Communication is the oil that keeps a project running smoothly.

Project communications management involves communications planning, information distribution, performance reporting, and stakeholder management. A communications management plan of some type should be created for all projects. A stakeholder analysis for project communications helps determine communications needs for different people involved in a project.

The various methods for distributing project information include formal, informal, written, and verbal. It is important to determine the most appropriate means for distributing different types of project information. Project managers and their teams should focus on the importance of building relationships as they communicate project information. As the number of people that need to communicate increases, the number of communications channels also increases.

Performance reporting involves collecting and disseminating information about how well a project is moving toward meeting its goals. Project teams can use earned value charts and other forms of progress information to communicate and assess project performance. Status review meetings are an important part of communicating, monitoring, and controlling projects.

Stakeholder management involves managing communications to satisfy the needs and expectations of project stakeholders. It also involves identifying and managing issues.

To improve project communications, project managers and their teams must develop good conflict management skills, as well as other communication skills. Conflict resolution is an important part of project communications management. The main causes of conflict during a project are schedules, priorities, staffing, technical opinions, procedures, cost, and personalities. A problem-solving approach to managing conflict is often the best approach. Other suggestions for improving project communications include learning how to run more effective meetings, how to use e-mail more effectively, how to use templates for project communications, and how to develop a communications infrastructure.

New hardware and software continues to become available to help improve communications. As more people work remotely, it is important to make sure they have the necessary tools to be productive. Enterprise project management software provides many features to enhance communications across the organization.

Discussion Questions

1. Think of examples in the media that poke fun at the communications skills of technical professionals. How does poking fun at technical professionals' communications skills influence the industry and educational programs?

2. What items should a communications management plan address? How can a stakeholder analysis assist in preparing and implementing parts of this plan?

3. Discuss the advantages and disadvantages of different ways of distributing project information.

4. What are some of the ways to create and distribute project performance information?

5. Discuss the use of issue logs to assist in managing stakeholders.

6. Explain why you agree or disagree with some of the suggestions provided in this chapter for improving project communications. What other suggestions do you have?

7. How can software assist in project communications? How can it hurt project communications?

Exercises

1. Create a stakeholder analysis for project communications. Assume your organization has a project to determine employees' training needs and then provide in-house and external sources for courses in developing communications skills for employees. Stakeholders could be various levels and types of employees, suppliers, the Human Resources department in charge of the project, and so on. Determine specific stakeholders for the project. List at least one type of project communication for each stakeholder and the format for disseminating information to each. Use Table 10-1 as an example.

2. Review the following scenarios, and then write a paragraph for each one describing what media you think would be most appropriate to use and why. See Table 10-2 for suggestions.

   a. Many of the technical staff on the project come in from 9:30 to 10:00 a.m. while the business users always come in before 9:00 a.m. The business users have been making comments. The project manager wants to have the technical people come in by 9:00, although many of them leave late.

   b. Your company is bidding on a project for the entertainment industry. You know that you need new ideas on how to put together the proposal and communicate your approach in a way that will impress the customer.

c. Your business has been growing successfully, but you are becoming inundated with phone calls and e-mails asking similar types of questions.

d. You need to make a general announcement to a large group of people and want to make sure they get the information.

3. How many different communications channels does a project team with six people have? How many more communications channels would there be if the team grew to ten people?

4. Review the templates for various project documents provided in this chapter. Pick one of them and apply it to a project of your choice. Make suggestions for improving the template.

5. Write a lessons-learned report for a project of your choice using the template provided on the companion Web site and sample in Chapter 3 as guides. Do you think it is important for all project managers and team members to write lessons-learned reports? Would you take the time to read them if they were available in your organization? Why or why not?

6. Research software products that assist in communications management for large projects. Write a one- to two-page paper summarizing your findings. Include Web sites for software vendors and your opinion of some of the products.

RUNNING CASE

Several issues have arisen on the Recreation and Wellness Intranet Project. The person from the HR department supporting the project left the company, and the team really needs more support from that group. One of the members of the user group supporting the project is extremely vocal and hard to work with, and other users can hardly get a word in at meetings. The project manager, Tony, is getting weekly status reports from all of his team members, but many of them do not address challenges people are obviously facing. The team is having difficulties deciding how to communicate various project reports and documents and where to store all of the information being generated. Recall that the team members include you, a programmer/analyst and aspiring project manager; Patrick, a network specialist; Nancy, a business analyst; and Bonnie, another programmer/analyst.

1. Prepare a partial communications management plan to address some of the challenges mentioned above.

2. Create an issue log for the project. List at least three issues and related information based on the scenario presented.

3. Prepare a template and sample of a good weekly status report that could be used for this project. Include a list of tips to help team members provide information on these reports.

4. Write a one-page paper describing two suggested approaches to managing the conflict presented by the hard-to-work-with user.

ADDITIONAL RUNNING CASES AND OTHER APPENDICES

Appendix C provides additional case studies and questions you can use to practice applying the concepts, tools, and techniques you are learning throughout this and subsequent chapters. Review the running cases provided in Appendix C and on the companion Web site (*www.course.com/mis/schwalbe4e*). Appendix D includes templates for various project management documents. For additional sample documents based on real projects, visit the author's Web site at *www.kathyschwalbe.com*. Appendix E and the CD-ROM included with this text include a computer simulation where you can also practice applying the project management process groups and knowledge areas.

SUGGESTED READINGS

1. Foti, Ross. "The Virtual Hand Shake," *PM Network* (March 2004).

   This article looks at the advantages and disadvantages of using various types of technology to improve performance of virtual teams. It addresses the organizational and personnel issues that must be addressed to take advantage of new technologies and minimize miscommunication.

2. Gates, Bill. *Business @ the Speed of Thought: Using a Digital Nervous System.* New York: Warner Books, 1999.

   The main premise of the Microsoft CEO's 1999 book is that the speed of business is accelerating at an ever-increasing rate, and to survive, organizations must develop a "digital nervous system" that allows for rapid movement of information. Gates suggests that how organizations gather, manage, and use information to empower people will determine whether they win or lose in the competitive business environment.

3. Information Technology Infrastructure Library (ITIL) (*www.itil.org*) (2004).

   The Information Technology Infrastructure Library (ITIL) provides guidelines developed for the British government. Today, ITIL is the defacto global standard in the area of service management. It contains comprehensive, publicly accessible documentation on the planning, provision, and support of information technology services.

4. Kirchof, Nicki S. and John R. Adams. *Conflict Management for Project Managers.* Upper Darby, PA: Project Management Institute, 1989.

   This 53-page booklet contains information on the theory of conflict management, conflict in project organizations, conflict management and the project manager, and two-party conflict management. It also provides more information on Blake and Mouton's five basic modes for handling conflict and many references on the subject of conflict management.

5. Pritchard, Carl. *The Project Management Communications Toolkit*. Norwood, MA: Artech House Publishers, 2004.

> *This book describes how to coordinate the efforts of project teams and maintain an open dialog with senior executives. The author presents tools to assist project managers in determining project scope, tracking and controlling projects, and closing projects. The CD-ROM contains documents, forms, reports, and templates to assist in project communications management.*

6. TenStep (*www.tenstep.com*) (2004).

> *TenStep, Inc, is a company that specializes in project management methodology development, training, and consulting. Its Web site includes easy-to-read and -follow instructions and templates related to various aspects of project management on small, medium, and large projects. For example, step four describes how to manage issues on projects, and step six describes how to manage communication.*

7. Whitten, Neal. *Whitten's No-Nonsense Advice for Project Managers*. Vienna, VA, Management Concepts (2004).

> *Whitten provides insights, lessons learned, and answers to important questions that all project managers want and need. Many suggestions relate to improving communications skills.*

KEY TERMS

- **blogs** — easy to use journals on the Web that allow users to write entries, create links, and upload pictures, while readers can post comments to journal entries

- **communications infrastructure** — a set of tools, techniques, and principles that provide a foundation for the effective transfer of information among people

- **communications management plan** — a document that guides project communications

- **communications planning** — determining the information and communications needs of the stakeholders: who needs what information, when will they need it, and how will the information be given to them

- **compromise mode** — using a give-and-take approach to resolving conflicts; bargaining and searching for solutions that bring some degree of satisfaction to all the parties in a dispute

- **confrontation mode** — directly facing a conflict using a problem-solving approach that allows affected parties to work through their disagreements

- **expectations management matrix** — a tool to help understand unique measures of success for a particular project

- **forcing mode** — using a win-lose approach to conflict resolution to get one's way

- **forecasts** — used to predict future project status and progress based on past information and trends
- **groupthink** — conformance to the values or ethical standards of a group
- **information distribution** — making needed information available to project stakeholders in a timely manner
- **issue** — a matter under question or dispute that could impede project success
- **issue log** — a tool to document and monitor the resolution of project issues
- **kickoff meeting** — a meeting held at the beginning of a project or project phase where all major project stakeholders discuss project objectives, plans, and so on
- **lessons-learned report** — reflective statements written by project managers and their team members to document important things they have learned from working on the project
- **performance reporting** — collecting and disseminating performance information, which includes status reports, progress measurement, and forecasting
- **progress reports** — reports that describe what the project team has accomplished during a certain period of time
- **project archives** — a complete set of organized project records that provide an accurate history of the project
- **smoothing mode** — deemphasizing or avoiding areas of differences and emphasizing areas of agreements
- **status reports** — reports that describe where the project stands at a specific point in time
- **withdrawal mode** — retreating or withdrawing from an actual or potential disagreement

11

Project Risk Management

Objectives

After reading this chapter, you will be able to:
1. Understand what risk is and the importance of good project risk management
2. Discuss the elements involved in risk management planning and contents of a risk management plan
3. List common sources of risks on information technology projects
4. Describe the risk identification process, tools and techniques to help identify project risks, and the main output of risk identification: a risk register
5. Discuss the qualitative risk analysis process and explain how to calculate risk factors, create probability/impact matrixes, apply the Top Ten Risk Item Tracking technique, and use expert judgment to rank risks
6. Explain the quantitative risk analysis process and how to apply decision trees, simulation, and sensitivity analysis to quantify risks
7. Provide examples of using different risk response planning strategies to address both negative and positive risks
8. Discuss what is involved in risk monitoring and control
9. Describe how software can assist in project risk management

OPENING CASE

*C*liff Branch was the president of a small information technology consulting firm that specialized in developing Internet applications and providing full-service support. The staff consisted of programmers, business analysts, database specialists, Web designers, project managers, and so on. The firm had 50 people and planned to hire at least 10 more in the next year. The company had done very well the past few years, but was recently having difficulty winning contracts. Spending time and resources to respond to various requests for proposals from prospective clients was becoming expensive. Many clients were starting to require presentations and even some prototype development before awarding a contract.

Cliff knew he had an aggressive approach to risk and liked to bid on the projects with the highest payoff. He did not use a systematic approach to evaluate the risks involved in various projects before bidding on them. He focused on the profit potentials and on how challenging the projects were. His strategy was now causing problems for the company because it was investing heavily in the preparation of proposals, yet winning few contracts. Several consultants who were not currently working on projects were still on the payroll. What could Cliff and his company do to get a better understanding of project risks? Should Cliff adjust his strategy for deciding what projects to pursue? How?

THE IMPORTANCE OF PROJECT RISK MANAGEMENT

Project risk management is the art and science of identifying, analyzing, and responding to risk throughout the life of a project and in the best interests of meeting project objectives. A frequently overlooked aspect of project management, risk management can often result in significant improvements in the ultimate success of projects. Risk management can have a positive impact on selecting projects, determining the scope of projects, and developing realistic schedules and cost estimates. It helps project stakeholders understand the nature of the project, involves team members in defining strengths and weaknesses, and helps to integrate the other project management knowledge areas.

Good project risk management often goes unnoticed, unlike crisis management. With crisis management, there is an obvious danger to the success of a project. The crisis, in turn, receives the intense interest of the entire project team. Resolving a crisis has much greater visibility, often accompanied by rewards from management, than successful risk management. In contrast, when risk management is effective, it results in fewer problems, and for the few problems that exist, it results in more expeditious resolutions. It may be difficult for outside observers to tell whether risk management or luck was responsible for the smooth development of a new system, but project teams will always know that their projects worked out better because of good risk management.

All industries, especially the software development industry, tend to neglect the importance of project risk management. As mentioned in Chapter 8, Project Quality Management, William Ibbs and Young H. Kwak performed a study to assess project management maturity. The 38 organizations participating in the study were divided into four industry groups: engineering and construction, telecommunications, information systems/software development, and high-tech manufacturing. Survey participants answered 148 multiple-choice questions to assess how mature their organization was in the project management knowledge areas of scope, time, cost, quality, human resources, communications, risk, and procurement. The rating scale ranged from 1 to 5, with 5 being the highest maturity rating. Table 11-1 shows the results of the survey. Notice that risk management was the only knowledge area for which all ratings were less than 3. This study shows that all organizations should put more effort into project risk management, especially the information systems/software development industry, which had the lowest rating of 2.75.[1]

Table 11-1: Project Management Maturity by Industry Group and Knowledge Area

KEY: 1 = Lowest Maturity Rating   5 = Highest Maturity Rating

KNOWLEDGE AREA	ENGINEERING/ CONSTRUCTION	TELECOMMUNICATIONS	INFORMATION SYSTEMS	HI-TECH MANUFACTURING
Scope	3.52	3.45	3.25	3.37
Time	3.55	3.41	3.03	3.50
Cost	3.74	3.22	3.20	3.97
Quality	2.91	3.22	2.88	3.26
Human Resources	3.18	3.20	2.93	3.18
Communications	3.53	3.53	3.21	3.48
Risk	**2.93**	**2.87**	**2.75**	**2.76**
Procurement	3.33	3.01	2.91	3.33

Ibbs, C. William and Young Hoon Kwak. "Assessing Project Management Maturity," *Project Management Journal* (March 2000).

KLCI Research Group surveyed 260 software organizations worldwide in 2001 to study software risk management practices. Below are some of their findings:

- 97 percent of the participants said they had procedures in place to identify and assess risk.
- 80 percent identified anticipating and avoiding problems as the primary benefit of risk management.

[1] Ibbs, C. William and Young Hoon Kwak, "Assessing Project Management Maturity," *Project Management Journal* (March 2000).

- 70 percent of the organizations had defined software development processes.
- 64 percent had a Project Management Office.

Figure 11-1 shows the main benefits from software risk management practices cited by survey respondents. In addition to anticipating/avoiding problems, risk management practices helped software project managers prevent surprises, improve negotiations, meet customer commitments, and reduce schedule slips cost overruns.[2]

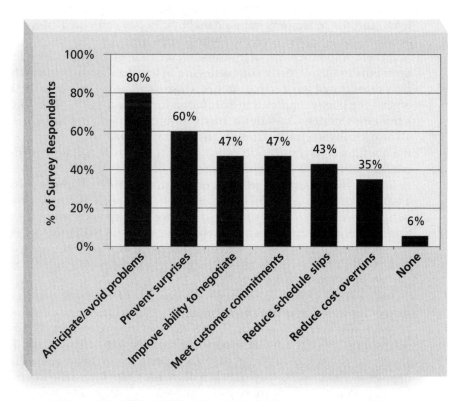

Figure 11-1. Benefits from Software Risk Management Practices

Before you can improve project risk management, you must understand what risk is. A basic dictionary definition says that risk is "the possibility of loss or injury." This definition highlights the negativity often associated with risk and suggests that uncertainty is involved. Project risk management involves understanding potential problems that might occur on the project and how they might impede project success. The *PMBOK® Guide 2004* refers to this type

[2] Kulik, Peter and Catherine Weber, "Software Risk Management Practices – 2001," KLCI Research Group (August 2001).

of risk as a negative risk. However, there are also positive risks, which can result in good things happening on a project. A general definition of a project **risk**, therefore, is an uncertainty that can have a negative or positive effect on meeting project objectives.

In many respects, negative risk management is like a form of insurance. It is an activity undertaken to lessen the impact of potentially adverse events on a project. Positive risk management is like investing in opportunities. It is important to note that risk management *is* an investment—there are costs associated with it. The investment an organization is willing to make in risk management activities depends on the nature of the project, the experience of the project team, and the constraints imposed on both. In any case, the cost for risk management should not exceed the potential benefits.

If there is so much risk in information technology projects, why do organizations pursue them? Many companies are in business today because they took risks that created great opportunities. Organizations survive over the long term when they pursue opportunities. Information technology is often a key part of a business's strategy; without it, many businesses might not survive. Given that all projects involve uncertainties that can have negative or positive outcomes, the question is how to decide which projects to pursue and how to identify and manage project risk throughout a project's life cycle.

Several risk experts suggest that organizations and individuals strive to find a balance between risks and opportunities in all aspects of projects and their personal lives. The idea of striving to balance risks and opportunities suggests that different organizations and people have different tolerances for risk. Some organizations or people have a neutral tolerance for risk, some have an aversion to risk, and others are risk-seeking. These three preferences for risk are part of the utility theory of risk.

Risk utility or **risk tolerance** is the amount of satisfaction or pleasure received from a potential payoff. Figure 11-2 shows the basic difference between risk-averse, risk-neutral, and risk-seeking preferences. The y-axis represents utility, or the amount of pleasure received from taking a risk. The x-axis shows the amount of potential payoff, opportunity, or dollar value of the opportunity at stake. Utility rises at a decreasing rate for a **risk-averse** person. In other words, when more payoff or money is at stake, a person or organization that is risk-averse gains less satisfaction from the risk, or has lower tolerance for the risk. Those who are **risk-seeking** have a higher tolerance for risk, and their satisfaction increases when more payoff is at stake. A risk-seeking person prefers outcomes that are more uncertain and is often willing to pay a penalty to take risks. A **risk-neutral** person achieves a balance between risk and payoff.

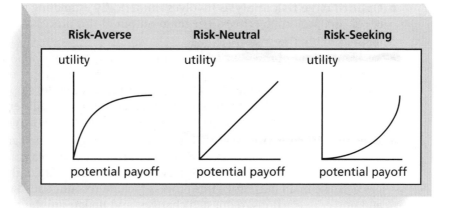

Figure 11-2. Risk Utility Function and Risk Preference

The goal of project risk management can be viewed as minimizing potential negative risks while maximizing potential positive risks. The term "known risks" is sometimes used to describe risks that the project team have identified and analyzed. Known risks can be managed proactively. However, unknown risks, or risks that have not been identified and analyzed, cannot be managed. As you can imagine, good project managers know it is good practice to take the time to identify and manage project risks. There are six major processes involved in risk management:

- **Risk management planning** involves deciding how to approach and plan the risk management activities for the project. By reviewing the project scope statement, project management plan, enterprise environmental factors, and organizational process assets, project teams can discuss and analyze risk management activities for their particular projects. The main output of this process is a risk management plan.

- **Risk identification** involves determining which risks are likely to affect a project and documenting the characteristics of each. The main output of this process is the start of a risk register, as described in more detail later in this chapter.

- **Qualitative risk analysis** involves prioritizing risks based on their probability and impact of occurrence. After identifying risks, project teams can use various tools and techniques to rank risks and update information in the risk register. The main output is updates to the risk register.

■ **Quantitative risk analysis** involves numerically estimating the effects of risks on project objectives. The main output of this process is also updates to the risk register.

■ **Risk response planning** involves taking steps to enhance opportunities and reduce threats to meeting project objectives. Using outputs from the preceding risk management processes, project teams can develop risk response strategies that often result in updates to the risk register and project management plan as well as risk-related contractual agreements.

■ **Risk monitoring and control** involves monitoring identified and residual risks, identifying new risks, carrying out risk response plans, and evaluating the effectiveness of risk strategies throughout the life of the project. The main outputs of this process include recommended corrective and preventive actions, requested changes, and updates to the risk register, project management plan, and organizational process assets.

The first step in project risk management is deciding how to address this knowledge area for a particular project by performing risk management planning.

RISK MANAGEMENT PLANNING

Risk management planning is the process of deciding how to approach and plan for risk management activities for a project, and the main output of this process is a risk management plan. A **risk management plan** documents the procedures for managing risk throughout the project. Project teams should hold several planning meetings early in the project's life cycle to help develop the risk management plan. The project team should review project documents as well as corporate risk management policies, risk categories, lessons-learned reports from past projects, and templates for creating a risk management plan. It is also important to review the risk tolerances of various stakeholders. For example, if the project sponsor is risk-averse, the project might require a different approach to risk management than if the project sponsor were a risk seeker.

A risk management plan summarizes how risk management will be performed on a particular project. Like other specific knowledge area plans, it becomes a subset of the project management plan. Table 11-2 lists the general topics that a risk management plan should address. It is important to clarify roles and responsibilities, prepare budget and schedule estimates for risk-related work, and identify risk categories for consideration. It is also important to describe how risk management will be done, including assessment of risk probabilities and impacts as well as creation of risk related documentation. The level of detail included in the risk management plan will vary with the needs of the project.

Table 11-2: Topics Addressed in a Risk Management Plan

- Methodology: How will risk management be performed on this project? What tools and data sources are available and applicable?

- Roles and Responsibilities: Who are the individuals responsible for implementing specific tasks and providing deliverables related to risk management?

- Budget and Schedule: What are the estimated costs and schedules for performing risk-related activities?

- Risk Categories: What are the main categories of risks that should be addressed on this project? Is there a risk breakdown structure for the project? (See the information on risk breakdown structures later in this section.)

- Risk Probability and Impact: How will the probabilities and impacts of risk items be assessed? What scoring and interpretation methods will be used for the qualitative and quantitative analysis of risks?

- Risk Documentation: What reporting formats and processes will be used for risk management activities?

In addition to a risk management plan, many projects also include contingency plans, fallback plans, and contingency reserves. **Contingency plans** are predefined actions that the project team will take if an identified risk event occurs. For example, if the project team knows that a new release of a software package may not be available in time for them to use it for their project, they might have a contingency plan to use the existing, older version of the software. **Fallback plans** are developed for risks that have a high impact on meeting project objectives, and are put into effect if attempts to reduce the risk are not effective. For example, a new college graduate might have a main plan and several contingency plans on where to live after graduation, but if none of those plans works out, a fallback plan might be to live at home for a while. Sometimes the terms contingency plan and fallback plan are used interchangeably. **Contingency reserves** or **contingency allowances** are provisions held by the project sponsor or organization to reduce the risk of cost or schedule overruns to an acceptable level. For example, if a project appears to be off course because the staff is inexperienced with some new technology and the team had not identified that as a risk, the project sponsor may provide additional funds from contingency reserves to hire an outside consultant to train and advise the project staff in using the new technology.

Before you can really understand and use the other project risk management processes on information technology projects, it is necessary to recognize and understand the common sources of risk.

COMMON SOURCES OF RISK ON INFORMATION TECHNOLOGY PROJECTS

Several studies have shown that information technology projects share some common sources of risk. For example, the Standish Group did a follow-up study to the 1995 CHAOS research, which they called Unfinished Voyages. This study brought together 60 information technology professionals to elaborate on how to evaluate a project's overall likelihood of being successful. Table 11-3 shows the Standish Group's success potential scoring sheet and the relative importance of the project success criteria factors. If a potential project does not receive a minimum score, the organization might decide not to work on it or to take actions to reduce the risks before it invests too much time or money.[3]

Table 11-3: Information Technology Success Potential Scoring Sheet

Success Criterion	Relative Importance
User Involvement	19
Executive Management Support	16
Clear Statement of Requirements	15
Proper Planning	11
Realistic Expectations	10
Smaller Project Milestones	9
Competent Staff	8
Ownership	6
Clear Visions and Objectives	3
Hardworking, Focused Staff	3
Total	**100**

The Standish Group provides specific questions for each success criterion to help decide the number of points to assign to a project. For example, the five questions related to user involvement include the following:

- Do I have the right user(s)?
- Did I involve the user(s) early and often?

---

[3] The Standish Group, "Unfinished Voyages" (1996) (*www.standishgroup.com*).

- Do I have a quality relationship with the user(s)?
- Do I make involvement easy?
- Did I find out what the user(s) need(s)?

The number of questions corresponding to each success criterion determines the number of points each positive response is assigned. For example, in the case of user involvement there are five questions. For each positive reply, you would get 3.8 (19/5) points; 19 represents the weight of the criterion, and 5 represents the number of questions. Therefore, you would assign a value to the user involvement criterion by adding 3.8 points to the score for each question you can answer positively.

Many organizations develop their own risk questionnaires. Broad categories of risks described on these questionnaires might include:

- *Market risk*: If the information technology project is to produce a new product or service, will it be useful to the organization or marketable to others? Will users accept and use the product or service? Will someone else create a better product or service faster, making the project a waste of time and money?

- *Financial risk*: Can the organization afford to undertake the project? How confident are stakeholders in the financial projections? Will the project meet NPV, ROI, and payback estimates? If not, can the organization afford to continue the project? Is this project the best way to use the organization's financial resources?

- *Technology risk*: Is the project technically feasible? Will it use mature, leading edge, or bleeding edge technologies? When will decisions be made on which technology to use? Will hardware, software, and networks function properly? Will the technology be available in time to meet project objectives? Could the technology be obsolete before a useful product can be produced? You can also break down the technology risk category into hardware, software, and network technology, if desired.

- *People risk*: Does the organization have or can they find people with appropriate skills to complete the project successfully? Do people have the proper managerial and technical skills? Do they have enough experience? Does senior management support the project? Is there a project champion? Is the organization familiar with the sponsor/customer for the project? How good is the relationship with the sponsor/customer?

- *Structure/process risk*: What is the degree of change the new project will introduce into user areas and business procedures? How many distinct user groups does the project need to satisfy? With how many other systems does the new project/system need to interact? Does the organization have processes in place to complete the project successfully?

What Went Wrong?

KPMG, a large consulting firm, published a study in 1995 that found that 55 percent of **runaway projects**—projects that have significant cost or schedule overruns—did *no* risk management at all, 38 percent did some (but half did not use their risk findings after the project was underway), and 7 percent did not know whether they did risk management or not.[4] This study suggests that performing risk management is important to improving the likelihood of project success and preventing runaway projects.

Risk management is also becoming the weak link for many organizations in their supply chains. Global events such as terrorist attacks, political instability in developing economies, and the 2004 explosion in global metal prices have awakened managers to the importance of assessing and managing risks for supply chain projects. Ironically, many of the new risks being identified by supply managers are the product of the cost reduction strategies that are sweeping business today. Advocates of new supply strategies spent too much time and energy on selling new supply strategies to management and too little effort on identifying and evaluating all of the risks and potential strategies for addressing them.[5]

Reviewing a proposed project in terms of the Standish Group's success criteria, a risk questionnaire, or any other similar tool is a good method for understanding common sources of risk on information technology projects. It is also useful to review the work breakdown structure (WBS) for a project to see if there might be specific risks by WBS categories. For example, if one item on the WBS involves preparing a press release and no one on the project team has ever done that, it could be a negative risk if it is not handled professionally.

A risk breakdown structure is a useful tool that can help project managers consider potential risks in different categories. Similar in structure to a work breakdown structure, a **risk breakdown structure** is a hierarchy of potential risk categories for a project. Figure 11-3 shows a sample risk breakdown structure that might apply to many information technology projects. The highest-level categories are business, technical, organizational, and project management. Competitors, suppliers, and cash flow are categories that fall under business risks. Under technical risks are the categories of hardware, software, and network. Notice how the risk breakdown structure provides a simple, one-page chart to help ensure a project team is considering important risk categories related to all information technology projects. For example, Cliff and his managers in the opening case could have benefited from considering several of the categories listed under project management—estimates, communication, and resources. They could have discussed these and other types of risks related to the projects their company bid on and developed appropriate strategies for optimizing positive risks and minimizing negative ones.

[4] Cole, Andy, "Runaway Projects—Cause and Effects," *Software World*, Vol. 26, no. 3, pp. 3–5 (1995).

[5] Morgan, James, "Poor risk management threatens supply chain," *Purchasing.com* (June 3, 2004).

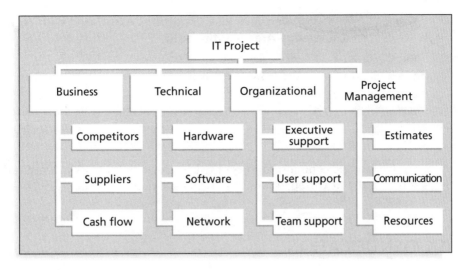

Figure 11-3. Sample Risk Breakdown Structure

In addition to identifying risk based on the nature of the project or products produced, it is also important to identify potential risks according to project management knowledge areas, such as scope, time, cost, and quality. Notice that one of the major categories in the risk breakdown structure in Figure 11-3 is project management. Table 11-4 lists potential negative risk conditions that can exist within each knowledge area.[6]

Table 11-4: Potential Negative Risk Conditions Associated with Each Knowledge Area

KNOWLEDGE AREA	RISK CONDITIONS
Integration	Inadequate planning; poor resource allocation; poor integration management; lack of post-project review
Scope	Poor definition of scope or work packages; incomplete definition
Time	Errors in estimating time or resource availability; errors in determining the critical path; poor allocation and management of float; early release of competitive products
Cost	Estimating errors; inadequate productivity, cost, change, or contingency
Quality	Poor attitude toward quality; substandard design/materials/workmanship; inadequate quality assurance program
Human Resources	Poor conflict management; poor project organization and definition of responsibilities; absence of leadership

[6] Wideman, R. Max, "Project and Program Risk Management: A Guide to Managing Project Risks and Opportunities", Upper Darby, PA: Project Management Institute, 1992, II–4.

Table 11-4: Potential Negative Risk Conditions Associated with Each Knowledge Area (continued)

KNOWLEDGE AREA	RISK CONDITIONS
Communications	Carelessness in planning or communicating; lack of consultation with key stakeholders
Risk	Ignoring risk; unclear analysis of risk; poor insurance management
Procurement	Unenforceable conditions or contract clauses; adversarial relations

Understanding common sources of risk is very helpful in risk identification, which is the next step in project risk management.

RISK IDENTIFICATION

Identifying risks is the process of understanding what potential events might hurt or enhance a particular project. It is important to identify potential risks early, but you must also continue to identify risks based on the changing project environment. Also remember that you cannot manage risks if you do not first identify them. By understanding common sources of risks and reviewing a project's risk management plan, project scope statement, project management plan, enterprise environmental factors, and organizational process assets, project managers and their teams can identify many potential risks.

Suggestions for Identifying Risks

There are several tools and techniques for identifying risks. Project teams often begin the risk identification process by reviewing project documentation, recent and historical information related to the organization, and assumptions that might affect the project. Project team members and outside experts often hold meetings to discuss this information and ask important questions about them as they relate to risk. After identifying potential risks at this initial meeting, the project team might then use different information-gathering techniques to further identify risks. Five common information-gathering techniques include brainstorming, the Delphi Technique, interviewing, root cause analysis, and SWOT analysis.

Brainstorming is a technique by which a group attempts to generate ideas or find a solution for a specific problem by amassing ideas spontaneously and without judgment. This approach can help the group create a comprehensive list of risks to address later in the qualitative and quantitative risk analysis processes. An experienced facilitator should run the brainstorming session and introduce new categories of potential risks to keep the ideas flowing. After the ideas are collected, the facilitator can group and categorize the ideas to make them more manageable. Care must be taken, however, not to overuse or misuse brain-storming. Although businesses use brainstorming widely to generate new ideas, the psychology literature shows that individuals, working alone, produce a greater number of ideas than the same individuals produce through brainstorming in small, face-to-face groups. Group effects, such as fear of social disapproval, the effects of authority hierarchy, and domination of the session by one or two very vocal people often inhibit idea generation for many participants.[7]

An approach to gathering information that helps prevent some of the nega-tive group affects found in brainstorming is the Delphi Technique. The basic concept of the **Delphi Technique** is to derive a consensus among a panel of experts who make predictions about future developments. Developed by the Rand Corporation for the U.S. Air Force in the late 1960s, the Delphi Technique is a systematic, interactive forecasting procedure based on indepen-dent and *anonymous* input regarding future events. The Delphi Technique uses repeated rounds of questioning and written responses, including feedback to earlier-round responses, to take advantage of group input, while avoiding the biasing effects possible in oral panel deliberations. To use the Delphi Technique, you must select a panel of experts for the particular area in question. For example, Cliff Branch from the opening case could use the Delphi Technique to help him understand why his company is no longer winning many contracts. Cliff could assemble a panel of people with knowledge in his business area. Each expert would answer questions related to Cliff's scenario, and then Cliff or a facilitator would evaluate their responses, together with opinions and jus-tifications, and provide that feedback to each expert in the next iteration. Cliff would continue this process until the group responses converge to a specific solution. If the responses diverge, the facilitator of the Delphi Technique needs to determine if there is a problem with the process.

Interviewing is a fact-finding technique for collecting information in face-to-face, phone, e-mail, or instant-messaging discussions. Interviewing people with similar project experience is an important tool for identifying potential risks. For example, if a new project involves using a particular type of hard-ware or software, someone with recent experience with that hardware or soft-ware could describe problems he or she had on a past project. If someone has worked with a particular customer, he or she might provide insight into poten-tial risks involved in working for that group again.

It is not uncommon for people to identify problems or opportunities without really understanding them. Before suggesting courses of action, it is important to identify the root cause of a problem or opportunity. Chapter 8, Project Quality Management, provides information on root cause analysis. Root cause analysis often results in identifying even more potential risks for a project.

Another technique described in Chapter 4, Project Integration Management, is a SWOT analysis (strengths, weaknesses, opportunities, and threats), which

---

[7] Couger, J. Daniel, *Creative Problem Solving and Opportunity Finding*, Boyd & Fraser Publishing Company, 1995.

is often used in strategic planning. SWOT analysis can also be used during risk identification by having project teams focus on the broad perspectives of potential risks for particular projects. For example, before writing a particular proposal, Cliff Branch could have a group of his employees discuss in detail what their company's strengths are, what their weaknesses are related to that project, and what opportunities and threats exist. Do they know that several competing firms are much more likely to win a certain contract? Do they know that winning a particular contract will likely lead to future contracts and help expand their business? Applying SWOT to specific potential projects can help identify the broad risks and opportunities that apply in that scenario.

Three other techniques for risk identification include the use of checklists, analysis of assumptions, and creation of diagrams.

- Checklists based on risks that have been encountered in previous projects provide a meaningful template for understanding risks in a current project. You can use checklists similar to those developed by the Standish Group or other groups to help identify risks on information technology projects.

- It is important to analyze project assumptions to make sure they are valid. Incomplete, inaccurate, or inconsistent assumptions might lead to identifying more risks.

- Diagramming techniques include using cause-and-effect diagrams or fishbone diagrams, flow charts, and influence diagrams. Recall from Chapter 8, Project Quality Management, that fishbone diagrams help you trace problems back to their root cause. System or process **flow charts** are diagrams that show how different parts of a system interrelate. For example, many programmers create flow charts to show programming logic. Another type of diagram, an **influence diagram**, represents decision problems by displaying essential elements, including decisions, uncertainties, causality, and objectives, and how they influence each other.

The Risk Register

The main output of the risk identification process is a list of identified risks and other information needed to begin creating a risk register. A **risk register** is a document that contains results of various risk management processes, often displayed in a table or spreadsheet format. It is a tool for documenting potential risk events and related information. **Risk events** refer to specific, uncertain events that may occur to the detriment or enhancement of the project. For example, negative risk events might include the performance failure of a product produced as part of a project, delays in completing work as scheduled, increases in estimated costs, supply shortages, litigation against the company, strikes, and so on. Examples of positive risk events include completing work sooner or cheaper than planned, collaborating with suppliers to produce better products, good publicity resulting from the project, and so on.

Table 11-5 provides a sample of the format for a risk register that Cliff and his managers from the opening case might use on a new project. Actual data that might be entered for one of the risks is included below the table. Notice the main headings often included in the register. Many of these items are described in more detail later in this chapter.

- *An identification number for each risk event*: The project team may want to sort or quickly search for specific risk events, so they need to identify each risk with some type of unique descriptor, like an identification number.
- *A rank for each risk event*: The rank is usually a number, with 1 being the highest ranked risk.
- *The name of the risk event*: For example, defective server, late completion of testing, reduced consulting costs, or good publicity.
- *A description of the risk event*: Because the name of a risk event is often abbreviated, it helps to provide a more detailed description. For example, reduced consulting costs might be expanded in the description to say that the organization might be able to negotiate lower-than-average costs for a particular consultant because the consultant really enjoys working for that company in that particular location.
- *The category under which the risk event falls*: For example, defective server might fall under the broader category of technology or hardware technology.
- *The root cause of the risk*: The root cause of the defective server might be a defective power supply.
- *Triggers for each risk*: **Triggers** are indicators or symptoms of actual risk events. For example, cost overruns on early activities may be symptoms of poor cost estimates. Defective products may be symptoms of a low-quality supplier. Documenting potential risk symptoms for projects also helps the project team identify more potential risk events.
- *Potential responses to each risk*: A potential response to the risk event of a defective server might be the inclusion of a clause in a contract with the supplier to replace a defective server within a certain time period at a negotiated cost.
- *The **risk owner** or person who will own or take responsibility for the risk*: For example, a certain person might be in charge of any server-related risk events and managing response strategies.
- *The probability of the risk occurring*: There might be a high, medium, or low probability of a certain risk event occurring. For example, the risk might be low that the server would actually be defective.
- *The impact to the project if the risk occurs*: There might be a high, medium, or low impact to project success if the risk event actually occurs. A defective server might have a high impact on successfully completing a project on time.

- *The status of the risk*: Did the risk event occur? Was the response strategy completed? Is the risk no longer relevant to the project? For example, a contract clause may have been completed to address the risk of a defective server.

Table 11-5: Sample Risk Register

No.	Rank	Risk	Description	Category	Root Cause	Triggers	Potential Responses	Risk Owner	Probability	Impact	Status
R44	1										
R21	2										
R7	3										

For example, the following data might be entered for the first risk in the register as follows. Notice that Cliff's team is taking a very proactive approach in managing this risk.

- No.: R44
- Rank: 1
- Risk: New customer
- Description: We have never done a project for this organization before and don't know too much about them. One of our company's strengths is building good customer relationships, which often leads to further projects with that customer. We might have trouble working with this customer since they are new to us.
- Category: People risk
- Root cause: We won a contract to work on a project without really getting to know the customer.
- Triggers: The project manager and other senior managers realize that we don't know much about this customer and could easily misunderstand their needs or expectations.
- Risk Responses: Make sure the project manager is sensitive to the fact that this is a new customer and takes the time to understand them. Have the PM set up a meeting to get to know the customer and clarify their expectations. Have Cliff attend the meeting, too.
- Risk owner: Our project manager
- Probability: Medium
- Impact: High
- Status: PM will set up the meeting within the week.

After identifying risks, the next step is to understand which risks are most important by performing qualitative risk analysis.

QUALITATIVE RISK ANALYSIS

Qualitative risk analysis involves assessing the likelihood and impact of identified risks, to determine their magnitude and priority. This section describes examples of using a probability/impact matrix to produce a prioritized list of risks. It also provides examples of using the Top Ten Risk Item Tracking technique to produce an overall ranking for project risks and to track trends in qualitative risk analysis. Finally, it discusses the importance of expert judgment in performing risk analysis.

Using Probability/Impact Matrixes to Calculate Risk Factors

People often describe a risk probability or consequence as being high, medium or moderate, or low. For example, a meteorologist might predict that there is a high probability, or likelihood, of severe rain showers on a certain day. If that day happens to be your wedding day and you are planning a large outdoor wedding, the consequences or impact of severe showers might also be high.

A project manager can chart the probability and impact of risks on a **probability/impact matrix** or **chart**. A probability/impact matrix or chart lists the relative probability of a risk occurring on one side of a matrix or axis on a chart and the relative impact of the risk occurring on the other. Many project teams would benefit from using this simple technique to help them identify risks that they need to pay attention to. To use this approach, project stakeholders list the risks they think might occur on their projects. They then label each risk as being high, medium, or low in terms of its probability of occurrence and its impact if it did occur.

The project manager then summarizes the results in a probability/impact matrix or chart, as shown in Figure 11-4. For example, Cliff Branch and some of his project managers in the opening case could each identify three negative and positive potential risks for their company or a particular project. They could then label each risk as being high, medium, or low in terms of probability and impact. For example, one project manager might identify a negative risk of a severe market downturn. He or she might think that particular risk is low in probability and high in impact. Cliff may have listed the risk of a severe market downturn as being medium in both probability and impact. The team could then plot all of the risks on a matrix or chart using sticky notes. As a group, they could then combine any common risks and decide where those risks should be on the matrix or chart, if there were differences of opinion. After agreeing on the risks and their placement on the chart, the team should then focus on any risks that fall in the high sections of the probability/impact matrix or chart. For example, Risks 1 and 4 are listed as high in both categories of probability and impact. Risk 6 is high in the probability category but low in the impact category. Risk 9 is high in the probability category and medium in the impact category. Risk 12 is low in the probability category and high in the impact category. The team should then discuss how they plan to respond to those risks if they occur, as discussed later in this chapter in the section on risk response planning.

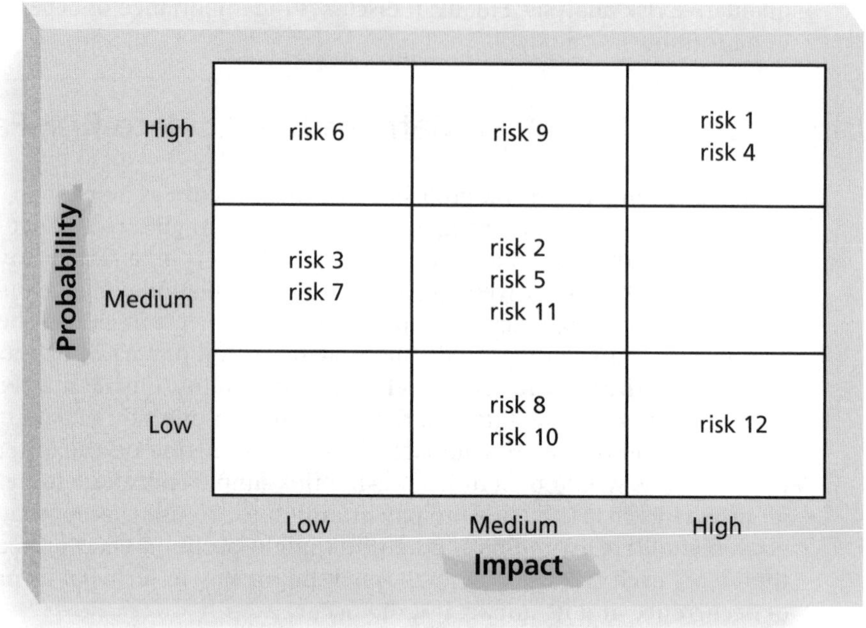

Figure 11-4. Sample Probability/Impact Matrix

It may be useful to create a separate probability/impact matrix or chart for negative risks and positive risks to make sure both types of risks are adequately addressed. Some project teams also collect data based on the probability and impact of risks in terms of negatively or positively affecting scope, time, and cost goals. Qualitative risk analysis is normally done quickly, so the project team has to decide what type of approach makes the most sense for their project.

Some project teams develop a single number for a risk score by simply multiplying a numeric score for probability by a numeric score for impact. A more sophisticated approach to using probability/impact information is to calculate risk factors. To quantify risk probability and consequence, the Defense Systems Management College (DSMC) developed a technique for calculating **risk factors**—numbers that represent the overall risk of specific events, based on their probability of occurring and the consequences to the project if they do occur. The technique makes use of a probability/impact matrix that shows the probability of risks occurring and the impact or consequences of the risks.

Probabilities of a risk occurring can be estimated based on several factors, as determined by the unique nature of each project. For example, factors evaluated for potential hardware or software technology risks could include the technology not being mature, the technology being too complex, and an inadequate support base for developing the technology. The impact of a risk occurring could include factors such as the availability of fallback solutions or the consequences of not meeting performance, cost, and schedule estimates.

Table 11-6 provides an example of two probability/impact matrixes that were used in assessing the risk of various technologies that could make aircraft more reliable. The U.S. Air Force sponsored a multimillion-dollar research project in the mid-1980s, the High Reliability Fighter study, to evaluate potential technologies for improving the reliability of fighter aircraft. Many aircraft were too often unfit for flying because of reliability problems with various components of the planes and problems with their maintenance. The purpose of the High Reliability Fighter study was to help the U.S. Air Force decide what technologies to invest in to make planes more reliable and maintainable, thereby decreasing the planes' downtime. The matrixes in Table 11-6 were used to assess the risk of proposed technologies, such as developing a sophisticated onboard computer system to monitor and adjust various systems in the aircraft. The top matrix in Table 11-6 was used to assign a Probability of Failure (Pf) value to each proposed technology, and the bottom matrix was used to assign a Consequence of Failure (Cf) value. Experts used their judgment to assign a value for both the probability and impact of each proposed technology. For example, an expert assigned a Pf value of .1 for the radial tires technology since the hardware existed, had a simple design, and had multiple programs and services that used it. Likewise, the expert assigned a Cf value of .9 to a risk that had no acceptable alternatives, would increase project life cycle costs, was unlikely to meet schedule dates, and might not reduce the aircraft downtime factor. These values were then used in a formula to calculate an overall risk factor. A risk factor is defined as the probability of failure (Pf) plus the consequence of failure (Cf) minus the product of the two.[8] For example, a technology with a Pf of .1 and a Cf of .9 would have a risk factor of .91, or $(.1 + .9) - (.1 * .9) = .91$.

[8] Defense Systems Management College, *Systems Engineering Management Guide,* Washington, DC, 1989.

Table 11-6: Sample Probability/Impact Matrixes for Qualitative Risk Assessment

PROBABILITY OF FAILURE (Pf) ATTRIBUTES OF SUGGESTED TECHNOLOGY

VALUE	MATURITY HARDWARE/SOFTWARE	COMPLEXITY HARDWARE/SOFTWARE	SUPPORT BASE
0.1	Existing	Simple design	Multiple programs and services
0.3	Minor redesign	Somewhat complex	Multiple programs
0.5	Major change feasible	Somewhat complex	Several parallel programs
0.7	Complex hardware design/new software similar to existing	Very complex	At least one other program
0.9	Some research completed/never done before	Extremely complex	No additional programs

CONSEQUENCE OF FAILURE (Cf) ATTRIBUTES OF SUGGESTED TECHNOLOGY

VALUE	FALLBACK SOLUTIONS	LIFE CYCLE COST (LCC) FACTOR	SCHEDULE FACTOR (INITIAL OPERATIONAL CAPABILITY = IOC)	DOWNTIME (DT) FACTOR
0.1	Several acceptable alternatives	Highly confident will reduce LCC	90–100% confident will meet IOC significantly	Highly confident will reduce DT significantly
0.3	A few known alternatives	Fairly confident will reduce LCC	75–90% confident will meet IOC	Fairly confident will reduce DT significantly
0.5	Single acceptable alternative	LCC will not change much	50–75% confident will meet IOC	Highly confident will reduce DT somewhat
0.7	Some possible alternatives	Fairly confident will increase LCC	25–50% confident will meet IOC	Fairly confident will reduce DT somewhat
0.9	No acceptable alternatives	Highly confident will increase LCC	0–25% confident will meet IOC	DT may not be reduced much

Figure 11-5 provides an example of how the risk factors were used to graph the probability of failure and consequence of failure for proposed technologies in this study. The figure classifies potential technologies (dots on the chart) as high-, medium-, or low-risk, based on the probability of failure and consequences of failure. The researchers for this study highly recommended that the U.S. Air Force invest in the low- to medium-risk technologies and suggested that it not pursue the high-risk technologies.[9] You can see that the rigor behind

---

[9] McDonnell Douglas Corporation, "Hi-Rel Fighter Concept," Report MDC B0642, 1988.

using the probability/impact matrix and risk factors provides a much stronger argument than simply stating that risk probabilities or consequences are high, medium, or low.

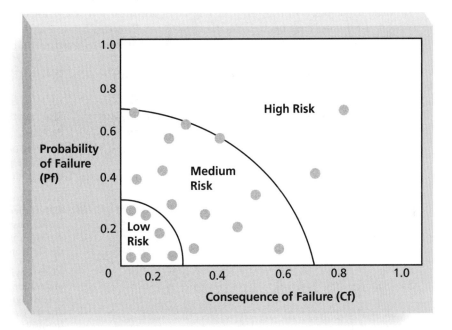

Figure 11-5. Chart Showing High-, Medium-, and Low-Risk Technologies

Top Ten Risk Item Tracking

Top Ten Risk Item Tracking is a qualitative risk analysis tool, and in addition to identifying risks, it maintains an awareness of risks throughout the life of a project. It involves establishing a periodic review of the project's most significant risk items with management and, optionally, with the customer. The review begins with a summary of the status of the top-ten sources of risk on the project. The summary includes each item's current ranking, previous ranking, number of times it appears on the list over a period of time, and a summary of progress made in resolving the risk item since the previous review. The Microsoft Solution Framework (MSF) includes a risk management model that includes developing and monitoring a top-ten master list of risks. MSF is the methodology Microsoft uses for managing projects. It combines aspects of software design and development, and building and deploying infrastructure, into a single project life cycle for guiding technology solutions of all kinds. Consult Microsoft's Web site for more information on MSF (*www.microsoft.com/msf*).

Table 11-7 provides an example of a Top Ten Risk Item Tracking chart that could be used at a management review meeting for a project. This particular example includes only the top five negative risk events. Notice that each risk event is ranked based on the current month, previous month, and how many months it has been in the top ten. The last column briefly describes the progress for resolving each particular risk item. You can have separate charts for negative and positive risks or combine them into one chart.

Table 11-7: Example of Top Ten Risk Item Tracking

	MONTHLY RANKING			
RISK EVENT	RANK THIS MONTH	RANK LAST MONTH	NUMBER OF MONTHS IN TOP TEN	RISK RESOLUTION PROGRESS
Inadequate planning	1	2	4	Working on revising the entire project management plan
Poor definition	2	3	3	Holding meetings with project customer and sponsor to clarify scope
Absence of leadership	3	1	2	After previous project manager quit, assigned a new one to lead the project
Poor cost estimates	4	4	3	Revising cost estimates
Poor time estimates	5	5	3	Revising schedule estimates

A risk management review accomplishes several objectives. First, it keeps management and the customer (if included) aware of the major influences that could prevent or enhance the project's success. Second, by involving the customer, the project team may be able to consider alternative strategies for addressing the risks. Third, it is a means of promoting confidence in the project team by demonstrating to management and the customer that the team is aware of the significant risks, has a strategy in place, and is effectively carrying out that strategy.

Expert Judgment

Many organizations rely on the intuitive feelings and experience of experts in performing qualitative risk analyses. They may use expert judgment in lieu of or in addition to other techniques for analyzing risks. For example, experts can categorize risks as being high, medium, or low with or without sophisticated techniques such as calculating risk factors, as described earlier.

Using sophisticated risk analysis techniques has a number of disadvantages. For example, the outputs are only as good as the inputs, and the people using the techniques may be using poor assumptions. Many people get confused when someone tries to explain the math and statistics behind the various techniques. Because of these disadvantages, it is important to include expert opinions when using both qualitative and quantitative risk assessment techniques.

The main output of qualitative risk analysis is updating the risk register. The ranking column of the risk register should be filled in, along with a numeric value or high/medium/low rating for the probability and impact of the risk event. Additional information is often added for risk events, such as identification of risks that need more attention in the near term or those that can be placed on a watch list. A **watch list** is a list of risks that are low priority, but are still identified as potential risks. Qualitative analysis can also identify risks that should be evaluated on a quantitative basis, as described in the next section.

QUANTITATIVE RISK ANALYSIS

Quantitative risk analysis often follows qualitative risk analysis, yet both processes can be done together or separately. On some projects, the team may only perform qualitative risk analysis. The nature of the project and availability of time and money affect the type of risk analysis techniques used. Large, complex projects involving leading-edge technologies often require extensive quantitative risk analysis. The main techniques for quantitative risk analysis include data gathering, quantitative risk analysis, and modeling techniques. Data gathering often involves interviewing, expert judgment, and collection of probability distribution information. This section focuses on using the quantitative risk analysis and modeling techniques of decision tree analysis, simulation, and sensitivity analysis.

Decision Trees and Expected Monetary Value

A **decision tree** is a diagramming analysis technique used to help select the best course of action in situations in which future outcomes are uncertain. A common application of decision tree analysis involves calculating expected monetary value. **Expected monetary value (EMV)** is the product of a risk event probability and the risk event's monetary value. Figure 11-6 uses the decision of which project(s) an organization might pursue to illustrate this concept. Suppose Cliff Branch's firm was trying to decide if it should submit a proposal for Project 1, Project 2, both projects, or neither project. The team could draw a decision tree with two branches, one for Project 1 and one for Project 2.

The firm could then calculate the expected monetary value to help make this decision.

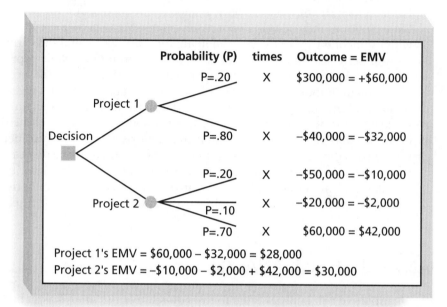

Figure 11-6. Expected Monetary Value (EMV) Example

To create a decision tree, and to calculate expected monetary value specifically, you must estimate the probabilities, or chances, of certain events occurring. For example, in Figure 11-6 there is a 20 percent probability or chance (P = .20) that Cliff's firm will win the contract for Project 1, which is estimated to be worth $300,000 in profits—the outcome of the top branch in the figure. There is an 80 percent probability (P = .80) that it will not win the contract for Project 1, and the outcome is estimated to be –$40,000, meaning that the firm will have to invest $40,000 into Project 1 with no reimbursement if it is not awarded the contract. The sum of the probabilities for outcomes for each project must equal one (for Project 1, 20 percent plus 80 percent). Probabilities are normally determined based on expert judgment. Cliff or other people in his firm should have some sense of their likelihood of winning certain projects.

Figure 11-6 also shows probabilities and outcomes for Project 2. Suppose there is a 20 percent probability that Cliff's firm will lose $50,000 on Project 2, a 10 percent probability that it will lose $20,000, and a 70 percent probability that it will earn $60,000. Again, experts would need to estimate these dollar amounts and probabilities.

To calculate the expected monetary value (EMV) for each project, multiply the probability by the outcome value for each potential outcome for each project and sum the results. To calculate expected monetary value for Project 1, going from left to right, multiply the probability by the outcome for each branch and sum the results. In this example, the EMV for Project 1 is $28,000.

.2($300,000) + .8(–$40,000) = $60,000 – $32,000 = $28,000

The EMV for Project 2 is $30,000.

.2(–$50,000) + .1(–$20,000) + .7($60,000) = –$10,000 – $2,000 + $42,000 = $30,000

Because the EMV provides an estimate for the total dollar value of a decision, you want to have a positive number; the higher the EMV, the better. Since the EMV is positive for both Projects 1 and 2, Cliff's firm would expect a positive outcome from each and could bid on both projects. If it had to choose between the two projects, perhaps because of limited resources, Cliff's firm should bid on Project 2 because it has a higher EMV.

Also notice in Figure 11-6 that if you just looked at the potential outcome of the two projects, Project 1 looks more appealing. You could earn $300,000 in profits from Project 1, but you can only earn $60,000 for Project 2. If Cliff were a risk seeker, he would naturally want to bid on Project 1. However, there is only a 20 percent chance of getting that $300,000 on Project 1, and there is a 70 percent chance of earning $60,000 on Project 2. Using EMV helps account for all possible outcomes and their probabilities of occurrence, thereby reducing the tendency to pursue overly aggressive or conservative risk strategies.

Simulation

A more sophisticated quantitative risk analysis technique is simulation. Simulation uses a representation or model of a system to analyze the expected behavior or performance of the system. Most simulations are based on some form of Monte Carlo analysis. **Monte Carlo analysis** simulates a model's outcome many times to provide a statistical distribution of the calculated results. Monte Carlo analysis can determine that a project will finish by a certain date only 10 percent of the time, and determine another date for which the project will finish 50 percent of the time. In another words, the Monte Carlo analysis can predict the probability of finishing by a certain date or the probability that the cost will be equal to or less than a certain value.

You can use several different types of distribution functions when performing a Monte Carlo analysis. The example below is a simplified approach.

The basic steps of a Monte Carlo analysis are:

1. Assess the range for the variables being considered. In other words, collect the most likely, optimistic, and pessimistic estimates for the variables in the model. For example, if you are trying to determine the likelihood of meeting project schedule goals, the project network diagram would be your model. You would collect the most likely, optimistic, and pessimistic

time estimates for each task. Notice that this step is similar to collecting data for performing PERT estimates. However, instead of applying the same PERT weighted average formula, you go on to the following steps when performing a Monte Carlo simulation.

2. Determine the probability distribution of each variable. What is the likelihood of that variable falling between the optimistic and most likely estimates? For example, if an expert assigned to do a particular task provides a most likely estimate of ten weeks, an optimistic estimate of eight weeks, and a pessimistic estimate of fifteen weeks, you then ask what the probability is of completing that task between eight and ten weeks. The expert might respond that there is a 20 percent probability.

3. For each variable, such as the time estimate for a task, select a random value based on the probability distribution for the occurrence of the variable. For example, using the above scenario, you would randomly pick a value between eight weeks and ten weeks 20 percent of the time and a value between ten weeks and fifteen weeks 80 percent of the time.

4. Run a deterministic analysis or one pass through the model using the combination of values selected for each one of the variables. For example, the one task described above might have a value of 12 on the first run. All of the other tasks would have one random value assigned to them on that first run, also, based on their estimates and probability distributions.

5. Repeat Steps 3 and 4 many times to obtain the probability distribution of the model's results. The number of iterations depends on the number of variables and the degree of confidence required in the results, but it typically lies between 100 and 1,000. Using the project schedule as an example, the final simulation results will show you the probability of completing the entire project within a certain time period.

Figure 11-7 illustrates the results from a Monte Carlo-based simulation of a project schedule. The simulation was done using Microsoft Project and Risk+ software. On the left side of Figure 11-7 is a chart displaying columns and an S-shaped curve. The height of each column, read by the scale on the left of the chart, indicates how many times the project was completed in a given time interval during the simulation run, which is the sample count. In this example, the time interval is two working days, and the simulation was run 250 times. The first column shows that the project was completed by 1/29/07 only two times during the simulation. The S-shaped curve, read from the scale on the right of the chart, shows the cumulative probability of completing the project on or before a given date. The right side of Figure 11-7 shows the information in tabular form. For example, there is a 10 percent probability that the project will be completed by 2/8/07, a 50 percent chance of completion by 2/17/07, and a 90 percent chance of completion by 2/25/07.

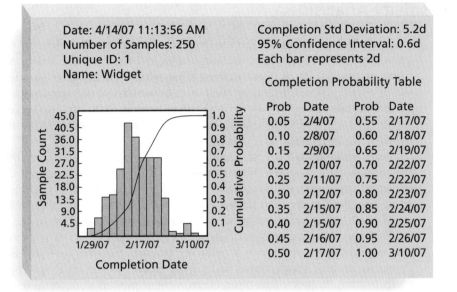

Figure 11-7. Sample Monte Carlo-Based Simulation Results for Project Schedule

As you can imagine, people use software to perform the steps required for a Monte Carlo analysis. Several PC-based software packages are available that perform Monte Carlo simulations. Many products will show you what the major risk drivers are, based on the simulation results. For example, a wide range for a certain task estimate might be causing most of the uncertainty in the project schedule. You will learn more about using simulation and other software related to project risk management later in this chapter.

☑ What Went Right?

A large aerospace company used Monte Carlo analysis to help quantify risks on several advanced-design engineering projects. The U.S. National Aerospace Plane (NASP) project involved many risks. The purpose of this multibillion-dollar project was to design and develop a vehicle that could fly into space using a single-stage-to-orbit approach. A single-stage-to-orbit approach meant the vehicle would have to achieve a speed of Mach 25 (25 times the speed of sound) without a rocket booster. A team of engineers and business professionals worked together in the mid-1980s to develop a software model for estimating the time and cost of developing the NASP project. This model was then linked with simulation software to determine the sources of cost and schedule risk for the project. The company then used the results of the Monte Carlo analysis to determine how it would invest its internal research and development funds. Although the NASP project was terminated, the resulting research has helped develop more advanced materials and propulsion systems used on many modern aircraft.

Sensitivity Analysis

Many people are familiar with using **sensitivity analysis** to see the effects of changing one or more variables on an outcome. For example, many people perform a sensitivity analysis to determine what their monthly payments will be for a loan given different interest rates or periods of the loan. What will your monthly mortgage payment be if you borrow $100,000 for 30 years at a 6 percent rate? What will it be if the interest rate is 7 percent? What if you pay off the loan in 15 years at 5 percent?

Many professionals use sensitivity analysis to help make several common business decisions, such as determining break-even points based on different assumptions. People often use spreadsheet software like Excel to perform sensitivity analysis. Figure 11-8 shows an example Excel file created to quickly show the break-even point for a product based on various inputs: the sales price per unit, the manufacturing cost per unit, and fixed monthly expenses. The current inputs result in a break-even point of 6,250 units sold. Users of this spreadsheet can change inputs and see the effects on the break-even point in chart format. Project teams often create similar models to determine the sensitivity of various project variables. For example, Cliff's team could develop sensitivity analysis models to estimate their profits on jobs by varying how many hours it would take them to do the job, costs per hour, and so on.

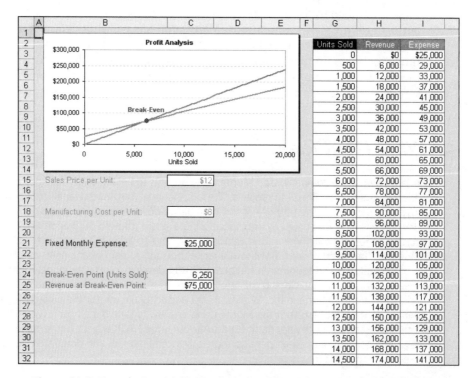

Figure 11-8. Sample Sensitivity Analysis for Determining Break-Even Point

The main outputs of quantitative risk analysis are updates to the risk register, such as revised risk rankings or detailed information behind those rankings. The quantitative analysis also provides high-level information in terms of the probabilities of achieving certain project objectives. This information might cause the project manager to suggest changes in contingency reserves. In some cases, projects may be redirected or canceled based on the quantitative analysis, or they might cause the initiation of new projects to help the current one succeed, such as the NASP project described in the What Went Right example.

RISK RESPONSE PLANNING

After an organization identifies and quantifies risks, it must develop an appropriate response to them. Developing a response to risks involves developing options and defining strategies for reducing negative risks and enhancing positive risks.

The four basic response strategies for negative risks are:

1. **Risk avoidance** or eliminating a specific threat, usually by eliminating its causes. Of course, not all risks can be eliminated, but specific risk events can be. For example, a project team may decide to continue using a specific piece of hardware or software on a project because they know it works. Other products that could be used on the project may be available, but if the project team is unfamiliar with them, they could cause significant risk. Using familiar hardware or software eliminates this risk.

2. **Risk acceptance** or accepting the consequences should a risk occur. For example, a project team planning a big project review meeting could take an active approach to risk by having a contingency or backup plan and contingency reserves if they cannot get approval for a specific site for the meeting. On the other hand, they could take a passive approach and accept whatever facility their organization provides them.

3. **Risk transference** or shifting the consequence of a risk and responsibility for its management to a third party. For example, risk transference is often used in dealing with financial risk exposure. A project team may purchase special insurance or warranty protection for specific hardware needed for a project. If the hardware fails, the insurer must replace it within an agreed-upon period of time.

4. **Risk mitigation** or reducing the impact of a risk event by reducing the probability of its occurrence. Suggestions for reducing common sources of risk on information technology projects were provided at the beginning of this chapter. Other examples of risk mitigation include using proven technology, having competent project personnel, using various analysis and validation techniques, and buying maintenance or service agreements from subcontractors.

Table 11-8 provides general mitigation strategies for technical, cost, and schedule risks on projects.[10] Note that increasing the frequency of project monitoring and using a work breakdown structure (WBS) and Critical Path Method (CPM) are strategies for all three areas. Increasing the project manager's authority is a strategy for mitigating technical and cost risks, and selecting the most experienced project manager is recommended for reducing schedule risks. Improving communication is also an effective strategy for mitigating risks.

Table 11-8: General Risk Mitigation Strategies for Technical, Cost, and Schedule Risks

TECHNICAL RISKS	COST RISKS	SCHEDULE RISKS
Emphasize team support and avoid stand-alone project structure	Increase the frequency of project monitoring	Increase the frequency of project monitoring
Increase project manager authority	Use WBS and CPM	Use WBS and CPM
Improve problem handling and communication	Improve communication, project goals understanding, and team support	Select the most experienced project manager
Increase the frequency of project monitoring	Increase project manager authority	
Use WBS and CPM		

The four basic response strategies for positive risks are:

1. **Risk exploitation** or doing whatever you can to make sure the positive risk happens. For example, suppose Cliff's company funded a project to provide new computer classrooms for a nearby school in need. The project manager might organize news coverage of the project, write a press release, or hold some other public event to ensure the project produces good public relations for the company, which could lead to more business.

2. **Risk sharing** or allocating ownership of the risk to another party. Using the same example of implementing new computer classrooms, the project manager could form a partnership with the school's principal, school board, or parent-teacher organization to share responsibility for achieving good public relations for the project. Or, the company might partner with a local training firm who agrees to provide free training for all of the teachers on how to use the new computer classrooms.

3. **Risk enhancement** or changing the size of the opportunity by identifying and maximizing key drivers of the positive risk. For example, an important driver of getting good public relations for the computer classrooms project

---

[10] Couillard, Jean, "The Role of Project Risk in Determining Project Management Approach," *Project Management Journal*, Project Management Institute (December 1995).

might be getting the students, parents, and teachers aware of and excited about the project. They could then do their own formal or informal advertising of the project and Cliff's company, which in turn might interest other groups and could generate more business.

4. Risk acceptance also applies to positive risks when the project team cannot or chooses not to take any actions toward a risk. For example, the computer classrooms project manager might just assume the project will result in good public relations for their company without doing anything extra.

The main outputs of risk response planning include risk-related contractual agreements, updates to the project management plan, and updates to the risk register. For example, if Cliff's company decided to partner with a local training firm on the computer classrooms project to share the opportunity of achieving good public relations, it could write a contract with that firm. The project management plan and its related plans might need to be updated if the risk response strategies require additional tasks, resources, or time to accomplish them. Risk response strategies often result in changes to the WBS and project schedule, so plans with that information must be updated. The risk response strategies also provide updated information for the risk register by describing the risk responses, risk owners, and status information.

Risk response strategies often include identification of residual and secondary risks as well as contingency plans and reserves, as described earlier. **Residual risks** are risks that remain after all of the response strategies have been implemented. For example, even though a more stable hardware product may have been used on a project, there may still be some risk of it failing to function properly. **Secondary risks** are a direct result of implementing a risk response. For example, using the more stable hardware may have caused a risk of peripheral devices failing to function properly.

🎥 Media Snapshot

A highly publicized example of a risk response to corporate financial scandals, such as those affecting Enron, Arthur Andersen, and WorldCom, was legal action. The Sarbanes-Oxley Act of 2002, signed into law on 30 July 2002, is considered the most significant change to federal securities laws in the United States since the New Deal. This Act has caused many organizations to initiate projects and other actions to avoid litigation. (For more detailed information on this law, see *www.sarbanes-oxley.com.*)

The Sarbanes-Oxley Act imposes fines and prison sentences of up to 20 years for anyone who knowingly alters or destroys a record or document with the intent to obstruct an investigation. While most provisions focus on financial records, they can also affect other corporate records. For example, during an investigation, lawyers can submit discovery requests to information technology departments. These requests could require access to all e-mail communication. The courts can treat e-mail messages and attachments as business records that must be retained to achieve regulatory compliance. Network administrators are often instructed to overwrite tapes, but that means their company knowingly allows potential evidence to be destroyed. Does a network administrator or anyone else in the organization want to spend up to 20 years in prison for unknowingly destroying important data?[11]

RISK MONITORING AND CONTROL

Risk monitoring and control involves executing the risk management processes to respond to risk events. Executing the risk management processes means ensuring that risk awareness is an ongoing activity performed by the entire project team throughout the entire project. Project risk management does not stop with the initial risk analysis. Identified risks may not materialize, or their probabilities of occurrence or loss may diminish. Previously identified risks may be determined to have a greater probability of occurrence or a higher estimated loss value. Similarly, new risks will be identified as the project progresses. Newly identified risks need to go through the same process as those identified during the initial risk assessment. A redistribution of resources devoted to risk management may be necessary because of relative changes in risk exposure.

Carrying out individual risk management plans involves monitoring risks based on defined milestones and making decisions regarding risks and their response strategies. It may be necessary to alter a strategy if it becomes ineffective, implement a planned contingency activity, or eliminate a risk from the list of potential risks when it no longer exists. Project teams sometimes use **workarounds**—unplanned responses to risk events—when they do not have contingency plans in place.

Risk reassessment, risk audits, variance and trend analysis, technical performance measurements, reserve analysis, and status meetings or periodic risk reviews such as the Top Ten Risk Item Tracking method are all tools and techniques for performing risk monitoring and control. Outputs of this process are requested changes, recommended corrective and preventive actions, and updates to the risk register, project management plan, and organizational process assets, such as lessons-learned information that might help future projects.

USING SOFTWARE TO ASSIST IN PROJECT RISK MANAGEMENT

As demonstrated in several parts of this chapter, you can use a variety of software tools to enhance various risk management processes. Most organizations use software to create, update, and distribute information in their risk registers. The risk register is often a simple Word or Excel file, but it can also be part of a

[11] Iosub, John C. "What the Sarbanes-Oxley Act means for IT managers," *TechRepublic,* (March 19, 2003) (*http://techrepublic.com.com/5100-6313-5034345.html*).

more sophisticated database. Spreadsheets can aid in tracking and quantifying risks, preparing charts and graphs, and performing sensitivity analysis. Software can be used to create decision trees and estimate expected monetary value.

More sophisticated risk management software, such as Monte Carlo simulation software, can help you develop models and use simulations to analyze and respond to various risks. You can purchase add-on software to perform Monte Carlo simulations using Excel (such as Decisioneering Crystal Ball or Palisade @Risk for Excel) or Project 2003 (such as CS Solutions Risk+ or Palisade @Risk for Project). There are also several software packages created specifically for project risk management. See the Project Management Institute Risk Specific Interest Group's Web site at *www.risksig.com* for a detailed list of software products available to assist in performing project risk management.

Although it has become easier to do sophisticated risk analysis with new software tools, project teams must be careful not to over-rely on using software when performing project risk management. If a risk is not identified, it cannot be managed, and it takes intelligent, experienced people to do a good job of identifying risks. It also takes hard work to develop and implement good risk response strategies. Software should be used as a tool to help make good decisions in project risk management, not as a scapegoat for when things go wrong.

Well-run projects, like a master violinist's performance, an Olympic athlete's gold medal win, or a Pulitzer Prize–winning book, appear to be almost effortless. Those on the outside—whether audiences, customers, or managers—cannot observe the effort that goes into a superb performance. They cannot see the hours of practice, the edited drafts, or the planning, management, and foresight that create the appearance of ease. To improve information technology project management, project managers should strive to make their jobs look easy—it reflects the results of a well-run project.

CASE WRAP-UP

Cliff Branch and two of his senior people attended a seminar on project risk management where the speaker discussed several techniques, such as estimating the expected monetary value of projects, Monte Carlo simulations, and so on. Cliff asked the speaker how these techniques could be used to help his company decide which projects to bid on, since bidding on projects often required up-front investments with the possibility of no payback. The speaker walked through an example of EMV and then ran a quick Monte Carlo simulation. Cliff did not have a strong math background and had a hard time understanding the EMV calculations. He thought the simulation was much too confusing to have any practical use for him. He believed in his gut instincts much more than any math calculation or computer output.

The speaker finally sensed that Cliff was not impressed, so she explained the importance of looking at the odds of winning project awards and not just at the potential profits. She suggested using a risk-neutral strategy by bidding on projects that the company had a good chance of winning (50 percent or so) and that had a good profit potential, instead of focusing on projects that they had a small chance of winning and that had a larger profit potential. Cliff disagreed with this advice, and he continued to bid on high-risk projects. The two other managers who attended the seminar now understood why the firm was having problems—their leader loved taking risks, even if it hurt the company. They soon found jobs with competing companies, as did several other employees.

CHAPTER SUMMARY

Risk is an uncertainty that can have a negative or positive effect on meeting project objectives. Projects, by virtue of their unique nature, involve risk. Many organizations do a poor job of project risk management, if they do any at all. Successful organizations realize the value of good project risk management.

Risk management is an investment; that is, there are costs associated with identifying risks, analyzing those risks, and establishing plans to address those risks. Those costs must be included in cost, schedule, and resource planning.

Risk utility or risk tolerance is the amount of satisfaction or pleasure received from a potential payoff. Risk seekers enjoy high risks, risk-averse people do not like to take risks, and risk-neutral people seek to balance risks and potential payoff.

Project risk management is a process in which the project team continually assesses what may negatively or positively impact the project, determines the probability of such events occurring, and determines the impact should such events occur. It also involves analyzing and determining alternate strategies to deal with risks. The six main processes involved in risk management are risk management planning, risk identification, qualitative risk analysis, quantitative risk analysis, risk response planning, and risk monitoring and control.

Risk management planning is the process of deciding how to approach and plan for risk management activities for a particular project. A risk management plan is a key output of the risk management planning process, and a risk register is a key output of the other risk management processes. Contingency plans are predefined actions that a project team will take if an identified risk event occurs. Fallback plans are developed for risks that have a high impact on meeting project objectives, and are implemented if attempts to reduce the risk are not effective. Contingency reserves or contingency allowances are provisions held by the project sponsor or organization to reduce the risk of cost or schedule overruns to an acceptable level.

Information technology projects often involve several risks: lack of user involvement, lack of executive management support, unclear requirements, poor planning, and so on. Lists developed by the Standish Group and other organizations can help you identify potential risks on information technology projects. A risk breakdown structure is a useful tool that can help project managers consider potential risks in different categories. Lists of common risk conditions in project management knowledge areas can also be helpful in identifying risks, as can information-gathering techniques such as brainstorming, the Delphi Technique, interviewing, and SWOT analysis. A risk register is a document that contains results of various risk management processes, often displayed in a table or spreadsheet format. It is a tool for documenting potential risk events and related information. Risk events refer to specific, uncertain events that may occur to the detriment or enhancement of the project.

Risks can be assessed qualitatively and quantitatively. Tools for qualitative risk analysis include a probability/impact matrix, the Top Ten Risk Item Tracking technique, and expert judgment. Tools for quantitative risk analysis include decision trees and Monte Carlo simulation. Expected monetary value (EMV) uses decision trees to evaluate potential projects based on their expected value. Simulations are a more sophisticated method for creating estimates to help you determine the likelihood of meeting specific project schedule or cost goals. Sensitivity analysis is used to show the effects of changing one or more variables on an outcome.

The four basic responses to risk are avoidance, acceptance, transference, and mitigation. Risk avoidance involves eliminating a specific threat or risk. Risk acceptance means accepting the consequences of a risk, should it occur. Risk transference is shifting the consequence of a risk and responsibility for its management to a third party. Risk mitigation is reducing the impact of a risk event by reducing the probability of its occurrence. The four basic response strategies for positive risks are risk exploitation, risk sharing, risk enhancement, and risk acceptance.

Risk monitoring and control involves executing the risk management processes and the risk management plan to respond to risks. Outputs of this process include requested changes, recommended corrective and preventive actions, and updates to the risk register, project management plan, and organizational process assets.

Several types of software can assist in project risk management. Monte Carlo–based simulation software is a particularly useful tool for helping get a better idea of project risks and top sources of risk or risk drivers.

DISCUSSION QUESTIONS

1. Discuss the risk utility function and risk preference chart in Figure 11-2. Would you rate yourself as being risk-averse, risk-neutral, or risk-seeking? Give examples of each approach from different aspects of your life, such as your current job, your personal finances, romances, and eating habits.

2. What are some questions that should be addressed in a risk management plan?

3. Discuss the common sources of risk on information technology projects and suggestions for managing them. Which suggestions do you find most useful? Which do you feel would not work in your organization? Why?

4. What is the difference between using brainstorming and the Delphi Technique for risk identification? What are some of the advantages and disadvantages of each approach? Describe the contents of a risk register and how the risk register is used in several risk management processes.

5. Describe how to use a probability/impact matrix, risk factors, and the Top Ten Risk Item Tracking approaches for performing qualitative risk analysis. How could you use each technique on a project?

6. Explain how to use decision trees and Monte Carlo analysis for quantifying risk. Give an example of how you could use each technique on an information technology project.

7. Provide realistic examples of each of the risk response strategies for both negative and positive risks.

8. List the tools and techniques for performing risk monitoring and control.

9. How can you use Excel to assist in project risk management? What other software can help project teams make better risk management decisions?

EXERCISES

1. Suppose your college or organization is considering a new project that would involve developing an information system that would allow all employees and students/customers to access and maintain their own human resources–related information, such as address, marital status, tax information, and so on. The main benefits of the system would be a reduction in human resources personnel and more accurate information. For example, if an employee, student, or customer had a new telephone number or e-mail address, he or she would be responsible for entering the new data in the new system. The new system would also allow employees to change their tax withholdings or pension plan contributions. Identify five potential risks for this new project, being sure to list some negative and positive risks. Provide a detailed description of each risk and propose strategies for addressing each risk.

2. Review a project risk management plan. Did the organization do a good job of answering the questions that should be addressed in a risk management plan? If you cannot find a risk management plan, write one of your own for an information technology project your organization is considering.

3. Research risk management software. (See the sixth Suggested Reading for information.) Are many products available? What are the main advantages of using them? What are the main disadvantages? Write a paper discussing your findings, and include at least three references.

4. Suppose your organization is deciding which of four projects to bid on. Information on each is in the table below. Assume that all up-front investments are not recovered, so they are shown as negative profits. Draw a diagram and calculate the EMV for each project. Write a few paragraphs explaining which projects you would bid on. Be sure to use the EMV information and your personal risk tolerance to justify your answer.

PROJECT	CHANCE OF OUTCOME	ESTIMATED PROFITS
Project 1	50 percent	$120,000
	50 percent	–$50,000
Project 2	30 percent	$100,000
	40 percent	$50,000
	30 percent	–$60,000
Project 3	70 percent	$20,000
	30 percent	–$5,000
Project 4	30 percent	$40,000
	30 percent	$30,000
	20 percent	$20,000
	20 percent	–$50,000

5. Find an example of a company that took a big risk on an information technology project and succeeded. In addition, find an example of a company that took a big risk and failed. Summarize each project and situation. Did anything besides luck make a difference between success and failure?

RUNNING CASE

Tony and his team identified some risks the first month of the Recreation and Wellness Intranet Project. However, all they did was document them in a list. They never ranked them or developed any response strategies. Since several problems have been occurring on the project, such as key team members leaving the company, users being uncooperative, and team members not providing good status information, Tony has decided to be more proactive in managing risks. He also wants to address positive as well as negative risks.

1. Create a risk register for the project, using Table 11-5 and the data below it as a guide. Identify six potential risks, including risks related to the problems described above. Include negative and positive risks.

2. Plot the six risks on a probability/impact matrix, using Figure 11-6. Also assign a numeric value for the probability and impact of each risk on meeting the main project objective. Use a scale of 1 to 10 in assigning the values, with 1 being low and 10 being high. For a simple risk factor calculation, multiply these two values (the probability score and the impact score). Add a column to your risk register to the right of the impact column called Risk Score. Enter the new data in the risk register. Write your rationale for how you determined the scores for one of the negative risks and one of the positive risks.

3. Develop a response strategy for one of the negative risks and one of the positive risks. Enter the information in the risk register. Also write a separate paragraph describing what specific tasks would need to be done to implement the strategy. Include time and cost estimates for each strategy, as well.

ADDITIONAL RUNNING CASES AND OTHER APPENDICES

Appendix C provides additional case studies and questions you can use to practice applying the concepts, tools, and techniques you are learning throughout this and subsequent chapters. Review the running cases provided in Appendix C and on the companion Web site (*www.course.com/mis/schwalbe4e*). Appendix D includes templates for various project management documents. For additional sample documents based on real projects, visit the author's Web site at *www.kathyschwalbe.com*. Appendix E and the CD-ROM included with this text include a computer simulation where you can also practice applying the project management process groups and knowledge areas.

SUGGESTED READINGS

1. Boehm, Barry W. "Software Risk Management: Principles and Practices." *IEEE Software* (January 1991).

    Boehm describes detailed approaches for several risk management techniques, including risk identification checklists, risk prioritization, risk management planning, and risk monitoring. This article includes an example of Top Ten Risk Item Tracking.

2. Defense Acquisition University. "Risk Management Guide for DoD Acquisition (Fifth Edition, Version 2.0)." (*www.dau.mil/pubs/gdbks/ risk_management.asp*) (June 2003).

    This free risk management guide is an excellent source of information on project risk management as applied to Department of Defense (DoD) projects. DoD has over 134,000 people working in acquisition, technology, and logistics who are involved in managing and supporting thousands of projects. Much of the information contained in this guide is directly applicable to information technology projects.

3. DeMarco, Tom and Timothy Lister. *Waltzing with Bears: Managing Risk on Software Projects*. Dorset House Publishing Company (March 2003).

   The authors, who also wrote Peopleware, *are also consultants in risk management. They say they have moved from a focus on risk control to a focus on risk taking, which takes a more proactive approach to risk management. In this book, they describe how to identify and embrace worthwhile risks on software development projects. They also offer strategies for common risks that software projects face, such as schedule flaws, requirements inflation, and specification problems.*

4. Hillson, David. "The Risk Breakdown Structure (RBS) as an Aid to Effective Risk Management" Presented at PMI Europe (June 2002).

   Although people have created work breakdown structures for a long time, the concept of a risk breakdown structure is fairly new. In his paper, Dr. Hillson shows how the RBS can be an invaluable tool in understanding and managing project risks.

5. Microsoft. White Papers on Microsoft Solutions Framework (MSF). *www.microsoft.com/enterpriseservices* (2001).

   Microsoft has published several white papers describing its approach to managing risks (search for risk management). The Microsoft Solutions Framework includes three core disciplines, one being risk management. The "Process of Risk Management" white paper describes techniques that project managers can use to create an environment in which teams successfully identify and manage risks.

6. Project Management Institute. Risk Specific Interest Group (SIG) Web Site. (*www.risksig.com*) (2004).

   The mission of PMI's Risk SIG is to establish and promote the principles of risk management as the foundation for effective project management. You can access several articles and learn more about project risk management from this Web site.

KEY TERMS

- **brainstorming** — a technique by which a group attempts to generate ideas or find a solution for a specific problem by amassing ideas spontaneously and without judgment

- **contingency plans** — predefined actions that the project team will take if an identified risk event occurs

- **contingency reserves** or **contingency allowances** — provisions held by the project sponsor or organization to reduce the risk of cost or schedule overruns to an acceptable level

- **decision tree** — a diagramming analysis technique used to help select the best course of action in situations in which future outcomes are uncertain
- **Delphi technique** — an approach used to derive a consensus among a panel of experts, to make predictions about future developments
- **expected monetary value (EMV)** — the product of the risk event probability and the risk event's monetary value
- **fallback plans** — plans developed for risks that have a high impact on meeting project objectives, to be implemented if attempts to reduce the risk are not effective
- **flowcharts** — diagrams that show how various elements of a system relate to each other
- **influence diagrams** — diagrams that represent decision problems by displaying essential elements, including decisions, uncertainties, and objectives, and how they influence each other
- **interviewing** — a fact-finding technique that is normally done face-to-face, but can also occur through phone calls, e-mail, or instant messaging
- **Monte Carlo analysis** — a risk quantification technique that simulates a model's outcome many times, to provide a statistical distribution of the calculated results
- **probability/impact matrix or chart** — a matrix or chart that lists the relative probability of a risk occurring on one side of a matrix or axis on a chart and the relative impact of the risk occurring on the other
- **qualitative risk analysis** — qualitatively analyzing risks and prioritizing their effects on project objectives
- **quantitative risk analysis** — measuring the probability and consequences of risks and estimating their effects on project objectives
- **residual risks** — risks that remain after all of the response strategies have been implemented
- **risk** — an uncertainty that can have a negative or positive effect on meeting project objectives
- **risk acceptance** — accepting the consequences should a risk occur
- **risk-averse** — having a low tolerance for risk
- **risk avoidance** — eliminating a specific threat or risk, usually by eliminating its causes
- **risk breakdown structure** — a hierarchy of potential risk categories for a project
- **risk enhancement** — changing the size of an opportunity by identifying and maximizing key drivers of the positive risk
- **risk events** — specific circumstances that may occur to the detriment of the project
- **risk exploitation** — doing whatever you can to make sure the positive risk happens

- **risk factors** — numbers that represent overall risk of specific events, given their probability of occurring and the consequence to the project if they do occur
- **risk identification** — determining which risks are likely to affect a project and documenting the characteristics of each
- **risk management plan** — a plan that documents the procedures for managing risk throughout a project
- **risk management planning** — deciding how to approach and plan the risk management activities for a project, by reviewing the project charter, WBS, roles and responsibilities, stakeholder risk tolerances, and the organization's risk management policies and plan templates
- **risk mitigation** — reducing the impact of a risk event by reducing the probability of its occurrence
- **risk monitoring and control** — monitoring known risks, identifying new risks, reducing risks, and evaluating the effectiveness of risk reduction throughout the life of the project
- **risk-neutral** — a balance between risk and payoff
- **risk owner** — the person who will take responsibility for a risk and its associated response strategies and tasks
- **risk register** — a document that contains results of various risk management processes, often displayed in a table or spreadsheet format
- **risk response planning** — taking steps to enhance opportunities and reduce threats to meeting project objectives
- **risk-seeking** — having a high tolerance for risk
- **risk sharing** — allocating ownership of the risk to another party
- **risk transference** — shifting the consequence of a risk and responsibility for its management to a third party
- **risk utility** or **risk tolerance** — the amount of satisfaction or pleasure received from a potential payoff
- **runaway projects** — projects that have significant cost or schedule overruns
- **secondary risks** — risks that are a direct result of implementing a risk response
- **sensitivity analysis** — a technique used to show the effects of changing one or more variables on an outcome
- **Top Ten Risk Item Tracking** — a qualitative risk analysis tool for identifying risks and maintaining an awareness of risks throughout the life of a project
- **triggers** — indications for actual risk events
- **watch list** — a list of risks that are low priority, but are still identified as potential risks
- **workarounds** — unplanned responses to risk events when there are no contingency plans in place

12

Project Procurement Management

Objectives

After reading this chapter, you will be able to:

1. *Understand the importance of project procurement management and the increasing use of outsourcing for information technology projects*
2. *Describe the work involved in planning purchases and acquisitions for projects, the contents of a procurement management plan and contract statement of work, and calculations involved in a make-or-buy analysis*
3. *Discuss what is involved in planning contracting, including the creation of various procurement documents and evaluation criteria for sellers*
4. *Understand the process of requesting seller responses and the difference between proposals and bids*
5. *Describe the seller selection process and recognize different approaches for evaluating proposals or selecting suppliers*
6. *Discuss the importance of good contract administration*
7. *Describe the contract closure process*
8. *Discuss types of software available to assist in project procurement management*

OPENING CASE

*M*arie McBride could not believe how much money her company was paying for outside consultants to help the company finish an important operating system conversion project. The consulting company's proposal said it would provide experienced professionals who had completed similar conversions, and that the job would be finished in six months or less with four consultants working full time. Nine months later her company was still paying high consulting fees, and half of the original consultants on the project had been replaced with new people. One new consultant had graduated from college only two months before. Marie's internal staff complained that they were wasting time training some of these "experienced professionals." Marie talked to her company's purchasing manager about the contract, fees, and special clauses that might be relevant to the problems they were experiencing.

Marie was dismayed at how difficult it was to interpret the contract. It was very long and obviously written by someone with a legal background. When she asked what her company could do since the consulting firm was not following its proposal, the purchasing manager stated that the proposal was not part of the official contract. Marie's company was paying for time and materials, not specific deliverables. There was no clause stating the minimum experience level required for the consultants, nor were there penalty clauses for not completing the work on time. There was a termination clause, however, meaning the company could terminate the contract. Marie wondered why her company had signed such a poor contract. Was there a better way to deal with procuring services from outside the company?

THE IMPORTANCE OF PROJECT PROCUREMENT MANAGEMENT

Procurement means acquiring goods and/or services from an outside source. The term procurement is widely used in government; many private companies use the term *purchasing*, and information technology professionals use the term *outsourcing*. Organizations or individuals who provide procurement services are referred to as suppliers, vendors, contractors, subcontractors, or sellers, with suppliers being the most widely used term. Many information technology projects involve the use of goods and services from outside the organization.

In the last few years, outsourcing has become a hot topic for research and debate, especially the implication of outsourcing to other countries, referred to as offshore outsourcing. The outsourcing statistics below are from a March 2004 Information Technology Association of America (ITAA)–sponsored report:

- Spending for global sources of computer software and services is expected to grow at a compound annual rate of about 26 percent, increasing from $10 billion in 2003 to $31 billion in 2008. These spending amounts represent 2.3 percent and 6.2 percent of the total information technology software and services spending of U.S. corporations in 2003 and 2008, respectively. India is the leading country for U.S. offshore outsourcing.

■ Total savings from offshore resources during the same time period are estimated to grow from $6.7 billion to $20.9 billion. The cost savings and use of offshore resources lower inflation, increase productivity, and lower interest rates, which boosts business and consumer spending and increases economic activity.[1]

Politicians debate on whether offshore outsourcing helps their own country or not. Andy Bork, chief operating officer of a computer network support service provider, describes outsourcing as an essential part of a healthy business diet. He describes good vs. bad outsourcing, similar to good vs. bad cholesterol. He says that most people view offshore outsourcing as being bad because it takes jobs away from domestic workers. However, many companies are realizing that they can use offshore outsourcing *and* create more jobs at home. For example, Atlanta-based Delta Air Lines created 1,000 call-center jobs in India in 2003, saving $25 million, which enabled them to add 1,200 job positions for reservations and sales agents in the United States.[2] Other companies, like Wal-Mart, successfully manage the majority of their information technology projects in-house with very little commercial software and no outsourcing at all. (See the sixth Suggested Reading for details on "Wal-Mart's Way.")

Most organizations do use some form of outsourcing to meet their information technology needs, spending most money within their own country. In fact, many information technology professionals are choosing to work for temporary job agencies so they can move from company to company while maintaining stability by working for the same temporary agency. In the United States, temporary workers numbered 2.25 million nationwide in 2003, up almost 50 percent over ten years, according to the U.S. Bureau of Labor Statistics. The bureau also projects that the temporary employment industry will create more jobs than any other in this country through 2012. Because projects are temporary and business needs can change quickly, demand for temporary workers will continue to grow.[3]

Because outsourcing is a growing area, it is important for project managers to understand project procurement management. Many organizations are turning to outsourcing to:

■ **Reduce both fixed and recurrent costs.** Outsourcing suppliers are often able to use economies of scale that may not be available to the client alone, especially for hardware and software. It can also be less expensive to outsource some labor costs to other organizations in the same country or offshore. Companies can also use outsourcing to reduce labor costs on projects by avoiding the costs of hiring, firing, and reassigning people to projects or paying their salaries when they are between projects.

---

[1] Global Insight, "Executive Summary: The Comprehensive Impact of Offshore IT Software and Services Outsourcing on the U.S. Economy and the IT Industry," Information Technology Association of America, *www.itaa.org* (March 2004).

[2] Bork, Andy, "Soft skills needed in a hard world," *Star Tribune* (May 24, 2004).

[3] Cummins, H.J. "Permanent Temps," *Star Tribune* (October 6, 2004).

- Allow the client organization to focus on its core business. Most organizations are not in business to provide information technology services, yet many have spent valuable time and resources on information technology functions when they should have focused on core competencies such as marketing, customer service, and new product design. By outsourcing many information technology functions, employees can focus on jobs that are critical to the success of the organization.

- Access skills and technologies. Organizations can gain access to specific skills and technologies when they are required by using outside resources. For example, a project may require an expert in a particular field or require the use of expensive hardware or software for one particular month on a project. Planning for this procurement will ensure that the needed skills or technology will be available for the project.

- Provide flexibility. Outsourcing to provide extra staff during periods of peak workloads can be much more economical than trying to staff entire projects with internal resources. Many companies cite quicker flexibility in staffing as a key reason for outsourcing.

- Increase accountability. A well-written **contract**—a mutually binding agreement that obligates the seller to provide the specified products or services and obligates the buyer to pay for them—can clarify responsibilities and sharpen focus on key deliverables of a project. Because contracts are legally binding, there is more accountability for delivering the work as stated in the contract.

A recent trend in outsourcing is the increasing size of contracts. The $6.9 billion Navy-Marine Corps Intranet contract awarded in October 2000 was the biggest technology outsourcing contract ever for the U.S. government.[4] Procter & Gamble recently opened its information technology and business operations outsourcing work for bids, and analysts have estimated the potential value of the contract at anywhere from $4 billion to $10 billion over ten years.[5] With so much money at stake, project managers must be well versed in procurement management.

Organizations must also consider reasons they might *not* want to outsource. When an organization outsources work, it often does not have as much control over those aspects of projects that suppliers carry out. In addition, an organization could become too dependent on particular suppliers. If those suppliers went out of business or lost key personnel, it could cause great damage to a project. Organizations must also be careful to protect strategic information that could become vulnerable in the hands of suppliers. According to Scott McNeally, CEO of Sun Microsystems, Inc., "What you want to handle in-house is the stuff that

---

[4] Chen, Yu-Che and James L. Perry, "IT Outsourcing: A Primer for Public Managers," *IBM Endowment for The Business of Government* (February 2002).

[5] Weiss, Todd R. and Juan Carlos Perez, "EDS Renews P&G Outsourcing Talks," *ComputerWorld* (September 23, 2002).

gives you an edge over your competition—your core competencies. I call it your 'secret sauce.' If you're on Wall Street and you have your own program for tracking and analyzing the market, you'll hang onto that. At Sun, we have a complex program for testing microprocessor designs, and we'll keep it."[6] Project teams must think carefully about procurement issues and make wise decisions based on the unique needs of their projects and organizations. They can also change their minds on outsourcing as business conditions change.

ⓧ *What Went Wrong?*

In December 2002, when the financial services company JPMorgan Chase announced a seven-year, $5 billion deal to outsource much of its data processing to IBM, both companies bragged that the contract was the largest of its kind. It seemed like a win-win situation—IBM would make money and reduce costs, and JPMorgan Chase could push for innovation. However, in September 2004, JPMorgan Chase revoked the contract less than two years into its existence. The 4,000 employees of JPMorgan Chase who moved to IBM as part of the deal are planned to return to the bank next year.

What seemed like a good procurement plan no longer fit with JPMorgan Chase's business strategy. According to Austin Adams, chief information officer at JPMorgan Chase, "We believe managing our own technology infrastructure is best for the long-term growth and success of our company as well as our shareholders." IBM said the canceled contract was simply a result of JPMorgan Chase's merger earlier this year with Bank One. They tried to shrug off the loss of a large business deal. "The combined firm found itself with an abundance of IT assets," IBM spokesperson James Sciales said. "This decision was like other business decisions related to the merger."[7]

Outsourcing can also cause problems in other areas for companies and nations as a whole. For example, many people in Australia are concerned about outsourcing software development work. "The Australian Computer Society says sending work offshore may lower the number of students entering IT courses, deplete the number of skilled IT professionals, and diminish the nation's strategic technology capability. Another issue is security, which encompasses the protection of intellectual property, integrity of data, and the reliability of infrastructure in offshore locations."[8]

The success of many information technology projects that use outside resources is often due to good project procurement management. **Project procurement management** includes the processes required to acquire

---

[6] McNeally, Scott, "The Future of the Net: Why We Don't Want You to Buy Our Software," Executive Perspectives, Sun Microsystems, Inc., *www.sun.com/dot-com/perspectives/stop.html* (1999): 1.

[7] Morgenson, Gretchen, "IBM Shrugs Off Loss of Big Contract," *TechNewsWorld* (September 17, 2004).

[8] Beer, Stan, "Is going offshore good for Australia?" *The Age* (*www.theage.com*) (September 21, 2004).

goods and services for a project from outside the performing organization. Organizations can be either the buyer or the seller of products or services under a contract.

There are six main processes of project procurement management:

- **Planning purchases and acquisitions** involves determining what to procure, when, and how. In procurement planning, one must decide what to outsource, determine the type of contract, and describe the work for potential sellers. **Sellers** are contractors, suppliers, or providers who provide goods and services to other organizations. Outputs of this process include a procurement management plan, contract statement of work, make-or-buy decisions, and requested changes to the project that might result from this process.

- **Planning contracting** involves describing requirements for the products or services desired from the procurement and identifying potential sources or sellers. Outputs include procurement documents, such as a Request for Proposal (RFP), evaluation criteria, and updates to the contract statement of work.

- **Requesting seller responses** involves obtaining information, quotes, bids, offers, or proposals from sellers, as appropriate. The main outputs of this process include a qualified sellers list, a procurement document package, and proposals.

- **Selecting sellers** involves choosing from among potential suppliers through a process of evaluating potential sellers and negotiating the contract. Outputs include selected sellers, contracts, a contract management plan, resource availability information, updates to the procurement management plan, and requested changes.

- **Administering the contract** involves managing the relationship with the selected seller. Outputs include contract documentation, requested changes, recommended corrective actions, and updates to organizational process assets and the project management plan.

- **Closing the contract** involves completion and settlement of each contract, including resolution of any open items. Outputs of this process include closed contracts and updates to organizational process assets.

PLANNING PURCHASES AND ACQUISITIONS

Planning purchases and acquisitions involves identifying which project needs can best be met by using products or services outside the organization. It involves deciding whether to procure, how to procure, what to procure, how much to procure, and when to procure. An important output of planning purchases and acquisitions is the make-or-buy decision. A **make-or-buy decision** is one in which an organization decides if it is in its best interests to make certain products or perform certain services inside the organization, or if it is better to buy them from an outside organization. If there is no need to buy any products or services from outside the organization, then there is no need to perform any of the other procurement management processes.

For many projects, properly outsourcing some information technology functions can be a great investment, as shown in the following examples of What Went Right.

☑ *What Went Right?*

The Boots Company PLC, a pharmacy and health care company in Nottingham, England, outsourced its information technology systems to IBM in October 2002. The Boots Company signed a ten-year contract worth about $1.1 billion and expected to save $203.9 million over that period compared with the cost of running the systems itself. IBM managed and developed The Boots Company's systems infrastructure "from the mainframes to the tills in our 1,400 stores, to the computer on my desk," said spokesperson Francis Thomas. More than 400 Boots employees were transferred to IBM's payroll but continued to work at Boots' head office, with extra IBM staff brought in as needed. Thomas added, "The great thing about this is that if IBM has an expert in Singapore and [if] we need that expertise, we can tap into it for three months. It keeps our costs on an even keel."[9]

Properly planning purchases and acquisitions and writing good contracts can also save organizations millions of dollars. Many companies centralize purchasing for products such as personal computers, software, and printers to earn special pricing discounts. For example, in the mid-1980s the U.S. Air Force awarded a five-year, multimillion-dollar contract to automate 15 Air Force Systems Command bases. The project manager and contracting officer decided to allow for a unit pricing strategy for some items required in the contract, such as the workstations and printers. By not requiring everything to be negotiated at a fixed cost, the winning supplier lowered its final bid by more than $40 million.[10]

Inputs needed for planning purchases and acquisitions include the project scope statement, WBS and WBS dictionary, project management plan, and information on enterprise environmental factors and organizational process assets. For example, a large clothing company might consider outsourcing the delivery of, maintenance of, and basic user training and support for laptops supplied to its international sales and marketing force. If there were suppliers who could provide this service well at a reasonable price, it would make sense to outsource, because this could reduce fixed and recurring costs for the clothing company and let them focus on their core business of selling clothes.

---

[9] Law, Gillian, "IBM wins $1.1B outsourcing deal in England," *ComputerWorld* (October 1, 2002).
[10] Schwalbe, Kathy, Air Force Commendation Medal Citation, 1986.

It is important to understand why a company would want to procure goods or services and what inputs are needed to plan purchases and acquisitions. In the opening case, Marie's company hired outside consultants to help complete an operating system conversion project because it needed people with specialized skills for a short period of time. This is a common occurrence in many information technology projects. It can be more effective to hire skilled consultants to perform specific tasks for a short period of time than to hire or keep employees on staff full time.

However, it is also important to define clearly the scope of the project, the products, services, or results required, market conditions, and constraints and assumptions. In Marie's case, the scope of the project and services required were relatively clear, but her company may not have adequately discussed or documented the market conditions or constraints and assumptions involved in using the outside consultants. Were there many companies that provided consultants to do operating conversion projects similar to theirs? Did the project team investigate the background of the company that provided their consultants? Did they list important constraints and assumptions for using the consultants, such as limiting the time that the consultants had to complete the conversion project or the minimum years of experience for any consultant assigned to their project? It is very important to answer these types of questions before going into an outsourcing agreement.

Tools and Techniques for Planning Purchases and Acquisitions

There are several tools and techniques to help project managers and their teams in planning purchases and acquisitions, including make-or-buy analysis, expert judgment, and contract types.

Make-or-Buy Analysis

Make-or-buy analysis is a general management technique used to determine whether an organization should make or perform a particular product or service inside the organization or buy from someone else. This form of analysis involves estimating the internal costs of providing a product or service and comparing that estimate to the cost of outsourcing. Consider a company that has 1,000 international salespeople with laptops. Using make-or-buy analysis, the company would compare the cost of providing those services using internal resources to the cost of buying those services from an outside source. If supplier quotes were less than its internal estimates, the company should definitely consider outsourcing the training and user support services.

Many organizations also use make-or-buy analysis to decide if they should either purchase or lease items for a particular project. For example, suppose you need a piece of equipment for a project that has a purchase price of $12,000. Assume it also had a daily operational cost of $400. Suppose you could lease the same piece of equipment for $800 per day, including the operational costs. You can set up an equation in which the purchase cost equals the lease cost to determine when it makes sense financially to lease or buy the equipment. In this example, d = the number of days you need the piece of equipment. The equation would then be:

$$800d = \$12{,}000 + \$400d$$

Subtracting $400d from both sides, you get:

$$\$400d = \$12{,}000$$

Dividing both sides by $400, you get:

$$d = 30$$

which means that the purchase cost equals the lease cost in 30 days. So, if you need the equipment for less than 30 days, it would be more economical to lease it. If you need the equipment for more than 30 days, you should purchase it.

Expert Judgment

Experts inside an organization and outside an organization could provide excellent advice in planning purchases and acquisitions. Project teams often need to consult experts within their organization as part of good business practice. Internal experts might suggest that the company in the above example could not provide quality training and user support for the 1,000 laptop users since the service involves so many people with different skill levels in so many different locations. Experts in the company might also know that most of their competitors outsource this type of work and know who the qualified outside suppliers are. It is also important to consult legal experts since contracts for outsourced work are legal agreements.

Experts outside the company, including potential suppliers themselves, can also provide expert judgment. For example, suppliers might suggest an option for salespeople to purchase the laptops themselves at a reduced cost. This option would solve problems during employee turnover—exiting employees would own their laptops and new employees would purchase a laptop through the program. An internal expert might then suggest that employees receive a technology bonus to help offset what they might view as an added expense.

Expert judgment, both internal and external, is an asset in making many procurement decisions.

Types of Contracts

Contract type is an important consideration. Different types of contracts can be used in different situations. Three broad categories of contracts are fixed price or lump sum, cost reimbursable, and time and material. A single contract can actually include all three of these categories, if it makes sense for that particular procurement. For example, you could have a contract with a seller that includes purchasing specific hardware for a fixed price or lump sum, some services that are provided on a cost reimbursable basis, and other services that are provided on a time and material basis. Project managers and their teams must understand and decide which approaches to use to meet their particular project needs. It is also important to understand when and how you can take advantage of unit pricing in contracts.

Fixed-price or **lump-sum contracts** involve a fixed total price for a well-defined product or service. The buyer incurs little risk in this situation since the price is predetermined. The sellers often pad their estimate somewhat to reduce their risk, realizing their price must still be competitive. For example, a company could award a fixed-price contract to purchase 100 laser printers with a certain print resolution and print speed to be delivered to one location within two months. In this example, the product and delivery date are well defined. Several sellers could create fixed price estimates for completing the job. Fixed-price contracts may also include incentives for meeting or exceeding selected project objectives. For example, the contract could include an incentive fee paid if the laser printers are delivered within one month. A firm-fixed price (FFP) contract has the least amount of risk for the buyer, followed by a fixed-price incentive (FPI) contract.

Cost-reimbursable contracts involve payment to the supplier for direct and indirect actual costs. Recall from Chapter 7 that direct costs are costs that can be directly related to producing the products and services of the project. They normally can be traced back to a project in a cost-effective way. Indirect costs are costs that are not directly related to the products or services of the project, but are indirectly related to performing the project. They normally cannot be traced back in a cost-effective way. For example, the salaries for people working directly on a project and hardware or software purchased for a specific project are direct costs, while the cost of providing a work space with electricity, a cafeteria, and so on, are indirect costs. Indirect costs are often calculated as a percentage of direct costs. Cost-reimbursable contracts often include fees such as a profit percentage or incentives for meeting or exceeding selected project objectives. These contracts are often used for projects that include providing goods and services that involve new technologies. The buyer absorbs more of the risk with cost-reimbursable contracts than they do with fixed-price contracts. Three types of cost-reimbursable contracts, in order of lowest to highest risk to the buyer, include cost plus incentive fee, cost plus fixed fee, and cost plus percentage of costs.

- With a cost plus incentive fee (CPIF) contract, the buyer pays the supplier for allowable performance costs along with a predetermined fee and an incentive bonus. If the final cost is less than the expected cost, both the buyer and the supplier benefit from the cost savings, according to a prenegotiated share formula. For example, suppose the expected cost of a project is $100,000, the fee to the supplier is $10,000, and the share formula is 85/15, meaning that the buyer absorbs 85 percent of the uncertainty and the supplier absorbs 15 percent. If the final price is $80,000, the cost savings are $20,000. The supplier would be paid the final cost and the fee plus an incentive of $3,000 (15 percent of $20,000), for a total reimbursement of $93,000.

- With a **cost plus fixed fee (CPFF) contract**, the buyer pays the supplier for allowable performance costs plus a fixed fee payment usually based on a percentage of estimated costs. This fee does not vary, however, unless the scope of the contract changes. For example, suppose the expected cost of a project is $100,000 and the fixed fee is $10,000. If the actual cost of the contract rises to $120,000 and the scope of the contract remains the same, the contractor will still receive the fee of $10,000.

- With a **cost plus percentage of costs (CPPC) contract**, the buyer pays the supplier for allowable performance costs along with a predetermined percentage based on total costs. From the buyer's perspective, this is the least desirable type of contract because the supplier has no incentive to decrease costs. In fact, the supplier may be motivated to increase costs, since doing so will automatically increase profits based on the percentage of costs. This type of contract is prohibited for U.S. federal government use, but is sometimes used in private industry, particularly in the construction industry. All of the risk is borne by the buyer.

Time and material contracts are a hybrid of both fixed-price and cost-reimbursable contracts. For example, an independent computer consultant might have a contract with a company based on a fee of $80 per hour for his or her services plus a fixed price of $10,000 for providing specific materials for the project. The materials fee might also be based on approved receipts for purchasing items, with a ceiling of $10,000. The consultant would send an invoice to the company each week or month, listing the materials fee, the number of hours worked, and a description of the work produced. This type of contract is often used for services that are needed when the work cannot be clearly specified and total costs cannot be estimated in a contract. Many contract programmers and consultants, such as those Marie's company hired in the opening case, prefer time and material contracts.

Unit pricing can also be used in various types of contracts to require the buyer to pay the supplier a predetermined amount per unit of product or service.

The total value of the contract is a function of the quantities needed to complete the work. Consider an information technology department that might have a unit price contract for purchasing computer hardware. If the company purchases only one unit, the cost might be $1,000. If it purchases ten units, the cost would be $10,000. This type of pricing often involves volume discounts. For example, if the company purchases between 10 and 50 units, the contracted cost might be $900 per unit. If it purchases over 50 units, the cost might go down to $800 per unit. This flexible pricing strategy is often advantageous to both the buyer and the seller. (See the second example in the What Went Right earlier in this chapter.)

Any type of contract should include specific clauses that take into account issues unique to the project. For example, if a company uses a time and material contract for consulting services, the contract should stipulate different hourly rates based on the level of experience of each individual contractor. The services of a junior programmer with no Bachelor's degree and less than three years' experience might be billed at $40 per hour, whereas the services of a senior programmer with at least a Bachelor's degree and more than ten years of experience might be billed at $80 per hour.

Figure 12-1 summarizes the spectrum of risk to the buyer and supplier for different types of contracts. Buyers have the lowest risk with firm-fixed price contracts because they know exactly what they will need to pay the supplier. Buyers have the most risk with cost plus percentage of costs (CPPC) contracts because they do not know what the supplier's costs will be in advance, and the suppliers may be motivated to keep increasing costs. From the supplier's perspective, there is the least risk with a CPPC contract and the most risk with the firm-fixed price contract.

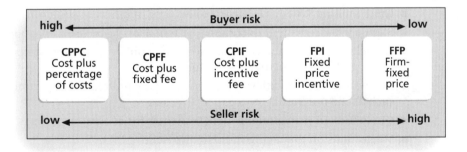

Figure 12-1. Contract Types Versus Risk

Time and material contracts and unit-price contracts can be high- or low-risk, depending on the nature of the project and other contract clauses. For example, if an organization is unclear on what work needs to be done, it cannot expect a supplier to sign a firm-fixed price contract. However, the buyer could find a consultant or group of consultants to work on specific tasks based on a predetermined hourly rate. The buying organization could evaluate the work produced every day or week to decide if it wants to continue using the consultants. In this case the contract would include a **termination clause**—a contract clause that allows the buyer or supplier to end the contract. Some termination clauses state that the buyer can terminate a contract for any reason and give the supplier only 24 hours' notice. Suppliers must often give a one-week notice to terminate a contract and must have sufficient reasons for the termination. The buyer could also include a contract clause specifying hourly rates that are based on education and experience of consultants. These contract clauses reduce the risk incurred by the buyer while providing flexibility for accomplishing the work.

In addition to make-or-buy decisions and requested changes to the project based on the procurement decision, two important outputs of planning purchases and acquisitions are a procurement management plan and a contract statement of work.

Procurement Management Plan

As stated earlier, every project management knowledge area includes some planning. The procurement management plan is a document that describes how the procurement processes will be managed, from developing documentation for making outside purchases or acquisitions to contract closure. Like other project plans, contents of the procurement management plan will vary with project needs. Below are some of the topics that can be included in a procurement management plan:

- Guidelines on types of contracts to be used in different situations
- Standard procurement documents or templates to be used, if applicable
- Guidelines for creating contract work breakdown structures, statements of work, and other procurement documents
- Roles and responsibilities of the project team and related departments, such as the purchasing or legal department
- Guidelines on using independent estimates for evaluating sellers
- Suggestions on managing multiple providers
- Processes for coordinating procurement decisions, such as make-or-buy decisions, with other project areas, such as scheduling and performance reporting
- Constraints and assumptions related to purchases and acquisitions
- Lead times for purchases and acquisitions
- Risk mitigation strategies for purchases and acquisitions, such as insurance contracts and bonds

- Guidelines for identifying prequalified sellers and organizational lists of preferred sellers
- Procurement metrics to assist in evaluating sellers and managing contracts

Contract Statement of Work

The **statement of work (SOW)** is a description of the work required for the procurement. Some organizations use the term statement of work to describe a document for describing internal work, as well. If a SOW is used as part of a contract to describe only the work required for that particular contract, it is called a contract statement of work. The contract SOW is a type of scope statement that describes the work in sufficient detail to allow prospective suppliers to determine if they are capable of providing the goods and services required and to determine an appropriate price. A contract SOW should be clear, concise, and as complete as possible. It should describe all services required and include performance reporting. It is important to use appropriate words in a contract SOW such as *must* instead of *may*. For example, *must* implies that something has to be done; *may* implies that there is a choice involved in doing something or not. The contract SOW should specify the products and services required for the project, use industry terms, and refer to industry standards.

Many organizations use samples and templates to generate SOWs. Figure 12-2 provides a basic outline or template for a contract SOW that Marie's organization could use when they hire outside consultants or purchase other goods or services. For example, for the operating system conversion project, Marie's company should specify the specific manufacturer and model number for the hardware involved, the former operating systems and new ones for the conversion, the number of pieces of each type of hardware involved (mainframes, midrange computers, or PCs), and so on. The contract SOW should also specify the location of the work, the expected period of performance, specific deliverables and when they are due, applicable standards, acceptance criteria, and special requirements. A good contract SOW gives bidders a better understanding of the buyer's expectations. A contract SOW should become part of the official contract to ensure that the buyer gets what the supplier bid on.

Statement of Work (SOW)

I.   **Scope of Work:** Describe the work to be done in detail. Specify the hardware and software involved and the exact nature of the work.

II.   **Location of Work:** Describe where the work must be performed. Specify the location of hardware and software and where the people must perform the work.

III.   **Period of Performance:** Specify when the work is expected to start and end, working hours, number of hours that can be billed per week, where the work must be performed, and related schedule information.

IV.   **Deliverables Schedule:** List specific deliverables, describe them in detail, and specify when they are due.

V.   **Applicable Standards:** Specify any company or industry-specific standards that are relevant to performing the work.

VI.   **Acceptance Criteria:** Describe how the buyer organization will determine if the work is acceptable.

VII.   **Special Requirements:** Specify any special requirements such as hardware or software certifications, minimum degree or experience level of personnel, travel requirements, and so on.

Figure 12-2. Statement of Work (SOW) Template

PLANNING CONTRACTING

Planning contracting involves preparing the documents needed for potential sellers to prepare their responses and determining the evaluation criteria for the contract award. The procurement management plan, contract statement of work, make-or-buy decisions, and project management plan are all important inputs for this process. The project team often uses standard forms and expert judgment as tools to help them create relevant procurement documents and evaluation criteria.

Two common examples of procurement documents include a Request for Proposal (RFP) and a Request for Quote (RFQ). A **Request for Proposal (RFP)** is a document used to solicit proposals from prospective suppliers. A **proposal** is a document prepared by a seller when there are different approaches for meeting buyer needs. For example, if an organization wants to automate its work practices or find a solution to a business problem, it can write and issue an RFP so suppliers can respond with proposals. Suppliers might propose various hardware, software, and networking solutions to meet the organization's need. Selections of winning sellers are often made on a variety of criteria, not just the lowest price.

A **Request for Quote (RFQ)** is a document used to solicit quotes or bids from prospective suppliers. A **bid**, also called a tender or quote (short for quotation), is a document prepared by sellers providing pricing for standard items that have been clearly defined by the buyer. Organizations often use an RFQ for solicitations that involve specific items. For example, if a company wanted to purchase 100 personal computers with specific features, it might issue an RFQ to potential suppliers. RFQs usually do not take nearly as long to prepare as RFPs, nor do responses to them. Selections are often made based on the lowest price bid.

Writing a good RFP is a critical part of project procurement management. Many people have never had to write or respond to an RFP. To generate a good RFP, expertise is invaluable. Many examples of RFPs are available within different companies, from potential contractors, and from government agencies. There are often legal requirements involved in issuing RFPs and reviewing proposals, especially for government projects. It is important to consult with experts familiar with the contract planning process for particular organizations. To make sure the RFP has enough information to provide the basis for a good proposal, the buying organization should try to put themselves in the suppliers' shoes. Could you develop a good proposal based on the information in the RFP? Could you determine detailed pricing and schedule information based on the RFP? Developing a good RFP is difficult, as is writing a good proposal.

Figure 12-3 provides a basic outline or template for an RFP. The main sections of an RFP usually include a statement of the purpose of the RFP, background information on the organization issuing the RFP, the basic requirements for the products and/or services being proposed, the hardware and software environment (usually important information for information technology related proposals), a description of the RFP process, the statement of work and schedule information, and possible appendices. A simple RFP might be three to five pages long, while an RFP for a larger, more complicated procurement might be hundreds of pages long. See the fifth Suggested Reading for information on finding samples of RFPs, such as the State University of New York Integrated Library Management System RFP. See Appendix D for information on accessing templates online.

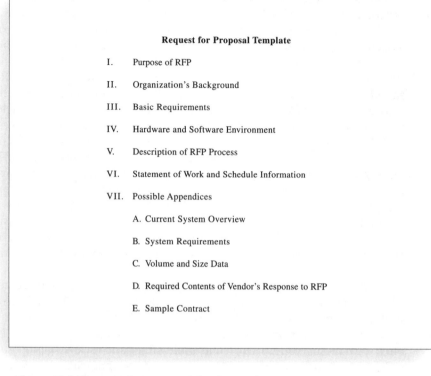

Figure 12-3. Request for Proposal (RFP) Template

Other terms used for RFQs and RFPs include invitations for bid, invitations for negotiation, and initial contractor responses. Regardless of what they are called, all procurement documents should be written to facilitate accurate and complete responses from prospective sellers. They should include background information on the organization and project, the relevant statement of work, a schedule, a description of the desired form of response, evaluation criteria, pricing forms, and any required contractual provisions. They should also be rigorous enough to ensure consistent, comparable responses, but flexible enough to allow consideration of seller suggestions for better ways to satisfy the requirements.

It is very important for organizations to prepare some form of evaluation criteria, preferably before they issue a formal RFP or RFQ. Organizations use criteria to rate or score proposals, and they often assign a weight to each criterion to indicate how important it is. Some examples of criteria include the technical approach (30 percent weight), management approach (30 percent weight), past performance (20 percent weight), and price (20 percent weight). The criteria should be specific and objective. For example, if the buyer wants the supplier's project manager to be a certified Project Management Professional (PMP), the procurement documents should state that requirement clearly and follow it during the award process. Losing bidders may pursue legal recourse if the buyer does not follow a fair and consistent evaluation process.

Organizations should heed the saying, "Let the buyer beware." It is critical to evaluate proposals based on more than the appearance of the paperwork submitted. A key factor in evaluating bids, particularly for projects involving information technology, is the past performance record of the bidder. The RFP should require bidders to list other similar projects they have worked on and provide customer references for those projects. Reviewing performance records and references helps to reduce the risk of selecting a supplier with a poor track record. Suppliers should also demonstrate their understanding of the buyer's need, their technical and financial capabilities, their management approach to the project, and their price for delivering the desired goods and services. It is also crucial to write the contract to protect the buyer's interests.

Some information technology projects also require potential sellers to deliver a technical presentation as part of their proposal. The proposed project manager should lead the potential seller's presentation team. When the outside project manager leads the proposal presentation, the organization can begin building a relationship with the potential seller from the beginning. Visits to contractor sites can also help the buyer get a better feeling for the seller's capabilities and management style.

REQUESTING SELLER RESPONSES

After planning for contracting, the next procurement management process involves deciding whom to ask to do the work, sending appropriate documentation to potential sellers, and obtaining proposals or bids. Prospective sellers do most of the work in this process, normally at no cost to the buyer or project. The buying organization is responsible for advertising the work, and for large procurements, they often hold some sort of bidders' conference to answer questions about the job. The main outputs of this process are a procurement document package, a qualified sellers list, and the proposals or bids received by potential sellers.

Organizations can advertise to procure outside goods and services in many different ways. Sometimes a specific supplier might be the number-one choice for the buyer. In this case, the buyer gives procurement information to just that company. If the preferred supplier responds favorably, both organizations proceed to work together. Many organizations have formed good working relationships with certain suppliers, so they want to continue working with them.

In many cases, however, there may be more than one supplier qualified to provide the goods and services. Providing information and receiving bids from multiple sources often takes advantage of the competitive business environment.

Offshore outsourcing, as described earlier, has increased tremendously as organizations find suitable sellers around the globe. As a result of pursuing a competitive bidding strategy, the buyer can receive better goods and services than expected at a lower price.

A bidders' conference, also called a supplier conference or pre-bid conference, is a meeting with prospective sellers prior to preparation of their proposals or bids. These conferences help ensure that everyone has a clear, common understanding of the buyer's desired products or services. In some cases, the bidders' conference might be held online via a Webcast or using other communications technology. Buyers will also post procurement information on a Web site and post answers to frequently asked questions. Before, during, or after the bidders' conference, the buyer may incorporate responses to questions into the procurement documents as amendments.

SELECTING SELLERS

Once buyers receive proposals or bids, they can select a supplier or decide to cancel the procurement. Selecting suppliers or sellers, often called source selection, involves evaluating proposals or bids from sellers, choosing the best one, negotiating the contract, and awarding the contract. It is often a long, tedious process, especially for large procurements. Several stakeholders in the procurement process should be involved in selecting the best supplier for the project. Often, teams of people are responsible for evaluating various sections of the proposals. There might be a technical team, a management team, and a cost team to focus on each of those major areas. Often, buyers develop a short list of the top three to five suppliers to reduce the work involved in selecting a source. The main outputs of this process are the selected sellers, a contract, a contract management plan, resource availability information, requested changes to the project based on the seller selected, and updates to the procurement management plan.

Experts in source selection highly recommend that buyers use formal proposal evaluation sheets during source selection. Figure 12-4 provides a sample proposal evaluation sheet that the project team might use to help create a short list of the best three to five proposals. Notice that this example is a form of a weighted scoring model as described in Chapter 4, Project Integration Management. The score for a criterion would be calculated by multiplying the weight of that criterion by the rating for that proposal. Adding up the scores would provide the total weighted score for each proposal. The proposals with the highest weighted scores should be included in the short list of possible sellers. Experts also recommend that technical criteria should not be given more weight than management or cost criteria. Many organizations have suffered the consequences of paying too much attention to the technical aspects of proposals. For example, the project might cost much more than expected or take longer to complete because the source selection team focused only on technical aspects of proposals. Paying too much attention to technical aspects of proposals is especially likely to occur on information technology projects. However, it is often the supplier's management team—not the technical team—that makes procurement successful.

Criteria	Weight	Proposal 1		Proposal 2		Proposal 3, etc.	
		Rating	Score	Rating	Score	Rating	Score
Technical approach	30%						
Management approach	30%						
Past performance	20%						
Price	20%						
Total score	100%						

Figure 12-4. Sample Proposal Evaluation Sheet

After developing a short list of possible sellers, organizations often follow a more detailed proposal evaluation process. For example, they might list more detailed criteria for important categories, such as the management approach. They might assign points for the potential project manager's educational background and PMP certification, his or her presentation (meaning the sellers had to give a formal presentation as part of the evaluation process), top management support for the project, and the organization's project management methodologies. If the criteria and evaluation are done well, the seller with the most points based on all of the criteria should be offered the contract.

It is customary to have contract negotiations during the source selection process. Sellers on the short list are often asked to prepare a best and final offer (BAFO). People who negotiate contracts for a living often conduct these negotiations for contracts involving large amounts of money. In addition, top managers from both the buying and selling organizations often meet before making final decisions. The final output from the process to select sellers is a contract that obligates the seller to provide the specified products or services and obligates the buyer to pay for them. It is also appropriate on some projects to also prepare a contract management plan to describe details about how the contract will be managed.

Media Snapshot

Many organizations realize that selecting appropriate sellers can often provide a win-win situation. Several companies, including those owned by famous celebrities, work closely with outside sources to help both parties come out ahead. For example, Oprah Winfrey celebrated the premiere of her show's nineteenth season by giving each of her 276 audience members a new car.

"We're calling this our wildest dream season, because this year on the Oprah show, no dream is too wild, no surprise too impossible to pull off," said Winfrey. Everyone in the audience was there because their friends or family had written about their need for a new car. Winfrey's organization did an excellent job of planning the event and keeping it a secret. They also worked closely with other organizations, including Pontiac, who donated all of the cars, which retail for $28,000 each. The audience members were obviously overjoyed. Many marketing experts say Pontiac's generosity was also outstanding for their business.

In other segments on the same show, Winfrey surprised a 20-year-old woman who had spent years in foster care and homeless shelters with a four-year college scholarship, sponsored by the telecommunications company SBC. The young woman also received a makeover and $10,000 in clothes provided by several other companies. Winfrey also showed how her organization worked with many other companies to provide a needy family with $130,000 and several new products to buy and repair their home. The Web site for the show includes a detailed list of sponsoring organizations and links to their Web sites. By collaborating with outside companies, everyone can come out ahead.[11]

ADMINISTERING THE CONTRACT

Administering the contract, or contract administration, ensures that the seller's performance meets contractual requirements. The contractual relationship is a legal relationship and as such is subject to state and federal contract laws. It is very important that appropriate legal and contracting professionals be involved in writing and administering contracts.

Good contracts and close contract administration can help organizations avoid problems. For example, Department 56, Inc., a giftware company in Minnesota, filed a $6 billion lawsuit against Arthur Andersen Worldwide in March 2001. Department 56, Inc. claimed that the well-known consulting firm badly managed the overhaul of their computer systems. Arthur Andersen assured Department 56 officials that the new system would not cost more than $3.3 million, but costs ultimately exceeded $14 million. Arthur Andersen claimed the lawsuit was retaliation for a lawsuit it had filed five months earlier against Department 56 for nonpayment. Arthur Andersen settled the lawsuit on March 1, 2002, when it paid Department 56 $11 million. Both companies learned important lessons from this procurement.[12]

---

[11] "Oprah Gives Audience 276 Cars," *CBSNEWS.com* (September 13, 2004).

[12] Girard, Kim, "Why Andersen Cracked an IT Suit," *Baseline Magazine* (March 6, 2002).

Many project managers know very little about contract administration. Many technical professionals would prefer not to look at contracts at all. However, many project managers and technical people do not understand or trust their own purchasing departments, if their organization even has one, because of problems they may have encountered in the past. Ideally, the project manager, a project team member, or an active user involved in the project should be actively involved in writing and administering the contract so that everyone understands the importance of good procurement management. The project team should also seek expert advice in working with contractual issues.

Project team members must be aware of potential legal problems they might cause by not understanding a contract. For example, most projects involve changes, and these changes must be handled properly for items under contract. Without understanding the provisions of the contract, a project manager may not realize he or she is authorizing the contractor to do additional work at additional costs. Therefore, change control is an important part of the contract administration process.

It is critical that project managers and team members watch for constructive change orders. **Constructive change orders** are oral or written acts or omissions by someone with actual or apparent authority that can be construed to have the same effect as a written change order. For example, if a member of the buyer's project team has met with the contractor on a weekly basis for three months to provide guidelines for performing work, he or she can be viewed as an apparent authority. If he or she tells the contractor to redo part of a report that has already been delivered and accepted by the project manager, that action can be viewed as a constructive change order and the contractor can legally bill the buyer for the additional work. Likewise, if this apparent authority tells the contractor to skip parts of a critical review meeting in the interests of time, the omission of that information is not the contractor's fault.

The following suggestions help ensure adequate change control and good contract administration:

- Changes to any part of the project need to be reviewed, approved, and documented by the same people in the same way that the original part of the plan was approved.
- Evaluation of any change should include an impact analysis. How will the change affect the scope, time, cost, and quality of the goods or services being provided? There must also be a baseline to understand and analyze changes.
- Changes must be documented in writing. Project team members should document all important meetings and telephone calls.
- When procuring complex information systems, project managers and their teams must stay closely involved to make sure the new system will meet business needs and work in an operational environment. Don't assume everything will go fine because you hired a reputable supplier. The buying organization needs to provide expertise as well.

- Have backup plans in case the new system doesn't work as planned when it's put into operation.
- Several tools and techniques can help in contract administration, such as a formal contract change control system, buyer-conducted performance reviews, inspections and audits, performance reporting, payment systems, claims administration, records management, and information technology to support contract administration.

CLOSING THE CONTRACT

The final process in project procurement management is closing the contract, or contract closure. Contract closure involves completion and settlement of contracts and resolution of any open items. The project team should determine if all work required in each contract was completed correctly and satisfactorily. They should also update records to reflect final results and archive information for future use.

Two tools to assist in contract closure include procurement audits and a records management system. Procurement audits are often done during contract closure to identify lessons learned in the entire procurement process. Organizations should strive to improve all of their business processes, including procurement management. A records management system provides the ability to easily organize, find, and archive procurement-related documents. It is often an automated system, or at least partially automated, since there can be a large amount of information related to project procurement.

Outputs from contract closure include closed contracts and updates to organizational process assets. The buying organization often provides the seller with formal written notice that the contract has been completed. The contract itself should include requirements for formal acceptance and closure.

USING SOFTWARE TO ASSIST IN PROJECT PROCUREMENT MANAGEMENT

Over the years, organizations have used various types of productivity software to assist in project procurement management. For example, most organizations use word-processing software to write proposals or contracts, spreadsheet software to create proposal evaluation worksheets, databases to track suppliers, and presentation software to present procurement-related information.

It was not until the late 1990s and early 2000s, however, that many companies started using more advanced software to assist in procurement management. In fact, the term "e-procurement" often describes various procurement functions that are now done electronically. "The purchasing function was neglected for many, many years and has only recently begun to be viewed as strategic," said Hank Adamany, of PricewaterhouseCoopers. "Purchasing was a place where you had people who spent 20 to 25 years in positions below the senior team, and as a result the department never got much attention or investment. That's odd when you consider that those departments can account for up to 70 percent of a company's total spending."[13]

Many different Web sites and software tools are now available to assist in various procurement functions. For example, most college students have discovered Web sites to help them purchase books, airline tickets, or even automobiles. Likewise, many organizations can now access free Web-based portals for purchasing numerous items, or they can buy specialized software to help streamline their procurement activities. Companies such as Commerce One, Ariba, and others started providing corporate procurement services over the Internet in the past few years. Other established companies, such as Oracle, SAS, and Baan, have developed new software products to assist in procurement management. Traditional procurement methods were very inefficient and costly, and new e-procurement services have proved to be very effective in reducing the costs and burdens of indirect procurement. As Hank Adamany of PricewaterhouseCoopers said, "In fact, most user organizations that we surveyed were able to realize more than a 300 percent return on investment in Internet procurement automation within the first year of deployment."[14]

Organizations can also take advantage of information available on the Web, in industry publications, or in various discussion groups offering advice on selecting suppliers. For example, many organizations invest millions of dollars in enterprise project management software. Before deciding which seller's software to use, organizations use the Internet to find information describing specific products provided by various suppliers, prices, case studies, and current customer information to assist in making procurement decisions. Buyers can also use the Internet to hold bidders' conferences, as described earlier in this chapter, or to communicate procurement-related information.

As with any information or software tool, organizations must focus on using the information and tools to meet project and organizational needs. Many nontechnical issues are often involved in getting the most value out of new technologies, especially new e-procurement software. For example, organizations must often develop partnerships and strategic alliances with other organizations to take advantage of potential cost savings. Organizations should practice good procurement management in selecting new software tools and managing relationships with the chosen suppliers.

The processes involved in project procurement management follow a clear, logical sequence. However, many project managers are not familiar with the many issues involved in purchasing goods and services from other organizations. If projects will benefit by procuring goods or services, then project managers and

[13] "What Tools These Portals Be," *Fortune* (*www.fortune.com*) (January 16, 2001).

[14] Ibid.

their teams must follow good project procurement management. As outsourcing for information technology projects increases, it is important for all project managers to have a fundamental understanding of this knowledge area.

CASE WRAP-UP

After reading the contract for their consultants carefully, Marie McBride found a clause giving her company the right to terminate the contract with a one-week notice. She met with her project team to get their suggestions. They still needed help completing the operating system conversion project. One team member had a friend who worked for a competing consulting firm. The competing consulting firm had experienced people available, and their fees were lower than the fees in the current contract. Marie asked this team member to help her research other consulting firms in the area who could work on the operating system conversion project. Marie then requested bids from these companies. She personally interviewed people from the top three suppliers' management teams and checked their references for similar projects.

Marie worked with the purchasing department to terminate the original contract and issue a new one with a new consulting firm that had a much better reputation and lower hourly rates. This time, she made certain the contract included a statement of work, specific deliverables, and requirements stating the minimum experience level of consultants provided. The contract also included incentive fees for completing the conversion work within a certain time period. Marie learned the importance of good project procurement management.

CHAPTER SUMMARY

Procurement, purchasing, or outsourcing is acquiring goods and/or services from an outside source. Information technology outsourcing continues to grow, both within an organization's own country and offshore. Organizations outsource to reduce costs, focus on their core business, access skills and technologies, provide flexibility, and increase accountability. It is becoming increasingly important for information technology professionals to understand project procurement management.

Project procurement management processes include planning purchases and acquisitions, planning contracting, requesting seller responses, selecting sellers, administering contracts, and closing contracts.

Planning purchases and acquisitions involves deciding what to procure or outsource, what type of contract to use, and how to describe the effort in a statement of work. Make-or-buy analysis helps an organization determine whether it can cost-effectively procure a product or service. Project managers should consult internal and external experts to assist them with procurement planning because many legal, organizational, and financial issues are often involved.

The basic types of contracts are fixed price, cost reimbursable, and time and material. Fixed-price contracts involve a fixed total price for a well-defined product and entail the least risk to buyers. Cost-reimbursable contracts involve payments to suppliers for direct and indirect actual costs and require buyers to absorb some of the risk. Time and material contracts are a hybrid of fixed-price and cost-reimbursable contracts and are commonly used by consultants. Unit pricing involves paying suppliers a predetermined amount per unit of service and imposes different levels of risk on buyers, depending on how the contract is written. It is important to decide which contract type is most appropriate for a particular procurement. All contracts should include specific clauses that address unique aspects of a project and that describe termination requirements.

A contract statement of work (SOW) describes the work required for the procurement in enough detail to allow prospective suppliers to determine if they are capable of providing the goods and services and to determine an appropriate price.

Planning contracting involves writing procurement documents, such as a Request for Proposal (RFP) or a Request for Quote (RFQ), and developing source selection evaluation criteria.

Requesting seller responses involves finalizing procurement documents, advertising the work, holding a bidders' conference, and receiving proposals or bids.

Selecting sellers is the process used to evaluate prospective suppliers and negotiate a contract. Organizations should use a formal proposal evaluation form when evaluating suppliers. Technical criteria should not be given more weight than management or cost criteria during evaluation.

Administering the contract involves finalizing and awarding a contract, monitoring performance, and awarding contract modifications. The project manager and key team members should be involved in writing and administering the contract. Project managers must be aware of potential legal problems they might cause when they do not understand a contract. Project managers and teams should use change control procedures when working with outside contracts and should be especially careful about constructive change orders.

Closing the contract includes completing and settling each contract and resolving open items. Procurement audits identify lessons learned during the procurement process.

Several types of software can assist in project procurement management. A new category of e-procurement software has been developed in the past few years to help organizations save money in procuring various goods and services. Organizations can also use the Web, industry publications, and discussion groups to research and compare various suppliers.

DISCUSSION QUESTIONS

1. List five reasons why organizations outsource. Why is there a growing trend in outsourcing, especially offshore?

2. Explain the make-or-buy decision process and describe how to perform the financial calculations involved in the simple lease-or-buy example provided in this chapter. What are the main types of contracts if you decide to outsource? What are the advantages and disadvantages of each?

3. Do you think many information technology professionals have experience writing RFPs and evaluating proposals for information technology projects? What skills would be useful for these tasks?

4. What is involved in the process of requesting seller responses? How do organizations decide whom to send RFPs or RFQs?

5. How can organizations use a weighted decision matrix to evaluate proposals as part of seller selection?

6. List two suggestions for ensuring adequate change control on projects that involve outside contracts.

7. What is the main purpose of a procurement audit?

8. How can software assist in procuring goods and services? What is e-procurement software?

EXERCISES

1. Research information on information technology outsourcing. Find at least two articles and summarize them. Answer the following questions:

   - What are the main types of goods and services being outsourced?
   - What are some of the largest companies that provide information technology outsourcing services?
   - Why has outsourcing grown so rapidly?
   - Have most organizations benefited from outsourcing? Why or why not?

2. Interview someone who was involved in an information technology procurement process and have him or her explain the process that was followed. Write a paper describing the procurement and any lessons learned by the organization.

3. Suppose your company is trying to decide whether it should buy special equipment to prepare some of its high-quality publications itself or lease the equipment from another company. Suppose leasing the equipment

costs $240 per day. If you decide to purchase the equipment, the initial investment is $6,800, and operations will cost $70 per day. After how many days will the lease cost be the same as the purchase cost for the equipment? Assume your company would only use this equipment for 30 days. Should your company buy the equipment or lease it?

4. Find an example of a contract for information technology services. Analyze the key features of the contract. What type of contract was used and why? Review the language and clauses in the contract. What are some of the key clauses? List questions you have about the contract and try to get answers from someone familiar with the contract.

5. Review the SUNY Library Automation Migration RFP mentioned in the fifth Suggested Reading (available on the companion Web site at *www.course.com/mis/schwalbe4e* under Chapter 12). Write a two-page paper summarizing the purpose of the RFP and how well you think it describes the work required.

6. Draft the source selection criteria that might be used for evaluating proposals for providing laptops for all students, faculty, and staff at your college or university. Use Figure 12-4 as a guide. Include at least five criteria, and make the total weights add up to 100. Write a one-to-two-page paper explaining and justifying the criteria and their weights.

RUNNING CASE

Senior management at Manage Your Health, Inc. decided that it would be best to outsource the work involved in training employees on the soon to be rolled-out Recreation and Wellness system and providing incentives for employees to use the system and improve their health. MYH feels that the right outside company could get people excited about the system and provide a good incentive program. As part of the seller selection process, MYH will require interviews and samples of similar work to be physically presented to a review team. Recall that MYH has more than 20,000 full-time employees and more than 5,000 part-time employees Assume the work would involve holding several instructor-led training sessions, developing a training video that could be viewed from the company's Intranet site, developing a training manual for the courses and for anyone to download from the Intranet site, developing an incentive program for using the system and improving health, creating surveys to assess the training and incentive programs, and developing monthly presentations and reports on the work completed. The initial contract would last one year, with annual renewal options.

1. Suppose that your team has discussed management's request. You agree that it makes sense to have another organization manage the incentive program for this new application, but you don't think it makes sense to outsource the training. Your company has a lot of experience doing internal training. You also know that your staff will have to support the system, so you want to develop the training to minimize future support calls.

Write a one-page memo to senior management stating why you think the training should be done in-house.

2. Assume the source selection criteria for evaluating proposals is as follows:

- Management approach 15%
- Technical approach 15%
- Past performance 20%
- Price 20%
- Interview results and samples 30%

Using Figure 12-4 as a guide and the weighted scoring model template, if desired, create a spreadsheet that could be used to enter ratings and calculate scores for each criterion and total weighted scores for three proposals. Enter scores for Proposal 1 as 80, 90, 70, 90, and 80, respectively. Enter scores for Proposal 2 as 90, 50, 95, 80, and 95. Enter scores for Proposal 3 as 60, 90, 90, 80, and 65. Add a paragraph summarizing the results and your recommendation on the spreadsheet. Print your results on one page.

3. Draft potential clauses you could include in the contract to provide incentives to the seller based on MYH achieving its main goal of improving employee health and lowering health care premiums as a result of this project. Be creative in your response, and document your ideas in a one-page paper.

ADDITIONAL RUNNING CASES AND OTHER APPENDICES

Appendix C provides additional case studies and questions you can use to practice applying the concepts, tools, and techniques you are learning throughout this and previous chapters. Review the running cases provided in Appendix C and on the companion Web site (*www.course.com/mis/schwalbe4e*). Appendix D includes templates for various project management documents. For additional sample documents based on real projects, visit the author's Web site at *www.kathyschwalbe.com*. Appendix E and the CD-ROM included with this text include a computer simulation where you can also practice applying the project management process groups and knowledge areas.

SUGGESTED READINGS

1. *CIO Magazine.* "Outsourcing Around the World," (*www.cio.com*) (July 15, 2004).

    This special issue of CIO Magazine *provides detailed information on the information technology outsourcing market. Information is provided on 26 countries that are the strongest contenders for outsourcing and ratings for each country. With nearly half of all CIOs using offshore providers today, it's crucial to understand the options and issues of offshore outsourcing.*

2. Fleming, Quentin. *Project Procurement Management: Contracting, Subcontracting, Teaming.* FMC Press (2003).

   In this 276-page book, the author uses PMBOK® Guide *terminology while providing more detailed information on project procurement management. Fleming focuses on the critical relationship between the project manager and buyers or other procurement specialists.*

3. *Information Week.* "Outsourcing," (*www.informationweek.com*) (2004).

   Information Week's *Web site has a whole section dedicated to the topic of outsourcing. You can access several articles and white papers from this site.*

4. Meyers, Mike. "Offshore Jobs Bring Gains at Home," *Star Tribune* (*www.startribune.com/outsource*) (September 5, 2004).

   The Minneapolis Star Tribune *ran a series of articles addressing various outsourcing issues. The entire series is available online, with new articles being adding every week.*

5. State University of New York Office of Library and Information Services. "SUNY Library Automation Migration RFP" (January 19, 1999).

   This document provides an excellent example of an RFP for a large information technology project. It includes specifications and related information for the delivery, installation, implementation, and maintenance of an integrated library management system for the libraries of the SUNY institutions. The proposal described requirements for the software, planned to be installed over a four- to five-year period as funds became available. You can access this detailed document from the online companion site for this text at www.course.com/mis/schwalbe/4e.

6. Sullivan, Laurie. "Wal-Mart's Way," *InformationWeek* (September 27, 2004).

   This article describes how Wal-Mart successfully manages 95 percent of its many information technology projects. Its "way" is to rely on very little commercial software and not outsource at all. Wal-Mart spends below the average on information technology for retailers at less than 1 percent of worldwide revenue. Linda Dillman, VP and CIO of Wal-Mart Stores Inc., says that her group can implement things faster than a third party could, so they have no reason to outsource.

KEY TERMS

- **bid** — also called a tender or quote (short for quotation), a document prepared by sellers providing pricing for standard items that have been clearly defined by the buyer
- **constructive change orders** — oral or written acts or omissions by someone with actual or apparent authority that can be construed to have the same effect as a written change order

- **contract** — a mutually binding agreement that obligates the seller to provide the specified products or services, and obligates the buyer to pay for them
- **cost plus fixed fee (CPFF) contract** — a contract in which the buyer pays the supplier for allowable performance costs plus a fixed fee payment usually based on a percentage of estimated costs
- **cost plus incentive fee (CPIF) contract** — a contract in which the buyer pays the supplier for allowable performance costs along with a predetermined fee and an incentive bonus
- **cost plus percentage of costs (CPPC) contract** — a contract in which the buyer pays the supplier for allowable performance costs along with a predetermined percentage based on total costs
- **cost-reimbursable contracts** — contracts involving payment to the supplier for direct and indirect actual costs
- **fixed-price** or **lump-sum contracts** — contracts with a fixed total price for a well-defined product or service
- **make-or-buy decision** — when an organization decides if it is in its best interests to make certain products or perform certain services inside the organization, or if it is better to buy them from an outside organization
- **procurement** — acquiring goods and/or services from an outside source
- **project procurement management** — the processes required to acquire goods and services for a project from outside the performing organization
- **proposal** — a document prepared by sellers when there are different approaches for meeting buyer needs
- **Request for Proposal (RFP)** — a document used to solicit proposals from prospective suppliers
- **Request for Quote (RFQ)** — a document used to solicit quotes or bids from prospective suppliers
- **sellers** — contractors, suppliers, or providers who provide goods and services to other organizations
- **statement of work (SOW)** — a description of the work required for the procurement
- **termination clause** — a contract clause that allows the buyer or supplier to end the contract
- **time and material contracts** — a hybrid of both fixed-price and cost-reimbursable contracts
- **unit pricing** — an approach in which the buyer pays the supplier a predetermined amount per unit of service, and the total value of the contract is a function of the quantities needed to complete the work

Appendix A

Guide to Using Microsoft Project 2003

INTRODUCTION

This appendix provides a concise guide to using Microsoft Office Project Standard 2003 (often referred to as Project Standard 2003 or, simply, Project 2003) to assist in performing project management. Project 2003 can help users manage different aspects of all nine project management knowledge areas. Most users, however, focus on using Project 2003 to assist with scope, time, cost, human resource, and communications management. This guide uses these project management knowledge areas as the context for learning how to use Project 2003. Refer to chapters on the other knowledge areas for suggestions on using software to assist in those areas. Also, note the basic order of the steps in this appendix. It is good practice to first determine the scope of a project, then the time, and then the resource and cost information. You can then set a baseline and enter actuals to track and communicate performance.

Hundreds of project management software products are available today. Microsoft Project is the clear market leader among midrange applications. A 2003 survey by TechRepublic showed that 94 percent of their 397 respondents used Microsoft Project compared to 86 percent of respondents in their 2000 study.[1]

Before you can use Project 2003 or any project management software effectively, you must understand the fundamental concepts of project management, such as creating work breakdown structures, linking tasks, entering duration estimates, and so on. See the Suggested Readings for recommendations on other resources to help you gain an even deeper understanding of Project 2003.

[1] Blakely, Beth. "Survey shows little change in consultants' use, satisfaction with PM software," TechRepublic, 2/3/03 (http://techrepublic.com.com/5100-6330_11-5031832.html).

NEW FEATURES OF PROJECT 2003

Because there have been several previous versions of Microsoft Project, it is useful to understand some of the new capabilities of Project 2003, especially if you are working with people who are upgrading from a previous version.

With the introduction of Project 2002, Microsoft provided two versions of Project—a standard version and an enterprise version. The same is true for Project 2003.

1. **Microsoft Office Project Standard 2003**, or simply Project 2003, is most similar to previous versions of Microsoft Project and is intended for individual desktop usage. Project 2002 Standard included a Client Access License (CAL) for connection to Project Server 2002 via the Web, but Project Standard 2003 does not. You must be using Project Professional 2003 to connect to Project Server 2003, as described below. Also note that Project 2002 can open, work with, and save Project 2003 files, so you don't have to save files in Project 2002 format from Project 2003. This appendix focuses on using the standard version of Project 2003.

2. **The Microsoft Enterprise Project Management (EPM) Solution** is a complete database and analysis software package that helps organizations plan and manage projects on an enterprise level. It includes several components, such as Microsoft Office Project Professional 2003 (which includes all of the capabilities of Project Standard 2003 plus more features), Project Web Access, Project Server 2003, Microsoft Windows SharePoint Services, and Microsoft SQL Server. Project Web Access is the Web portal to Project Server 2003 that provides an easy-to-use interface with a customized home page and functions that are based on each user's role on the project. Project Web Access users do not have to install Project Professional 2003; they just need a Client Access License (CAL) and a Web browser. Organizations should develop and apply many standards, templates, codes, and procedures before using the enterprise version of Project 2003 to make the best use of its capabilities. In fact, Microsoft provides a template file for implementing their EPM solution, and the schedule goal in the template file is about six months.

Figures A-1, A-2, and A-3 provide sample screens from the Microsoft EPM Solution taken from an online tour available from Microsoft's Web site. Figure A-1 shows the customized home page that an executive would see when using his or her Client Access License. Figure A-2 shows an executive summary view of all of the projects at this company. Figure A-3 shows a sample screen available from the portfolio analyzer. The reporting and analysis capabilities in the EPM solution help organizations use information from many projects and team members to optimize resources, prioritize work, and align project portfolios with overall business objectives. For more information on the Microsoft Office EPM Solution, contact Microsoft or visit their Web site at *www.microsoft.com/project*.

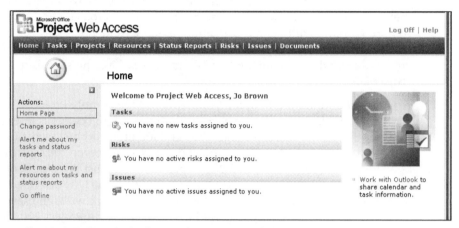

Figure A-1. Sample Project Web Access Home Page

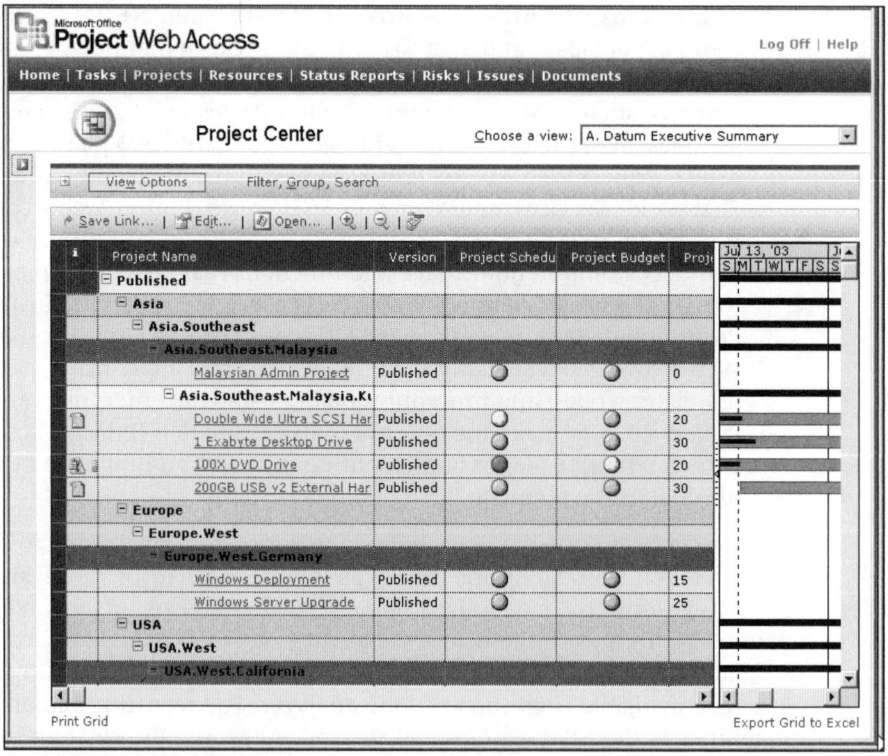

Figure A-2. Sample Project Web Access Project Center Screen

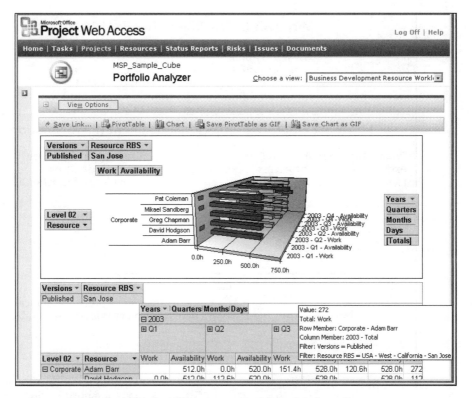

Figure A-3. Sample Project Web Access Portfolio Analyzer Screen

Figure A-4 provides an overview of using either Microsoft Office Project Standard 2003 (Project 2003) or the Microsoft Office EPM Solution. Remember that Project 2003 is a stand-alone product that enables users to plan and manage projects with familiar, easy-to-use tools. Project Professional 2003, Project Web Access, Project Server 2003, and Microsoft SQL Server work together to make up the Microsoft Office EPM Solution, which should have many users.

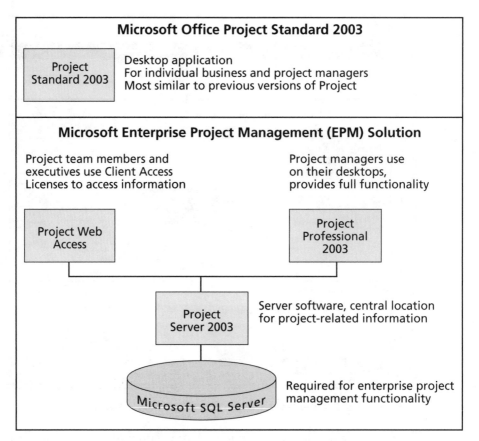

Figure A-4. Microsoft Office Project Standard 2003 Versus Microsoft Enterprise Project Management (EPM) Solution

There are also several new and improved features available in Project 2003:

- New ability to manage proposed and committed resources
- New shared workspace for storing non-Project files (including check-in and versioning)
- Backward and forward compatibility so users of previous and future versions of Project can easily share files
- New features for sharing data between Project and Excel, PowerPoint, and Visio, including the new Copy Picture to Office Wizard
- New online Help for up-to-date information and assistance via the Web
- New Euro currency converter toolbar

- Improved ability to create one-page printouts of project plans easily
- New calendar and tracking setup wizards
- Improved Project Guide, which is an interactive Help tool
- No more requirement that a developer hard-code paths for files used in customizing the Project Guide
- Enhanced developer tools for integration with Microsoft Office

BEFORE YOU BEGIN

This appendix assumes you are using Project 2003 with Windows 2000 or XP and are familiar with other Windows-based applications. You can check your work using the solution files that are available for download from the companion Web site for this text (*www.course.com/mis/schwalbe4e*) or from your instructor.

The CD-ROM in the back of this text includes a 120-day trial version of Project 2003. You need to be running Windows 2000 or Windows XP to use this software. If you have any technical difficulties, contact Course Technology's Support Services at *www.course.com/support*. You can also download or order evaluation copies of Project 2003 from *www.microsoft.com/project*. Also note that students and faculty can purchase software at deep discounts from several sources, such as *www.journeyed.com*.

This appendix uses a fictitious information technology project—the Project Tracking Database project—to illustrate how to use the software. The goal of this project is to create a database to keep track of all the projects a company is working on. Each section of the appendix includes hands-on activities through which you will work. To complete the hands-on activities in the appendix, you will need to create a Data Disk or copy files to a location on your hard drive or network drive. You will need to copy a set of files from a file server or from the companion Web site for this text (*www.course.com/mis/schwalbe4e*). When you begin each set of steps, make sure you are using the correct file from your Data Disk, hard drive, or network drive. You should have the following files before you begin your work:

- PRJ03DPL.mpp, the Project 2003 Deployment template file that comes with the software
- resource.mpp
- tracking.mpp
- kickoffmeeting.doc

In addition, you will create the following files from scratch as you work through the steps:

- scope.mpp
- time.mpp
- tracking.mpp
- baseline.mpp
- level.mpp
- taskinfo.html

Now that you understand project management concepts and the basic project management terminology, you will learn how to start Project 2003, review the Help facility and a template file, and begin to plan the Project Tracking Database project.

OVERVIEW OF PROJECT 2003

The first step to mastering Project 2003 is to become familiar with the Help facility, online tutorials, major screen elements, views, and filters. This section describes each of these features.

Starting Project 2003 and Using the Getting Started and Project Guide Features

To start Project 2003:

1. *Open Project 2003.* Click the **Start** button on the taskbar, click **Programs**, and then click **Microsoft Office Project 2003**. (If you are using Windows XP, point to **All Programs**, point to **Microsoft Office**, and then click **Microsoft Office Project 2003**.) Alternatively, a shortcut or icon might be available on the desktop; in this case, double-click the icon to start the software.

2. *Maximize Project 2003.* If the Project 2003 window does not fill the entire screen as shown in Figure A-5, click the **Maximize** button ▢ in the upper-right corner of the window.

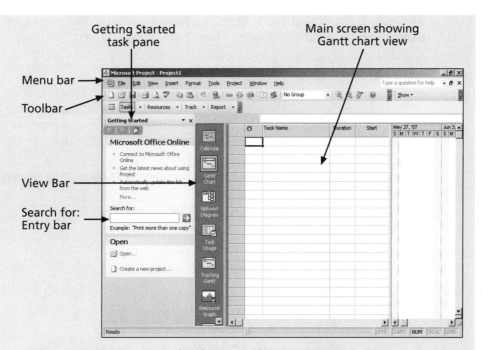

Getting Started task pane

Main screen showing Gantt chart view

Menu bar

Toolbar

View Bar

Search for: Entry bar

Figure A-5. Getting Started Task Pane and Project 2003 Main Screen

If the View Bar is not open, click View on the menu bar, and then click View Bar.

Do not worry if your toolbars or other screen elements look slightly different from Figure A-5. Focus on the Project 2003 main screen and the Getting Started task pane.

Project 2003 is now running and ready to use. Notice some of the elements of the Project 2003 screen. The default view is the Gantt Chart view, which shows tasks and other information in the Entry table as well as a calendar display. You can access other views by clicking each button in the View Bar on the left side of the screen. Also notice that when Project 2003 starts, it opens a new file named Project1, as shown in the title bar. If you open a second file, the name will be Project2, and so on, until you save and rename the file.

When you start Project 2003, the Getting Started task pane automatically opens, as shown in Figure A-5. You can access Microsoft Office Online, search for help on various topics, or open files from this pane. If you were used to seeing the Project Guide when starting Project 2002, you may notice that you now access the new Project Guide from the Project Guide toolbar, as shown in

Figure A-6. The Project Guide helps you learn Project 2003 by providing instructions for completing various steps in building a Project 2003 file.

To use the Project Guide:

1. *Open the Project Guide.* Click the **Tasks** list arrow on the Project Guide toolbar, as shown in Figure A-6. Notice the various options listed. The Resources, Track, and Report buttons on the Project Guide toolbar list additional options. You'll use some of these options later in this appendix.

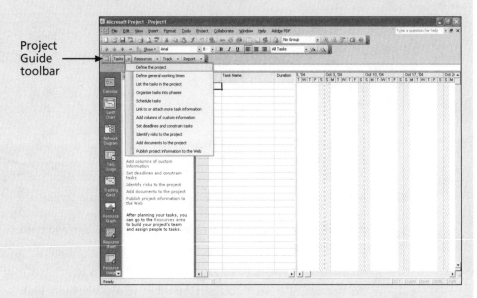

Figure A-6. Using the Project Guide

2. *Set the project start date.* Click **Define the project**.
3. *Review the Project Guide instructions.* Review the information that displays in the pane on the left side of the screen. As you select different options in the Project Guide, the task pane displays specific information related to that option. Notice that the default project start date is set at the current date. Click the **date** list arrow on the pop-up calendar to select a date. Select the last day of the month that displays.
4. *Close the Project Guide.* Click the **Close** button ⊠ in the upper-right corner of the Project Guide window, and then click **Yes** when prompted to exit the wizard.

If the Project Guide toolbar is not visible, click View on the menu bar, select Toolbars, and then click Project Guide.

In addition to using the Project Guide window, you can access other information to help you learn how to use Project 2003 under the Help menu. Figure A-7 displays the Help menu options available, and Figure A-8 provides a description of each option.

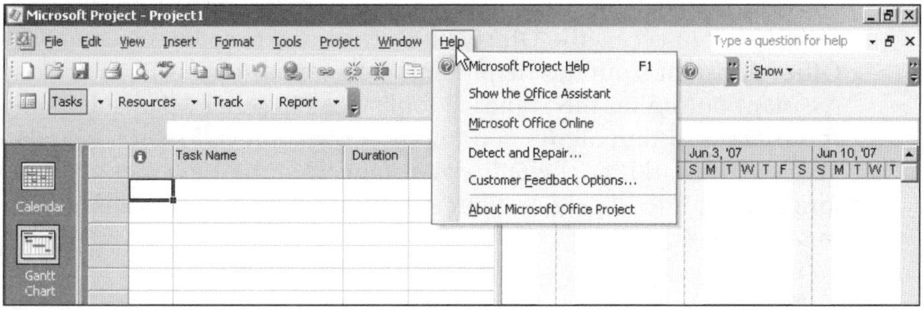

Figure A-7. Project 2003 Help Menu Options

HELP MENU OPTION	DESCRIPTION
Microsoft Project Help	Displays a Search for: textbox for assistance, the Table of Contents for Help, and options for using Microsoft Office Online.
Show the Office Assistant	Acts as a toggle to display and hide the Office Assistant, a context-sensitive, interactive guide to finding information on Project 2003.
Microsoft Office Online	Takes you directly to the Web site for Microsoft Office Online.
Customer Feedback Option	Provides information about Microsoft's Customer Experience Improvement Program. This program allows Microsoft to collect anonymous information about your hardware configuration and how you use their software and services to identify trends and usage patterns.
Detect and Repair	Automatically finds and fixes errors in all Office files.
About Microsoft Office Project	Displays the version of Microsoft Office Project, licensed use, and other information about Project 2003.

Figure A-8. Descriptions of Project 2003 Help Menu Options

Microsoft realizes that Project 2003 can take some time to learn and provides a number of resources on its Web site. Select MS Office Online from the Project 2003 Help menu to go to Microsoft's Web site for Office products. Microsoft's Web site for Project 2003 (*www.microsoft.com/project*) provides files for users to download, case studies, articles, and other useful materials.

Within Project 2003, the Office Assistant helps answer questions as they arise. You can access the Office Assistant at any time by selecting Show the Office Assistant from the Help menu. You can also press F1 or click the Office Assistant button on the Standard toolbar. As in other Microsoft applications, the Office Assistant allows you to type a question and search for related information. You can hide the Office Assistant from the Help menu if you no longer want it to appear on your screen. Feel free to experiment with the various Help tools.

Main Screen Elements

The Project 2003 default main screen is called the Gantt Chart view. At the top of the main screen, the menu bar and Standard toolbar are similar to those in other Windows programs. Note that the order and appearance of buttons on the Standard toolbar may vary, depending on the features you are using and how the Standard toolbar is customized. When you use a button, it automatically becomes visible on the toolbar it's associated with. You can display the Standard toolbar in one row or two, or you can set it to always show full menus so that all buttons are visible all the time. Additional toolbars in Project 2003 include the Formatting toolbar, the Custom Forms toolbar, and the Tracking toolbar, among others.

To customize the Standard toolbar:

1. *Open the Customize dialog box.* Click **Tools** on the menu bar, select **Customize**, and then click **Toolbars** to display the Customize dialog box. Click the **Options tab**.
2. *Show full menus.* Click the option to **Always show full menus**, as shown in Figure A-9, and then click the **Close** button to close the dialog box.

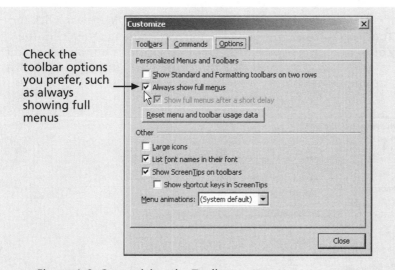

Check the toolbar options you prefer, such as always showing full menus

Figure A-9. Customizing the Toolbar

3. *If needed, add the Link Tasks button to the toolbar.* Click the **Toolbar Options** button on the toolbar 🔧, and then click the **Link Tasks** button 🔗, as shown in Figure A-10.

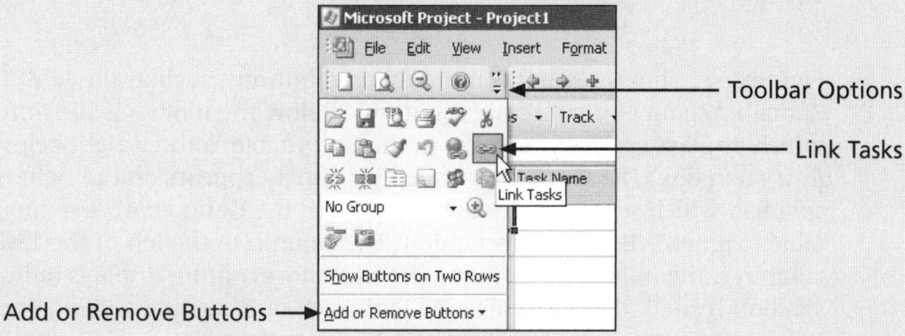

Toolbar Options

Link Tasks

Add or Remove Buttons

Figure A-10. Adding the Link Tasks Button to the Toolbar

4. *If needed, add the Zoom In and Zoom Out buttons to the toolbar.* Click the **Toolbar Options** button on the toolbar again, and then click the **Zoom In** button 🔍. Click the **Toolbar Options** button again, and then click the **Zoom Out** button 🔍, as shown in Figure A-11.

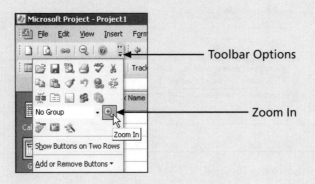

Figure A-11. Adding the Zoom In and Zoom Out Buttons to the Toolbar

💡 You can select the Add or Remove Buttons option to add or remove several toolbar buttons at once.

Tip

Figure A-12 shows several commonly used buttons, such as the Link Tasks, Zoom In, Zoom Out, and Indent buttons. Below the toolbar is the Entry bar, which displays entries you make in the Entry table, located right below the Entry text box. The Gantt chart calendar display appears on the right of the split bar, which separates the Entry table and the Gantt chart. Use the Entry table to enter WBS task information. The column to the left of the Task Name column is the Indicators column. The Indicators column displays indicators or symbols related to items associated with each task, such as task notes or hyperlinks to other files.

Entry bar   Link Tasks   Zoom In   Zoom Out   Indent

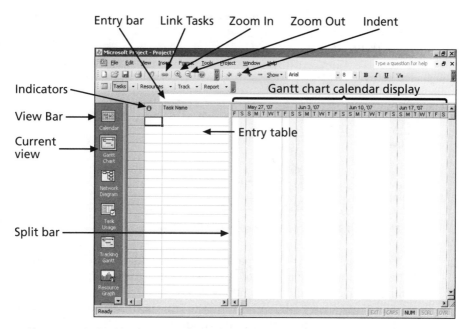

Indicators

View Bar

Current view

Split bar

Gantt chart calendar display

Entry table

Figure A-12. Project 2003 Main Screen Elements

Many features in Project 2003 are similar to features in other Windows programs. For example, to collapse or expand tasks, click the appropriate symbols to the left of the task name. To access shortcut items, right-click in either the Entry table area or the Gantt chart. Many of the Entry table operations in Project 2003 are very similar to operations in Excel. For example, to adjust a column width, click and drag or double-click between the column heading titles.

If you select another view and want to return to the Gantt Chart view, you can click the Gantt Chart button on the View Bar on the left of the screen, or select View from the menu bar, and click Gantt Chart, as shown in Figure A-13. If the Entry table on the left appears to be different, select View from the menu bar, point to Table: Entry, and then click Entry to return to the default Entry table view.

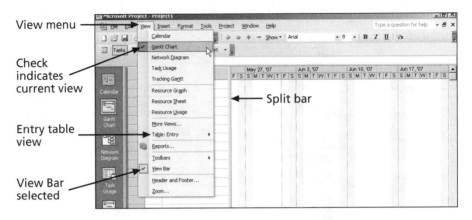

Figure A-13. Project 2003 View Menu Options

Notice the split bar that separates the Entry table from the Gantt chart. When you move the mouse over the split bar, your cursor changes to the resize pointer. Clicking and dragging the split bar to the right reveals other task information in the Entry table, including the Duration, Start date, Finish date, Predecessors, and Resource Names columns.

Next, you'll open a template file to explore more screen elements. Project 2003 comes with several template files, and you can also access templates on Microsoft's Web site. To open template files on your computer, click File on the menu bar, and then click New. In the New Project task pane, click the "On my computer" option under Templates, as shown in Figure A-14. You can also access templates on Office Online or from your own Web sites.

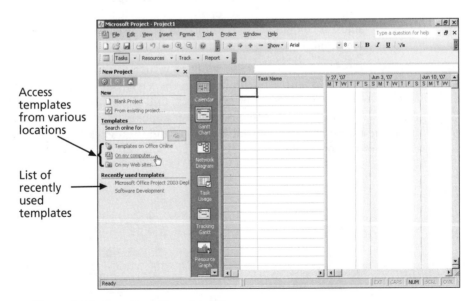

Figure A-14. Opening a Template File

To access template files on your computer, you must do a full or custom install of Project 2003.

Figure A-15 shows results from searching for templates that contain the word "Project" using the "Templates on Office Online" option. Note that the templates found will vary based on the date of access.

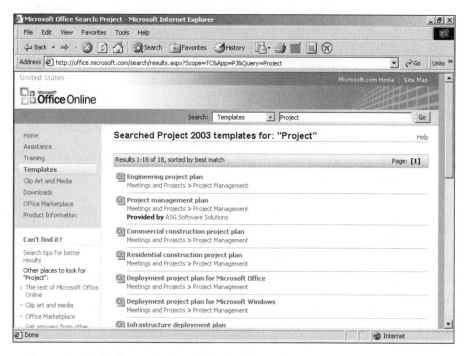

Figure A-15. Finding Templates Using Office Online

To open a template file and adjust Project 2003 screen elements:

1. *Open a template file.* Click the **Open button** 📂 on the Standard toolbar, browse to find the file named **PRJ03DPL.mpp** that you copied from the companion Web site for this text, and then double-click the filename to open the file. This file provides a template for a project to deploy Microsoft Office Project 2003 in a corporate environment. Note that you are simply opening this file instead of accessing it from the template options shown in Figure A-14.

2. *View the Note.* Move your mouse over the **Notes icon** in the Indicators column and read its contents. It's a good idea to provide a short note describing the purpose of project files. Your screen should resemble Figure A-16.

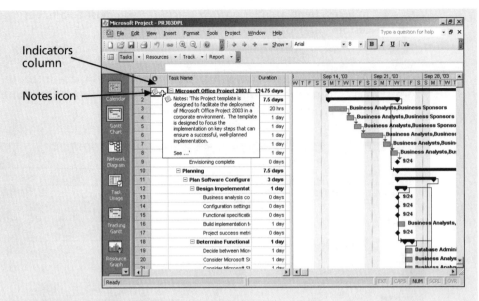

Figure A-16. The PRJ03DPL.mpp Template File

> **?** **Help** After you open template files, those filenames will appear on your screen under the New from Template section. For example, Figure A-14 shows that the template file for Software Deployment was recently accessed on this user's screen. You can also see recently opened files from the File menu.

3. *Adjust the timescale.* Click the **Zoom Out** button 🔍 on the toolbar *three times* to display the timescale in quarters. If you cannot find the Zoom Out button, you can add it to the Standard toolbar by clicking the **Toolbar Options** button, as described earlier.

4. *Select Outline Level 2 to display WBS level 1 tasks.* On the Standard toolbar, click the **Show** button's list arrow Show ▾, and then click **Outline Level 2**, as shown in Figure A-17. Notice that only the level 1 WBS items display in the Entry table after you select Outline Level 2, and the timescale shows quarters. The black bars on the Gantt chart represent the summary tasks. This file included a summary task for the entire project as the Outline Level 1 task. Recall from Chapter 5, Project Scope Management that the entire project is normally referred to as WBS level zero.

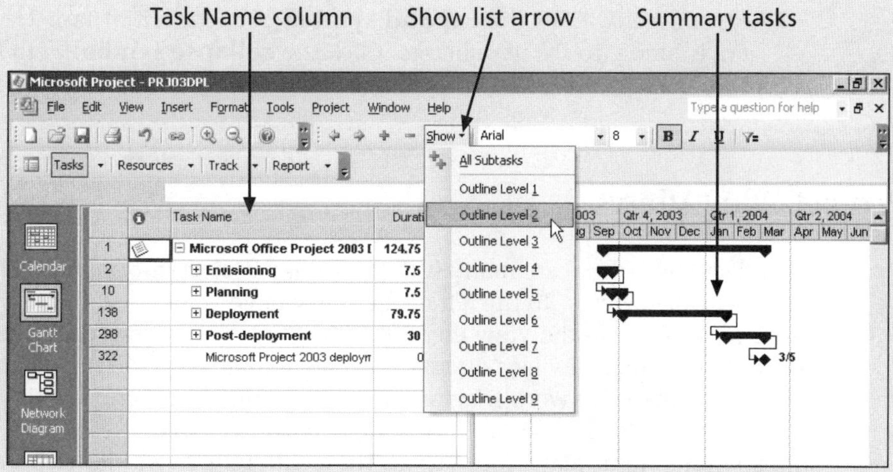

Task Name column   Show list arrow   Summary tasks

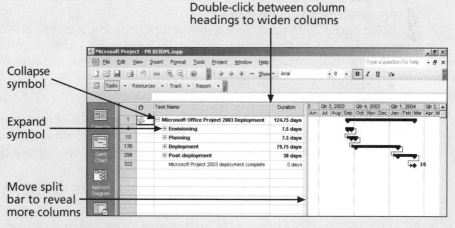

Figure A-17. Viewing Summary Tasks Using the Show Button

5. *Widen the Task Name column.* Move the cursor over the right-column gridline in the Task Name column heading until you see the resize pointer ⬌, and then double-click the **left mouse** button to resize the column width automatically to show all the text.

6. *Move the split bar to reveal more Entry table columns.* Move the split bar to the right to reveal the entire Duration column. Your screen should resemble Figure A-18.

Double-click between column
headings to widen columns

Collapse symbol

Expand symbol

Move split
bar to reveal
more columns

Figure A-18. Adjusting Screen Elements

7. *Expand a task*. Click the **expand symbol** ⊞ to the left of Task 2, Envisioning, to see its subtasks. Click the **collapse symbol** ⊟ to hide its subtasks.

Project 2003 Views

In Project 2003, there are many ways to display project information. These displays are called views. To the left of the Entry table is the View Bar. Instead of using commands on the View menu to change views, you can click buttons on the View Bar. Several views discussed in the main text are on the default View Bar: Gantt Chart, Network Diagram, Tracking Gantt, and Resource Graph. Other views include Calendar, Task Usage, Resource Sheet, Resource Usage, and an option to display More Views. These different views allow you to examine project information in different ways, which helps you analyze and understand what is happening on your project.

To save screen space, you can hide the View Bar by selecting View from the menu bar and deselecting View Bar. When the View Bar is not visible, a gray line (or blue line in Windows XP) with the name of the current view appears to the far left of the main screen. When you right-click this gray line, a shortcut menu appears, which gives you quick access to the other views.

The View menu also provides access to different tables and reports that display information in a variety of ways. Some tables that you can access from the View menu include Schedule, Cost, Tracking, and Earned Value. Some reports you can access from the View menu include Overview, Current Activities, Costs, Assignments, and Workload.

Some Project 2003 views, such as the Gantt Chart view, present a broad look at the entire project, whereas others, such as the Form views, focus on specific pieces of information about each task. Three main categories of views are available:

- Graphical: A chart or graphical representation of data using bars, boxes, lines, and images.
- Task Sheet or Table: A spreadsheet-like representation of data in which each task appears as a new row, and each piece of information about the task is represented by a column. Different tables are applied to a task sheet to display different kinds of information.
- Form: A specific view of information for one task. Forms are used to focus on the details of one task.

Table A-1 describes some of the predesigned views within each category that Project 2003 provides to help you display the project or task information that you need.

Table A-1: Common Project 2003 Views

CATEGORY OF VIEW	VIEW NAME	DESCRIPTION OF VIEW
Graphical	Gantt Chart	Standard format for displaying project schedule information that lists project tasks and their corresponding start and finish dates in a calendar format. Shows each task as a horizontal bar with the length and position corresponding to the timescale at the top of the Gantt chart.
	Network Diagram	Schematic display of the logical relationships or sequencing of project activities. Shows each task as a box with linking lines between tasks to show sequencing. Critical tasks appear in red.
Task Sheet or Table	Entry Table	Default table view showing columns for Task Name and Duration. By revealing more of the Entry table, you can enter start and end dates, predecessors, and resource names.
	Schedule Table	Displays columns for Task Name, Start, Finish, Late Start, Late Finish, Free Slack, and Total Slack.
	Cost Table	Displays columns for Task Name, Fixed Cost, Fixed Cost Accrual, Total Cost, Baseline, Variance, Actual, and Remaining.
	Tracking Table	Displays columns for Task Name, Actual Start, Actual Finish, % Complete, Physical % Complete, Actual Duration, Remaining Duration, Actual Cost, and Actual Work.
	Earned Value	Displays columns for Task Name, BCWS, BCWP, ACWP, SV, CV, EAC, BAC, and VAC. See Chapter 7, Project Cost Management, for descriptions of these earned value terms.
Form	Task Details Form	Displays detailed information about a single task in one window.
	Task Name Form	Displays columns for Task Name, Resources, and Predecessors for a single task.

Next, you will use the same Microsoft Office Project 2003 Deployment template file (PRJ03DPL.mpp) to access and explore some of the views available in Project 2003. As mentioned earlier, you can configure the toolbars and toolbar buttons in different ways, as with other Office 2003 products.

To access and explore different views:

1. *Show all subtasks*. Click the **Show** button on the toolbar, and then click **All Subtasks**.

2. *Explore the Network Diagram view*. Click the **Network Diagram** button on the View Bar, and then click the **Zoom Out** button on the toolbar *three times*. Your screen should resemble Figure A-19.

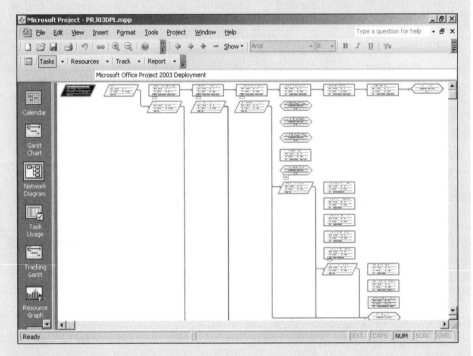

Figure A-19. Network Diagram View of Project 2003 Deployment File

3. *Explore the Calendar view*. Click the **Calendar** button on the View Bar. Your screen should resemble Figure A-20.

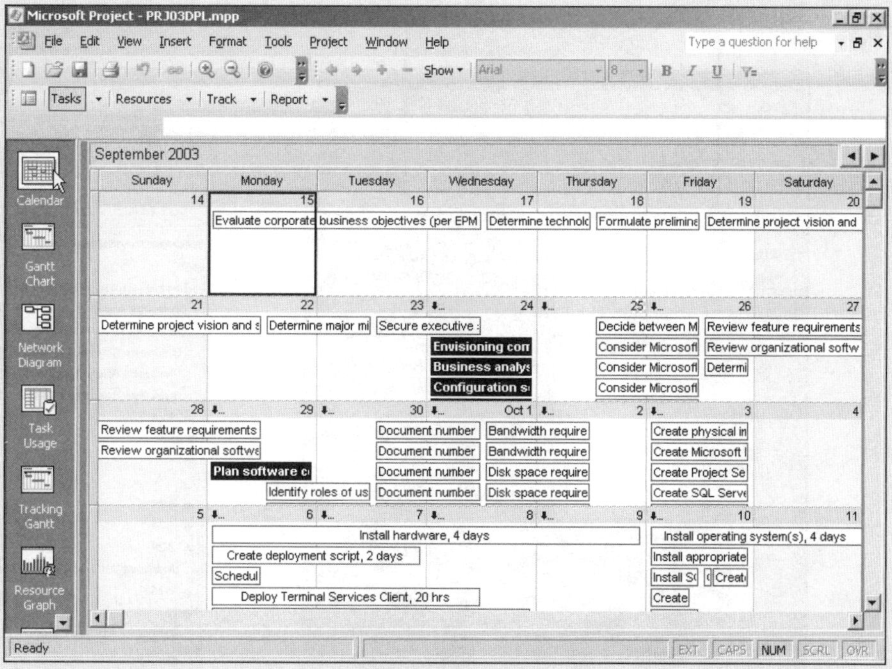

Figure A-20. Calendar View of Project 2003 Deployment File

4. *Examine columns in the Entry table.* Click the **Gantt Chart** button on the View Bar, and then move the split bar by moving your mouse between the Entry table and Gantt chart until you see the resize pointer. Click and drag your mouse to the right almost to the end of the screen to reveal other columns in the Entry table.

5. *Examine the Table: Schedule view.* Click **View** on the menu bar, move your mouse to **Table: Entry**, and then click **Schedule** in the cascading menu to the right. Notice that the columns to the left of the Gantt chart now display more detailed schedule information.

6. *Right-click the Select All button to access different table views.* Move your mouse to the **Select All** button to the left of the Task Name column symbol, and then **right-click** with your mouse. A shortcut menu to different table views displays, as shown in Figure A-21. Experiment with other table views, then return to the Table: Entry view.

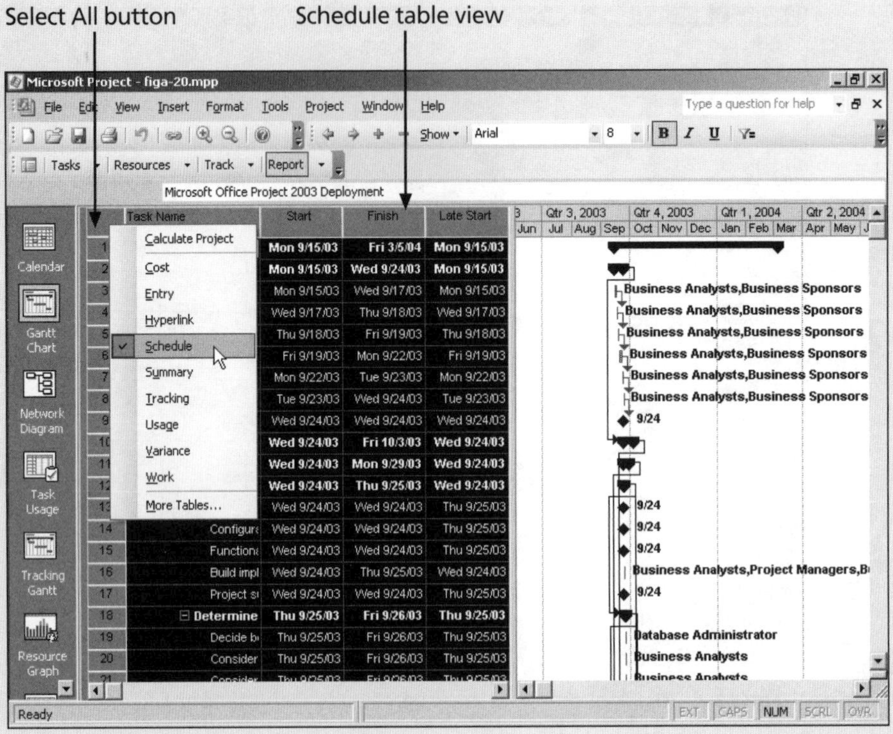

Select All button   Schedule table view

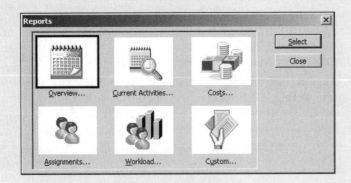

Figure A-21. Changing Table Views

7. *Explore the Reports feature.* Click **View** on the menu bar, and then click **Reports**. The Reports dialog box displays, as shown in Figure A-22.

Figure A-22. Reports Dialog Box

8. *View the Project Summary report.* Double-click **Overview** from the Reports dialog box, and then double-click **Project Summary** in the Overview Reports dialog box. Notice that the insertion point now resembles a magnifying glass. Click inside the report to Zoom In or Zoom Out. Click the **Close** button to close this report, and then experiment with viewing other reports. You will use several reports and other views throughout this appendix.

9. *Close the Reports feature.* Click **Close** to close the Reports dialog box and return to the Gantt Chart view.

Project 2003 Filters

Project 2003 uses an underlying relational database to filter, sort, store, and display information. This database is compatible with Microsoft Access, and it is relatively easy to import information from and export information to other applications. Project 2003 is an Object Linking and Embedding for Databases (OLE-DB) provider, meaning that it has low-level connectivity between multiple types of data sources. For example, you can import or export data to other applications such as Excel, Access, Oracle, and so on.

You can filter information by clicking the Filter text box list arrow, as shown in Figure A-23. You can access more filter options by using the scroll bar.

Filter text box list arrow

Figure A-23. Filter List Options

Filtering project information is very useful. For example, if a project includes hundreds of tasks, you might want to view only summary or milestone tasks to get an overview of the project. To get this type of overview of a project, select the Milestones or Summary Tasks filter from the Filter list. If you are concerned about the schedule, you can select a filter that shows only tasks on the critical path. Other filters include Completed Tasks, Late/Overbudget Tasks, and Date Range, which displays tasks based on dates you provide. As shown earlier, you can also click the Show list arrow on the toolbar to display different levels in the WBS quickly. For example, Outline Level 1 shows the highest-level items in the WBS, Outline Level 2 shows the next level of detail in the WBS, and so on.

To explore Project 2003 filters:

1. *Apply a filter to see only milestone tasks.* From the Table: Entry view in the Microsoft Office Project 2003 Deployment file, click the **Toolbar Options** button ▣ on the toolbar, and then click the **Filter** list arrow, as shown in Figure A-24.

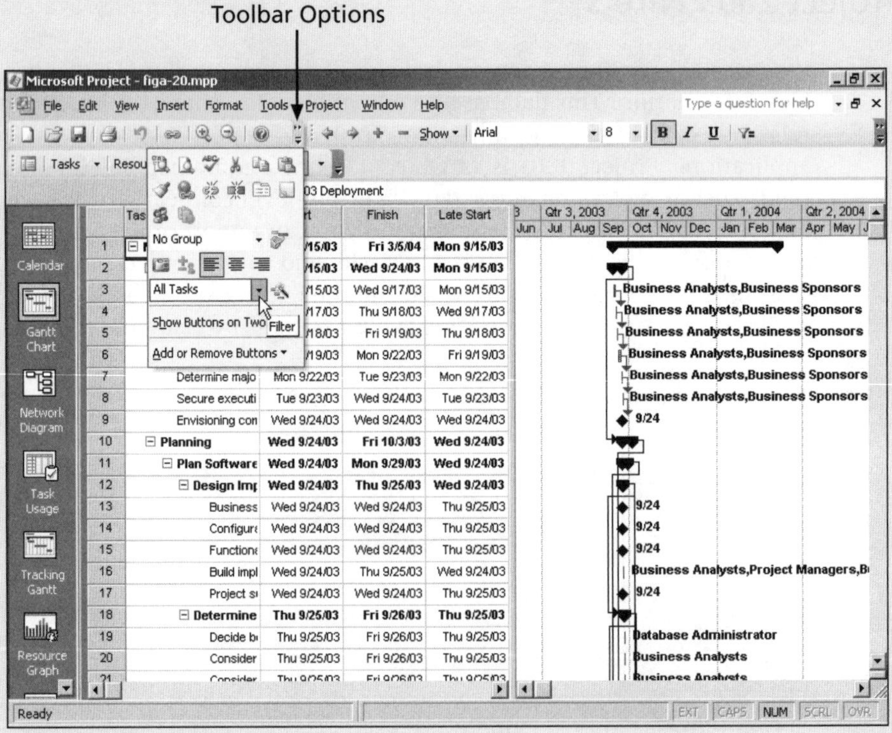

Figure A-24. Using a Filter

2. *Filter to show specific tasks.* Click **Milestones** in the list of filters. Your screen should resemble Figure A-25. Recall from Chapter 6, Project Time Management, that the black diamond symbol represents a milestone, a significant event on a project.

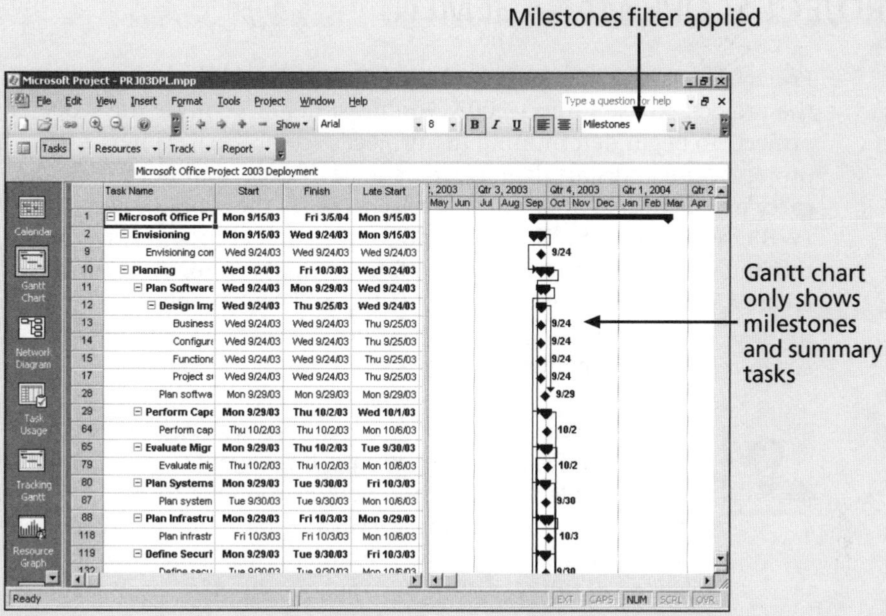

Figure A-25. Milestone Tasks Filter for Project 2003 Deployment File

3. *Show summary tasks.* Select **All Tasks** from the Filter list box to reveal all the tasks in the WBS again. Click the **Filter** list arrow, scroll down until you see Summary Tasks, and then click **Summary Tasks**. Now only the summary tasks appear in the WBS. Experiment with other outline levels and filters.

4. *Close the file.* When you are finished reviewing the Project 2003 Deployment file, click **Close** from the File menu or click the **Close** button. A dialog box appears asking if you want to save changes. Click **No**.

5. *Exit Project 2003.* Select **Exit** from the File menu or click the **Close** button for Project 2003.

Now that you are familiar with Project 2003 main screen elements, views, and filters, you will learn how to use Project 2003 to assist in project scope management by creating a new project file, developing a WBS, and setting a baseline.

PROJECT SCOPE MANAGEMENT

Project scope management involves defining the work to perform to carry out the project. To use Project 2003, you must first determine the scope of the project. To begin determining the project's scope, create a new file with the project name and start date. Develop a list of tasks that need to be done to carry out the project. This list of tasks becomes the work breakdown structure (WBS; see Chapter 5). If you intend to track actual project information against the initial plan, you must set a baseline. In this section, you will learn how to create a new project file, develop a WBS, and set a baseline to help plan and manage the Project Tracking Database project. To start, you will enter the scope-related information.

In this section, you will go through several steps to create a Project 2003 file named scope.mpp. If you want to download the completed file to check your work or continue to the next section, a copy of scope.mpp is available on the companion Web site for this text or from your instructor. You should try to complete an entire section of this appendix (project scope management, project time management, and so on) in one sitting to create the complete file.

Creating a New Project File

To create a new project file:

1. *Create a blank project.* Open Project 2003, and close the Getting Started pane and the following Tasks pane. A blank project file automatically opens when you start Project 2003. The default filenames are Project1, Project2, and so on. If Project 2003 is already open and you want to open a new file, click the **New** button on the toolbar, as with other Office programs.

2. *Open the Project Information dialog box.* Click **Project** on the menu bar, and then click **Project Information** to display the Project Information dialog box, as shown in Figure A-26. The Project Information dialog box enables you to set dates for the project, select the calendar to use, and view project statistics. The project start date will default to today's date.

Start date
text box

Project Information for 'Project1'   ×

Start date: Mon 6/4/07   ▼   Current date: Mon 6/4/07   ▼

Finish date: Mon 6/4/07   ▼   Status date: NA   ▼

Schedule from: Project Start Date ▼   Calendar: Standard   ▼

          All tasks begin as soon as possible.   Priority: 500 ▲▼

Enterprise Custom Fields

Custom Field Name   Value

      Help   Statistics...   OK   Cancel

Figure A-26. Project Information Dialog Box

3. *Enter the project start date.* In the Start date text box, enter **6/4/07**. Setting your project start date to 6/4/07 will ensure that your work matches the results that appear in this appendix. Leave the Finish date, Current date, and other information at the default settings. Click **OK**.

Project 2003 uses American date formats. For example, 6/4/07 represents June 4, 2007. Be sure to enter dates in this format.

4. *Enter project properties.* Click **File** on the menu bar, and then click **Properties**.

Any command that you click in the expanded menu is added immediately to the personalized (short) version of the menu. If you stop using a command for a while, Project 2003 stops showing it on the short version of the menu.

5. *Enter project information.* Type **Project Tracking Database** in the Title text box, type **Terry Dunlay** in the Author text box, as shown in Figure A-27, and then click **OK**. Note that you may have some default information entered in the Project Properties dialog box, such as your company's name. Keep this file open for the next set of steps.

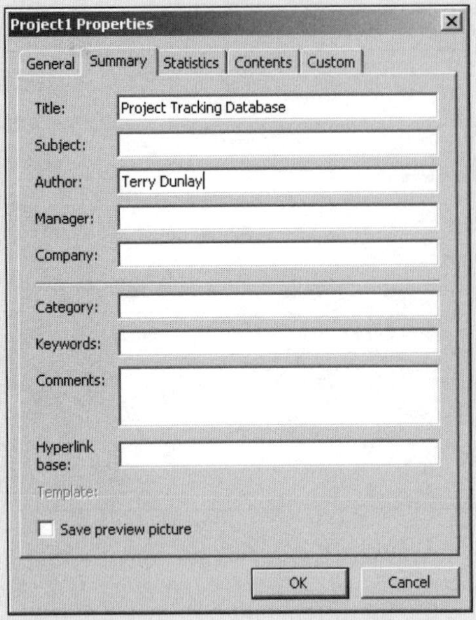

Figure A-27. Project Properties Dialog Box

Developing a Work Breakdown Structure

Before using Project 2003, you must develop a work breakdown structure (WBS) for your project. Developing a good WBS takes time, and it will make entering tasks into the Entry table easier if you develop the WBS first. It is also a good idea to establish milestones before entering tasks in Project 2003. See Chapter 5, Project Scope Management, for more information on creating a WBS and determining milestones. You will use the information in Table A-2 to enter tasks for the Project Tracking Database project. Be aware that this example is much shorter and simpler than most WBSs.

To develop a work breakdown structure and enter milestones for the Project Tracking Database project:

1. *Enter task names.* Enter the twenty tasks in Table A-2 into the Task Name column in the order shown. Do not worry about durations or any other information at this time. Type the name of each task into the Task Name column of the Entry table, beginning with the first row. Press **Enter** or the **down arrow** key on your keyboard to move to the next row.

If you accidentally skip a task, highlight the task row and select **Insert** from the menu bar, then select **New Task**, to insert a blank row. To edit a task entry, click the text for that task, click the Entry bar under the formatting toolbar, and either type over the old text or edit the existing text.

Table A-2: Project Tracking Database Tasks

ORDER	TASKS	ORDER	TASKS
1	Initiating	11	Design
2	Kickoff meeting	12	Implementation
3	Develop project charter	13	System implemented
4	Charter signed	14	Controlling
5	Planning	15	Report performance
6	Develop project plans	16	Control changes
7	Review project plans	17	Closing
8	Project plans approved	18	Prepare final project report
9	Executing	19	Present final project
10	Analysis	20	Project completed

Entering tasks into Project 2003 and editing the information is similar to entering and editing data in an Excel spreadsheet. Project 2003, like Project 2002, includes a feature called a Smart Tag, which appears, for example, when you delete a row. The Smart Tag clarifies whether you want to delete the entire task or only clear the contents of a cell.

2. *Move the split bar to reveal more columns.* If necessary, move the split bar to the right to reveal the entire Task Name and Duration columns.

3. *Adjust the Task Name column width, as needed.* To make all the text display in the Task Name column, move the mouse over the right-column gridline in the Task Name column heading until you see the resize pointer ↔, then click the **left mouse** button and drag the line to the right to make the column wider, or double-click to adjust the column width automatically.

This WBS separates tasks according to the project management process groups of initiating, planning, executing, controlling, and closing. These tasks will be the level 1 items in the WBS for this project. It is a good idea to include all of these process groups, because there are important tasks that must be done under each of them. Recall that the WBS should include *all* of the work required for the project. In the Project Tracking Database WBS, the tasks will purposefully be left at a high WBS level (level 2). You will create these levels, or the WBS hierarchy, next when you create summary tasks. For a real project, you would usually break the WBS into even more levels to provide more details to describe all the work involved in the project. For example, analysis tasks for a database project might be broken down further to include preparing entity relationship diagrams for the database and developing guidelines for the user interface. Design tasks might be broken down to include preparing prototypes, incorporating user feedback, entering data, and testing the database. Implementation tasks might include more levels, such as installing new hardware or software, training the users, fully documenting the system, and so on.

Creating Summary Tasks

After entering the WBS tasks listed in Table A-2 into the Entry table, the next step is to show the WBS levels by creating summary tasks. The summary tasks in this example are Tasks 1 (initiating), 5 (planning), 9 (executing), 14 (controlling), and 17 (closing). You create summary tasks by highlighting and indenting their respective subtasks.

To create the summary tasks:

1. *Select lower-level or subtasks.* Highlight **Tasks 2** through **4** by clicking the cell for Task 2 and dragging the mouse through the cells to Task 4.

2. *Indent subtasks.* Click the **Indent** button ⟶ on the Formatting toolbar, so your screen resembles Figure A-28. After the subtasks (Tasks 2 through 4) are indented, notice that Task 1 automatically becomes boldface, which indicates that it is a summary task. A collapse symbol ⊟ appears to the left of the new summary task name. Clicking the collapse symbol will collapse the summary task and hide the subtasks beneath it. When subtasks are hidden, an expand symbol ⊞ appears to the left of the summary task name. Clicking the expand symbol will expand the summary task. Also, notice that the symbol for the summary task on the Gantt chart has changed from a blue line to a black line with arrows indicating the start and end dates.

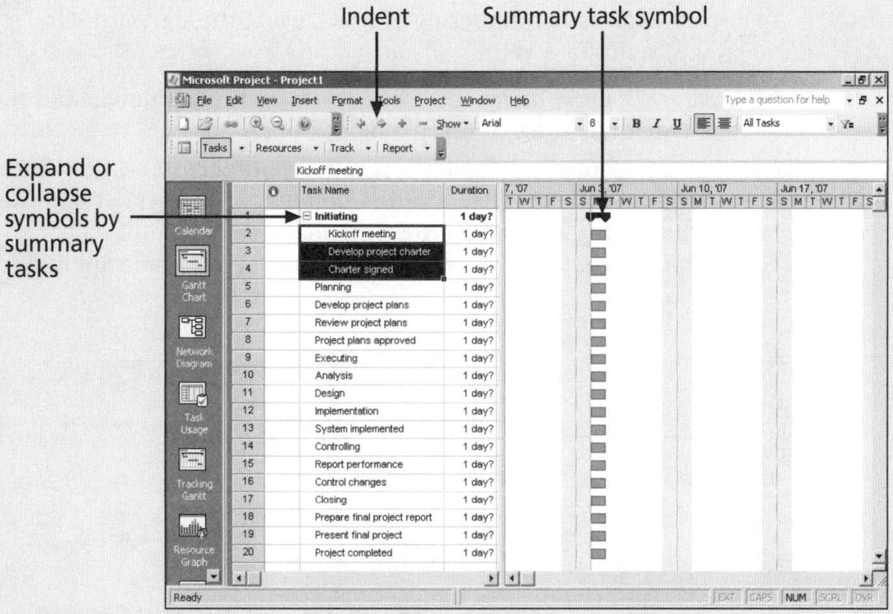

Indent   Summary task symbol

Expand or
collapse
symbols by
summary
tasks

Figure A-28. Indenting Tasks to Create the WBS Hierarchy

3. *Create other summary tasks and subtasks.* Create subtasks and summary
   tasks for Tasks 5, 9, 14, and 17 by following the same steps. Indent
   Tasks 6 through **8** to make Task 5 a summary task. Indent **Tasks 10**
   through **13** to make Task 9 a summary task. Indent **Tasks 15** through
   16 to make Task 14 a summary task. Indent **Tasks 18** through **20** to
   make Task 17 a summary task.

To change a task from a subtask to a summary task or to change
its level in the WBS, you can "outdent" the task. Click the cell of
the task or tasks you want to change, and then click the Outdent
button ⬚ on the Formatting toolbar. Remember that the tasks
in Project 2003 should be entered in an appropriate WBS format
with several levels in the hierarchy.

Numbering Tasks

Depending on how Project 2003 is set up on your computer, you may or may
not see numbers associated with tasks as you enter and indent them.

To display automatic numbering of tasks, using the standard tabular numbering system for a WBS:

1. *Display the Options dialog box.* Click **Tools** on the menu bar, and then click **Options**. The Options dialog box opens.

2. *Show outline numbers.* Click the **View** tab, if necessary. Check **Show outline number** in the "Outline options for Project1" section of the View tab. Click **OK** to close the dialog box. Figure A-29 shows the Options dialog box and resulting outline numbers applied to the WBS tasks.

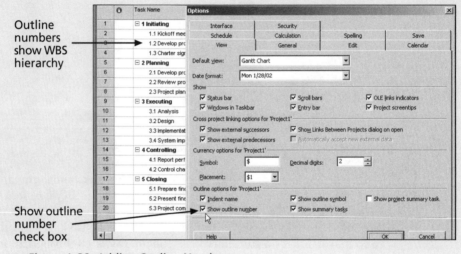

Outline numbers show WBS hierarchy

Show outline number check box

Figure A-29. Adding Outline Numbers

Saving Project Files With or Without a Baseline

An important part of project management is tracking performance against a baseline, or approved plan. Project 2003, like Project 2002, does not prompt you to save a file with or without a baseline each time you save it, as previous versions did. The default is to save without a baseline. It is important to wait until you are ready to save your file with a baseline, because Project 2003 will show changes against a baseline. Because you are still developing your project file for the Project Tracking Database project, you want to save the file without a baseline. Later in this appendix, you will save the file with a baseline by selecting Tools, Tracking, and then Set Baseline. You will then enter actual information to compare planned and actual performance data.

To save a file without a baseline:

1. *Save your file.* Click **File** on the menu bar and then click **Save**, or click the **Save** button 🖫 on the Standard toolbar.

2. *Enter a filename.* In the Save As dialog box, type **scope** in the File name text box. Save the file in your personal folder, on a floppy disk, or in the C:\temp folder, depending on how your computer is set up. Click **Save**. Your Project 2003 file should look like Figure A-30. (You can adjust the timescale to show months by clicking Zoom In or Zoom Out.)

 If you want to download the Project 2003 file scope.mpp to check your work or continue to the next section, a copy is available on the companion Web site for this text or from your instructor.

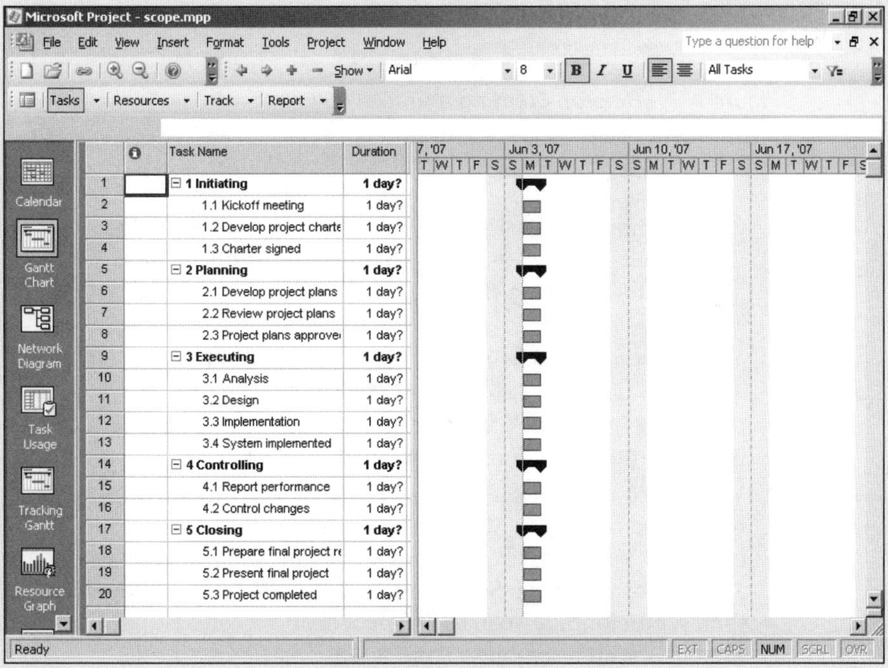

Figure A-30. Project 2003 Scope File

Project 2003 now includes the Visio WBS Chart Wizard. This wizard helps you to create a graphical view of your WBS, similar to the samples provided in Chapter 5. If you want to create a WBS in a chart format first and then import it into Project 2003, you can use several add-in products that provide that capability. Figure A-31 shows the Help information available from Project 2003 on creating a Visio WBS chart.

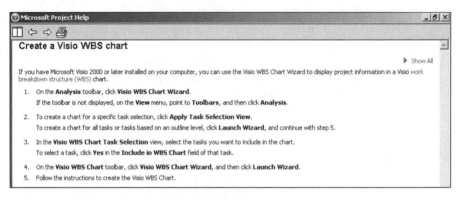

Figure A-31. Help on Creating a Visio WBS Chart

To create a WBS using the Visio WBS Chart Wizard:

 You must have Microsoft Visio 2000 or later installed on your computer to use this wizard. If you do not have Visio installed, skip the following steps.

1. *Show the analysis toolbar.* Click **View** on the menu bar, select **Toolbars**, and then click **Analysis**.
2. *Open the wizard.* Click **Visio WBS Chart Wizard** on the Analysis toolbar, and then click **Launch Wizard**.
3. *View the wizard options.* Click **Next** in Step 1 of the wizard, and then review the options in Step 2, as shown in Figure A-32. Notice that you can select which tasks or level of tasks you want to include, and include a second display field. Select the second radio button, if needed, and then set the outline level to 2. Uncheck the second Display Field option, so your screen resembles Figure A-32.

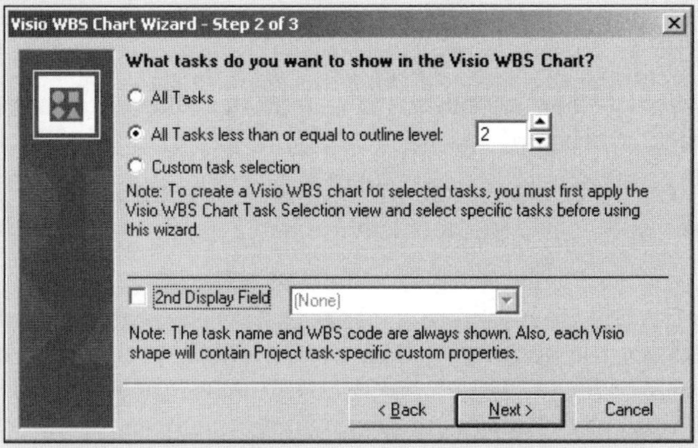

Figure A-32. Visio WBS Chart Wizard - Step 2 of 3

4. *Run the wizard.* Click **Next** to go to Step 3, click **Finish**, and then click **OK**. If you are prompted to automatically download and display the links, click No.

5. *Open the resulting file in Visio.* Click the **Visio** button that should now appear on your Windows 2000 or Windows XP Task Launch bar, and then close the Organization Chart toolbar, if necessary. Your screen should resemble Figure A-33, showing the WBS in Visio.

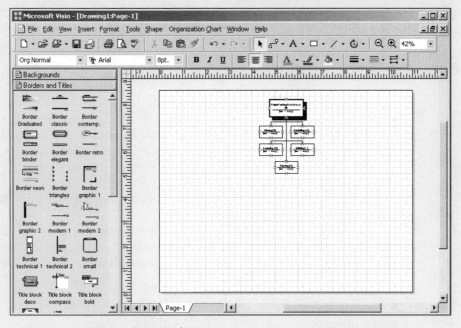

Figure A-33. WBS in Microsoft Visio

6. *Close Visio*. Click the **Close** button, and then click **No** when prompted to save the file to close Microsoft Visio.

7. *Hide the Analysis toolbar*. Click **View** on menu bar, select **Toolbars**, and then click **Analysis**.

8. *Save the file*. Click **File** on the menu bar and then click **Save**, or click the **Save** button 🖫 on the Standard toolbar.

Now that you have finished entering all twenty tasks, created the summary tasks and subtasks, set the options to show the standard WBS tabular numbering system, saved your file, and explored the Visio WBS Chart Wizard, you will learn how to use the Project 2003 time management features.

PROJECT TIME MANAGEMENT

Many people use Project 2003 for its time management features. The first step in using these features, after entering the WBS for the project, is to enter durations for tasks or specific dates when tasks will occur. Entering durations or specific dates will automatically update the Gantt chart. To use Project 2003 to adjust schedules automatically and to do critical path analysis, you must also enter task dependencies. After entering durations and task dependencies, you can view the network diagram and critical path information. This section describes how to use each of these time management features.

Entering Task Durations

When you enter a task, Project 2003 automatically assigns to it a default duration of one day, followed by a question mark. To change the default duration, type a task's estimated duration in the Duration column. If you are unsure of an estimate and want to review it again later, enter a question mark after it. For example, you could enter 5d? for a task with an estimated duration of five days that you want to review later. You can then use the Tasks With Estimated Durations filter to quickly see the tasks for which you need to review duration estimates.

To indicate the length of a task's duration, you must type both a number and an appropriate duration symbol. If you type only a number, Project 2003 automatically enters days as the duration unit. Duration unit symbols include:

- d = days (default)
- w = weeks
- m = minutes
- h = hours

- mo or mon = months
- ed = elapsed days
- ew = elapsed weeks

For example, to enter one week for a task duration, type 1w in the Duration column. (You can also type wk, wks, week, or weeks, instead of just w.) To enter two days for a task duration, type 2d in the Duration column. The default unit is days, so if you enter 2 for the duration, it will be entered as 2 days. You can also enter elapsed times in the Duration column. For example, 2ed means two elapsed days, and 2ew means two elapsed weeks. You would use an elapsed duration for a task like "Allow paint to dry." The paint will dry in exactly the same amount of time regardless of whether it is a workday, a week-end, or a holiday.

 If the Duration column is not visible, drag the split bar to the right until the Duration column is in view.

Entering time estimates or durations might seem like a straightforward process. However, you must follow a few important procedures:

- As you saw in the previous section, when you enter a task in the Task Name column, 1 day? appears in the Duration column. The question mark means that you are unsure of the duration or want to go back to it. Be sure to enter the durations you want to appear.
- Do not enter durations for summary tasks. Summary task durations are calculated automatically, based on the subtasks. If you enter a duration for a task and then make it a summary task, its duration will automatically change to match the durations of its subtasks. Project 2003 will not allow you to enter or change the duration of a summary task. Think of it as entering durations for work packages, the lowest-level items in your WBS. The other items are really the WBS hierarchy. Let Project 2003 do the duration calculations for you.
- To mark a task as a milestone, enter 0 for the duration. You can also mark tasks that have a non-zero duration as milestones by checking the Mark task as milestone option in the Task Information dialog box on the Advanced tab. The milestone symbol for those tasks will appear at their start date. Double-click a task to access the Task Information dialog box.
- To enter recurring tasks, such as weekly meetings or monthly status reports, select Recurring Task from the Insert menu. Enter the task name, the duration, and when the task occurs. Project 2003 will automatically insert appropriate subtasks based on the length of the project and the number of tasks required for the recurring task. For example, if you enter a recurring task for monthly review meetings that occur on the first day of

every month for a twelve-month project, Project 2003 will enter a summary task for monthly review meetings and twelve subtasks—one meeting for each month.

- You can enter the exact start and finish dates for activities, instead of entering durations. To enter start and finish dates, move the split bar to the right to reveal the Start and Finish columns. Enter start and finish dates only when those dates are certain. If you want task dates to adjust according to any other task dates, do not enter exact start and finish dates. Instead, you should enter a duration and then establish a dependency to related tasks. *The real scheduling power of Project 2003 comes from setting up dependencies or relationships among tasks, as described in the next section.*

- Project 2003 uses a default calendar with standard workdays and hours. Duration estimates will vary according to the calendar you use. For example, entering 5d in the standard calendar may result in more than five days on the Gantt chart if the time period includes Saturday or Sunday. You can change specific working and nonworking days, or the entire project calendar, by selecting Change Working Time from the Tools menu.

- You often need to adjust the timescale on the Gantt chart to see your project's schedule in different time frames, such as weeks, months, quarters, or years. You expand the timescale by clicking the Zoom Out button. The Zoom In button collapses the timescale. You can also change the format of the timescale by having three tiers instead of two or by formatting dates differently by double-clicking the timescale section of the Gantt chart.

Next, you will set task durations in your Project Tracking Database file (scope.mpp), created in the previous section. If you did not create the file named scope.mpp, you can download it from the companion Web site for this text. You will create a new recurring task and enter its duration, and then you will enter other task durations. First, create a new recurring task called Status Reports above Task 15, Report Performance.

To create a new recurring task:

1. *Insert a recurring task above Task 15, Report performance.* Open scope.mpp, if necessary, and then click **Report performance** (Task 15) in the Task Name column to select that task. Click **Insert** on the menu bar, and then click **Recurring Task**. The Recurring Task Information dialog box opens.

2. *Enter task and duration information for the recurring task.* Type **Status Reports** as the task title in the Task Name text box. Type **1h** in the Duration text box. Select the **Weekly** radio button under Recurrence Pattern. Click the **week on** list arrow and select **every** from the list. Select the **Wednesday** check box. In the Range of Recurrence section, type **6/13/07** in the Start text box, click the **End by** radio button, and

then type **10/31/07** in the End by text box. Click the **End by** list arrow to see the calendar, as shown in Figure A-34. You can use the calendar to enter the Start and End by dates. The new recurring task will appear above Task 15, Report Performance, when you are finished.

 You can also enter a number of occurrences instead of an End by for a recurring task. You might need to adjust the End by date after you enter all of your task durations and dependencies. Remember that the date on your computer determines the date listed as Today in the calendar.

Recurring Task Information

Task Name: Status Reports   Duration: 1h

Recurrence pattern
- ○ Daily   every ▾   week on:
- ● Weekly
- ○ Monthly   □ Sunday □ Monday □ Tuesday ☑ Wednesday
- ○ Yearly   □ Thursday □ Friday □ Saturday

Range of recurrence
Start: Wed 6/13/07 ▾   ○ End after: 21 ▾ occurrences
                          ● End by: Wed 10/31/07 ▾

Calendar for scheduling this task
Calendar: None ▾   □ Sched...

	October, 2007					
Sun	Mon	Tue	Wed	Thu	Fri	Sat
30	1	2	3	4	5	6
7	8	9	10	11	12	13
14	15	16	17	18	19	20
21	22	23	24	25	26	27
28	29	30	31	1	2	3
4	5	6	7	8	9	10

Today: 6/10/2007

Figure A-34. Recurring Task Information Dialog Box

3. *View the new summary task and its subtasks.* Click **OK**. Project 2003 inserts a new Status Reports subtask in the Task Name column. Note that this new subtask is boldface and has an expand symbol ⊞ to the left of the task name. Expand the new subtask by clicking ⊞ to the left of the words Status Reports. To collapse the recurring task, click the **collapse symbol** ⊟.

4. *Adjust column widths and the timescale.* Notice the # signs that appear in the Duration column for the Status Reports task. Like Excel, Project 2003 uses these symbols to denote that the column width needs to be increased. With the recurring task expanded, increase the Task Name column heading by moving the mouse to the right of the Task Name column heading until you see a resize pointer ↔. Double-click to adjust the column width and display the entire task name. Move the split bar to the right to reveal the entire Duration column, and then increase the Duration column's width by double-clicking to the right of the Duration column heading. Click the **Zoom Out** 🔍 button on the toolbar three times to display the Gantt chart in quarters. Your screen should resemble Figure A-35. Notice that the recurring task appears on the appropriate dates on the Gantt chart.

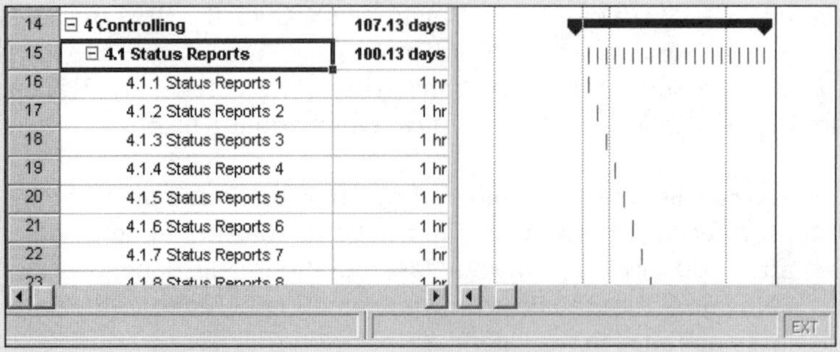

Figure A-35. Expanding a Recurring Task

Use the information in Table A-3 to enter durations for the other tasks for the Project Tracking Database project. The Project 2003 row number is shown to the left of each task name in the table. Remember that you already entered a duration for the recurring task. Also, remember that you should *not* enter durations for summary tasks. Durations for summary tasks are automatically calculated to match the durations and dependencies of subtasks, as described further in the next section, "Establishing Task Dependencies."

Table A-3: Durations for Project Tracking Database Tasks

TASK NUMBER/ROW	TASK NAME	DURATION
2	Kickoff meeting	2h
3	Develop project charter	10d
4	Charter signed	0d
6	Develop project plans	3w
7	Review project plans	4mo
8	Project plans approved	0d
10	Analysis	1mo
11	Design	2mo
12	Implementation	1mo
13	System implemented	0d
37	Report performance	5mo
38	Control changes	5mo
40	Prepare final project report	2w
41	Present final project	1w
42	Project completed	0d

To enter task durations for the other tasks:

1. *Enter the duration for Task 2.* Click the **Duration** column for row 2, Kickoff meeting, type **2h**, and then press **Enter**.

2. *Enter the duration for Task 3.* In the **Duration** column for row 3, Develop project charter, type **10d**, then press **Enter**. You can also just type 10, since d or days is the default duration.

3. *Enter remaining task durations.* Continue to enter the durations, using information in Table A-3.

4. *Save your file and name it.* Click **File** on the menu bar, and then click **Save As**. Enter **time** as the filename, and then save the file to the desired location on your computer or network. Keep this file open for the next set of steps.

Establishing Task Dependencies

To use Project 2003 to adjust schedules automatically and to do critical path analysis, you *must* determine the dependencies or relationships among tasks (see Chapter 6, Project Time Management). Project 2003 provides three methods for creating task dependencies: using the Link Tasks button; using the Predecessors column of the Entry table; or clicking and dragging the Gantt chart symbols for tasks with dependencies.

To create dependencies using the Link Tasks button, highlight tasks that are related and then click the Link Tasks button 🔗 on the toolbar. For example, to create a finish-to-start dependency between Task 1 and Task 2, click any cell in row 1, drag down to row 2, and then click the Link Tasks button. The default type of link is finish-to-start. In the Project Tracking Database example, all the tasks use this default relationship. You will learn about other types of dependencies later in this appendix.

 Selecting tasks is similar to selecting cells in Microsoft Excel. To select adjacent tasks, you can click and drag the mouse. You can also click the first task, hold down the Shift key, and then click the last task. To select nonadjacent tasks, hold down the Control (Ctrl) key as you click tasks in order of their dependencies.

When you use the Predecessors column of the Entry table to create dependencies, you must manually enter the information. To create dependencies manually, type the task row number of the preceding task in the Predecessors column of the Entry table. For example, Task 3 in Table A-3 has Task 2 as a predecessor, which can be entered in the Predecessors column, meaning that

Task 3 cannot start until Task 2 is finished. To see the Predecessors column of the Entry table, move the split bar to the right.

You can also create task dependencies by clicking the Gantt chart symbol for a task and then dragging to the Gantt chart symbol for a task that succeeds it. For example, you could click the Milestone symbol ◆ for Task 4, hold down the left mouse button, and drag to the Task Bar symbol for Task 6 to create a dependency, as shown in the Gantt chart in Figure A-36. Note the Finish-to-Start Link dialog box that appears when you use this method.

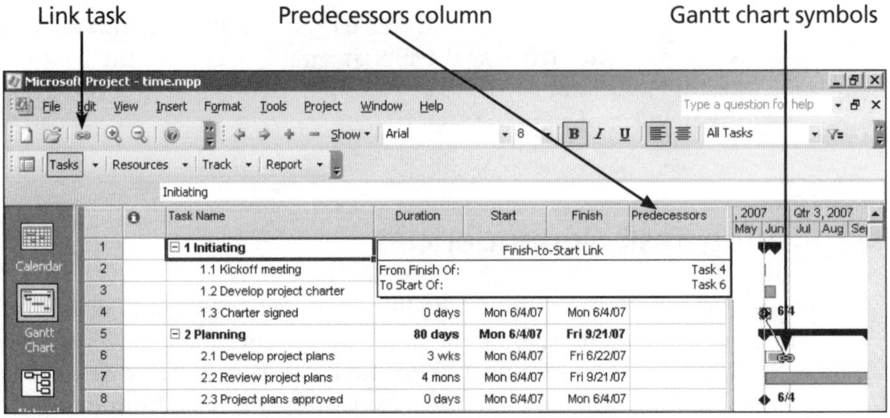

Figure A-36. Creating Task Dependencies

Next, you will use information from Figure A-37 to enter the predecessors for tasks as indicated. You will create some dependencies by manually typing the predecessors in the Predecessors column, some by using the Link Tasks button, some by using the Gantt chart symbols, and the remaining dependencies by using whichever method you prefer.

To link tasks or establish dependencies for the Project Tracking Database project:

1. *Display the Predecessors column in the Entry table.* Move the split bar to the right to reveal the Predecessors column in the time.mpp file. Widen the Task Name or other columns, if needed.

2. *Highlight the cell where you want to enter a predecessor, and then type the task number for the preceding task.* Click the **Predecessors cell** for Task 3, type **2**, and press **Enter**.

3. *Enter predecessors for Task 4.* Click the **Predecessors cell** for Task 4, type **3**, and press **Enter**. Notice that as you enter task dependencies, the Gantt chart changes to reflect the new schedule.

4. *Establish dependencies using the Link Tasks button.* To link Tasks 10 through 13, click the task name for Task 10 in the Task Name column and drag down through Task 13. Then, click the **Link Tasks** button on the toolbar. Alternately, you can also click **Edit** on the menu bar, and then click **Link Tasks**.

5. *Create a dependency using Gantt chart symbols.* Click the **Milestone** symbol ◆ for Task 4 on the Gantt chart, hold down the **left mouse** button, and drag to the **Task Bar** symbol for Task 6. (See Figure A-36 to see this step in progress.)

6. *Enter remaining dependencies.* Link the other tasks by manually entering the predecessors into the Predecessors column, or by using the Link Tasks button, or clicking and dragging the Gantt chart symbols. You can view the dependencies in Figure A-37. For example, Task 8 has Task 6 as a predecessor, Task 10 has Task 8 as a predecessor, and so on. If you have entered all data correctly, the project should end on 11/19/07.

7. *Adjust screen elements.* Click **View** from the menu bar, select **View Bar** to collapse the View Bar, and then drag the resize line between the Indicator column and the Task Name column to the left to remove the Indicator column, if needed. Click **View** on the menu bar, select **Toolbars**, and then click **Project Guide** to hide the Project Guide toolbar. When you finish, your screen should resemble Figure A-37. Double-check your screen to make sure you entered the dependencies correctly.

	Task Name	Duration	Start	Finish	Predeces
1	⊟ **1 Initiating**	**10.25 days**	**Mon 6/4/07**	**Mon 6/18/07**	
2	1.1 Kickoff meeting	2 hrs	Mon 6/4/07	Mon 6/4/07	
3	1.2 Develop project charter	10 days	Mon 6/4/07	Mon 6/18/07	2
4	1.3 Charter signed	0 days	Mon 6/18/07	Mon 6/18/07	3
5	⊟ **2 Planning**	**95 days**	**Mon 6/18/07**	**Mon 10/29/07**	
6	2.1 Develop project plans	3 wks	Mon 6/18/07	Mon 7/9/07	4
7	2.2 Review project plans	4 mons	Mon 7/9/07	Mon 10/29/07	6
8	2.3 Project plans approved	0 days	Mon 7/9/07	Mon 7/9/07	6
9	⊟ **3 Executing**	**80 days**	**Mon 7/9/07**	**Mon 10/29/07**	
10	3.1 Analysis	1 mon	Mon 7/9/07	Mon 8/6/07	8
11	3.2 Design	2 mons	Mon 8/6/07	Mon 10/1/07	10
12	3.3 Implementation	1 mon	Mon 10/1/07	Mon 10/29/07	11
13	3.4 System implemented	0 days	Mon 10/29/07	Mon 10/29/07	12
14	⊟ **4 Controlling**	**107.13 days**	**Mon 6/4/07**	**Wed 10/31/07**	
15	⊞ **4.1 Status Reports**	**100.13 days**	**Wed 6/13/07**	**Wed 10/31/07**	
37	4.2 Report performance	5 mons	Mon 6/4/07	Fri 10/19/07	
38	4.3 Control changes	5 mons	Mon 6/4/07	Fri 10/19/07	
39	⊟ **5 Closing**	**15 days**	**Mon 10/29/07**	**Mon 11/19/07**	
40	5.1 Prepare final project report	2 wks	Mon 10/29/07	Mon 11/12/07	13
41	5.2 Present final project	1 wk	Mon 11/12/07	Mon 11/19/07	40
42	5.3 Project completed	0 days	Mon 11/19/07	Mon 11/19/07	41

Figure A-37. Project Tracking Database File with Durations and Dependencies Entered

8. *Preview and save your file.* Click **File** on the menu bar, and then select **Print Preview**, or click the **Print Preview** button 🔍 on the Standard toolbar. When you are finished, close the Print Preview and save your file again by clicking the **Save** button 💾 on the Standard toolbar. If you desire, print your file by clicking the **Print** button 🖨 on the Standard toolbar, or by selecting Print from the File menu. Keep the file open for the next set of steps.

> **?**
> **Help**
>
> The entire Predecessors column must be visible in the Gantt Chart view for it to show up in the Print Preview. Close the Print Preview and move the split bar to reveal the column fully, and then select Print Preview again. Also, check your timescale to make sure you are not wasting paper by having a timescale too detailed. You should not print your file until the Print Preview displays the desired information.

Changing Task Dependency Types and Adding Lead or Lag Time

A task dependency or relationship describes how a task is related to the start or finish of another task. Project 2003 allows for four task dependencies: finish-to-start (FS), start-to-start (SS), finish-to-finish (FF), and start-to-finish (SF). You can find detailed descriptions of these dependencies in Chapter 6, Project Time Management. By using these dependencies effectively, you can modify the critical path and shorten your project schedule. The most common type of dependency is finish-to-start (FS). All of the dependencies in the Project Tracking Database example are FS dependencies. However, sometimes you need to establish other types of dependencies. This section describes how to change task dependency types. It also explains how to add lead or lag times between tasks. You will shorten the duration of the Project Tracking Database project by adding lead time between some tasks.

To change a dependency type, open the Task Information dialog box for that task by double-clicking the task name. On the Predecessors tab of the Task Information dialog box, select a new dependency type from the Type column list arrow.

The Predecessors tab also allows you to add lead or lag time to a dependency. You can enter both lead and lag time using the Lag column on the Predecessors tab. Lead time reflects an overlap between tasks that have a dependency. For example, if Task B can start when its predecessor, Task A, is half-finished, you can specify a finish-to-start dependency with a lead time of 50 percent for the successor task. Enter lead times as negative numbers. In this example, enter –50% in the first cell of the Lag column. Adding lead times is also called fast tracking and is one way to compress a project's schedule (see Chapter 6).

Lag time is the opposite of lead time—it is a time gap between tasks that have a dependency. If you need a two-day delay between the finish of Task C and the start of Task D, establish a finish-to-start dependency between Tasks C and D and specify a two-day lag time. Enter lag time as a positive value. In this example, type 2d in the Lag column.

In the Project Tracking Database example, notice that work on design tasks does not begin until all the work on the analysis tasks has been completed (see Rows 10 and 11 on the WBS), and work on implementation tasks does not begin until all the work on the design tasks has been completed (see Rows 11 and 12 on the WBS). In reality, it is rare to wait until all of the analysis work is complete before starting any design work, or to wait until all of the design work is finished before starting any implementation work. It is also a good idea to add some additional time, or a buffer, before crucial milestones, such as a system being implemented. To create a more realistic schedule, add lead times to the design and implementation tasks and lag time before the system implemented milestone.

To add lead and lag times:

1. *Open the Task Information dialog box for Task 11, Design.* In the Task Name column, double-click the text for Task 11, **Design**. The Task Information dialog box opens. Click the **Predecessors** tab.

2. *Enter lead time for Task 11.* Type **–10%** in the Lag column, as shown in Figure A-38. Click **OK**. You could also type a value such as –5d to indicate a five-day overlap. In the resulting Gantt chart, notice that the bar for this task has moved slightly to the left. Also, notice that the project completion date has moved from 11/19 to 11/15.

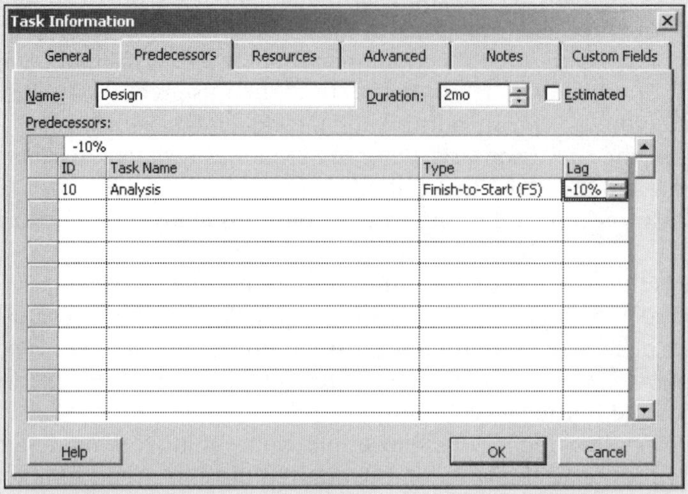

Figure A-38. Adding Lead or Lag Time to Task Dependencies

3. *Enter lead time for Task 12.* Double-click the text for Task 12,
   Implementation, type **–3d** in the Lag column, and click **OK**. Notice
   that the project now ends on 11/12 instead of 11/15.

4. *Enter lag time for Task 13.* Double-click the text for Task 13, **System
   Implemented**, type **5d** in the Lag column for this task, and click
   OK. Move the split bar to the right, if necessary, to reveal the
   Predecessors column. When you are finished, your screen should
   resemble Figure A-39. Notice the slight overlap in the taskbars for
   Tasks 10 and 11 and the short gap between the taskbars for Tasks
   12 and 13. You can zoom in to see the gap more clearly. The proj-
   ect completion date should be 11/19.

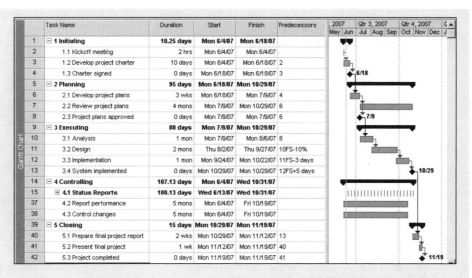

		Task Name	Duration	Start	Finish	Predecessors	2007	Qtr 3, 2007	Qtr 4, 2007	
							May Jun	Jul Aug Sep	Oct Nov Dec	J
1		□ 1 Initiating	10.25 days	Mon 6/4/07	Mon 6/18/07					
2		1.1 Kickoff meeting	2 hrs	Mon 6/4/07	Mon 6/4/07					
3		1.2 Develop project charter	10 days	Mon 6/4/07	Mon 6/18/07	2				
4		1.3 Charter signed	0 days	Mon 6/18/07	Mon 6/18/07	3	6/18			
5		□ 2 Planning	95 days	Mon 6/18/07	Mon 10/29/07					
6		2.1 Develop project plans	3 wks	Mon 6/18/07	Mon 7/9/07	4				
7		2.2 Review project plans	4 mons	Mon 7/9/07	Mon 10/29/07	6				
8		2.3 Project plans approved	0 days	Mon 7/9/07	Mon 7/9/07	6	7/9			
9		□ 3 Executing	80 days	Mon 7/9/07	Mon 10/29/07					
10		3.1 Analysis	1 mon	Mon 7/9/07	Mon 8/6/07	8				
11		3.2 Design	2 mons	Thu 8/2/07	Thu 9/27/07	10FS-10%				
12		3.3 Implementation	1 mon	Mon 9/24/07	Mon 10/22/07	11FS-3 days				
13		3.4 System implemented	0 days	Mon 10/29/07	Mon 10/29/07	12FS+5 days			10/29	
14		□ 4 Controlling	107.13 days	Mon 6/4/07	Wed 10/31/07					
15		⊞ 4.1 Status Reports	100.13 days	Wed 6/13/07	Wed 10/31/07			|||||||||||||||||||||		
37		4.2 Report performance	5 mons	Mon 6/4/07	Fri 10/19/07					
38		4.3 Control changes	5 mons	Mon 6/4/07	Fri 10/19/07					
39		□ 5 Closing	15 days	Mon 10/29/07	Mon 11/19/07					
40		5.1 Prepare final project report	2 wks	Mon 10/29/07	Mon 11/12/07	13				
41		5.2 Present final project	1 wk	Mon 11/12/07	Mon 11/19/07	40				
42		5.3 Project completed	0 days	Mon 11/19/07	Mon 11/19/07	41			11/19	

Figure A-39. Schedule for Project Tracking Database File with Lead and Lag Times

5. *Save your file.* Click **File** on the menu bar, and then click **Save**. Or, click the **Save** button 🖫 on the Standard toolbar. Keep this file open for the next set of steps.

You can enter or modify lead or lag times directly in the Predecessors column of the Entry table. Notice how the predecessors appear for Tasks 11, 12, and 13. For example, the Predecessors column for Task 11 shows 10FS-10%. This notation means that Task 11 has a finish-to-start (FS) dependency with Task 10 and a 10 percent lead. You can enter lead or lag times to task dependencies directly into the Predecessors column of the Entry table by using this same format: task row number, followed by type of dependency, followed by the amount of lead or lag.

Gantt Charts

Project 2003 shows a Gantt chart as the default view to the right of the Entry table. Gantt charts show the timescale for a project and all of its activities. In Project 2003, dependencies between tasks are shown on the Gantt chart by the arrows between tasks. Many Gantt charts, however, do not show any dependencies. Instead, as you might recall, network diagrams or PERT charts are used to show task dependencies. See Chapter 6 for more information about Gantt charts and network diagrams. This section explains important information about

Gantt charts and describes how to make critical path information more visible in the Gantt Chart view.

You must follow a few important procedures when working with Gantt charts:

- To adjust the timescale, click the Zoom Out button 🔍 or the Zoom In button 🔍. Clicking these buttons automatically makes the dates on the Gantt chart show more or less information. For example, if the timescale for the Gantt chart is showing months and you click the Zoom Out button, the timescale will adjust to show quarters. Clicking Zoom Out again will display the timescale in years. Similarly, each time you click the Zoom In button, the timescale changes to display more detailed time information— from years to quarters, quarters to months, and months to weeks.

- You can also adjust the timescale and access more formatting options by selecting Timescale from the Format menu. Adjusting the timescale enables you to see the entire Gantt chart on one screen and in the time increments that you desire.

- A Gantt Chart Wizard is available on the Format menu. This wizard helps you adjust the format of the Gantt chart. For example, you can select an option to display critical path information on the Gantt chart, and those items on the critical path will automatically display using a red bar.

- You can view a tracking Gantt chart by setting a baseline for your entire project or for selected tasks and then entering actual durations for tasks. The Tracking Gantt view displays two taskbars, one above the other, for each task. One taskbar shows planned or baseline start and finish dates, and the other taskbar shows actual start and finish dates. You will find a sample tracking Gantt chart later in this appendix, after you enter actual information for the Project Tracking Database project.

Because you have already created task dependencies, you can now find the critical path for the Project Tracking Database project. You can view the critical tasks by changing the color of those items on the critical path in the Task Name column. You can also change the color of bars on the Gantt chart. Tasks on the critical path will automatically be red in the Network Diagram view, as described in the following section.

To make the text for the critical path tasks appear in red in the Entry table and on the Gantt chart:

1. *Open the Text Styles dialog box.* Click **Format** on the menu bar, and then click **Text Styles**. The Text Styles dialog box opens.

2. *Change the critical tasks color option to red.* Click the **Item to Change** list arrow, and then select **Critical Tasks**. Click the **Color** list arrow, and then select **Red**, as shown in Figure A-40. Click **OK** to accept the changes to the text styles.

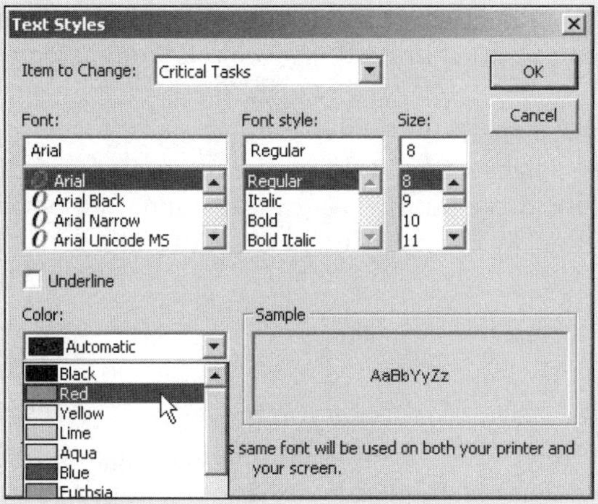

Figure A-40. Text Styles Dialog Box Settings to Display Critical Tasks in Red

3. *Open the Gantt Chart Wizard.* Click **Format** on the menu bar and then select **Gantt Chart Wizard**. Alternately, you can click the Gantt Chart Wizard button 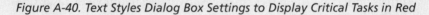 on the Formatting toolbar. The Gantt Chart Wizard opens. Click **Next**, and the next step of the wizard appears.

4. *Select the Critical Path option.* Click the **Critical path** radio button, as shown in Figure A-41. Notice that the sample Gantt chart in the Gantt Chart Wizard displays some bars in red, representing tasks on the critical path. Click **Finish** to proceed to the next step in the Gantt Chart Wizard.

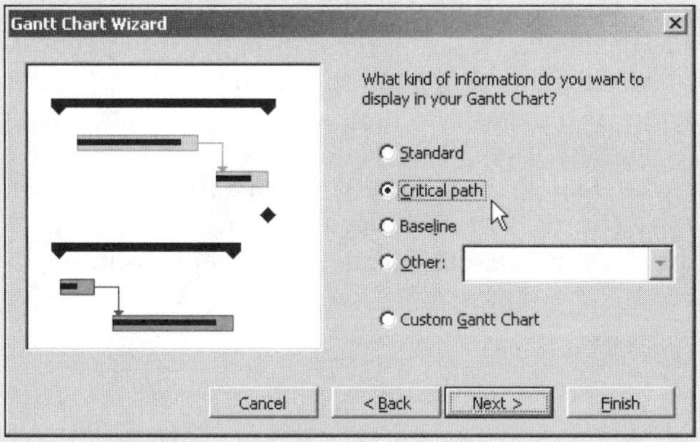

Figure A-41. Gantt Chart Wizard Showing Critical Path Format

5. *Format the Gantt chart.* Click **Format It**, and then click **Exit Wizard**. Critical taskbars on the Gantt chart will now be red. Your screen should resemble Figure A-39, but now task information and bars for all the critical tasks are displayed in red.

6. *Save your file without a baseline.* Click **File** on the menu bar, and then click **Save**. Or, click the **Save** button 🔲 on the Standard toolbar to save your file. Keep this file open for the next set of steps.

Network Diagrams

The network diagrams in Project 2003 use the precedence diagramming method, with tasks displayed in rectangular boxes and relationships shown by lines connecting the boxes. In the Network Diagram view, tasks on the critical path are automatically shown in red.

To view the network diagram for the Project Tracking Database project:

1. *View the network diagram.* Click **View** on the menu bar, and then click **View Bar**. Click the **Network Diagram** button 🔳 on the View Bar or select **Network Diagram** from the View menu.

2. *Adjust the Network Diagram view.* To see more tasks in the Network Diagram view, click the **Zoom Out** button 🔍 twice. You can also use the scroll bars to see different parts of the network diagram. Figure A-42 shows several of the tasks in the Project Tracking Database network diagram. Note that milestone tasks, such as Charter signed, appear as pointed rectangular boxes, while other tasks appear as rectangles. Tasks on the critical path automatically appear in red, while noncritical tasks appear in blue. Each task in the network diagram also shows information such as the start and finish dates, task ID, and duration. Move your mouse over any of the boxes to see them in a larger view. A dashed line on a network diagram represents a page break. You often need to change some of the default settings for the Network Diagram view before printing it.

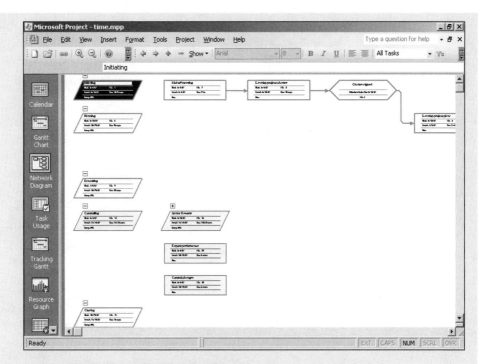

Figure A-42. Network Diagram View

3. *View the Help topic on rearranging network diagram boxes.* Click the
   Microsoft Project Help button 🔘 on the Standard toolbar or press
   F1 to display the Office Assistant. Type **How do I rearrange the
   network diagram?** in the text box and then press **Enter**. Click the
   resulting Help topic called **Rearrange Network Diagram boxes**.
   Figure A-43 shows the resulting Project 2003 Help window. Note that
   you can change several layout options for the network diagram, hide
   fields, and manually position boxes. Close the Help window, and then
   close the Search Results window.

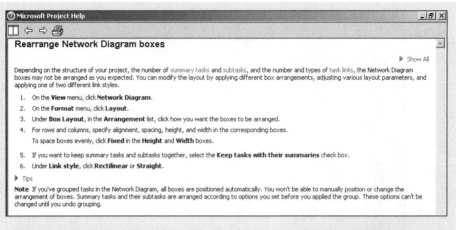

Figure A-43. Project 2003 Help Window for Rearranging the Network Diagram Boxes

4. *Return to Gantt Chart view.* Click the **Gantt Chart** button [icon] on the View Bar to return to Gantt Chart view. Alternately, you can select **Gantt Chart** from the View menu on the menu bar. Keep this file open for the next set of steps.

> **Tip:** Some people create new Project 2003 files or add tasks to existing files in Network Diagram view instead of Gantt Chart view. To add a new task or node in Network Diagram view, select New Task from the Insert menu, or press the Insert key on your keyboard. Double-click the new node to add a task name and other information. Create finish-to-start dependencies between tasks in Network Diagram view by clicking the preceding node and dragging to the succeeding node. To modify the type of dependency and add lead or lag time, double-click the arrow between the dependent nodes.

Critical Path Analysis

Recall that the critical path is the path through the network diagram with no slack or the least amount of slack; it represents the shortest possible time to complete the project (see Chapter 6). If a task on the critical path takes longer than planned, the project schedule will slip unless time is reduced on a task later on the critical path. Sometimes you can shift resources between tasks to help keep a project on schedule. Project 2003 has several views and reports to help analyze critical path information.

Two particularly useful features are the Schedule table view and the Critical Tasks report. The Schedule table view shows the early and late start dates for each task, the early and late finish dates for each task, and the free and total slack for each task. This information shows how flexible the schedule is and helps in making schedule compression decisions. The Critical Tasks report lists only tasks that are on the critical path for the project. If meeting schedule deadlines is essential for a project, project managers will want to monitor tasks on the critical path closely.

To access the Schedule table view and view the Critical Tasks report for a file:

1. *View the Schedule table.* Right-click the **Select All** button to the left of the Task Name column heading and select **Schedule**. Alternatively, you can click **View** on the menu bar, point to **Table: Entry**, and then click **Schedule**. The Schedule table replaces the Entry table to the left of the Gantt chart.

2. *Reveal all columns in the Schedule table.* Move the split bar to the right until you see the entire Schedule table. Your screen should resemble Figure A-44. This view shows the start and finish (meaning the early start and early finish) and late start and late finish dates for each task, as well as free and total slack. Right-click the **Select all** button and select **Entry** to return to the Entry table view.

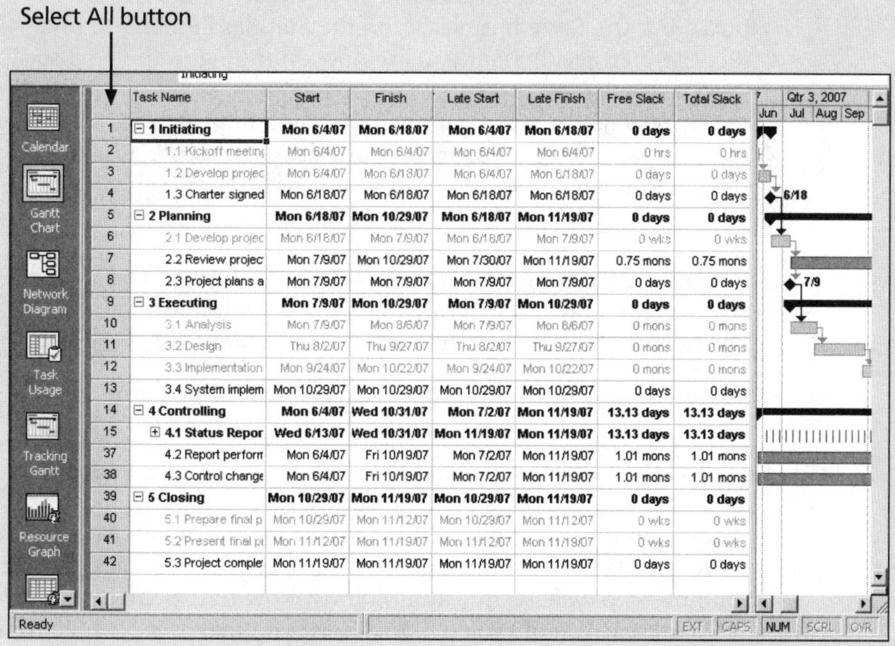

Select All button

Figure A-44. Schedule Table View

3. *Open the Reports dialog box.* Click **View** on the menu bar, and then click **Reports**. Double-click **Overview** to open the Overview Reports dialog box. Your screen should resemble Figure A-45.

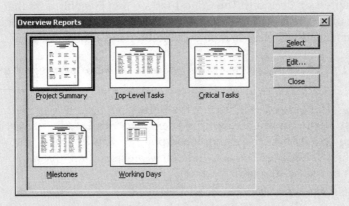

Figure A-45. Accessing the Critical Tasks Report

4. *Display the Critical Tasks report.* Double-click **Critical Tasks**. A Critical Tasks report as of today's date is displayed.

5. *Close the report and save your file.* When you are finished examining the Critical Tasks report, click **Close**. Click **Close** on the Reports dialog box. Click the **Save** button on the Standard toolbar to save your file. Close Project 2003 if you are not continuing to the next section.

?
Help
If you want to download the completed time.mpp file to check your work or continue to the next section, a copy is available on the companion Web site for this text or from your instructor.

Now that you have entered task durations, established task dependencies, and reviewed the network diagram and critical path information, you are ready to explore some of the Project 2003 cost management features.

PROJECT COST MANAGEMENT

Many people do not use Project 2003 for cost management. Most organizations have more established cost management software products and procedures in place, and many people simply do not know how to use the cost features. Using the cost features of Project 2003, however, makes it possible to integrate total project information more easily. This section offers brief instructions for entering fixed and variable cost estimates and actual cost and time information after establishing a baseline plan. It also explains how to use Project 2003 for earned value management. More details on these features are available in Project 2003 Help, online tutorials, or other books.

 To complete the steps in this section, you need to download the Project 2003 file resource.mpp from the companion Web site for this book and save it on a floppy disk or with your other project files on your hard drive.

Fixed and Variable Cost Estimates

The first step to using the cost features of Project 2003 is entering cost-related information. You enter costs as fixed or variable based on per-use material costs, or variable based on the type and amount of resources used. Costs related to personnel are often a significant part of project costs.

Entering Fixed Costs in the Cost Table

The Cost table allows you to enter fixed costs related to each task. To access the Cost table, right-click the Select All button and select Cost or select Table: Cost from the View menu. Figure A-46 shows the resulting view for the Project Tracking Database project using time.mpp. You can also assign a per-use cost to a resource that represents materials or supplies and use it as a base for calculating the total materials or supplies cost of a task. See Project 2003 Help for details on this feature.

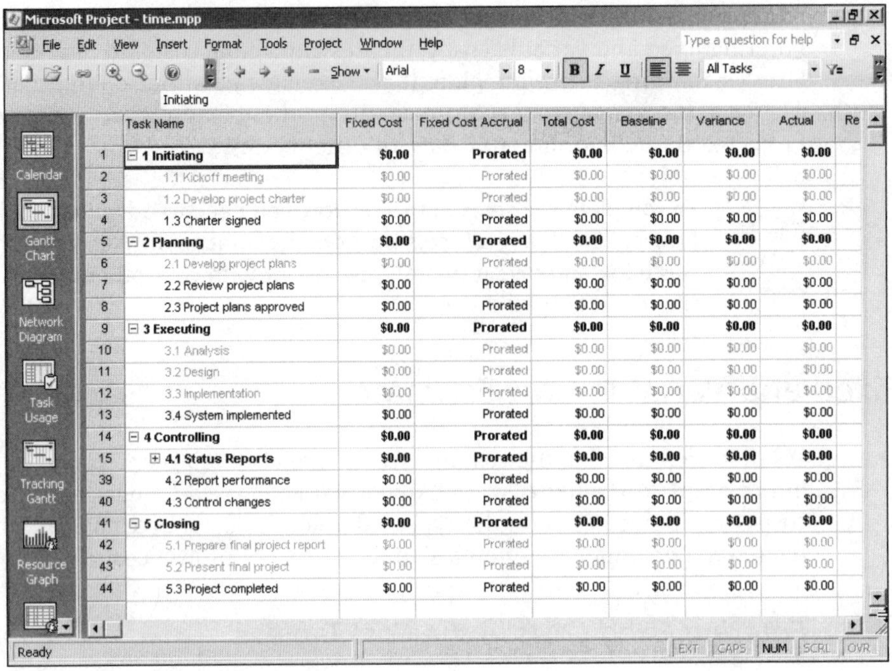

	Task Name	Fixed Cost	Fixed Cost Accrual	Total Cost	Baseline	Variance	Actual	Re
1	1 Initiating	$0.00	Prorated	$0.00	$0.00	$0.00	$0.00	
2	1.1 Kickoff meeting	$0.00	Prorated	$0.00	$0.00	$0.00	$0.00	
3	1.2 Develop project charter	$0.00	Prorated	$0.00	$0.00	$0.00	$0.00	
4	1.3 Charter signed	$0.00	Prorated	$0.00	$0.00	$0.00	$0.00	
5	2 Planning	$0.00	Prorated	$0.00	$0.00	$0.00	$0.00	
6	2.1 Develop project plans	$0.00	Prorated	$0.00	$0.00	$0.00	$0.00	
7	2.2 Review project plans	$0.00	Prorated	$0.00	$0.00	$0.00	$0.00	
8	2.3 Project plans approved	$0.00	Prorated	$0.00	$0.00	$0.00	$0.00	
9	3 Executing	$0.00	Prorated	$0.00	$0.00	$0.00	$0.00	
10	3.1 Analysis	$0.00	Prorated	$0.00	$0.00	$0.00	$0.00	
11	3.2 Design	$0.00	Prorated	$0.00	$0.00	$0.00	$0.00	
12	3.3 Implementation	$0.00	Prorated	$0.00	$0.00	$0.00	$0.00	
13	3.4 System implemented	$0.00	Prorated	$0.00	$0.00	$0.00	$0.00	
14	4 Controlling	$0.00	Prorated	$0.00	$0.00	$0.00	$0.00	
15	4.1 Status Reports	$0.00	Prorated	$0.00	$0.00	$0.00	$0.00	
39	4.2 Report performance	$0.00	Prorated	$0.00	$0.00	$0.00	$0.00	
40	4.3 Control changes	$0.00	Prorated	$0.00	$0.00	$0.00	$0.00	
41	5 Closing	$0.00	Prorated	$0.00	$0.00	$0.00	$0.00	
42	5.1 Prepare final project report	$0.00	Prorated	$0.00	$0.00	$0.00	$0.00	
43	5.2 Present final project	$0.00	Prorated	$0.00	$0.00	$0.00	$0.00	
44	5.3 Project completed	$0.00	Prorated	$0.00	$0.00	$0.00	$0.00	

Figure A-46. Cost Table View

Entering Human Resource Costs

Human resources represent a major part of the costs on many projects. By defining and then assigning human resources and their related costs to tasks in Project 2003, you can calculate human resource costs, track how people are used, identify potential resource shortages that could force you to miss deadlines, and identify underutilized resources. It is often possible to shorten a project's schedule by reassigning underutilized resources. This section focuses on entering human resource costs and assigning resources to tasks. The following section describes other features of Project 2003 related to human resource management.

There are several methods for entering resource information in Project 2003. One of the easiest is to enter basic resource information in the Resource Sheet. Access the Resource Sheet from the View Bar, or by selecting Resource Sheet from the View menu. The Resource Sheet allows you to enter the resource name, initials, resource group, maximum units, standard rate, overtime rate, cost/use, accrual method, base calendar, and code. Entering data into the Resource Sheet is similar to entering data into an Excel spreadsheet, and you can easily sort items by selecting Sort from the Project menu. In addition, you can use the Filter list on the Formatting toolbar to filter resources. Once you have established resources in the Resource Sheet, you can assign those resources to tasks in the Entry table with the list arrow that appears when you click a cell in the Resource Names column. The Resource Names column is the last column of the Entry table. You can also use other methods for assigning resources, as described below.

Next, you will use the Project 2003 file time.mpp, which you saved in the preceding section, to assign resources to tasks. (If you did not save your file, download it from the companion Web site for this text.) Assume that there are four people working on the Project Tracking database project. Assume that the only costs for this project are for these human resources: Kathy is the project manager; John is the business analyst; Mary is the database analyst; and Chris is an intern or trainee.

To enter basic information about each person into the Resource Sheet:

1. *Display the Resource Sheet view.* Open your Project 2003 file, time.mpp, if necessary. Click the **Resource Sheet** button 🗔 on the View Bar (you may need to scroll to see it), or click **View** on the menu bar, and then click **Resource Sheet**.

2. *Enter resource information.* Enter the information from Table A-4 into the Resource Sheet. Type the information as shown and press the **Tab** key to move to the next field. Notice that you are putting abbreviated job titles in the Initials column: PM stands for project manager; BA stands for business analyst; DA stands for database analyst; and IN stands for intern. When you type the standard and overtime rates, you can just type 50, and Project 2003 will automatically enter $50.00/hr. The standard and overtime rates entered are based on hourly rates. You can also enter annual salaries by typing the annual salary number followed by /y for "per year." Leave the default values for the other columns in the Resource Sheet as they are. Your screen should resemble Figure A-47 when you are finished entering the resource data.

Table A-4: Project Tracking Database Resource Data

RESOURCE NAME	INITIALS	GROUP	STAND. RATE	OVT. RATE
Kathy	PM	1	$50.00/h	$60.00/h
John	BA	1	$40.00/h	$50.00/h
Mary	DA	1	$40.00/h	$50.00/h
Chris	IN	1	$20.00/h	$25.00/h

		ⓘ	Resource Name	Type	Material Label	Initials	Group	Max. Units	Std. Rate	Ovt. Rate	Cos
	1		Kathy	Work		PM	1	100%	$50.00/hr	$60.00/hr	
	2		John	Work		BA	1	100%	$40.00/hr	$50.00/hr	
	3		Mary	Work		DA	1	100%	$40.00/hr	$50.00/hr	
	4		Chris	Work		IN	1	100%	$20.00/hr	$25.00/hr	

Figure A-47. Resource Sheet View with Resource Data Entered

> **Tip**
>
> If you know that some people will be available for a project only part-time, enter their percentage of availability in the Max Units column of the Resource Sheet. Project 2003 will then automatically assign those people based on their maximum units. For example, if someone can work only 25 percent of their time on a project throughout most of the project, enter 25% in the Max Units column for that person. When you enter that person as a resource for a task, his or her default number of hours will be 25 percent of a standard eight-hour workday, or two hours per day.

Adjusting Resource Costs

To make a resource cost adjustment, such as a raise, for a particular resource, double-click the person's name in the Resource Name column, select the Costs tab in the Resource Information dialog box, and then enter the effective date and raise percentage. You can also adjust other resource cost information, such as standard and overtime rates.

To give the project manager a 10 percent raise starting 8/1/07:

1. *Open the Resource Information dialog box.* In Resource Sheet view, double-click **Kathy** in the Resource Name column. The Resource Information dialog box opens.

2. *Enter an effective date for a raise.* Select the **Costs** tab, and then select **tab A**, if needed. Type **8/1/07** in the second cell in the Effective Date column and press **Enter**. Alternately, click the **list arrow** in the second cell and use the calendar that appears to enter the effective date.

3. *Enter the raise percentage.* Type **10%** in the second cell for the Standard Rate column, and then press **Enter**. The Resource Information screen should resemble Figure A-48. Notice that Project 2003 calculated the 10% raise to be $55.00/h. Click **OK**.

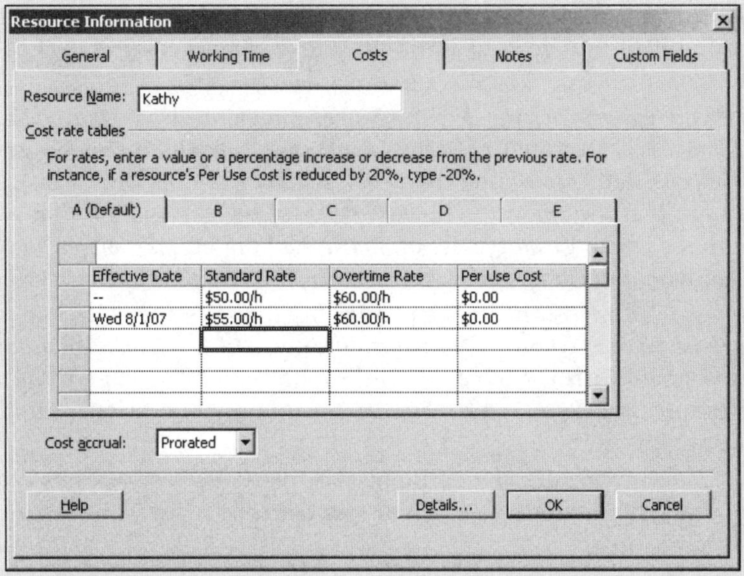

Figure A-48: Adjusting Resource Costs

Assigning Resources to Tasks

For Project 2003 to calculate resource costs, you must assign the appropriate resources to tasks in your WBS. There are several methods for assigning resources. The Resources column in the Entry table allows you to select a resource using a list. You can, however, use this method to assign only one resource to a task full-time. You often need to use other methods to assign resources, such as the Assign Resources button 🗐 on the Resource Management toolbar or the Resource Cost window, to ensure that Project 2003 captures resource assignments as you intend them. Next, you will use these three methods for assigning resources to the Project Tracking Database project.

Assigning Resources Using the Entry Table

To assign resources using the Entry table:

1. *Select the task to which you want to assign resources.* Click the **Gantt Chart** button on the View Bar. Click **View** on the menu bar, and then click **Table: Entry**, if necessary, to return to the Entry table.

2. *Reveal the Resource Names column of the Entry table.* Move the split bar to the right to reveal the entire Resource Names column in the Entry table.

3. *Select a resource from the Resource Names column.* In the Resource Names column, click the cell associated with Task 2, **Kickoff meeting.** Click the cell's **list arrow**, and then click **Kathy** to assign her to Task 2. Notice that the resource choices are based on information that you entered in the Resource Sheet. If you had not entered any resources, you would not have a list arrow or any choices to select.

4. *Try to select another resource for Task 2.* Again, click the cell's **list arrow** in the Resource Names column for Task 2, click **John**, and then press **Enter.** Notice that only John's name appears in the Resource Names column. You can assign only one resource to a task using this method.

5. *Clear the resource assignment.* Right-click the **Resource Names** column for Task 2, and then select **Clear Contents** to remove the resource assignments.

Assigning Resources Using the Toolbar

To assign resources using the Resource Management toolbar:

1. *Select the task to which you want to assign resources.* Click the task name for Task 2, **Kickoff meeting,** in the second row of the Task Name column.

2. *Open the Assign Resources dialog box.* Click the **Assign Resources** button on the Resource Management toolbar. (If the Assign Resources button is not visible, click the **Toolbar Options** button on the Resource Management toolbar, and then click the **Assign Resources** button.) The Assign Resources dialog box, which lists the names of the people assigned to the project, is displayed, as shown in Figure A-49. This dialog box remains open while you move from task to task to assign resources. Click the **expand symbol** to the left of **Resource list options**.

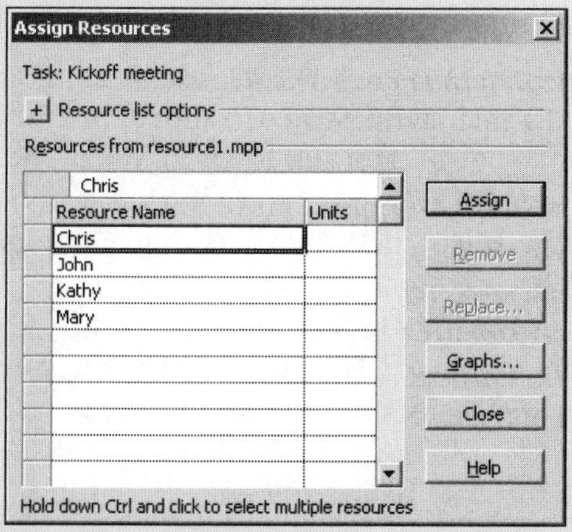

Figure A-49. Assign Resources Dialog Box

3. *Assign Kathy to Task 2.* Click **Kathy** in the Resource Name column of the Assign Resources dialog box, and then click **Assign**. Notice that the duration estimate for Task 2 remains at 2 hours, and Kathy's name appears on the Gantt chart by the bar for Task 2.

4. *Assign John to Task 2.* Click **John** in the Resource Name column of the Assign Resources dialog box, and then click **Assign**. The duration for Task 2 changes to 1 hour, *but you do not want this change to occur.* Click **Close** in the Assign Resources dialog box.

5. *Clear the resource assignments.* Right-click the **Resource Names** column for Task 2, and then select **Clear Contents** to remove the resource assignments.

6. *Reenter the duration for Task 2.* Type **2h** in the Duration column for Task 2 and press **Enter**.

Assigning Resources Using the Split Window and Resource Cost View

Even though using the Assign Resources button seems simpler, it is often better to use a split window and the Resource Cost view when assigning resources. When you assign resources using the split view, the durations of tasks will not automatically change when you assign more resources, and you have more control over how you enter information. Project 2003 Help offers more information on different options for resource assignment.

To assign both Kathy and John to attend the two-hour kickoff meeting:

1. *Split the window to reveal more information.* Click **Window** on the menu bar, and then click **Split**. The Gantt Chart view is displayed at the top of the screen and a resource information table is displayed at the bottom of the screen.

2. *Open the Resource Cost window.* Right-click anywhere in the bottom window and select **Resource Cost**.

3. *Assign Kathy to Task 2.* Select Task 2, **Kickoff meeting** in the top window, and then click the first cell in the **Resource Name** column in the Resource Cost window. Click the cell's **list arrow** and select **Kathy**. Press **Enter**.

4. *Assign John to Task 2.* Click the cell's **list arrow** and select **John**. Press **Enter**.

5. *Enter the resource assignment and review the Gantt chart.* Click **OK** in the Resource Cost window. Your screen should resemble Figure A-50. Notice that the duration for Task 2 is still two hours, and both Kathy and John are assigned to that task 100%. You can also enter a different percentage under the Units column if resources will not work 100 percent of their available time on a task. Note that the default settings are that tasks are effort driven and the Task type is Fixed Units. You will learn more about these settings later.

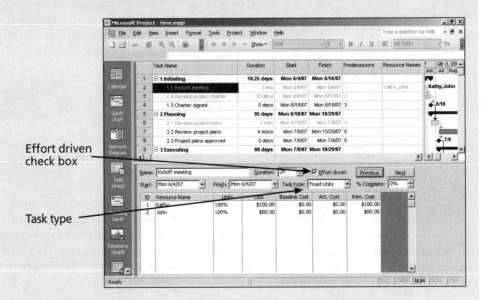

Figure A-50. Split Screen View for Entering Resource Information

6. *Open the Resource Schedule window.* Right-click anywhere in the lower window and select **Resource Schedule**. Notice that Kathy and John are assigned to Task 2 for two hours, as you intended. You can also enter resource information using the Resource Schedule option.

7. *Close the file and do not save it.* Close the file, but do not save the changes you made. Other resource information has been entered for you in the Project 2003 file named resource.mpp, available on the companion Web site for this text.

A copy of the Project 2003 file resource.mpp is available on the companion Web site for this text or from your instructor. You must use this file to continue the steps in the next section.

As you can see, you must be careful when assigning resources. Project 2003 assumes that the durations of tasks are not fixed, but are effort-driven, and this assumption can create some problems when you are assigning resources. Figure A-51 shows the results of adding resources with different settings. The first four task entries assume the task, Design, is effort-driven. When you assign one resource to the Design task, John, the duration is one month. If you later assign two resources, John and Mary, to that same task without changing any settings, Project 2003 automatically adjusts the duration of that task to half its original duration, or 0.5 months. If you change the task type to be Fixed Duration and assign two resources, the duration remains at one month and each resource is assigned 50 percent to the task. If you change the type to Fixed Work and assign two resources, the duration is cut in half and the resources are assigned full-time. The bottom part of Figure A-51 shows the results when a task is not effort-driven or when you assign the Fixed Work task type. You must be very careful when entering resource information to make sure the task duration and work hours reflect what you want.

Fixed Work forces the
Effort Driven field to "yes"

Note the effects of type
on resource assignments

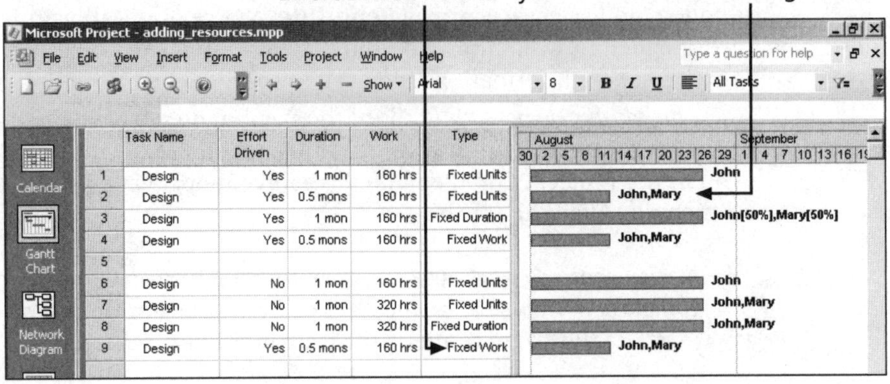

Figure A-51. Results of Adding Resources With Various Task Types

Viewing Project Cost Information

Once you enter resource information, Project 2003 automatically calculates resource costs for the project. There are several ways to view project cost information. You can view the Cost table to see cost information, or you can run various cost reports. Next, you will view cost information for the Project Tracking Database project.

To view cost information:

1. *Open resource.mpp*. Download **resource.mpp** from the companion Web site for this text or get it from your instructor. Open the file.

2. *Open the Cost table*. Right-click the **Select All** button and click **Cost**, or click **View** on the menu bar, point to **Table**: **Entry**, and then click **Cost**. The Cost table displays cost information and the Gantt chart. Move the split bar to the right to reveal all of the columns. Your screen should resemble Figure A-52. Note that by assigning resources, costs have been automatically calculated for tasks. You could also enter fixed costs for tasks in the cost table, but there are no fixed costs associated with this project.

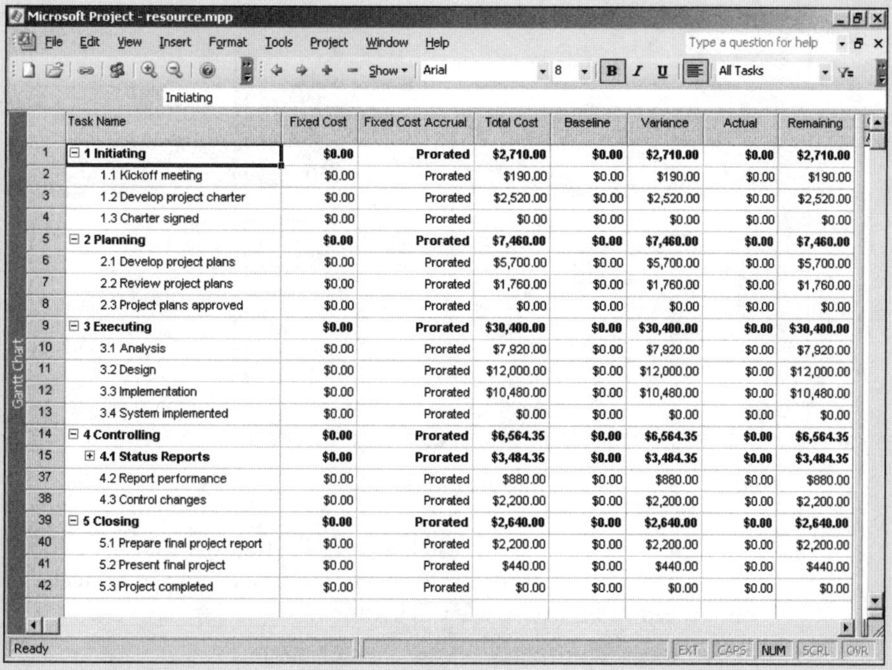

	Task Name	Fixed Cost	Fixed Cost Accrual	Total Cost	Baseline	Variance	Actual	Remaining
1	**1 Initiating**	**$0.00**	**Prorated**	**$2,710.00**	**$0.00**	**$2,710.00**	**$0.00**	**$2,710.00**
2	1.1 Kickoff meeting	$0.00	Prorated	$190.00	$0.00	$190.00	$0.00	$190.00
3	1.2 Develop project charter	$0.00	Prorated	$2,520.00	$0.00	$2,520.00	$0.00	$2,520.00
4	1.3 Charter signed	$0.00	Prorated	$0.00	$0.00	$0.00	$0.00	$0.00
5	**2 Planning**	**$0.00**	**Prorated**	**$7,460.00**	**$0.00**	**$7,460.00**	**$0.00**	**$7,460.00**
6	2.1 Develop project plans	$0.00	Prorated	$5,700.00	$0.00	$5,700.00	$0.00	$5,700.00
7	2.2 Review project plans	$0.00	Prorated	$1,760.00	$0.00	$1,760.00	$0.00	$1,760.00
8	2.3 Project plans approved	$0.00	Prorated	$0.00	$0.00	$0.00	$0.00	$0.00
9	**3 Executing**	**$0.00**	**Prorated**	**$30,400.00**	**$0.00**	**$30,400.00**	**$0.00**	**$30,400.00**
10	3.1 Analysis	$0.00	Prorated	$7,920.00	$0.00	$7,920.00	$0.00	$7,920.00
11	3.2 Design	$0.00	Prorated	$12,000.00	$0.00	$12,000.00	$0.00	$12,000.00
12	3.3 Implementation	$0.00	Prorated	$10,480.00	$0.00	$10,480.00	$0.00	$10,480.00
13	3.4 System implemented	$0.00	Prorated	$0.00	$0.00	$0.00	$0.00	$0.00
14	**4 Controlling**	**$0.00**	**Prorated**	**$6,564.35**	**$0.00**	**$6,564.35**	**$0.00**	**$6,564.35**
15	**4.1 Status Reports**	**$0.00**	**Prorated**	**$3,484.35**	**$0.00**	**$3,484.35**	**$0.00**	**$3,484.35**
37	4.2 Report performance	$0.00	Prorated	$880.00	$0.00	$880.00	$0.00	$880.00
38	4.3 Control changes	$0.00	Prorated	$2,200.00	$0.00	$2,200.00	$0.00	$2,200.00
39	**5 Closing**	**$0.00**	**Prorated**	**$2,640.00**	**$0.00**	**$2,640.00**	**$0.00**	**$2,640.00**
40	5.1 Prepare final project report	$0.00	Prorated	$2,200.00	$0.00	$2,200.00	$0.00	$2,200.00
41	5.2 Present final project	$0.00	Prorated	$440.00	$0.00	$440.00	$0.00	$440.00
42	5.3 Project completed	$0.00	Prorated	$0.00	$0.00	$0.00	$0.00	$0.00

Figure A-52. Cost Table for the Resource File

3. *Open the Cost Reports dialog box.* Click **View** on the menu bar, and then click **Reports**. Double-click **Costs** to display the Cost Reports dialog box.
4. *Set the time units for the report.* Click **Cash Flow**, if necessary, and then click **Edit**. The Crosstab Report dialog box is displayed. Click the **Column** list arrow and select **Months**, as shown in Figure A-53. Click **OK**.

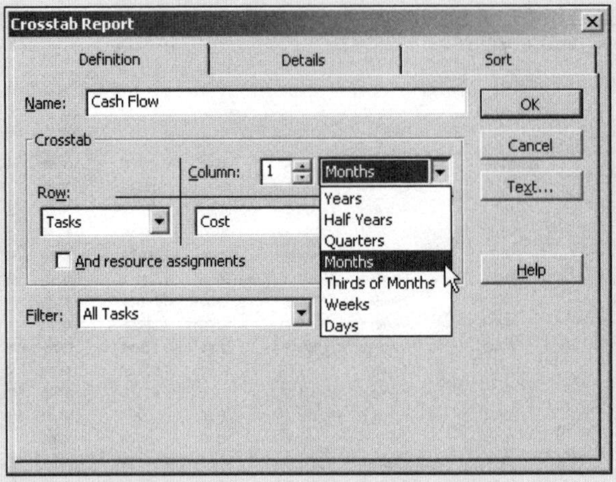

Figure A-53. Crosstab Report Dialog Box

5. *View the Cash Flow report.* Click **Select** in the Cost Reports dialog box. A Cash Flow report for the Project Tracking Database project appears, as shown in Figure A-54. Click **Close**.

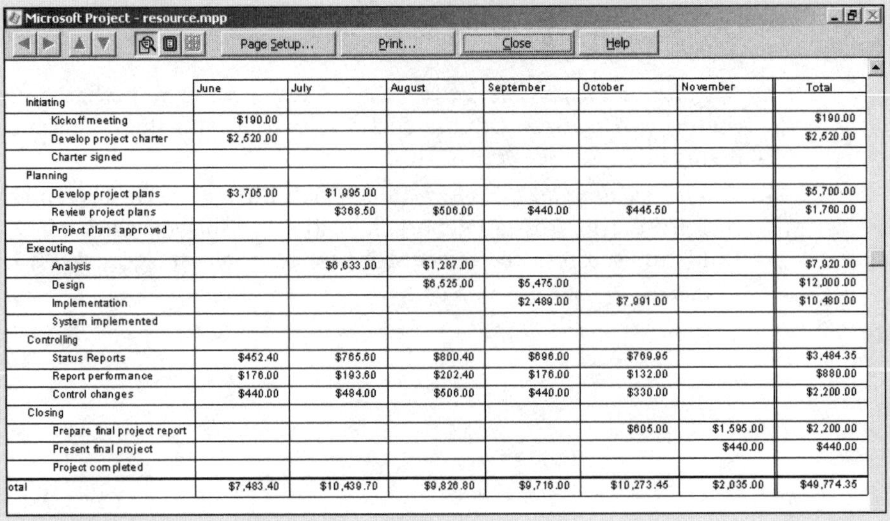

	June	July	August	September	October	November	Total
Initiating							
Kickoff meeting	$190.00						$190.00
Develop project charter	$2,520.00						$2,520.00
Charter signed							
Planning							
Develop project plans	$3,705.00	$1,995.00					$5,700.00
Review project plans		$368.50	$506.00	$440.00	$445.50		$1,760.00
Project plans approved							
Executing							
Analysis		$6,633.00	$1,287.00				$7,920.00
Design			$6,525.00	$5,475.00			$12,000.00
Implementation				$2,489.00	$7,991.00		$10,480.00
System implemented							
Controlling							
Status Reports	$452.40	$765.60	$800.40	$696.00	$769.95		$3,484.35
Report performance	$176.00	$193.60	$202.40	$176.00	$132.00		$880.00
Control changes	$440.00	$484.00	$506.00	$440.00	$330.00		$2,200.00
Closing							
Prepare final project report					$605.00	$1,595.00	$2,200.00
Present final project						$440.00	$440.00
Project completed							
Total	$7,483.40	$10,439.70	$9,826.80	$9,716.00	$10,273.45	$2,035.00	$49,774.35

Figure A-54. Cash Flow Report

6. *View the Project Summary report.* In the Reports dialog box, double-click **Overview**, and then double-click **Project Summary**. A Project Summary report for the Project Tracking Database project is displayed, listing information such as project baseline start and finish dates; actual start and finish dates; summaries of duration, work hours, and costs; as well as variance information. Since you have not yet saved the file as a baseline, much of this information is blank in the report. The Project Summary report provides a high-level overview of a project. Click **Close** to close the report after reviewing it. Click **Close** on the Reports dialog box. Keep this file open for the next set of steps.

 You can edit many of the report formats in Project 2003. Instead of double-clicking the report, highlight the desired report, and then click **Edit**.

The total estimated cost for this project, based on the information entered, should be $49,774.35, as shown in the Cash Flow and Project Summary reports. In the next section, you will save this data as a baseline plan and enter actual information.

Baseline Plan, Actual Costs, and Actual Times

Once you complete the initial process of creating a plan—entering tasks, establishing dependencies, assigning costs, and so on—you are ready to set a baseline plan. By comparing the information in your baseline plan to an updated plan during the course of the project, you can identify and solve problems. After the project ends, you can use the baseline and actual information to plan similar, future projects more accurately. To use Project 2003 to help control projects, you must establish a baseline plan, enter actual costs, and enter actual durations.

Establishing a Baseline Plan

An important part of project management is setting a baseline plan. If you plan to compare actual information such as durations and costs, you must first save the Project 2003 file as a baseline. The procedure involves selecting Tracking, and then Save Baseline from the Tools menu. Before setting a baseline, you must complete the baseline plan by entering time, cost, and human resources information. Be careful not to save a baseline until you have completed the baseline plan. If you do save a baseline before completing the baseline plan, Project 2003 allows you to save up to ten baselines. You can then clear unwanted baseline plans.

Even though you can clear a baseline plan or save multiple baselines, it is a good idea to save a separate, backup file of the original baseline for your project. Enter actuals and save that information in the main file, but always keep a backup baseline file without actuals. Keeping separate baseline and actual files will allow you to return to the original file in case you ever need to use it again.

To rename resource.mpp and then save it as a baseline plan in Project 2003:

1. *Save resource.mpp as a new file named baseline.mpp.* Click **File** on the menu bar, and then click **Save As**. Type **baseline** as the filename, and then click **Save**.

2. *Open the Save Baseline dialog box.* Click **Tools** on the menu bar, select **Tracking**, and then click **Save Baseline**. The Save Baseline dialog box is displayed. Click the **Baseline** list arrow to reveal multiple baselines, as shown in Figure A-55.

Save Baseline	✕
⦿ Save baseline	
	Baseline ▼
○ Save inte	Baseline
Copy:	Baseline 1
Into:	Baseline 2
	Baseline 3
For:	Baseline 4
⦿ Entire	Baseline 5
○ Select	Baseline 6
Roll u	Baseline 7
☐ To	Baseline 8
☐ En	Baseline 9
	Baseline 10
	Set as Default
Help	OK Cancel

Figure A-55. Save Baseline Dialog Box Showing Ability to Save Multiple Baselines

3. *Review the Save Baseline options, and save the entire project as a baseline.* If necessary, click the **Save baseline** radio button and the **Entire project** radio button. Use the default option of the first Save baseline name of just Baseline. These options should be the default settings in Project 2003. Click **OK**.

Notice that there are several options in the Save Baseline dialog box, as shown in Figure A-55. You can save the file as an interim plan if you anticipate several versions of the plan. You can also select the entire project or selected tasks when saving a baseline or interim plan. As mentioned above, you can save up to ten baselines with Project 2003.

Entering Actual Costs and Times

After you set the baseline plan, it is possible to track information on each of the tasks as the project progresses. You can also adjust planned task information for tasks still in the future. The Tracking table displays tracking information, and the Tracking toolbar helps you enter this information. Figure A-56 describes each button on the Tracking toolbar. The 100% button and the Update as Scheduled button are the most commonly used buttons for entering actual information. Note that the functions of these buttons have not changed from Project 2002 except that the Collaborate Toolbar button is no longer available.

Button	Name	Description
	Project Statistics	Provides summary information about the project baseline, the actual project start and finish dates, and overall project duration, costs, and work
	Update as Scheduled	Updates the selected tasks to indicate that actual dates, costs, and work match the scheduled dates, costs, and work
	Reschedule Work	Schedules the remaining duration for a task that is behind schedule so that it will continue from the status date
	Add Progress Line	Displays a progress line on the Gantt chart from a date that you select on the time scale
	0% Complete	Marks the selected tasks as 0% complete as of the status date (Actual date, work, and duration data is updated)
	25% Complete	Marks the selected tasks as 25% complete as of the status date (Actual date, work, and duration data is updated)
	50% Complete	Marks the selected tasks as 50% complete as of the status date (Actual date, work, and duration data is updated)
	75% Complete	Marks the selected tasks as 75% complete as of the status date (Actual date, work, and duration data is updated)
	100% Complete	Marks the selected tasks as 100% complete as of the status date (Actual date, work, and duration data is updated)
	Update Tasks	Displays the Update Tasks dialog box for the selected tasks so that you can enter their percentages completed, actual durations, remaining durations, or actual start or finish dates

Bunin, Rachel, (2004) *New Perspectives on Microsoft Office Project 2003, Introductory*, Course Technology (253).

Figure A-56. Buttons on the Tracking Toolbar

To practice entering actual information, enter just a few changes to the baseline. Assume that Tasks 1 through 8 were completed as planned, but that Task 10 took longer than planned.

To enter actual information for tasks that were completed as planned:

1. *Display the Tracking toolbar.* Click **View** on the menu bar, select **Toolbars**, and then click **Tracking**. Move the toolbar as desired.

2. *Display the Tracking table.* Right-click the **Select All** button, and then click **Tracking** to see more information as you enter actual data. Widen the Task Name column to see all of the text, and then move the split bar to reveal all the columns in the Tracking table.

3. *Mark Tasks 1 though 8 as 100% complete.* Click the Task Name for Task 1, **Initiating**, and drag down through Task 8 to highlight the first eight tasks. Click the **100% Complete** button ![100%] on the Tracking toolbar. The columns with dates, durations, and cost information should now contain data instead of the default values, such as NA or 0. The % Comp column should display 100%. Adjust column widths if needed. Your screen should resemble Figure A-57.

Tracking toolbar →

	Task Name	Act. Start	Act. Finish	% Comp.	Phys. % Comp.	Act. Dur.	Rem. Dur.	Act. Cost	Act. Wo
1	⊟ 1 Initiating	Mon 6/4/07	Mon 6/18/07	100%	0%	10.25 days	0 days	$2,710.00	52 h
2	1.1 Kickoff meeting	Mon 6/4/07	Mon 6/4/07	100%	0%	2 hrs	0 hrs	$190.00	4 h
3	1.2 Develop project charter	Mon 6/4/07	Mon 6/18/07	100%	0%	10 days	0 days	$2,520.00	48 h
4	1.3 Charter signed	Mon 6/18/07	Mon 6/18/07	100%	0%	0 days	0 days	$0.00	0 h
5	⊟ 2 Planning	Mon 6/18/07	Mon 10/29/07	100%	0%	95 days	0 days	$7,460.00	164 h
6	2.1 Develop project plans	Mon 6/18/07	Mon 7/9/07	100%	0%	3 wks	0 wks	$5,700.00	132 h
7	2.2 Review project plans	Mon 7/9/07	Mon 10/29/07	100%	0%	4 mons	0 mons	$1,760.00	32 h
8	2.3 Project plans approved	Mon 7/9/07	Mon 7/9/07	100%	0%	0 days	0 days	$0.00	0 h
9	⊟ 3 Executing	NA	NA	0%	0%	0 days	80 days	$0.00	0 h
10	3.1 Analysis	NA	NA	0%	0%	0 mons	1 mon	$0.00	0 h
11	3.2 Design	NA	NA	0%	0%	0 mons	2 mons	$0.00	0 h
12	3.3 Implementation	NA	NA	0%	0%	0 mons	1 mon	$0.00	0 h
13	3.4 System implemented	NA	NA	0%	0%	0 days	0 days	$0.00	0 h
14	⊟ 4 Controlling	NA	NA	0%	0%	0 days	107.13 days	$0.00	0 h
15	⊞ 4.1 Status Reports	NA	NA	0%	0%	0 days	100.13 days	$0.00	0 h
37	4.2 Report performance	NA	NA	0%	0%	0 mons	5 mons	$0.00	0 h
38	4.3 Control changes	NA	NA	0%	0%	0 mons	5 mons	$0.00	0 h
39	⊟ 5 Closing	NA	NA	0%	0%	0 days	15 days	$0.00	0 h
40	5.1 Prepare final project report	NA	NA	0%	0%	0 wks	2 wks	$0.00	0 h
41	5.2 Present final project	NA	NA	0%	0%	0 wks	1 wk	$0.00	0 h

Figure A-57. Tracking Table Information

4. *Enter actual completion dates for Task 10.* Click the Task Name for Task 10, **Analysis**, and then click the **Update Tasks** button ![icon] on the far right side of the Tracking toolbar. The Update Tasks dialog box opens. For Task 10, enter the Actual Start date as **7/11/07** and the Actual Finish date as **8/16/07**, as shown in Figure A-58. Click **OK**.

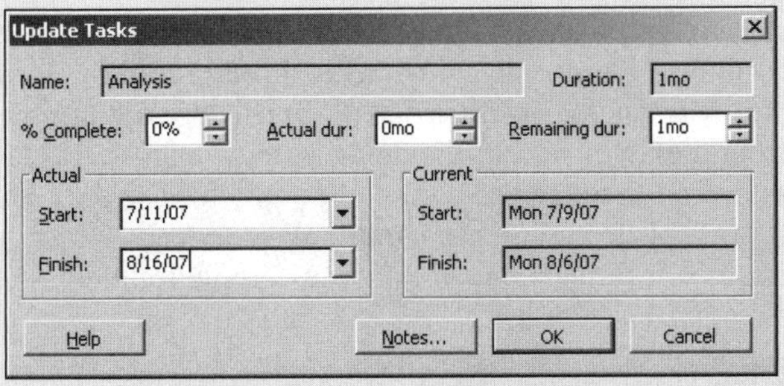

Figure A-58. Update Tasks Dialog Box

5. *Display the Indicators column.* Click **Insert** on the menu bar, and then click **Column**. The Column Definition dialog box opens. Click the **Field name** list arrow, select **Indicators**, and then click **OK**. The Indicators column is displayed, showing a check mark by completed tasks.

6. *Review changes in the Gantt chart.* Move the split bar to the left to reveal more of the Gantt chart. Notice that the Gantt chart bars for completed tasks have changed. Gantt chart bars for completed tasks appear with a black line drawn through the middle.

 You can hide the Indicators column by dragging the right border of the column heading to the left.

You can also view actual and baseline schedule information more clearly with the Tracking Gantt view. See Chapter 6 for descriptions of the symbols on a Tracking Gantt chart.

To display the Tracking Gantt chart:

1. *View the Tracking Gantt chart.* Click **View** on the menu bar, and then click **Tracking Gantt**. Alternatively, you can click the **Tracking Gantt** button on the View Bar, if it is displayed. Use the horizontal scroll bar in the Gantt chart window, if necessary, to see symbols on the Tracking Gantt chart.

2. *Display Gantt chart information in months.* Click **Zoom Out** 🔍 twice to display information in months. Your screen should resemble Figure A-59. (The Tracking toolbar was turned off in this figure to display the entire schedule). Notice that the delay in this one task on the critical path has caused the planned completion date for the entire project to slip. A good project manager would take corrective action to make up for this lost time or renegotiate the completion date.

Notice the analysis task started late and took longer than planned to complete

This one delay could cause the project completion date to slip unless corrective action is taken

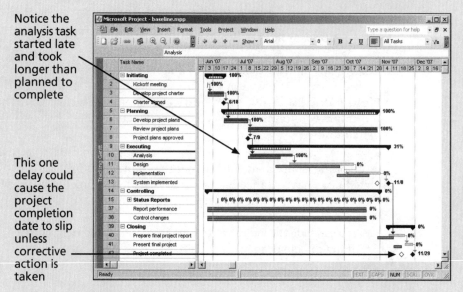

Figure A-59. Tracking Gantt Chart View

3. *Save your file as a new file named mytracking.mpp.* Click **File** on the menu bar, and then click **Save As**. Name the file **mytracking**, and then click **Save**.

Notice the additional information available on the Tracking Gantt chart. Completed tasks have 100% by their symbols on the Tracking Gantt chart. Tasks that have not started yet display 0%. Tasks in progress, such as Task 9, show the percentage of the work completed (31% in this example). Also, note that the project completion date has moved to 11/29 since several tasks depended on the completion of Task 10, which took longer than planned. Recall from Chapter 6 that a white diamond represents a slipped milestone. Viewing the Tracking Gantt chart allows you to easily see your schedule progress against the baseline plan.

After you have entered some actuals, you can review earned value information for the initiating tasks of this project.

Earned Value Management

Earned value management (see Chapter 7, Project Cost Management) is an important project management technique for measuring project performance. Because you have entered actual information for the initiating tasks in the Project Tracking Database project, you can now view earned value information in Project 2003.

To view earned value information:

1. *View the Earned Value table.* Open **tracking.mpp**, which you downloaded from the companion Web site. Click **View** on the menu bar, point to **Table: Entry**, and then select **More Tables**. The More Tables dialog box opens. Double-click **Earned Value**.

2. *Display all the Earned Value table columns.* Move the split bar to the right to reveal all of the columns. Note that the Earned Value table includes columns for each earned value term described in Chapter 7. Also note that the EAC (Estimate at Completion) is higher than the BAC (Budget at Completion) starting with Task 9, where the task took longer than planned to complete (Figure A-60). Task 9 shows a VAC (Variance at Completion) of ($2,772.00), meaning the project is projected to cost $2,772 more than planned at completion at that point in time. Remember that not all of the actual information has been entered yet.

	Task Name	BCWS	BCWP	ACWP	SV	CV	EAC	BAC	VAC
1	⊟ **Initiating**	**$1,387.00**	**$1,387.00**	**$1,387.00**	**$0.00**	**$0.00**	**$2,710.00**	**$2,710.00**	**$0.00**
2	Kickoff meeting	$190.00	$190.00	$190.00	$0.00	$0.00	$190.00	$190.00	$0.00
3	Develop project charter	$1,197.00	$1,197.00	$1,197.00	$0.00	$0.00	$2,520.00	$2,520.00	$0.00
4	Charter signed	$0.00	$0.00	$0.00	$0.00	$0.00	$0.00	$0.00	$0.00
5	⊟ **Planning**	**$0.00**	**$0.00**	**$0.00**	**$0.00**	**$0.00**	**$7,460.00**	**$7,460.00**	**$0.00**
6	Develop project plans	$0.00	$0.00	$0.00	$0.00	$0.00	$5,700.00	$5,700.00	$0.00
7	Review project plans	$0.00	$0.00	$0.00	$0.00	$0.00	$1,760.00	$1,760.00	$0.00
8	Project plans approved	$0.00	$0.00	$0.00	$0.00	$0.00	$0.00	$0.00	$0.00
9	⊟ **Executing**	**$0.00**	**$0.00**	**$0.00**	**$0.00**	**$0.00**	**$33,172.00**	**$30,400.00**	**($2,772.00)**
10	Analysis	$0.00	$0.00	$0.00	$0.00	$0.00	$10,692.00	$7,920.00	($2,772.00)
11	Design	$0.00	$0.00	$0.00	$0.00	$0.00	$12,000.00	$12,000.00	$0.00
12	Implementation	$0.00	$0.00	$0.00	$0.00	$0.00	$10,480.00	$10,480.00	$0.00
13	System implemented	$0.00	$0.00	$0.00	$0.00	$0.00	$0.00	$0.00	$0.00
14	⊟ **Controlling**	**$154.00**	**$0.00**	**$0.00**	**($154.00)**	**$0.00**	**$6,564.35**	**$6,564.35**	**$0.00**
15	⊞ **Status Reports**	**$0.00**	**$0.00**	**$0.00**	**$0.00**	**$0.00**	**$3,484.35**	**$3,484.35**	**$0.00**
37	Report performance	$44.00	$0.00	$0.00	($44.00)	$0.00	$880.00	$880.00	$0.00
38	Control changes	$110.00	$0.00	$0.00	($110.00)	$0.00	$2,200.00	$2,200.00	$0.00
39	⊟ **Closing**	**$0.00**	**$0.00**	**$0.00**	**$0.00**	**$0.00**	**$2,640.00**	**$2,640.00**	**$0.00**
40	Prepare final project report	$0.00	$0.00	$0.00	$0.00	$0.00	$2,200.00	$2,200.00	$0.00
41	Present final project	$0.00	$0.00	$0.00	$0.00	$0.00	$440.00	$440.00	$0.00
42	Project completed	$0.00	$0.00	$0.00	$0.00	$0.00	$0.00	$0.00	$0.00

Figure A-60. Earned Value Table

Project 2003 uses earned value terms from the 1996 PMBOK® Guide. The *PMBOK® Guides 2000* and *2004* use simpler acronyms for earned value terms. See Chapter 7 for explanations of all the earned value terms.

3. *Save and close the file.* Click the **Save** button on the Standard toolbar, and then close the tracking.mpp file.

If you want to check your work or continue to the next section, copies of the files baseline.mpp and tracking.mpp are available on the companion Web site for this text or from your instructor.

Further information on earned value is available from Project 2003 Help or in Chapter 7, Project Cost Management. To create an earned value chart, you must export data from Project 2003 into Excel. There are also add-in products you can use with Project 2003 to do more with earned value information. Consult Project 2003 Help and Microsoft's Web site for Project 2003 (*www.microsoft.com/project*) for more details.

Now that you have entered and analyzed various project cost information, you will examine some of the human resource management features of Project 2003.

PROJECT HUMAN RESOURCE MANAGEMENT

In the project cost management section, you learned how to enter resource information into Project 2003 and how to assign resources to tasks. Two other helpful human resource features include resource calendars and histograms. In addition, it is important to know how to use Project 2003 to assist in resource leveling.

Resource Calendars

When you created the Project Tracking Database file, you used the standard Project 2003 calendar. This calendar assumes that standard working hours are Monday through Friday, from 8:00 a.m. to 5:00 p.m., with an hour for lunch starting at noon. Rather than using this standard calendar, you can create a different calendar that takes into account each project's unique requirements. You can create a new calendar by using the Tasks pane or by changing the working time under the Tools menu.

To create a new base calendar:

1. *Open a new file*. Click the **New** button 🗋 on the Standard toolbar to open a new Project 2003 file. The Tasks pane opens on the left side of your screen.

2. *Use the Tasks pane to create a new calendar*. Click **Define general working times** in the Tasks pane. A calendar wizard appears, as shown in Figure A-61.

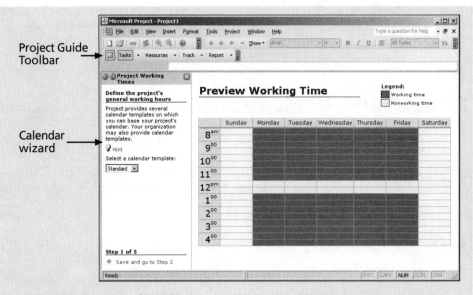

Project Guide Toolbar

Calendar wizard

Figure A-61. Using the Tasks Pane *to Change Working Time*

? **Help**

If the Tasks pane does not appear on your screen, click **View** on the menu bar, select **Toolbars,** and then click **Project Guide** to see the Project Guide toolbar. Click the **Tasks** list arrow, and then click **Define general working times.**

3. *Make changes to the calendar using the wizard.* Click **Save and go to Step 2** at the lower part of the Project Working Times pane, select the check box for **Saturday** to make it a working day, and then click the **I want to adjust the working hours shown for one or more days of the week** radio button. Scroll down to read the instructions for adjusting working hours, and make changes as desired. Click through the remaining steps to close the wizard.

4. *Create a new base calendar.* Click **Tools** on the menu bar, and then click **Change Working Time.** The Change Working Time dialog box opens.

5. *Name the new base calendar.* In the Change Working Time dialog box, click **New**. The Create New Base Calendar dialog box opens. Click the **Create new base calendar** radio button, type **Mine** as the name of the new calendar in the **Name** text box, and then click **OK**. Click **OK** to close the Change Working Time dialog box.

You can use this new calendar for the whole project, or you can assign it to specific resources on the project.

To assign the new calendar to the whole project:

1. *Open the Project Information dialog box.* Click **Project** on the menu bar, and then click **Project Information**. The Project Information dialog box opens.
2. *Select a new calendar.* Click the **Calendar** list arrow to display a list of available calendars. Select your new calendar named **Mine** from this list, and then click **OK**.

To assign a specific calendar to a specific resource:

1. *Assign a new calendar.* Click **View** on the menu bar, and then click **Resource Sheet**. Type **Me** in the Resource Name column, press **Enter**, and then select the word **Me**.
2. *Select the calendar.* Click the **Base Calendar** cell for that resource name. If the Base Calendar column is not visible, click the horizontal scroll bar to view more columns. Click the **Base Calendar** list arrow, and then select **Mine**.
3. *Block off vacation time.* Double-click the resource name **Me** to display the Resource Information dialog box, and then click the **Working Time** tab. You can block off vacation time for people by selecting the appropriate days on the calendar and marking them as nonworking days, as shown in Figure A-62. Close the Resource Information dialog box.
4. *Close the file without saving it.* Click **File** on the menu bar, and then click **Close**. Click **No** when you are prompted to save the file.

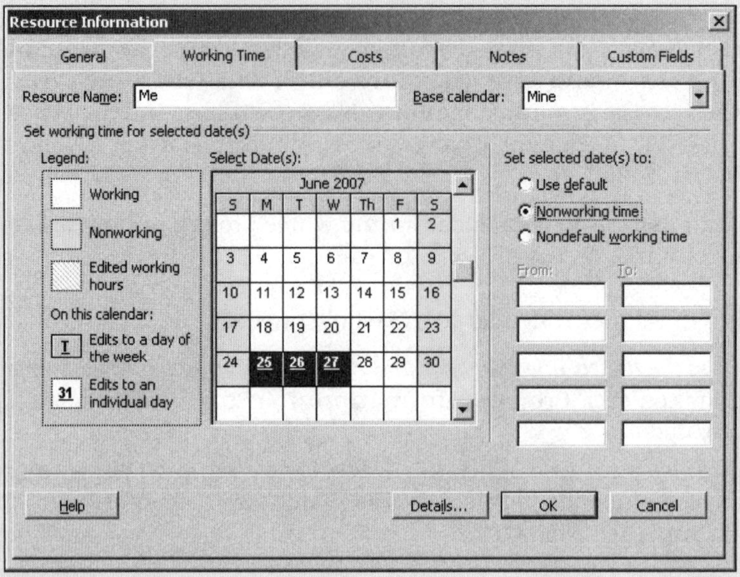

Figure A-62. Blocking Off Vacation Time

Resource Histograms

A resource histogram is a type of chart that shows the number of resources assigned to a project over time (see Chapter 9, Project Human Resource Management). A histogram by individual shows whether a person is over- or underallocated during certain periods. To view histograms in Project 2003, select Resource Graph from the View Bar or select Resource Graph from the View menu. The Resource Graph helps you see which resources are overallocated, by how much, and when. It also shows you the percentage of capacity each resource is scheduled to work, so you can reallocate resources, if necessary, to meet the needs of the project.

To view resource histograms for the Project Tracking Database project:

1. *View the Resource Graph.* Open the **baseline.mpp** file, and open the View Bar, if necessary. Click the **Resource Graph** button 🔳 on the View Bar. If you cannot see the Resource Graph button, you may need to click the up or down arrows on the View Bar. Alternatively, you can click **View** on the menu bar, and then click **Resource Graph**. A histogram for Kathy appears, as shown in Figure A-63. Notice that the screen is divided into two sections. The left pane

displays a person's name and the right pane displays a resource histogram for that person. You may need to click the right scroll bar to get the histogram to appear on your screen. Notice that Kathy is overallocated slightly in the month of June, since the column for that month goes above the 100% line.

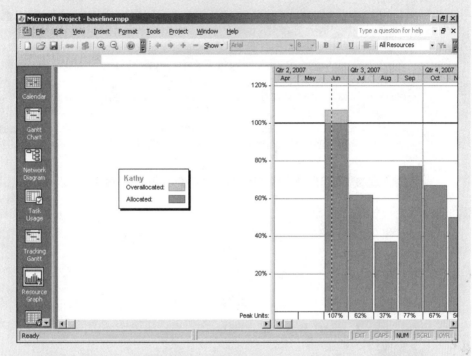

Figure A-63. Resource Histogram View

2. *Adjust the histogram's timescale.* Click **Zoom Out** 🔍 or **Zoom In** 🔍, if necessary, to adjust the timescale for the histogram so that it appears in quarters and then months, as shown in Figure A-63.

3. *View the next resource's histogram.* Click the **right scroll arrow** at the bottom of the resource name pane on the left side of the screen. The resource histogram for the next person appears. View the resource histograms for all four people, and then go back to the first one for Kathy.

Notice that Kathy's histogram has a partially red bar in June 2007. This red portion of the bar means that she has been overallocated during that month. The percentages at the bottom of each bar show the percentage each resource is assigned to work. For example, Kathy is scheduled to work 107% of her available time in June, 62% of her available time in July, and so on. Project 2003 has two tools that enable you to see more details about resource overallocation: the Resource Usage view and the Resource Management toolbar.

To see more details about an overallocated resource using the Resource Usage view:

1. *Display the Resource Usage view.* Click the **Resource Usage** button ⧉ from the View Bar or click **View** on the menu bar, and then click **Resource Usage**.

2. *Adjust the information displayed.* On the right side of the screen, click the **right scroll arrow** to display the hours Kathy is assigned to work each day, starting on June 3rd. You may also need to click the scroll down arrow to see all of Kathy's hours. If you need to adjust the timescale to display weeks, click **Zoom Out** or **Zoom In**. When you are finished, your screen should resemble Figure A-64.

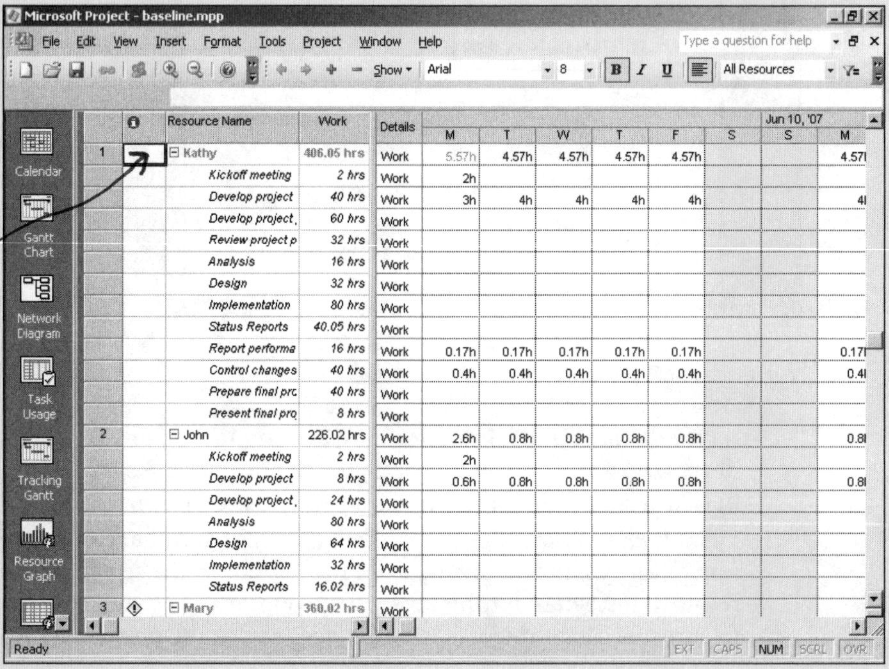

Figure A-64. Resource Usage View

3. *Examine overallocation information.* Notice that Kathy's name appears in red, as does the value 5.57h in the column for Monday. Although Kathy is not scheduled to work more than eight hours on Monday, some of the tasks were entered as hours, and Project 2003 assumes those hours start as soon as possible, thus causing the scheduling conflict. You'll remove the conflict using hour by hour leveling later in this section.

To learn more about an overallocated resource, you can also use the Resource Management toolbar for resource allocation.

To see more resource allocation information:

1. *View the Resource Management toolbar.* Click **View** on the menu bar, select **Toolbars**, and then click **Resource Management**. The Resource Management toolbar displays below the Formatting toolbar.

2. *Select the Resource Allocation view.* Click the **Resource Allocation View** button on the Resource Management toolbar. The Resource Allocation view appears, showing the Resource Usage view at the top of the screen and the Gantt Chart view at the bottom of the screen. Figure A-65 shows this view with Kathy's information highlighted.

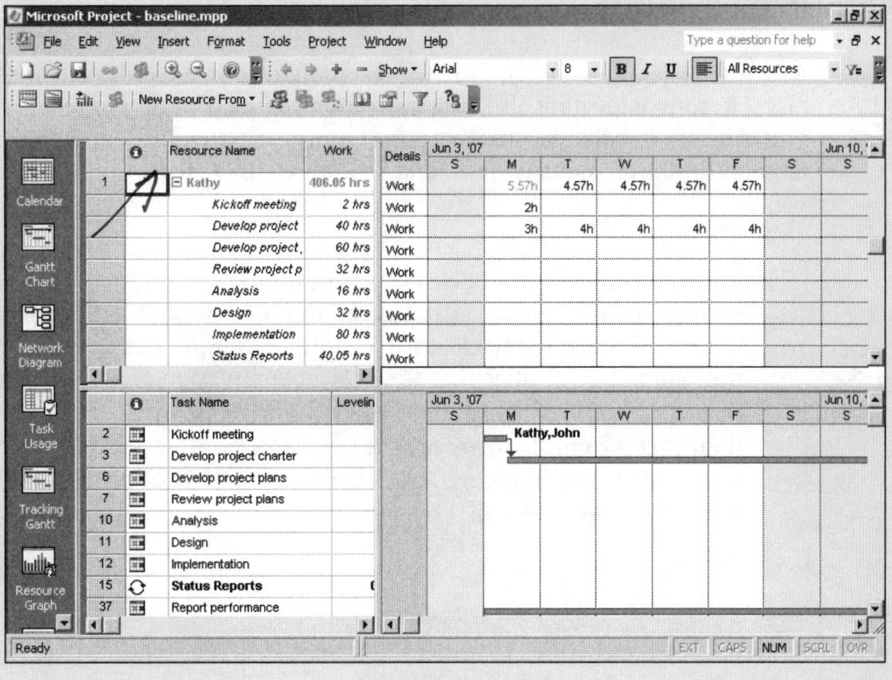

Figure A-65. Resource Allocation View

3. *Close the file, but do not save any changes.* Click **File** on the menu bar, and then click **Close**. Click **No** when you are prompted to save changes.

The Resource Allocation view can help you identify the source of a resource overallocation. It is not obvious that Kathy has an overallocation the first day of the project because she is only assigned to work a total of 5.57 hours. The problem in this case is that she is assigned to several tasks that day that were entered with durations of hours, and Project 2003 assumes each of those tasks starts as soon as possible, thus resulting in the overallocation on an hourly basis. You could ignore this problem or fix it with resource leveling, as shown in the next section. By using the scroll bar, you can view information for other resources. If you scroll down to look at Mary's resource allocation information and scroll to the right to reveal August and September, you can see that overlapping the design and implementation tasks may have caused another overallocation problem. To fix the problem, you can have Mary work overtime, assign another resource to help, or reschedule the implementation task to reduce the overlap. You can also see if resource leveling, as described in the next section, will help solve the problem.

Resource Leveling

Resource leveling is a technique for resolving resource conflicts by delaying tasks. Resource leveling also creates a smoother distribution of resource usage. You can find detailed information on resource leveling in Chapter 9, Project Human Resource Management, and in Project 2003 Help.

To use resource leveling:

1. *Reopen the baseline file.* Click the **Open** button 📂 on the Standard toolbar, navigate to **baseline.mpp**, and open it. Right-click the **Select All** button and click **Entry**. Move the split bar to see more of the Gantt chart. Notice that the project completion date is 11/19/07.

2. *Open the Resource Leveling dialog box.* Click **Tools** on the menu bar, and then click **Level Resources**. The Resource Leveling dialog box opens, as shown in Figure A-66. Uncheck the **Level only within available slack** option, if necessary.

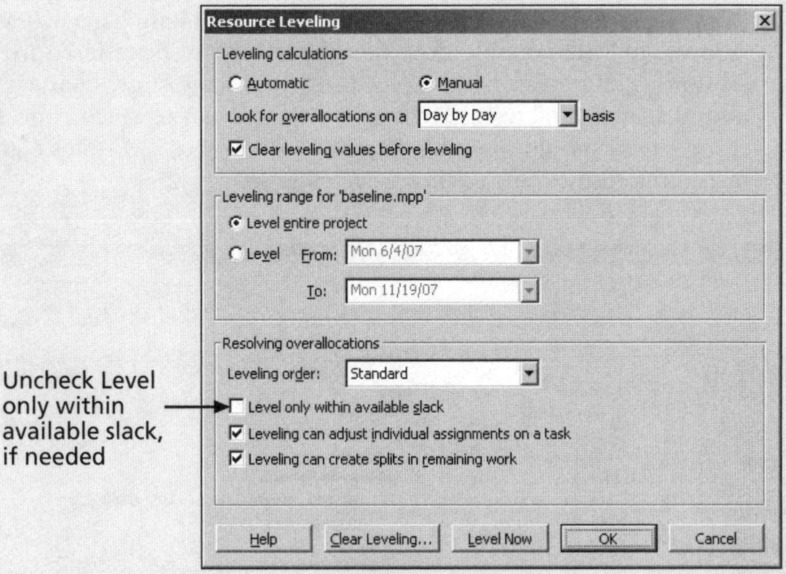

Uncheck Level
only within
available slack,
if needed

Figure A-66. Resource Leveling Dialog Box

3. *Level the file and review the date changes.* Click the **Level Now** button.
   In this example, resource leveling moved the project completion date
   from 11/19 to 11/22, as shown on the Gantt chart milestone symbol
   for Task 42, Project completed.

4. *View Kathy's Resource Graph again.* Click the **Resource Graph** button
   on the View Bar. If necessary, use the scroll bar in the resource
   histogram pane to display June 2007. Notice that Kathy is still overallocated in June 2007.

5. *Level using the Hour by Hour option.* Click **Tools** on the menu bar, and
   then click **Level Resources**. Change the basis for leveling from Day
   by Day to **Hour by Hour**, click the **Level Now** button, keep the
   default option of Entire pool in the Level Now dialog box, and then
   click **OK**. Notice that Kathy's overallocation disappears. Click the
   right scroll arrow in the resource name pane twice to reveal information about Mary. Remember that the left pane displays different
   resources, and the right pane reveals the resource histogram over
   time. Mary's overallocations are also gone, because of this resource
   leveling.

6. *View the Gantt chart again.* Click the **Gantt Chart** button on the
   View Bar and move the split bar to review the Gantt chart, if necessary. The project completion date has been moved back to 11/26
   because of the resource leveling.

7. *View the Leveling Gantt chart.* Click **View** on the menu bar, click **More Views**, and then double-click **Leveling Gantt**. Click the **Zoom Out** button, if necessary, to adjust the timescale to show quarters and months, and scroll to the right in the Gantt chart to reach June 2007. Your screen should resemble Figure A-67. Project 2003 adds a green bar to the Gantt chart to show leveled tasks.

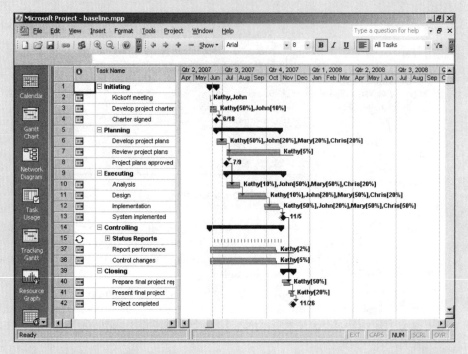

Figure A-67. Leveling Gantt Chart View

8. *Save the file as level.mpp.* Click **File** on the menu bar, click **Save As**, type **level** as the filename, and then click **Save**. Close the file.

 If you want to undo the leveling, immediately click the Undo button on the Standard toolbar. Alternatively, you can return to the Resource Leveling dialog box, and click the Clear Leveling button.

 If you want to check your work, a copy of level.mpp is available on the companion Web site for this text or from your instructor.

Consult the Project 2003 Help feature and use the key word "level" for more information on resource leveling. Also, be careful when setting options for this feature, so that the software adjusts resources only when it should. For example, the end date for the Project Tracking Database project was pushed back because you set the leveling options to allow the dates to slip. In this case, the project manager might prefer to ask her team to work a little overtime to remain on schedule.

Now that you have learned how to change resource calendars, view resource histograms, and level resources, you are ready to learn how to use Project 2003 to assist in project communications management.

PROJECT COMMUNICATIONS MANAGEMENT

Project 2003 can help you generate, collect, disseminate, store, and report project information. There are many different tables, views, and reports to aid in project communications, as you have seen in the previous sections. This section highlights some common reports and views. It also describes how to use templates and insert hyperlinks from Project 2003 into other project documents, how to save Project 2003 files as Web pages, and how to use Project 2003 in a workgroup setting.

Common Reports and Views

To use Project 2003 to enhance project communications, it is important to know when to use the many different ways to collect, view, and display project information. Table A-5 provides a brief summary of Project 2003 features and their functions, which will help you understand when to use which feature. Examples of most of these features are provided as figures in this appendix.

You can see from Table A-5 that many different reports are available in Project 2003. The overview reports provide summary information that top management might want to see, such as a project summary or a report of milestone tasks. Current activities reports help project managers stay abreast of and control project activities. The reports of unstarted tasks and slipping tasks alert project managers to problem areas. Cost reports provide information related to the cash flow for the project, budget information, overbudget items, and earned value management.

Assignment reports help the entire project team by providing different views of who is doing what on a project. You can see who is overallocated by running the Overallocated Resources report or by viewing the two workload reports. You can also create custom reports based on any project information you have entered into Project 2003.

Table A-5: Functions of Project 2003 Features

FEATURE	FUNCTION
Gantt Chart view, Entry table	Enter basic task information.
Network Diagram view	View task dependencies and critical path graphically.
Schedule table	View schedule information in a tabular form.
Cost table	Enter fixed costs or view cost information.
Resource Sheet view	Enter resource information.
Resource Information and Gantt Chart split view	Assign resources to tasks.
Save Baseline	Save project baseline plan.
Tracking toolbar	Enter actual information.
Earned Value table	View earned value information.
Resource Graph	View resource allocation.
Resource Usage	View detailed resource usage.
Resource Management	View resource usage and Gantt chart to find toolbar overallocation problems.
Resource Leveling	Level resources.
Overview Reports	View project summary, top-level tasks, critical tasks, milestones, working days.
Current Activities Reports	View unstarted tasks, tasks starting soon, tasks in progress, completed tasks, should have started tasks, slipping tasks.
Cost Reports	View cash flow, budget, overbudget tasks, overbudget resources, earned value.
Assignment Reports	View who does what, who does what when, to-do list, overallocated resources.
Workload Reports	View task usage, resource usage.
Custom Reports	Allow customization of each type of report.
Save as Web page	Save files as HTML documents.
Insert Hyperlink	Insert hyperlinks to other files or Web sites.

Using Templates and Inserting Hyperlinks and Comments

Chapter 10, Project Communications Management, and several other chapters provide examples of templates that you can use to improve project communications. Because it is often difficult to create good project plans, some organizations,

such as Northwest Airlines, Microsoft, and various government agencies, keep a
repository of sample Project 2003 files. As shown earlier in this appendix, Project
2003 includes several template files, and you can also access several templates
online. You must load the template files when installing Project 2003 or access
templates via the Internet.

To access the Project 2003 templates:

1. *Open the New Project task pane.* Start Project 2003, if necessary, click
   File on the menu bar, and then click **New** to display the New Project
   task pane, as shown in Figure A-68.

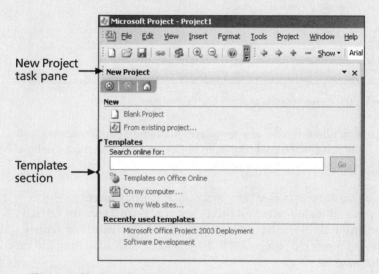

Figure A-68. Opening Templates

2. *Open the Templates dialog box.* Click the **On my computer** link below
   the **Templates** section in the New Project task pane. The Templates
   dialog box is displayed. Click the **Project Templates** tab, as shown
   in Figure A-69. As you can see, there are many templates available.
   You can also access additional templates from the **Templates on
   Office Online** link from the New Project task pane, from other Web
   sites, or from your own organization.

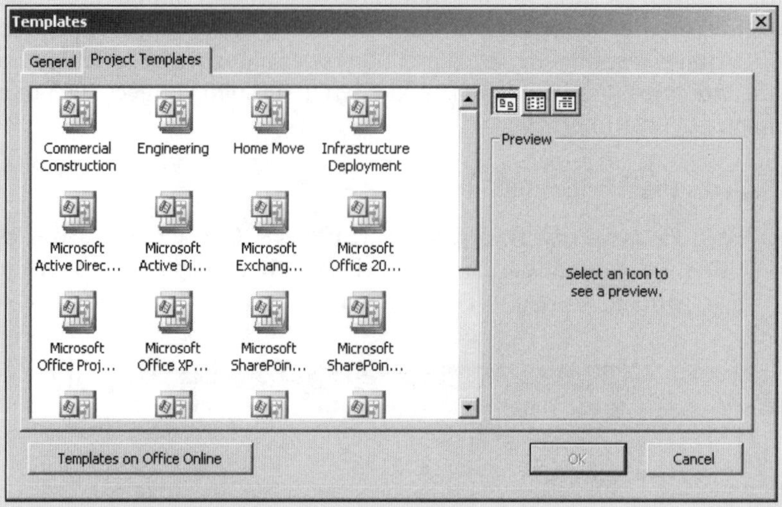

Figure A-69. Project Templates on My Computer

? Help
If you do not have any templates loaded on your computer, you will not see the files displayed in Figure A-69. You must load the template files when installing Project 2003.

3. *Open the Microsoft Office Templates Gallery.* Click **Cancel** in the Templates dialog box, and then click the **Templates on Office Online** link in the New Project task pane. The resulting screen should resemble Figure A-70. Experiment with searching for various template files.

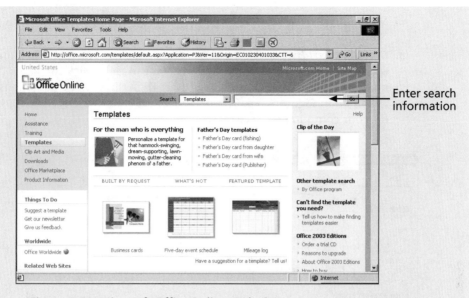

Figure A-70. Microsoft Office Online Web Site

4. *Close your browser and the open Project file.* Click the **Close** button ☒ on your browser and then close the file in Project 2003.

Using templates can help you prepare your project files, but be careful to address the unique needs of your project and organization. For example, even though the Home Move template file can provide guidance in preparing a Project 2003 file for a specific home move project, you need to tailor the file for your particular situation and schedule. You can also create your own Project 2003 templates by saving your file as a template using the Save As option from the File menu. Simply change the Save As Type option to Template. Project 2003 then gives the file an .mpt extension (Microsoft Project Template) to identify the file as a template.

In addition to using templates for Project 2003 files, it is helpful to use templates for other project documents and insert hyperlinks to them from Project 2003. For example, your organization might have templates for meeting agendas, project charters, status reports, and project plans. See Appendix D and the companion Web site for examples of other template files. Next, you will create hyperlinks to some template files created in other applications.

To insert a hyperlink within a Project 2003 file:

1. *Display the Entry table.* Open the **baseline.mpp** file. Right-click the **Select All** button and click **Entry** to return to the Entry table view.

2. *Select the task in which you want to insert a hyperlink.* Click the Task Name for Task 2, **Kickoff meeting**.

3. *Open the Insert Hyperlink dialog box.* Click **Insert** on the menu bar, and then click **Hyperlink**. Alternatively, you can right-click the task name and select **Hyperlink**. The Insert Hyperlink dialog box opens, as shown in Figure A-71. You may have different folders visible based on your computer's directory structure.

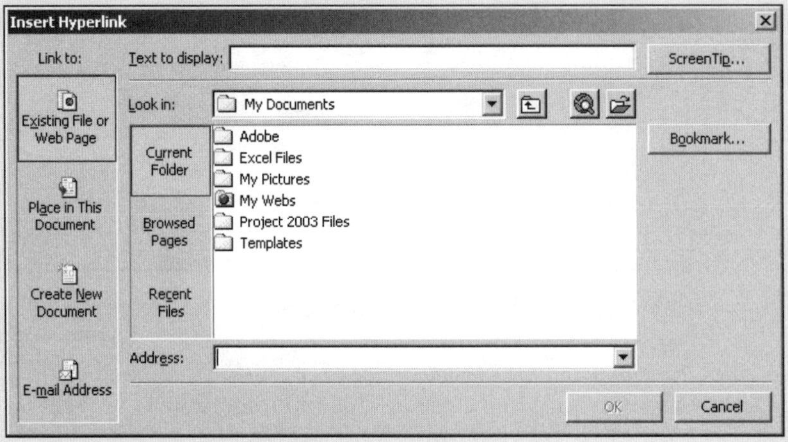

Figure A-71. Insert Hyperlink Dialog Box

4. *Enter the filename of the hyperlink file.* Click the **Browse for File** button to browse for the kickoffmeeting.doc file you downloaded from the companion Web site for this text. Double-click **kickoffmeeting.doc**, and then click **OK**.

5. *Reveal the Indicators column.* Click **Insert** on the menu bar, and then click **Column**. Click the **Field Name** list arrow, select **Indicators**, and then click **OK**. A Hyperlink button appears in the Indicators column to the left of the Task Name for Task 2. Move your mouse over the hyperlink button until the mouse pointer changes to the Hand symbol to reveal the name of the hyperlinked file.

Clicking the Hyperlink button in the Indicators column will automatically open the hyperlinked file. Using hyperlinks is a good way to keep all project documents organized.

It is also a good idea to insert notes or comments into Project 2003 files to provide more information on specific tasks.

To insert a note for Task 3:

1. *Open the Task Information dialog box for Task 3.* Double-click the Task Name for Task 3, **Develop project charter**. Click the **Notes** tab.

2. *Enter your note text.* Type **The charter was developed as a joint effort.** in the Notes text box, as shown in Figure A-72.

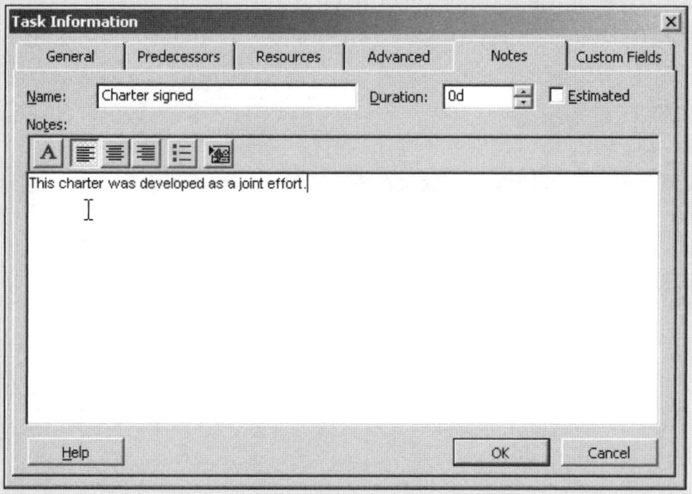

Figure A-72. Task Information Dialog Box Notes Tab

3. *See the resulting Notes icon.* Click **OK** to enter the note. The Notes icon appears in the Indicators column next to Task 3.

4. *Open the note.* Double-click the **Notes** icon in the Indicators column for Task 3 to view the note, and then click **OK**.

5. *Close the file without saving it.* Click **File** on the menu bar, and then click **Close**. Click **No** when you are prompted to save the file.

Saving Files as Web Pages

Many organizations use the Web to disseminate information to employees and to communicate with customers. Project 2003 includes a feature that enables you to save project information as a Web page. As with most Microsoft applications, this feature in Project 2003 is available by selecting Save As Web Page from the File menu. You can create Web pages in several standard formats, including Compare to Baseline, Cost Data By Task, Earned Value Information, Top Level Tasks List, a "Who Does What" Report, and more. You can also design your own formats for information to be saved as Web pages.

To save project information as a Web page:

1. *Open the Save As dialog box for Web pages.* Open **baseline.mpp**. Click **File** on the menu bar, and then click **Save As Web Page**.

2. *Name the Web page file.* Type **taskinfo** in the File name text box as the new filename, select the location where you'll save the file, and then click **Save**. The Export Wizard dialog box appears. Click **Next** to go to the next step of the Export Wizard, click the option to **Use existing map**, and then click **Next**, so your screen resembles Figure A-73.

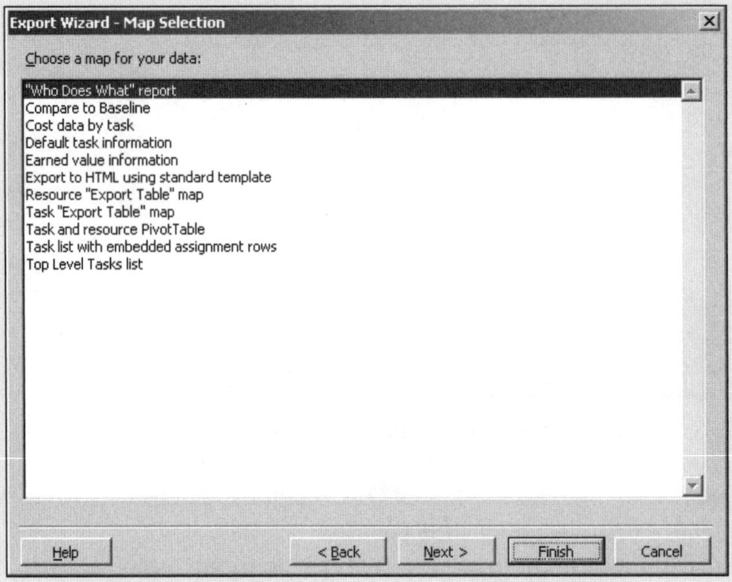

Figure A-73. Export Wizard — Map Selection Dialog Box

3. *Select the import/export map.* Click **Default task information** from the import/export map list, and then click **Finish**. Project 2003 saves an HTML version of task information from baseline.mpp. The name of the new Web page is taskinfo.html.

To see the new Web page, you must open the new HTML file you created.

To open the taskinfo.html file:

1. *Open Windows Explorer.* Click the **Start** button on the taskbar, point to **Programs** (or **All Programs** in Windows XP), point to **Accessories**, and then click **Windows Explorer**. Navigate your folders to find the taskinfo.html file. You can also use other methods to open Windows

Explorer or use My Computer to find the desired file. Alternatively, start your Web browser, and use the Open Page feature from the File menu to open the file.

2. *Open the taskinfo.html file in your Web browser.* Double-click **taskinfo.html** to open it in your Web browser. The file extension for the file may not appear, depending on your settings. Figure A-74 shows part of the taskinfo.html file in Microsoft Internet Explorer.

| | Microsoft Office Project Exported Information - Microsoft Internet Explorer | _ |8| x| |

File Edit View Favorites Tools Help

Back → ⊗ ⊡ ⌂ Search Favorites History ⊟ - ⬚ ⬚ - ⬚ ⊗

Address C:\Kathy\4th edition\Project2003_files\taskinfo.html   ▼ Go   Links »

Project Tracking Database

Project Start Date: Mon 6/4/07
Project Finish Date: Mon 11/19/07

Task_Data

ID	Task_Name	Duration	Start_Date	Finish_Date	Predecessors	Resource_Names
1	**Initiating**	10.25 days	Mon 6/4/07	Mon 6/18/07		
2	Kickoff meeting	2 hrs	Mon 6/4/07	Mon 6/4/07		Kathy,John
3	Develop project charter	10 days	Mon 6/4/07	Mon 6/18/07	2	Kathy[50%],John[10%]
4	Charter signed	0 days	Mon 6/18/07	Mon 6/18/07	3	
5	**Planning**	95 days	Mon 6/18/07	Mon 10/29/07		
6	Develop project plans	3 wks	Mon 6/18/07	Mon 7/9/07	4	Kathy[50%],John[20%],Mary [20%],Chris[20%]
7	Review project plans	4 mons	Mon 7/9/07	Mon 10/29/07	6	Kathy[5%]
	Project plans					

Done   My Computer

Figure A-74. Default Task Information HTML File

3. *Close the taskinfo.html and baseline.mpp files.* Click **Close** to close your browser, and then go back to your Project 2003 window and close the baseline.mpp file without saving it.

There are several ways to change or customize the standard HTML documents in Project 2003. The Export Wizard allows you to change formats for existing maps, and you can also create your own map. Consult the Project 2003 Help topic "HTML" to find more information on HTML features of Project 2003 and saving your files as Web pages.

Saving project information as Web pages includes text and numerical information, but not graphic information. You can create Web displays of graphic views, such as Gantt charts or network diagrams, but the process often requires knowledge of how to include images in Web pages.

Using the Copy Picture to Office Wizard

The Copy Picture to Office Wizard is a new feature in Project 2003 that helps you easily create and copy graphical images. After answering a few questions in the wizard, you can quickly display project data as a static picture in Office applications, including PowerPoint, Word, and Visio. You can access this wizard on the Analysis toolbar.

To use the Copy Picture to Office Wizard:

1. *Prepare the file you want to copy.* Open **baseline.mpp**. Drag the split bar to the left so that only the entire Task Name column is showing. Click the **collapse symbol** ⊟ for Task 1, Initiating. You should be able to see the entire Gantt chart on your screen.

2. *View the Analysis Toolbar.* Click **View** on the menu bar, select **Toolbars**, and then click **Analysis** to display the Analysis toolbar.

3. *Open the Copy Picture to Office Wizard.* Click the **Copy Picture to Office Wizard** button 🖺 on the Analysis toolbar. Click **Next** at each step of the wizard until you reach Step 4. You have now accepted the default settings to copy the entire picture of the screen to PowerPoint. Click **Finish**, and then click **Close** when the wizard is complete. Your screen should resemble Figure A-75. The wizard has created a picture of your screen in Project 2003 and copied it into PowerPoint.

Figure A-75. Copy Picture to Office Wizard Results in PowerPoint

4. *Close all open windows.* Click the Close button on any open windows. You do not need to save any of the open files.

As you can see, Project 2003 is a very powerful tool. If used properly, it can greatly help users successfully manage projects.

DISCUSSION QUESTIONS

1. What are some new features of Project 2003 Standard and how do they differ from previous versions of Microsoft Project?

2. How do you use Project 2003 to create a WBS hierarchy?

3. Summarize how you use Project 2003 to assist in time management. How do you enter durations, link tasks, and view critical path information?

4. How can Project 2003 assist you in project cost management? How do you enter fixed costs? How do you enter resources and assign them to tasks? How can you view earned value information?

5. Briefly describe how to change resource calendars, view resource histograms, and level resources.

6. Summarize different ways to communicate information with Project 2003. How do you link to other files from within your Project 2003 file? How can you use templates and share information on the Web? How can you copy pictures into other applications?

SUGGESTED READINGS

1. Bunin, Rachel. *New Perspectives on Microsoft Office Project 2003, Introductory.* Boston: Course Technology (2004).

    This book walks you through step-by-step tutorials on how to use Project 2003 on one continuous running case. It includes six tutorials on planning a project, creating a project schedule, communicating project information, assigning resources and costs, tracking progress and closing a project, and sharing information across projects, applications, and the World Wide Web. There are review activities and several cases for each chapter. There are also supporting files and other information on the publisher's Web site at www.course.com. You can also run an interactive tutorial of Project 2002 from www.scsite.com/selabs.

2. Marmel, Elaine. *Microsoft Office Project 2003 Bible*, Hungry Minds, Inc. (2003).

    This comprehensive text (960 pages) includes detailed information on how to use Project 2003. It also includes a CD-ROM with an evaluation copy of Project, add-on tools, and sample cases.

3. Pyron, Tim et al. *Special Edition: Using Microsoft Office Project 2003*, Que Publishing (2004).

    This text includes detailed coverage of Project 2003. It includes 1248 pages of text and additional information on Que's Web site. The companion CD-ROM contains all the examples developed for the book, a third-party software library, and Web resources. Tim Pyron is a Microsoft Project consultant, trainer, and editor of Woody's Project Watch, a free newsletter for Project users (www.woodyswatch.com/project).

4. Stover, Teresa S. *Microsoft Office Project 2003 Inside Out*. Microsoft Press (2003).

    This text includes 1152 pages of text and a CD-ROM that contains a trial version of Microsoft Office Project 2003 Standard Edition, Microsoft resources and demos, a computer dictionary, and information about third-party utilities and tools.

EXERCISES

The more you practice using the different features of Project 2003, the more quickly you will master the application and be able to use it to manage projects. This section includes exercises based on three examples of information technology projects that could benefit from using Project 2003 to assist in project scope, time, cost, human resource, and communications management. It also includes an exercise you can use to apply Project 2003 on a real project. There are simpler exercises that use Project 2003 at the end of several other chapters in this text. You can also re-create several screen shots in this text to practice your skills, or you can open and then make slight changes to template files or the files that come with this text.

Exercise A-1: Web Site Development

A nonprofit organization would like you to lead a Web site development project for them. The organization has Internet access that includes space on a Web server, but has no experience developing Web pages or Web sites. In addition to creating their Web site, they would like you to train two people on their staff to do simple Web page updates. The organization wants their Web site to include the following basic information, as a minimum: description of the organization (mission, history, and recent events), list of services, and contact information. They want the Web site to include graphics (photographs and other images) and have an attractive, easy-to-use layout.

1. Project Scope Management

   Create a WBS for this project and enter the tasks in Project 2003. Create milestones and summary tasks. Assume that some of the project management tasks you need to do are similar to tasks from the Project Tracking Database project. Some of the specific analysis, design, and implementation tasks will be to:

   a. Collect information on the organization in hardcopy and digital form (brochures, reports, organization charts, photographs, and so on).

   b. Research Web sites of similar organizations.

   c. Collect detailed information about the customer's design preferences and access to space on a Web server.

   d. Develop a template for the customer to review (background color for all pages, position of navigation buttons, layout of text and images, typography, including basic text font and display type, and so on).

   e. Create a site map or hierarchy chart showing the flow of Web pages.

   f. Digitize the photographs and find other images for the Web pages; digitize hard-copy text.

   g. Create the individual Web pages for the site.

   h. Test the pages and the site.

   i. Implement the Web site on the customer's Web server.

   j. Get customer feedback.

   k. Incorporate changes.

    l. Create training materials for the customer on how to update the Web pages.

    m. Train the customer's staff on updating the Web pages.

2. Project Time Management

    a. Enter realistic durations for each task, and then link the tasks as appropriate. Be sure that all tasks are linked (in some fashion) to the start and end of the project. Assume that you have four months to complete the entire project. (*Hint*: Use the Project Tracking Database as an example.)

    b. Print the Gantt Chart view and Network Diagram view for the project.

    c. Print the Schedule table to see key dates and slack times for each task.

3. Project Cost Management

    a. Assume that you have three people working on the project, and each of them would charge $20 per hour. Enter this information in the Resource Sheet.

    b. Estimate that each person will spend an average of about five hours per week for the four-month period. Assign resources to the tasks, and try to make the final cost in line with this estimate.

    c. Print the budget report for your project.

4. Project Human Resource Management

    a. Assume that one project team member will be unavailable (due to vacation) for two weeks in the middle of the project. Make adjustments to accommodate this vacation so that the schedule does not slip and the costs do not change. Document the changes from the original plan and the new plan.

    b. Use the Resource Usage view to see each person's work each month. Print a copy of the Resource Usage view.

5. Project Communications Management

    a. Print a Gantt chart for this project. Use a timescale that enables the chart to fit on one page. Then use the Copy Picture to Office Wizard to show the same Gantt chart in PowerPoint.

    b. Print a "To-do List" report for each team member.

    c. Create a "Who Does What Report" HTML document. Print it from your Web browser.

Exercise A-2: Software Training Program

ABC Company has 50,000 employees in its headquarters. The company wants to increase employee productivity by setting up an internal software applications training program for its employees. The training program will teach employees how to use software programs such as Windows, Word, Excel, PowerPoint, Access, and Project 2003. Courses will be offered in the evenings and on

Saturdays and will be taught by qualified volunteer employees. The instructors will be paid $40 per hour. In the past, various departments sent employees to courses offered by local vendors during company time. In contrast to local vendors' programs, this internal training program should save the company money on training as well as make its people more productive. The Human Resources department will manage the program, and any employee can take the courses. Employees will receive a certificate for completing courses, and a copy of the certificate will be put in their personnel files. The company has decided to use off-the-shelf training materials, but is not sure which vendor's materials to use. The company needs to set up a training classroom, survey employees on what courses they want to take, find qualified volunteer instructors, and start offering courses. The company wants to offer the first courses within six months. One person from Human Resources is assigned full-time to manage this project, and top management has pledged its support of the project.

1. Project Scope Management

   Create a WBS for this project and enter the tasks in Project 2003. Create milestones and summary tasks. Assume that some of the project management tasks you need to do are similar to tasks from the Project Tracking Database example. Some of the tasks specific to this project will be to:

   a. Review off-the-shelf training materials from three major vendors and decide which materials to use.

   b. Negotiate a contract with the selected vendor for their materials.

   c. Develop communications information about this new training program. Disseminate the information via department meetings, e-mail, the company's intranet, and flyers to all employees.

   d. Create a survey to determine the number and type of courses needed and employees' preferred times for taking courses.

   e. Administer the survey.

   f. Solicit qualified volunteers to teach the courses.

   g. Review resumes, interview candidates for teaching the courses, and develop a list of preferred instructors.

   h. Coordinate with the Facilities department to build two classrooms with twenty personal computers in each, a teacher station, and an overhead projection system (assume that the Facilities department will manage this part of the project).

   i. Schedule courses.

   j. Develop a fair system for signing up for classes.

   k. Develop a course evaluation form to assess the usefulness of each course and the instructor's teaching ability.

   l. Offer classes.

2. Project Time Management

    a. Enter realistic durations for each task and then link appropriate tasks. Be sure that all tasks are linked in some fashion to the start and end of the project. Use the Project Tracking Database as an example. Assume that you have six months to complete the entire project.

    b. Print the Gantt Chart view and Network Diagram view for the project.

    c. Print the Schedule table to see key dates and slack times for each task.

3. Project Cost Management

    a. Assume that you have four people from various departments available part-time to support the full-time Human Resources person, Terry, on the project. Assume that Terry's hourly rate is $40. Two people from the Information Technology department will each spend up to 25 percent of their time supporting the project. Their hourly rate is $50. One person from the Marketing department is available 25 percent of the time at $40 per hour, and one person from Corporate is available 30 percent of the time at $35 per hour. Enter this information about time and hourly wages into the Resource Sheet. Assume that the cost to build the two classrooms will be $100,000, and enter it as a fixed cost.

    b. Using your best judgment, assign resources to the tasks.

    c. View the Resource Graphs for each person. If anyone is overallocated, make adjustments.

    d. Print the budget report for the project.

4. Project Human Resource Management

    a. Assume that the Marketing person will be unavailable for one week, two months into the project, and for another week, four months into the project. Make adjustments to accommodate this unavailability, so the schedule does not slip and costs do not change. Document the changes from the original plan and the new plan.

    b. Add to each resource a 5 percent raise that starts three months into the project. Print a new budget report.

    c. Use the Resource Usage view to see each person's work each month. Print a copy.

5. Project Communications Management

    a. Print a Gantt chart for this project. Use a timescale that enables the chart to fit on one page. Then use the Copy Picture to Office Wizard to show the same Gantt chart in PowerPoint.

    b. Print a "To-do List" report for each team member.

    c. Create an HTML document using the "Default task information" map. Print it from your Web browser.

Exercise A-3: Project Tracking Database

Expand the Project Tracking Database example. Assume that XYZ Company wants to create a history of project information, and the Project Tracking Database example is the best format for this type of history. The company wants to track information on twenty past and current projects and wants the database to be able to handle one hundred projects total. It wants to track the following project information:

- Project name
- Sponsor name
- Sponsor department
- Type of project
- Project description
- Project manager
- Team members
- Date project was proposed
- Date project was approved or denied
- Initial cost estimate
- Initial time estimate
- Dates of milestones (for example, project approval, funding approval, project completion)
- Actual cost
- Actual time
- Location of project files

1. Project Scope Management

   Open scope.mpp and add more detail to Executing tasks, using the information in Table A-6.

Table A-6: Company XYZ Project Tracking Database Executing Tasks

ANALYSIS TASKS	DESIGN TASKS	IMPLEMENTATION TASKS
Collect list of 20 projects	Gather detailed requirements for desired outputs from the database	Enter project data
Gather information on projects	Create fully attributed, normalized data model	Test database
Define draft requirements	Create list of edit rules for fields, determine default values, develop queries, and determine formats for reports	Make adjustments, as needed
Create entity relationship diagram for database	Develop list of queries required for database	Conduct user testing
Create sample entry screen	Review design information with customer	Make adjustments based on user testing
Create sample report	Make adjustments to design based on customer feedback	Create online Help, user manual, and other documentation
Develop simple prototype of database	Create full table structure in database prototype	Train users on the system
Review prototype with customer	Create data input screens	
Make adjustments based on customer feedback	Write queries	
	Create reports	
	Create main screen	
	Review new prototype with customer	

2. Project Time Management

  a. Enter realistic durations for each task, and then link appropriate tasks. Make the additional tasks fit the time estimate: 20 days for analysis tasks, 30 days for design tasks, and 20 days for implementation tasks. Assume that all of the executing tasks have a total duration of 70 days. Do not overlap analysis, design, and implementation tasks.

  b. Print the Gantt Chart view and Network Diagram view for the project.

  c. Print the Schedule table to see key dates and slack times for each task.

3. Project Cost Management

  a. Use the resource and cost information provided in resource.mpp.

    b. Assign resources to the new tasks. Try to make the final cost about the same as that shown in the Project Tracking Database example: about $50,000.

    c. Print the budget report for the project.

4. Project Human Resources Management

    a. Two months after the project begins, give everyone on the team a 10 percent raise. Document the increase in costs that these raises cause.

    b. Use the Resource Usage view to see each person's work each month. Print a copy of the Resource Usage view.

5. Project Communications Management

    a. Print a Gantt chart for this project. Use a timescale that enables the chart to fit on one page. Then use the Copy Picture to Office Wizard to show the same Gantt chart in PowerPoint.

    b. Print a "Top-Level Tasks" report.

    c. Create a "Cost Data by Task" HTML document. Print it from your Web browser.

Exercise A-4: Real Project Application

If you are doing a group project as part of your class or for a project at work, use Project 2003 to create a detailed file describing the work you plan to do. Enter a sufficient WBS, estimate task durations, link tasks, enter resources and costs, assign resources, and so on. Then save your file as a baseline and track your progress. View earned value information when you are about halfway done with the project or course. Export the earned value information into Excel and create an earned value chart. Continue tracking your progress until the project or course is finished. Print your Gantt chart, Resource Sheet, Project Summary report, and relevant information. In addition, write a two- to three-page paper describing your experience. What did you learn about Project 2003 from this exercise? What did you learn about managing a project? How do you think Project 2003 helps in managing a project? You may also want to interview people who use Project 2003 and ask them about their experiences and suggestions.

ADDITIONAL RUNNING CASES

Review the additional running cases provided in Appendix C. Several questions involve using Project 2003.

Appendix B

Advice for the Project Management Professional (PMP) Exam and Related Certifications

INTRODUCTION TO PROJECT MANAGEMENT CERTIFICATION PROGRAMS

This appendix provides information on project management certification programs and offers some advice for earning them. It briefly describes various certification programs and provides detailed information on PMI's PMP and CompTIA's Project+ certifications, the structure and content of these exams, suggestions on preparing for the exams, tips for taking the exams, sample questions, information on related certifications, and final advice on certification and project management in general.

WHAT IS PMP CERTIFICATION?

The Project Management Institute (PMI) offers certification as a Project Management Professional (PMP). As mentioned in Chapter 1, the number of people earning PMP certification grew rapidly in the past ten years, and there were 84,302 PMPs by the end of June 2004. There are PMPs in more than 120 countries throughout the world. A major milestone occurred in 1999 when PMI's Certification Program Department became the first professional certification program department in the world to attain ISO 9001 recognition. Detailed information about PMP certification, the PMP Certification Handbook, and an online application is available from PMI's Web site (*www.pmi.org*) under Professional Development & Careers. The following information is quoted from PMI's Web site:

> The Project Management Institute (PMI®) stands as a global leader in the field of project management. It is well known that PMI certification involves a rigorous, examination-based process that represents the highest caliber in professional standards. Therefore, PMI's professional certification is universally accepted and

recognized. As a demonstration of our commitment to professional excellence, the PMI program also maintains ISO 9001 certification in Quality Management Systems.

If you enjoy the prestige that comes from being the best in your field, then you will appreciate the professional advantages derived from becoming a PMP. PMP certification is the profession's most globally recognized and respected certification credential. The PMP designation following your name tells current and potential employers that you have a solid foundation of project management knowledge that you can readily apply in the workplace.[1]

Many companies and organizations are recommending or even requiring PMP certification for their project managers. A February 2003 newsletter reported that Microsoft chose PMI's PMP Certification Program as the certification of choice for its Microsoft Services Operation. Microsoft chose the PMP certification because of its global recognition and proven record in professional development for project managers.[2]

Certification Magazine published its annual review of how certification affects salaries of information technology professionals. This industry-wide study uses real-world numbers to show how education and experience affect a person's bottom-line salary. During a down market, people might ask why they should seek additional technical certifications. According to Gary Gabelhouse of *Certification Magazine*, "Perhaps it is best expressed in two words: job security. In boom times, one constantly reviews the rate of growth in salary as a key personal-success measurement. However, in down times, job security is paramount."[3]

The October 2004 issue of *PM Network* included an article entitled "The Rise of PMP," which provided several examples of companies and countries that have made concerted efforts to increase their number of PMPs. Hewlett-Packard had only six registered PMPs in 1997, but by August 2004, it had more than 1,500 and was adding 500 per year. Although most PMPs are in the U.S. (51,498) and Canada (7,444), the PMP credential is growing in popularity in several countries, such as Japan (6,001), China (4,472), and India (2,281). Thomas Walenta, PMP, a senior project manager for IBM Germany, said, "The PMP credential lends portability to a career plan. It has significantly shaped careers in the IT industry, with global companies creating career models for project managers based on PMI certification requirements."[4]

PMI also offers certification as a Certified Associate in Project Management (CAPM). PMI developed this newer certification as a stepping-stone to PMP certification. Candidates for the CAPM certification must also meet specific education

[1] Project Management Institute, "PMI® Certifications," (*www.pmi.org/info/PDC_CertificationsOverview.asp?nav=0401*) (October 2004).

[2] Project Management Institute Information Systems Specific Interest Group (ISSIG), *ISSIG Bits* (*www.pmi-issig.org*) (February 2003).

[3] Global Knowledge, "2003 Certification Salary by Certmag," *Global Knowledge E-Newsletter,* Issue #56 (March 2003).

[4] Rewi, Adrienne, "The Rise of PMP," *PM Network* (October 2004) p. 18.

and experience requirements and pass an exam. As you can imagine, the requirements are not as rigorous as they are for the PMP exam. Consult PMI's Web site for detailed information. You might want to consider getting the CAPM, or just wait until you have enough experience to earn the more popular PMP certification. As described below, you need 4,500 hours of work experience if you have a bachelor's degree to qualify for PMP certification. This amounts to working full-time for two or three years in a project environment, or part-time for a longer time period.

What Are the Requirements for Earning and Maintaining PMP Certification?

You can now apply to take the PMP exam online. You can also fill out the application forms and mail them in with a check, if desired. *Before* you apply to take the exam, you must meet the following four requirements:

1. Have experience working in the field of project management. When you apply to take the exam, you enter the role you played on one or several projects and how many hours you worked in each of the five project management process groups for each project. Roles include:

   - Project contributor
   - Supervisor
   - Manager
   - Project leader
   - Project manager
   - Educator
   - Consultant
   - Administrator
   - Other

   Note that you do not have to have experience as a project manager to take the PMP exam; any of these roles will suffice. PMP certification requires all applicants with a bachelor's degree to have at least 4,500 hours of project management experience within the six-year period prior to application. Applicants without a bachelor's degree are required to have 7,500 hours of experience within an eight-year period prior to application. You must fill out a simple form online (or by paper, if you choose not to apply online) listing the title of a project or projects you worked on; the start and end date for when you worked on the project(s); your role on the project; the number of hours you spent in initiating, planning, executing, controlling, and closing processes; and a list of the deliverables that you managed on the project. PMI staff will review your qualifications and let you know if you are qualified to take the PMP exam. *You cannot take the exam without this experience qualification.*

2. Document at least 35 contact hours of project management education. There is no time frame for this requirement. A university or college, a training company or independent consultant, a PMI chapter, a PMI Registered Education Provider (REP), a company-sponsored program, or a distance-learning company can provide the education. The hours must include content on project quality, scope, time, cost, human resources, communications, risk, procurement, and integration management. You must list the course title, institution, date, and number of hours for each course. For example, if you took a "principles of management" course at a university 20 years ago, you could list that on the education form along with a one- or two-day PMP exam preparation class.

3. Agree to the PMP certificant and candidate agreement and release statement. This form certifies that application information is accurate and complete and that candidates will conduct themselves in accordance with the PMP Code of Professional Conduct, professional development requirements, and other PMI certification program policies and procedures. You can simply check a box saying you agree to this information when applying online.

4. Pay the appropriate exam fee. As of October 2004, the PMP certification fee is $405 for PMI members and $555 for non-members. The re-examination fee (if you don't pass the exam) is $275 for PMI members and $375 for non-members. The annual individual PMI membership fee is $129 (including the application fee). Note that students, or anyone enrolled in a degree-granting program at an accredited, or globally equivalent, college or university, can join PMI at the student member rate of only $40. If you want to earn your PMP certification, it makes sense to join PMI, not only for the financial savings, but also for other benefits. Consult PMI's Web site for membership information.

The last step in earning PMP certification is passing the exam! After PMI sends you an eligibility letter to take the PMP exam, you can sign up to take it at several different testing sites. You must take the exam within six months of receiving your eligibility letter. The eligibility letter will include complete details for scheduling your exam. As of October 2004, the PMP exam consists of 200 four-option multiple-choice questions. Although you cannot use any study aids during the exam, you can bring a nonprogrammable calculator to assist in performing calculations required to answer some of the questions. You are also given two blank pieces of paper, so you can write down formulas or other information when you enter the exam room, but you cannot bring in any notes or other materials. The questions on each test are randomly selected from a large test bank, so each person taking the exam receives different questions. Test takers have 4.5 hours to take this computerized exam, and a passing score is 68.5 percent, or at least 137 correct answers out of 200. PMI reviews and revises the exam annually. PMI has announced that the exam will change based on changes made to the *PMBOK® Guide 2004* in late 2005. *Be sure you study from the correct version of the* PMBOK® Guide.

Starting in January 1999, PMI instituted a new professional development program for maintaining the PMP certification. To maintain your PMP status, you must earn at least 60 Professional Development Units within three years, pay a recertification fee every three years when you renew your certification ($60 as of October 2004), and agree to continue to adhere to PMI's PMP Code of Professional Conduct. The Continuing Certification Requirements Handbook, available from PMI's Web site, provides more details on maintaining your PMP status. For detailed information, contact PMI's Web site under Professional Development & Careers, certifications, and then continuing certification requirements. Many people are confused about maintaining PMP certification, so be sure to read the handbook.

What Is the Structure and Content of the PMP Exam?

The PMP exam is based on information from the entire project management body of knowledge (PMBOK) as well as the area of professional responsibility. Essentially, the exam reviews concepts and terminology in PMI's *PMBOK® Guide 2004*, and texts such as this one will help to reinforce your understanding of key topics in project management. Table B-1 shows the approximate breakdown of questions on the PMP exam by process group as of October 2004. Candidates should review updated exam information on PMI's Web site to make sure they are using the latest information.

Table B-1: Breakdown of Questions on the PMP Exam by Process Groups

PROCESS GROUP	PERCENT OF QUESTIONS ON PMP EXAM	NUMBER OF QUESTIONS ON PMP EXAM
Initiating	8.5	17
Planning	23.5	47
Executing	23.5	47
Controlling	23	46
Closing	7	14
Professional Responsibility	14.5	29

Study Table 3-1 from Chapter 3 of this text to understand the relationships among project management process groups, activities, and knowledge areas. Table 3-1 briefly outlines which activities are performed during each of the project management process groups and what is involved in each of the knowledge areas. It is also important to understand what each of the project management activities includes. Several questions on the certification exam require an understanding of this framework for project management, and many questions require an understanding of the various tools and techniques described in the *PMBOK® Guide 2004* and this text.

The PMP exam includes three basic types of questions:

1. *Conceptual questions* test your understanding of key terms and concepts in project management. For example, you should know basic definitions such as what a project is, what project management is, and what key activities are included in project scope management.

2. *Application questions* test your ability to apply techniques to specific problems. For example, a question might provide information for constructing a network diagram and ask you to find the critical path or determine how much slack is available on another path. You might be given cost and schedule information and be asked to find the schedule or cost performance index by applying earned value formulas.

3. *Evaluative questions* provide situations or scenarios for you to analyze, and your response will indicate how you would handle them. For example, a project might have many problems. A question might ask what you would do in that situation, given the information provided. Remember that all questions are multiple-choice, so you must select the best answer from the options provided.

How Should You Prepare for the PMP Exam?

To prepare for the PMP exam, it is important to understand *your* learning and testing style, and to use whatever resources and study techniques work best for *you*. Below are some important questions to consider:

■ *Are you a good test taker?* Some people are very good at studying and do well on multiple-choice exams, but others are not. Several people can and have passed the PMP exam right after completing one college course or workshop using this text or another text, by reading the *PMBOK® Guide* once, or just by trusting their experience. Other people know they need to study more to understand many of the basic concepts of project management, or they need to practice taking sample tests.

■ *How confident do you need to be before taking the exam?* To pass the exam, you need to answer only 68.5 percent of the questions correctly. It's all multiple-choice, so just random guessing should theoretically result in a score of 25 percent. Your PMP certificate will not display your final score, so it does not matter if you get 70, 80, 90, or even 100 percent correct. Some people won't take the exam until they consistently score 80 or 90 percent or higher on sample tests. Are you that risk averse, or are you willing to take the exam when you consistently score 70 percent or more on sample tests? Remember that your ultimate goal is to be a good project manager, not to brag about your score or to have more letters behind your name.

■ *How much information do you need to review before taking the exam?* The *PMBOK® Guide 2004* and this text should be enough content information, but many people want to review even more information before taking the PMP exam. Several companies sell books, sample tests, CD-ROMs, and audiotapes, or provide courses designed to help people pass the PMP exam. There's even a *PMP Certification for Dummies* book available with a CD-ROM of sample questions. In October 2004, performing a search using the keywords "PMP exam" at *www.amazon.com* resulted in 621 different books and related products! Performing a search using the keywords "PMP exam" at *www.google.com* resulted in 52,300 hits. See the Suggested Readings section for a short list of suggested resources, many at no cost at all.

■ *How much time and money do you want to spend studying for the PMP exam?* Some people with little free time or money to spend on extra courses or materials would rather just take the PMP exam with little preparation; this will tell them what they need to study further if they don't pass on their first attempt. Even though there is a reexamination fee of $275 (for PMI members), this cost is generally less than what you would have to pay for most exam preparation courses. Spend the time and money you need to feel confident enough to pass the exam, but don't over-extend yourself. Remember that there is no letter grade assigned—just a pass or fail. And no one needs to know if you don't pass the first time.

■ *Do you know PMI's language?* Even if you think you know about project management, studying the material in the *PMBOK® Guide 2004* before taking the exam will help you. Volunteer PMPs created the exam, and they often refer to information from the *PMBOK® Guide 2004* when writing questions. Many outstanding project managers might fail the exam if they don't use PMI's terminology or processes in their jobs.

■ *Do you really understand the triple constraint of project management?* Many questions on the PMP exam are based on the scope, time, and cost knowledge areas. You should be familiar with project charters, WBSs, network diagrams, critical path analysis, cost estimates, earned value, and so on before you take the exam.

■ *Do you want to meet other people in the field of project management as you study for your PMP exam?* Several chapters of PMI offer PMP exam review courses. These courses are often a good way to network with other local project managers or soon to be project managers. Many other organizations provide online and instructor-led courses in PMP preparation where you can also meet people in the field.

■ *Do you need extra support, peer pressure, or incentives to pass the PMP exam?* Having some support and positive peer pressure might help to ensure that you actually take and pass the exam in a timely manner. It often helps to verbalize information when you are studying, and explaining something to someone else in a study group can help you understand and remember it. If you don't want to be part of a study group, even just telling a friend, colleague, or loved one that you have set a goal to pass the PMP exam by a certain date might provide motivation to actually do it. You could also reward yourself after you pass the exam.

■ *Can you make earning your PMP a project and give it high priority?* This is probably the most important question to ask yourself. If passing the exam is really important to you, you *can* make it happen.

■ *How much are you willing to invest in this project?* You can view earning certification as a personal project and set your own goals and strategies for success. Your scope goal is to pass the PMP exam, your time goal might be within three or six months, and your cost goal is how much time and money you plan to spend. If you have the time and money, you could take one of the immersion courses several companies offer, like Cheetah Learning, Velociteach, Project Management Training Institute, or mScholar. These courses generally last four to five days and cost between two and three thousand dollars. The company offering the course will have you come to class with your qualification to take the PMP exam already completed, and you actually take the PMP exam on the fourth or fifth day of the class. If you don't pass, they'll let you take the class again at no additional cost. If you don't have any extra money to spend, you can find several free resources (see the Suggested Readings) and form your own study group, or just go take the exam and see how you do.

Ten Tips for Taking the PMP Exam

1. The PMP exam is computer based and begins with a short tutorial on how to use the testing software. The software makes it easy to mark questions you want to review later, so learning how to mark questions is helpful. Using this feature can give you a feel for how well you are doing on the test. It is a good idea to go through every question somewhat quickly and mark those questions on which you want to spend more time. If you mark 63 questions or less (the total number you can miss), you should pass the exam.

2. The time allotted for the exam is 4.5 hours, and each multiple-choice question has four answer choices. You should have plenty of time to complete the exam. Use the first two to three hours to go through all of the questions. Do not spend too much time on any one question. As you

work, mark each question that you would like to return to for further consideration. Then use the remaining time to check the questions you are not sure of. If you're a morning person, schedule your exam in the morning. If you work better after lunch, schedule an afternoon exam. Make sure you are alert and well rested when you go in to take the exam.

3. Some people believe it is better to change answers you were originally unsure of. If you think that a different answer is better, after reading the question again, then change your answer. If you are still unsure, choose your best guess and leave the question marked so you can easily review it again later. Don't get hung up on any questions. Move on and focus on answering the ones you can answer correctly.

4. Do not try to read more into the questions than what is stated. There are no trick questions, but some may be poorly worded or just bad questions. Remember that they were written by volunteers and are part of a huge test bank. Most of the questions are relatively short, and there are only four options from which to choose the answer. Few of the options include "all of the above" or "none of the above." Some questions might include the option "There is not enough information."

5. To increase your chances of getting the right answer, first eliminate obviously wrong options, and then choose among the remaining options. Take the time to read all of the options before selecting an answer. Remember you have to pick the *best* answer available.

6. Some questions require doing calculations such as earned value management. It is worthwhile to memorize the earned value formulas to make answering these questions easier. You may use a nonprogrammable calculator while taking the exam, so be sure to bring one to make performing calculations easier.

7. You will be given two pieces of blank paper to use during the exam. Before starting the test, you might want to write down important equations so that you do not have to rely on your memory. When you come to a question involving calculations, write the calculations down so you can check your work for errors.

8. Read all questions carefully. A few sections of the test require that you answer three to four questions about a scenario. These questions can be difficult; it can seem as if two of the choices could be correct, although you can choose only one. Read the directions for these types of questions several times to be sure you know exactly what you are supposed to do. Also, remember important concepts such as the importance of using a WBS, emphasizing teamwork, and practicing professional integrity. You might want to skip the longer or more difficult questions and answer the shorter or easier ones first.

9. If you do not know an answer and need to guess, be wary of choices that include words such as always, never, only, must, and completely. These extreme words usually indicate incorrect answers because there are many exceptions to rules.

10. After an hour or two, take a short break to clear your mind. You might want to bring a snack to have during your break. You might also consider bringing earplugs if you're easily distracted by noises in the room.

Sample PMP Exam Questions

A few sample questions similar to those you will find on the PMP exam are provided below. You can check your answers at the end of this appendix. If you miss less than 7 out of these 20 questions, you are probably ready to take the PMP exam. You can find additional sample questions and their answers on the companion Web site for this test.

1. A document that formally recognizes the existence of a project is a
    A. Gantt chart
    B. WBS
    C. project charter
    D. scope statement

2. Decomposition is used in developing
    A. the management plan
    B. the communications plan
    C. the earned value
    D. the WBS

3. The critical path on a project represents
    A. the shortest path through a network diagram
    B. the longest path through a network diagram
    C. the most important tasks on a project
    D. the highest-risk tasks on a project

4. If the earned value (EV) for a project is $30,000, the actual cost (AC) is $33,000, and the planned value (PV) is $25,000, what is the cost variance?
    A. $3,000
    B. -$3,000
    C. $5,000
    D. -$5,000

5. If the earned value (EV) for a project is $30,000, the actual cost (AC) is $33,000, and the planned value (PV) is $25,000, how is the project performing?

   A. The project is over budget and ahead of schedule.

   B. The project is over budget and behind schedule.

   C. The project is under budget and ahead of schedule.

   D. The project is under budget and behind schedule.

6. What is the target goal for defects per million opportunities using Six Sigma?

   A. 1

   B. 3.4

   C. 34

   D. 100

7. Project human resource management does not include which of the following processes?

   A. acquiring the project team

   B. developing the project team

   C. managing the project team

   D. estimating activity resources

8. If a project team goes from three to five people, how many more communications channels are there?

   A. 7

   B. 6

   C. 5

   D. 4

9. Your project team has identified several risks related to your project. You decide to take actions to reduce the impact of a particular risk event by reducing the probability of its occurrence. What risk response strategy are you using?

   A. risk avoidance

   B. risk acceptance

   C. risk mitigation

   D. contingency planning

10. Which type of contract provides the least amount of risk for the buyer?

    A. firm fixed price

    B. fixed price incentive

    C. cost plus incentive fee

    D. cost plus fixed fee

11. Suppose you have a project with four tasks as follows:
    - Task 1 can start immediately and has an estimated duration of 1.
    - Task 2 can start after Task 1 is completed and has an estimated duration of 4.
    - Task 3 can start after Task 2 is completed and has an estimated duration of 5.
    - Task 4 can start after Task 1 is completed and must be completed when Task 3 is completed. Its estimated duration is 8.

    What is the length of the critical path for this project?

    A. 9

    B. 10

    C. 11

    D. 12

12. In which of the following project management process groups are the most time and money typically spent?

    A. initiating

    B. planning

    C. executing

    D. controlling

13. Creating a probability/impact matrix is part of which risk management process?

    A. risk management planning

    B. risk identification

    C. qualitative risk analysis

    D. quantitative risk analysis

14. It is crucial that your project team finish your project on time. Your team is using a technique to account for limited resources. You have also added a project buffer before the end date and feeding buffers before each critical task. What technique are you using?

    A. critical path analysis

    B. PERT

    C. critical chain scheduling

    D. earned value management

15. One of your senior technical specialists informs you that a major design flaw exists in a systems development project you are managing. You are already testing the system and planned to roll it out to more than 5,000 users in a month. You know that changing the design now will cause several cost and schedule overruns. As project manager, what should you do first?

    A. Issue a stop work order until you understand the extent of the flaw.

    B. Notify your project sponsor immediately to see if there are additional funds available to work on this problem.

C. Notify your senior management and let them decide what to do.

D. Hold a meeting as soon as possible with key members of your project team to discuss possible solutions to the problem.

16. You are a member of a large government project. You know that the contract insists that all equipment be manufactured in the United States. You see a senior member of your team replacing a company etching on a piece of equipment that was made in a foreign country. You confront this person, and he says he is following the project manager's orders. What should you do?

   A. Nothing; the project manager made the decision.

   B. Immediately report the violation to the government.

   C. Update your resume and look for another job.

   D. Talk to the project manager about the situation, and then decide what to do.

17. Which of the following is not an output of the integrated change control process?

   A. project management plan updates

   B. approved corrective action

   C. forecasts

   D. deliverables

18. Which type of dependency is inherent in the nature of the work being done?

   A. mandatory dependency

   B. discretionary dependency

   C. external dependency

   D. internal dependency

19. Obtaining quotes, bids, offers, or proposals is part of which project procurement management process?

   A. plan purchases and acquisitions

   B. request seller responses

   C. select sellers

   D. contract administration

20. Your boss believes that all of your project team members avoid work as much as possible. He or she often uses threats and various control schemes to make sure people are doing their jobs. Which approach to managing people does your boss follow?

   A. Maslow's hierarchy of needs

   B. Theory X

   C. Theory Y

   D. Herzberg's motivation and hygiene factors

WHAT IS PROJECT+ CERTIFICATION?

More than 600,000 people worldwide have earned Computing Technology Industry Association (CompTIA) certifications in PC service, networking, document imaging, Internet, and PC Server technologies. In April 2001, CompTIA started offering its IT Project+ certification, which was purchased from Prometric-Thomson Learning, and recognized as the Gartner Institute Certification Program. The certification was renamed Project+ in August 2004. According to CompTIA's manager of public relations, there are more than 6,000 people with CompTIA's Project+ certification as of October 2004. Detailed information about Project+ certification is available from CompTIA's Web site (*www.comptia.org*) under certification. The following information is quoted from a press release on their site:

> "Individuals managing any type of project for their company—whether or not it has an IT component to it—can benefit enormously from using the certification as a means of validating project management competency," said Lisa-Ann Barnes, president, IreeTec, Inc. and chair of the CompTIA Project+ advisory committee.
>
> Individuals passing the certification exam validate knowledge mastery equivalent to 2,000 hours of on-the-job practical experience. CompTIA Project+ can serve as a standalone certificate for individuals or as a stepping-stone toward the Project Management Institute PMP certification, which requires 4,500 hours of experience.[5]

What Are the Requirements for Earning and Maintaining Project+ Certification?

You can register to take the Project+ exam online from CompTIA's Web site. Testing sites include Thomson Prometric and Pearson VUE. Unlike the PMP exam, there are very few requirements you need to meet to take the Project+ exam. CompTIA does not require that you have any work experience or formal education in project management before you can take the Project+ exam, but they do recommend 2,000 hours of work experience. The main requirements include paying a fee and passing the exam. Important information is summarized below:

1. As of October 2004, the cost for taking the Project+ exam is $155 for CompTIA members and $207 for non-members.
2. To pass this 90-minute, 80-question exam, you must score at least 63 percent.
3. You do not need to renew your Project+ certification.
4. Project+ certification is one of the prerequisites or equivalents to Novell's Certified Novell Engineer (CNE), Certified Novell Administrator (CNA),

---

[5] CompTIA Web site (*www.comptia.org/pressroom/get_pr.aspx?prid=489*) (October 2004).

and Certified Novell Instructor (CNI) certifications. CompTIA's Project+ certification is also a Continuing Certification Requirement (CCR) to maintaining a Master CNE certification.

5. You can earn college credit with the Project+ certification. For example, CompTIA's Web site says that Capella University will provide six credit hours to someone with the Project+ certification. Many colleges and universities will grant credit based on your experience or other certifications. There is usually some fee involved, but it's often much less than the cost to take the courses.

Additional Information on the Project+ Exam

Because there are no experience or education requirements for taking the Project+ exam, you might want to take it very early in your career. Once you have enough experience and education to take the PMP exam, you might want to earn and maintain PMP certification.

As stated above, the Project+ exam consists of 80 questions. Table B-2 shows the approximate breakdown of questions on the Project+ exam by four domain areas. Candidates should review updated exam information on CompTIA's Web site to make sure they are prepared for the exams. For example, CompTIA provides a detailed list of objectives you should understand before taking the Project+ exam. Studying information in this book will also help you prepare for the Project+ exam. You might also want to purchase an exam guide to get specific information and access to more sample questions.

Table B-2: Breakdown of Questions on the Project+ Exam by Domain Areas

DOMAIN AREA	PERCENT OF QUESTIONS ON PROJECT+ EXAM
Project Initiation and Scope Definition	20
Project Planning	30
Project Execution, Control, and Coordination	43
Project Closure, Acceptance, and Support	7

Much of the advice for taking the PMP exam also applies to taking the Project+ exam. Many people find the questions to be similar on both exams, although the PMP exam is longer and more comprehensive. Below are some of the main differences between the content and types of questions you will find on the Project+ exam:

- The Project+ exam includes some scenarios and content specific to the information technology industry. For example, you should understand the various systems development life cycles and issues that often occur on information technology projects.

- You should understand the various roles of people on information technology projects, such as business analysts, database analysts, programmers, and so on.
- Although many of the questions are multiple-choice like the PMP exam questions, several questions involve choosing two or more correct answers. Several questions also involve matching or putting items in order, called drag-and-drop questions.
- The Project+ exam is not based on the *PMBOK® Guide*, so you do not need to know the processes involved in the various knowledge areas. However, much of the terminology, concepts, tools, and techniques are the same on both exams.

Sample Project+ Exam Questions

A few sample questions similar to those you will find on the Project+ exam are provided below. CompTIA also provides sample questions on their Web site at *www.comptia.org/certification/project/samplequestions.aspx*. You can check your answers at the end of this appendix.

1. Two software developers on your project disagree on how to design an important part of a system. There are several technologies and methodologies they could use. What should be the primary driver in deciding how to proceed?
    A. following corporate standards
    B. following industry standards
    C. meeting business needs
    D. using the lowest-cost approach

2. Match the following items to their descriptions:

    Stakeholder   A. Acts as a liaison between the business area and developers

    Project manager   B. Writes software code

    Business analyst   C. Person involved in or affected by project activities

    Programmer   D. Responsible for managing project activities

3. For a project to be successful, the project manager should strive to understand and meet certain goals. What are the three main project goals to meet? Select three answers.
    A. scope or performance goals
    B. time goals
    C. political goals
    D. cost goals
    E. stock price goals

4. You have received an incomplete project scope definition. Put the following actions in order of how you should proceed to complete them.

   A. Incorporate additional changes to the scope definition document.

   B. Review the draft scope definition document with your project team.

   C. Get signatures on the completed scope definition document.

   D. Rewrite the draft scope definition document with your users and project team.

5. What term is used to describe the process of reaching agreement on a collective decision?

   A. collaboration

   B. cooperation

   C. coordination

   D. consensus

6. What is a variance?

   A. a buffer in a duration estimate

   B. a small amount of money set aside for contingencies

   C. a form of risk management

   D. a deviation from the project plan

7. Which of the following would be legitimate reasons for a vendor to request a delay in delivering a product? Select two answers.

   A. The vendor may have underestimated the amount of time required to produce and deliver the product.

   B. The project contact from the vendor's organization may be going on vacation.

   C. The vendor might be able to provide a better product by delivering the product late.

   D. The vendor might lose money by delivering the product late.

8. When should you involve stakeholders in the change control process on information technology projects?

   A. before a change is submitted

   B. after a change is submitted

   C. when a change is submitted

   D. throughout the life of a project

9. Which of the following techniques can be used to help manage requirements? Select three answers.

   A. prototyping

   B. use case modeling

   C. JAD

   D. worst case modeling

10. Match the following items to their descriptions:

Lessons learned	A. Leave a clear and complete history of a project
Project audits	B. Review project progress and results
Project archives	C. Document what went right or wrong on a project

WHAT OTHER EXAMS OR CERTIFICATIONS RELATED TO PROJECT MANAGEMENT ARE AVAILABLE?

In recent years, several organizations have been developing more certifications related to project management and information technology project management, in particular. Some involve taking exams while others require coursework or attendance at workshops. Below are brief descriptions of other existing exams or certifications related to project management:

- Microsoft provides certification as a Microsoft Office Specialist (MOS) to recognize the ability to demonstrate proficiency in using its software products. You can earn certification for Microsoft Project at the core or comprehensive level. Consult Microsoft's Web site for detailed information on the MOS certification.

- The International Project Management Association (IPMA) offers a four-level certification program. The main requirements for each level are derived from typical activities, responsibilities, and requirements from practice. The IPMA four-level certification system, in descending order, includes the Certified Project Director, Certified Project Manager, Certified Project Management Professional, and Certified Project Management Practitioner. See IPMA's Web site (*www.ipma.org*) for more details.

- Certified IT Project Manager (CITPM): In 1998, the Singapore Computer Society collaborated with the Infocomm Development Authority of Singapore to establish an Information Technology Project Management Certification Program. PMI signed a Memorandum of Understanding with the Singapore Computer Society to support and advance the global credentialing of project management and information technology expertise.

- PMI's additional certifications. As mentioned earlier, PMI also offers certification as a Certified Associate in Project Management (CAPM). Candidates for the CAPM certification must also meet education and experience requirements and pass an exam. Applicants with a bachelor's degree need 1,500 hours of experience working on projects, and applicants without a bachelor's degree (but with a high school diploma or equivalent) need 2,500 hours of experience. Applicants also need at least 23 contact hours of project management education. For a while, PMI also offered Certificates of Added Qualification (CAQs) in specific industry areas, like the automotive and information systems fields, but PMI has cancelled the CAQs. Consult PMI's Web site for updated information on their certification programs.

- Several colleges, universities, and corporate training companies now provide their own certificate programs or entire degrees in project management. Typing "project management certificate" into google.com in October 2004 resulted in 17,800 hits. Some of the certificate courses apply toward bachelor's or advanced degrees, while many do not. Like any other educational program, it is important to research the quality of the program and find one that will meet your specific needs.

FINAL ADVICE ON CERTIFICATION AND PROJECT MANAGEMENT IN GENERAL

Now that you have read this text and discussed project management with others, I hope that you have grown to see project management as a valuable skill, especially in the information technology field. Project management certification can be your first step toward advancing your career in the fast-paced world of technology and business, for no matter how much things change, the need for projects and project managers is a constant.

The knowledge and experience I have gained working on and managing projects continues to help me in my career and my personal life. I was fortunate to step into a project management role very early in my career as an Air Force officer. My first real job at the age of 22 was as a project manager. I have held several job titles since then—systems analyst, senior engineer, technical specialist, information technology management consultant, independent consultant, college professor, and now author. All of these jobs included working on or managing projects. As a wife and mother of three, I can also attest to the fact that project management skills help in planning and executing social activities (weddings, birthday parties, fundraisers, and so on) and in dealing with the joys and challenges of everyday life.

DISCUSSION QUESTIONS

1. What is PMP certification, and why do you think the number of people earning it has grown so much in the past ten years?

2. What do you need to do before you can take the PMP exam? What is the exam itself like? What do you need to do to maintain PMP certification? What do you need to do to take the Project+ exam? How does the Project+ exam differ from the PMP exam? Do you need to renew Project+ certification?

3. What is the difference between conceptual, application, and evaluative questions? Which project management process groups have the most questions on the PMP exam? What are the four domain areas tested on the Project+ exam?

4. Which tips for taking the PMP exam do you think would be most helpful for you?

5. If you plan to take the Project+ or PMP exam soon, what should you do to prepare?

6. Briefly describe project management certification programs other than the PMP or Project+ certifications.

EXERCISES

1. Go to PMI's Web site and review the information about taking the PMP exam. Write a one- to two-page paper summarizing what you found.

2. Go to CompTIA's Web site and review the information about taking the Project+ exam. Write a one- to two-page paper summarizing what you found.

3. Answer the 20 sample PMP questions in this text or take another sample test (see the Suggested Readings or do an Internet search to find some sample PMP tests). Take the sample test, and then score your results. Summarize how you did and areas you would need to study before you could take the PMP exam.

4. Interview someone who has PMP or Project+ certification. Ask him or her why he or she earned the certification and how it has affected his or her career. Write your findings in a one- to two-page paper.

5. Do an Internet search on earning PMP or Project+ certification. Be sure to search for Yahoo! Groups related to these topics. What are some of the options you found to help people prepare for either exam? If you were to take one of the exams, what do you think you would do to help study for it? Do you think you would need additional information beyond what is in this text to help you pass? Write a one- to two-page paper describing your findings and opinions.

6. Read a recent issue of PMI's *PM Network* magazine. You can order a copy from PMI's Web site, or contact someone from a local PMI chapter to obtain a copy. Summarize all of the ads you find in the magazine related to earning PMP or other project management related certification in a one- to two-page paper. Also, include your opinion on which ad, course, book, CD-ROM, or other media appeals to you the most.

SUGGESTED READINGS

1. Amazon.com, search all products for "PMP" or "Project+" (*www.amazon.com*) (2004).

   There are hundreds of books, CD-ROMs, and audiotapes available for sale from Amazon.com or similar sites, with new offerings coming out every year. The "Dummy" series even published a book called PMP Certification for Dummies *in 2003. Read the descriptions and decide which resources best meet your needs and budget.*

2. CompTIA's Web Site (*www.comptia.org*) (2004).

> *If you're interested in earning Project+ certification, read the latest information and access sample questions from CompTIA's Web site.*

3. Free Simulated Practice Test (*www.pmstudy.com*) (2004).

> *This Web site provides a free sample PMP exam, including 200 questions you take online in 4 hours or less. You can also purchase additional practice exams or enroll in courses from this site.*

4. VLP's PMP Exam Online Self Assessment Test - 75 Sample Questions (*http://vl-p.net/pmp-online-self-test/75-free-questions.htm*).

> *This site provides 75 free sample questions you can complete online in 90 minutes or less. You can also download their pdf file called "25 Tips and Tricks for the PMP Exam."*

5. Project Management Institute (PMI) Standards Committee, *A Guide to the Project Management Body of Knowledge (PMBOK® Guide)*, 2004.

> *You get this guide on CD-ROM for free as part of your PMI membership. Students enrolled in an accredited, degree-granting program can join PMI at a discounted rate ($40 versus $129 for regular annual membership as of October 2004). Be sure to study the PMBOK® Guide 2004 since it provides the basis for what is on the PMP exam. Also, read recent information on the PMP certification from www.pmi.org. PMI also provides links to several sites that provide project management and PMP-related materials and courses.*

6. Yahoo Groups (*www.yahoo.group.com*) (2004).

> *There are 277 Yahoo Groups related to PMP certification as of December 2004. You can also start your own group at no cost, if you like. One of the most popular PMP groups (over 7,200 members as of December 2004) was at http://groups.yahoo.com/group/PMPCert.*

ANSWERS TO SAMPLE PMP EXAM QUESTIONS

1. C
2. D
3. B
4. B
5. A
6. B
7. D
8. A
9. C
10. A
11. B

12. C
13. C
14. C
15. D
16. D
17. C
18. A
19. B
20. B

ANSWERS TO SAMPLE PROJECT+ EXAM QUESTIONS

1. C
2. C, D, A, B
3. A, B, D
4. B, D, A, C
5. D
6. D
7. A, C
8. D
9. A, B, C
10. C, B, A

Appendix C
Additional Running Cases

INTRODUCTION

This appendix provides an additional running case with tasks covering each of the nine knowledge areas covered in Chapters 4-12. Additional running cases are available on the companion Web site (*www.course.com/mis/schwalbe/4e*). This particular case should be interesting to the video game enthusiast as well as to anyone interested in marketing and providing products and services primarily over the Internet. The main purpose of this and other running cases is to help you practice some of the project management skills you are developing as part of your course. Several of the tasks involve using templates provided in Appendix D or on the companion Web site. Instructors can download the suggested solutions for this and additional running cases from the password-protected section on Course Technology's Web site (*www.course.com*).

ADDITIONAL CASE 1: VIDEO GAME DELIVERY PROJECT

Part 1: Project Integration Management

An international marketing and distribution company has decided to provide a monthly video game rental program as a result of its market research. The Video Game Delivery Project involves developing a Web-based application and support structure to provide customers with video games on a monthly rental basis. For example, a customer would pay a monthly fee and then be able to order several video games over the Internet, receive the games via express mail, return those games via express mail, and keep receiving additional games. Several companies already provide this type of service for movie rentals. Market research and corporate values suggested your company focus on educational and sports-related video games only, and games would be available for all types of platforms, including popular gaming systems, computers, and learning systems used in preschools and elementary schools. You also plan to serve an international market, providing information and products in several different languages. This system must be very user-friendly, providing customers the ability to search for specific games by platform, age and gender appropriateness, customer reviews, sport (for sports-related games), language, and so on. Customers must be able to order the video games, track delivery and return of video games, pay online or via other payment methods (including credit or debit card, check, school payment systems, other electronic payment systems, etc.), and write reviews of the games they rent. The system must also be able to track referrals to the site from corporate partners and customer referrals, display advertisements, and track customer usage patterns.

Tasks

1. Research similar Web-based applications. Based on the scenario description of similar Web sites and your personal opinions, summarize the main functions that this system could provide. Rate the functions as mandatory, optional, and nice-to-have. Then provide an initial assessment of how difficult it would be to provide each function. Write your results in a one- to two-page paper, citing important references.

2. Prepare a weighted decision matrix using the template from Appendix D to evaluate providing the functions you identified in Task 1 above. Develop at least four criteria, assign weights to each criterion, assign scores, and then calculate the weighted scores for each function. Print the spreadsheet and bar chart with the results. Write a one-page paper describing this weighted decision matrix and summarize the results.

3. Prepare the financial section of a business case for the Video Game Delivery Project. Assume this project will take 12 months to complete and cost about $500,000, and monthly operating costs would be about $50,000 per month for year one and $60,000 per month for years two and three. Estimated benefits are about $1 million the first year after implementation and $2 million the following two years. Use the business case spreadsheet template from Appendix D to help calculate the NPV, ROI, and year in which payback occurs. Assume a 7 percent discount rate.

4. Prepare a project charter for the Video Game Delivery Project. Assume the project will take 12 months to complete and cost about $500,000. Use the project charter template and examples of project charters in Chapters 3 and 4 as guidelines.

5. You know that people will be requesting changes to the project and want to make sure you have a good integrated change control process in place. You also know that you want to address change requests as quickly as possible. Review the template for a change request form provided on the companion Web site (change_request.doc). Write a one- to two-page paper describing how you plan to manage changes on this project in a timely manner. Address who will be involved in making change control decisions, what paperwork/electronic systems will be used to collect and respond to changes, and other related issues.

Part 2: Project Scope Management

Congratulations! You have been selected as the project manager for the Video Game Delivery Project. The company's VP of marketing, Lori, is the project sponsor. Now you need to put together your project team and get to work on this high-visibility project. Top management has told you that you can hand pick your team. In addition, you will be working with several other companies on this project. Instead of developing all of the software yourselves, you'll use a Web-based application developed by ABC Corp. Of course, you'll need to customize the application somewhat to meet requirements for this project. ABC Corp.'s senior consultant, Gaurav, is your main contact with that company. You'll also be working with Edsys, an educational systems consulting firm that will help you in determining user requirements and developing partnership programs. Your company has never sold directly to schools, and senior management thinks it makes sense to get professional help in selling to that market segment. One of Edsys's top consultants, Julie, and some of her colleagues will assist you. Initial estimates suggest that about half of the $500,000 budgeted for this project will go to hardware costs and the outsourced software and consulting services. You will need two information technology professionals, two marketing specialists, and one purchasing specialist on your internal project team.

Tasks

1. Develop a scope statement for the project. Use the template provided in Appendix D and the example in Chapter 3 as guides. Be as specific as possible in describing product characteristics, requirements, and deliverables.

2. Develop a work breakdown structure (WBS) for the project. Break down the work to level 2 or level 3, as appropriate. Use the template in Appendix D and samples in Chapters 3 and 5 as guides. Print the WBS in list form as a Word file. Be sure to base your WBS on the project charter, scope statement, and other relevant information.

3. Use the WBS you developed in Task 2 above to create a Gantt chart in Project 2003 for the project. Use the outline numbering feature to display the outline numbers (click Tools on the menu bar, click Options, and then click Show outline number). Do not enter any durations or dependencies. Print the resulting Gantt chart on one page, being sure to display the entire Task Name column.

4. Develop a strategy for scope verification and change control for this project. Write a two-page paper summarizing key points of the strategy.

Part 3: Project Time Management

As project manager, you are actively leading the Video Game Delivery Project team in developing a schedule. The two information technology professionals on your team are Matt and Najwa, the marketing specialists are Magda and John, and the purchasing specialist is Nora. Recall that the project is expected to be completed in one year for $500,000. Assume that all of your internal team members are available to work up to 75 percent of their time on this project. Your project sponsor, Lori, has made it clear that it is important to meet or beat the one-year schedule goal. Your team has agreed to add a one-month buffer at the end of the project to ensure that you finish on time or early.

Tasks

1. Review the WBS and Gantt chart you created for Tasks 2 and 3 in Part 2. Propose three to five additional activities you think should be added to help you estimate resources and durations. Write a one-page paper describing these new activities.

2. Identify at least eight milestones for this project. Write a one-page paper describing each milestone using the SMART criteria.

3. Using the Gantt chart you created for Task 3 in Part 2, and the new activities and milestones you proposed in Tasks 1 and 2 above, estimate the task durations and enter dependencies, as appropriate. Remember that your schedule goal for the project is one year. Print the Gantt chart and network diagram.

4. Write a one-page paper summarizing how you would assign people to each activity. Include a table or matrix listing how many hours each person would work on each task. These resource assignments should make sense given the duration estimates made in Task 3 above.

5. Assume that your project team starts falling behind schedule. You know that it is crucial to meet or beat the one-year schedule goal. Describe strategies for making up lost time and avoiding schedule slips in the future.

Part 4: Project Cost Management

Your project sponsor has asked you and your team to refine the existing cost estimate for the project so that there is a solid cost baseline for evaluating project performance. Recall that your schedule and cost goals are to complete the project in one year or less for under $500,000. Also recall that you plan to purchase a Web-based application developed by ABC Corp. and customize it to meet requirements for this project. You'll also have consultants from Edsys, who will help you determine user requirements and develop partnership programs. Initial estimates suggested that about half of the $500,000 budgeted for this project would go to hardware costs and the outsourced software and consulting services.

Tasks

1. Prepare and print a one-page cost estimate for the project, similar to the one provided in Figure 7-1 in Chapter 7. Use the WBS level 1 categories provided below, and be sure to document assumptions you make in preparing the cost estimate. Assume a labor rate of $100/hour for the project manager and $60/hour for other project team members. Assume twice those numbers for outsourced labor.

   a. Project Management (internal)

   b. Requirements Definition

   c. Off-the-Shelf System Installation

   d. System Customization (for transferring/entering product and customer information and developing new capabilities like managing referrals, displaying advertisements, and tracking customer usage patterns)

   e. Testing

   f. Training, Roll Out, and Support

2. Using the cost estimate you created above, prepare a cost baseline by allocating the costs by WBS for each month of the project.

3. Assume you have completed six months of the project. The BAC was $500,000 for this 12-month project. Also assume the following:

   PV= $300,000

   EV= $280,000

   AC= $250,000

   Using this information, write a one- to two-page report that answers the following questions

   a. What is the cost variance, schedule variance, cost performance index (CPI), and schedule performance index (SPI) for the project?

   b. Use the CPI to calculate the estimate at completion (EAC) for this project. Use the SPI to estimate how long it will take to finish this project. Sketch an earned value chart using the above information, including the EAC point. See Figure 7-5 as a guide. Write a paragraph explaining what this chart shows.

   c. How is the project doing? Is it ahead of schedule or behind schedule? Is it under budget or over budget? Should you alert your senior management and ask for assistance? Should you talk to your sponsor before the meeting to discuss this data?

4. You notice that several of the tasks that involve getting inputs and approvals from stakeholders inside your own company (specifically the sales and legal departments) have taken longer to complete than planned. You have talked to the people in those departments several times, but everyone says they are doing the best they can. Write a one-page paper describing corrective action you could take to addresses this problem.

Part 5: Project Quality Management

The Video Game Delivery Project team is working hard to ensure that the new system meets expectations. The team has a detailed scope statement, schedule, and so on, but as the project manager, you want to make sure you'll satisfy key stakeholders, especially Lori, the project sponsor. Lori is also very sensitive to customer needs, so you have to make sure the new system is stable and easy to use. You know Lori wants the system to provide the most important capabilities within a year so the company can start making money from this new venture. You also know that you might need to make some tradeoffs in terms of cost and scope to meet the schedule goal.

Tasks

1. Develop a list of quality standards or requirements related to meeting the stakeholder expectations described above, especially for customers. Also provide a brief description of each requirement. For example, a requirement might be that the new system is available 24 hours a day, 7 days a week, and that any normal 12-year-old should be able to use it.

2. You have decided to expand the testing program to observe several potential customers using early versions of the system to get their feedback. Write a one- to two-page paper describing how this testing might work. Try to find information on how real companies get customer feedback such as this.

3. After analyzing results of early customer testing of the new system, you decide to create a Pareto diagram to easily see what problems or suggestions users reported. First, create a spreadsheet in Excel, using the data in the table below. List the most frequently described problems or suggestions first. Add a column called "% of Total" and another one called "Cumulative %". Then enter formulas to calculate those items. Next, use the Excel Chart Wizard to create a Pareto diagram based on this data. Use the Line – Column on 2 Axis custom type chart so your resulting chart looks similar to the one in Figure 8-1.

PROBLEMS/SUGGESTIONS	# OF TIMES REQUESTED
Shopping cart is not like others	18
No easy way to sort products by age, rating, cost, etc.	22
No product pictures	15
Can't use Paypal or similar payment options	3
System is too slow	10
Interface is boring	4

Part 6: Project Human Resource Management

Now that you have expanded the testing for the Video Game Delivery Project, you need to clarify who needs to do what for the customer-testing portion of the project. Recall that the team members include you, the project manager; Lori, your project sponsor; Matt and Najwa from the information technology department; Magda and John from the marketing department; Nora from the purchasing department; Gaurav from ABC Corp., who's providing the Web-based application and helping you customize it; and Julie, your main contact from Edsys, the educational systems consulting firm.

1. Prepare a responsibility assignment matrix based on the following information: The main tasks that need to be done for early customer testing are to determine the demographics for the testers, plan how many people and how many testing sessions will be held, find the testers, develop the test, prepare the facilities, create the survey and other methods for getting customer feedback, and analyze the results. Prepare a RACI chart to help clarify roles and responsibilities for these customer-testing tasks. Document key assumptions you make in preparing the chart.

2. You've decided to temporarily add additional people to the project to help run the customer testing. Based on feedback from your marketing specialists and the outside consultants, you decide that the customer testing will occur over a ten-day period. Assume that you'll need two marketing specialists and three technical specialists each day, yourself and one administrative assistant half time each day, one facilities expert for the first two days, and two assessment experts the last two days. Create a resource histogram, similar to the one in Figure 9-6, based on this information.

3. Customer testing has just started, but you realize that there isn't enough physical space or access to computers and other items everyone assumed would be available. You cannot add more people to address these problems, but you can suggest overtime, find additional computers, or provide other amenities to improve the work environment. Write a one-page paper describing what you can do quickly to address this situation.

Part 7: Project Communications Management

Several issues have arisen on the Video Game Delivery Project. Four months have passed since the project started. Gaurav and his company are complaining about not being paid appropriately. You initially thought you would only need a small amount of customization of the outsourced software, but now the supplier says the customization will cost another $100,000 over their initial budget. The purchasing specialist on your internal project team, Nora, does not like this supplier at all and gets very hostile at team meetings. Everyone else has been fairly happy with the supplier, and you weren't that surprised when they raised the customization estimate. Nora went to senior management (without telling you), and suggested terminating that supplier. Your two information technology specialists, Matt and Najwa, came and told you that they both feel that they are being underutilized on the project. As project manager, you have been getting short, weekly status reports from all of your team members, but many of them did not address challenges people are obviously facing.

1. Create an issue log for the project. List at least three issues and related information based on the scenario presented.
2. In addition to written weekly status reports, what else might you suggest to improve project communications? Summarize your ideas in a one-page paper.
3. Write a one- to two-page paper describing how you might approach two of the conflicts described above.

Part 8: Project Risk Management

Since several problems have been occurring on the Video Game Delivery project (see the running case information in Part 7 above), you have decided to be more proactive in managing risks. You also want to address positive and negative risks.

1. Create a risk register for the project, using Table 11-6 as a guide. Identify six potential risks, including at least two positive risks.
2. Plot the six risks on a probability/impact matrix, using Figure 11-6 as a guide. Assign a numeric value for the probability of each risk, and its impact on meeting the main project objectives. Use a scale of 1 to 10 to assign the values, with 1 being lowest and 10 being highest. For a simple risk factor calculation, multiply these two values (the probability score and the impact score). Enter the new data in the risk register. Write your rationale for how you determined the scores for one of the negative risks and one of the positive risks.

3. Develop a response strategy for one of the negative risks and one of the positive risks. Enter the information in the risk register. Write a separate paragraph describing what specific tasks would need to be done to implement the strategy. In addition, include time and cost estimates for each strategy.

Part 9: Project Procurement Management

After a monthly progress review meeting six months into your project, top management approved adding $100,000 to the budget for software customization. Lori, your project sponsor, was not happy with this increase in costs, and she suggested that a procurement audit be performed as soon as possible to avoid any future contractual problems. You have also decided to replace Nora, your internal purchasing specialist.

Tasks

1. Research information on procurement audits. Write a one- to two-page paper with your findings, and describe how a procurement audit might help your company.

2. Draft part of an RFP to have an outside expert/firm perform the procurement audit for this project. Assume it would take one month or less to complete, and you would require a report and presentation on the main findings and suggestions. Write a paragraph or two describing the purpose of the RFP, the organization's background, the basic requirements, and a description of the RFP process. Limit your answer to two or three pages. Also, document at least three major questions you would need answered before sending out the RFP.

3. Write a one-page position description for a new purchasing specialist for this project. Be specific in terms of educational and work experience requirements, personal traits, and so on. Include a requirement that this person work closely with project teams to help them know when contractors are not living up to expectations, how they can provide evidence to support their claims, and what they can do to prevent this type of problem from occurring.

4. Prepare a lessons-learned report for what you may have learned so far as project manager for this project. Use the template provided on the companion Web site (lessons_learned_report.doc) and be creative in your response.

Appendix D

Templates

INTRODUCTION

As mentioned throughout this text, using templates can help you prepare various project management documents, spreadsheets, charts, and other files. This appendix provides an organized listing of the templates used or described in this text and information on where to find additional templates. It also discusses how you can create your own templates to meet your specific project needs. The companion Web site for this text (*www.course.com/mis/schwalbe/4e*) and the author's home page (*www.kathyschwalbe.com*) both include a link to download all of the template files in one compressed file. To open the individual template files, you must have the application software that a file was created in. The companion Web site also includes links to many other Web sites containing templates based on results of a student project to find useful templates.

TEMPLATES FROM THIS TEXT

This text describes several project management templates. Table D-1 lists all the templates available to download from the companion Web site, and includes the template name, chapter number, project management process group(s) where you normally use the template, application software used to create it, and the file name for the template. Note that you could use other software to create some of the templates, and some of the templates can be used in other project management process groups. For example, the stakeholder analysis template was created using Word, but you could also create it using Excel. Several templates listed under planning could also be used for monitoring and controlling. The templates are listed in the order in which you might use them on a real project. Be careful to enter information into the templates carefully, and feel free to modify the templates to meet your particular project needs. For example, you could modify the template for a Software Project Management Plan to use it for any type of project's project management plan. All of the template files listed in Table D-1 are contained in the file named templates.zip on the companion Web site and author's site.

Table D-1: Templates Available for Download on the Companion Web Site

TEMPLATE NAME	PROCESS GROUP	CHAPTER(S) WHERE USED	APPLICATION SOFTWARE	FILENAME
Kickoff Meeting	Initiating	Appendix A	Word	kickoffmeeting.doc
Business Case	Initiating	3	Word	business_case.doc
Business Case Financial Analysis	Initiating	3,4	Excel	business_case_financials.xls
Payback Chart	Initiating	4	Excel	payback.xls
Weighted Decision Matrix	Initiating	4, 12	Excel	wtd_decision_matrix.xls
Project Charter	Initiating	3, 4, 5	Word	charter.doc
Team Contract	Planning	3	Word	team_contract.doc
Scope Statement	Planning	3, 4, 5	Word	scope_statement.doc
Statement of Work	Planning	12	Word	statement_of_work.doc
Request for Proposal	Planning	12	Word	rfp_outline.doc
Stakeholder Analysis	Planning	4	Word	stakeholder_analysis.doc
Software Project Management Plan	Planning	4	Word	sw_project_mgt_plan.doc
Work Breakdown Structure	Planning	3, 5, 6	Word	wbs.doc
Gantt Chart	Planning, Executing	3, 5, 6	Project	Gantt_chart.mpp
Network Diagram	Planning, Executing	3, 6	Project	network_diagram.mpp
Project Cost Estimate	Planning	7	Excel	cost_estimate.xls
Earned Value Data and Chart	Monitoring and Controlling	7	Excel	earned_value.xls
Quality Assurance Plan	Planning, Executing	8	Word	quality_assurance_plan.doc
Pareto Diagram	Monitoring and Controlling	8	Excel	pareto_diagram.xls
Project Organizational Chart	Planning, Executing	9	PowerPoint	project_org_chart.ppt
Responsibility Assignment Matrix	Planning, Executing	9	Excel	ram.xls
Resource Histogram	Planning, Executing	9	Excel	resource_histogram.xls

Table D-1: Templates Available for Download on the Companion Web Site (continued)

Template Name	Process Group	Chapter(s) Where Used	Application Software	Filename
Communications Management Plan	Planning	10	Word	comm_plan.doc
Project Description (text)	Planning	10	Word	project_desc_text.doc
Project Description (Gantt chart)	Planning	10	Project	project_desc_Gantt.mpp
Milestone Report	Executing	3, 6	Word	milestone_report.doc
Change Request Form	Planning, Monitoring and Controlling	4	Word	change_request.doc
Status/Progress Report	Monitoring and Controlling	3, 10	Word	status_report.doc
Expectations Management Matrix	Monitoring and Controlling	10	Word	expectations.doc
Issue Log	Monitoring and Controlling	10	Word	issue_log.doc
Probability/ Impact Matrix	Planning, Executing, Monitoring and Controlling	11	PowerPoint	prob_impact_matrix.ppt
List of Prioritized Risks	Planning, Executing, Monitoring and Controlling	3, 11	Word	list_of_risks.doc
Risk Register	Planning, Monitoring and Controlling	11	Excel	risk_register.xls
Top 10 Risk Item Tracking	Planning, Monitoring and Controlling	11	Excel	top_10.xls
Breakeven/Sensitivity Analysis	Planning	11	Excel	breakeven.xls
Client Acceptance Form	Closing	4, 10	Word	client_acceptance.doc
Lessons-Learned Report	Closing	3, 4, 10	Word	lessons_learned_report.doc
Final Project Documentation	Closing	3, 4, 10	Word	final_documentation.doc

TEMPLATES FROM OTHER SOURCES

In the spring of 2003, four students in a project management course at the University of Minnesota did a group project to help find and evaluate templates. Bill Hunt was the project manager, and his team members included Heather Fox, Kinwai Chan, and Roy Musselman. All four students took the class in a virtual environment. I was the sponsor for the project and their instructor for the course, which met one night a week from mid-January through early May 2003. You can access the team's Web site with links to templates found from the companion Web site for this text.

In their preliminary project report, one of the first product-related deliverables for the project, the students reported finding 54 unique Web sites that offered project management templates. Thirty-five of these sites offered free templates with no required registration or membership, 11 sites required a registration or free membership to access the templates, and 18 sites required a fee for either membership or for direct purchase of the templates. The students used several search engines and entered keywords such as "free project management templates" and similar variations.

Table D-2 provides a count of the templates the students found by knowledge area and process group. Note that the students found 333 templates. Several of these templates fulfilled the same purpose, such as handling a change request. Figure D-1 provides a bar chart showing the number of templates found by knowledge area. Figure D-2 shows the percentage of templates found by progress group in a pie chart. Note that most of the templates—41 percent—fit under the planning process group, followed by the monitoring and controlling process group with 32 percent. The majority of the templates fell under the communications and scope management knowledge areas, with 73 (22 percent) and 70 (21 percent) each, respectively. These results make sense given the nature of the documentation involved in managing projects. It is difficult to understand the scope of many projects and communicate project information, so many templates are needed in those areas. There are also a wide variety of ways you can provide scope and communications information.

Table D-2: Number of Templates Found by Knowledge Area and Process Group

KNOWLEDGE AREA	PROJECT PROCESS GROUPS					
	INITIATING	PLANNING	EXECUTING	MONITORING AND CONTROLLING	CLOSING	TOTAL
Integration		15	7	30		52
Scope	35	23		12		70
Time		10		8		18
Cost		14		4		18
Quality		11	14	5		30
Human Resource		13	4			17
Communications		12	11	35	15	73
Risk		38		12		50
Procurement		4	1		0	5
Total	**35**	**140**	**37**	**106**	**15**	**333**

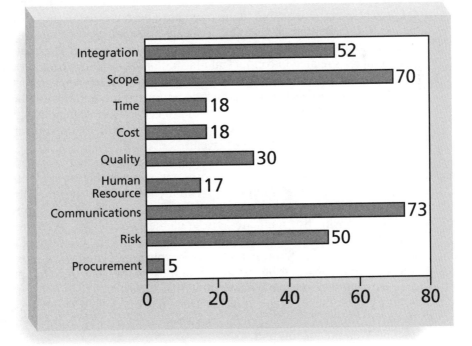

Figure D-1. Number of Templates Found by Knowledge Area

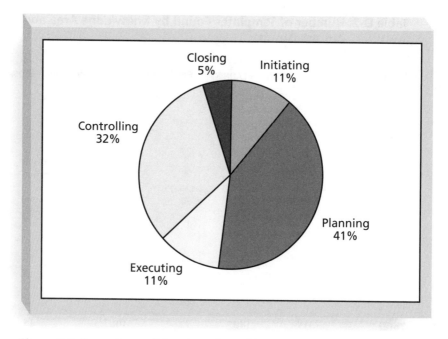

Figure D-2. Percentage of Templates Found by Project Management Process Group

As you can see, there are plenty of templates available to assist in most areas of project management, and many more have been created in the last year alone. This student project was by no means an exhaustive search. For example, typing "contract templates" into google.com produces several sites that provide free templates related to contract administration, and typing "contract closeout templates" produces several sites with free templates related to contract closeout.

Many software packages also include template files. You learned how to access several template files to use with Project 2003 in Appendix A. Word, Excel, and PowerPoint also include several template files, and you can save your own files as template files as well.

Table D-3 lists the Web site names for the sites found that provided 20 or more free templates. To help avoid information overload, the students evaluated all of the sites they found and recommended which ones they thought were most valuable. Table D-4 lists the students' ranking of the "top ten" Web sites for accessing free templates. The students evaluated the Web sites based on the overall quality of the free templates as well as the quantity of templates available. You can determine for yourself which templates are of most value in meeting your particular needs. Note that most of the Web sites that provide free templates are created by government agencies, which do not profit from selling project management services. They may make their templates available as part of their public service. You can access all of these sites through the companion Web site for this text.

Table D-3: Web Sites with 20 or More Free Templates

NUMBER OF FREE TEMPLATES	WEB SITE NAME
50	allPM.com
44	TenStep.com
40	California Department of Finance Statewide Information Technology Web Site
40	University of Waterloo Information Systems and Technology—Project Templates
30	Minnesota Office of Technology
23	New York State Office for Technology
22	Michigan Department of Information Technology
20	smartdraw.com
20	U.S. Department of Energy, Office of the Chief Information Officer
20	Process Impact
20	Army Training Information Management Program
20	State of Texas Department of Information Resources
20	Project Management—Tasmanian State Government
20	ProjectWare

Table D-4: Students' "Top Ten" Web Sites with Free Templates

NUMBER OF FREE TEMPLATES	WEB SITE NAME
22	Michigan Department of Information Technology
28	Minnesota Office of Technology
23	New York State Office for Technology
12	Tasmanian State Government
28	U.S. Department of Energy
52	University of Waterloo Information Systems & Technology
12	Method 123
19	ProjectWare
14	State of Texas Department of Information Resources
8	U.S. Defense Logistics Agency

CREATING YOUR OWN TEMPLATES

If you cannot find an existing template to meet your needs, you can create your own. It often helps to review other templates to get ideas from them and then prepare several draft templates for people in your organization to review. You can save files as template files in most software applications, or you can simply save them as regular files and make them easily available in your organization. Consult the Help facility of your software application to learn how to save files as templates. Be sure to find out if you need formal approval to create or use new templates in your organization and work within the proper channels. Remember that the main goal of using templates is to make it easier for people to manage projects.

In addition, be sure to follow copyright laws. All of the templates created for this text (available on the companion Web site) have no copyright restrictions. You can use them freely. Most of the free templates listed on the companion Web site also have no copyright restrictions. If there is a copyright statement on a template, you must ask for permission to use the form from that source first.

I also create templates to help me teach my project management classes. For example, I use a survey template to collect information on students in every class I teach. The template includes important contact information; asks students about their strengths, weaknesses, and what they want to learn in the class; and provides an easy means for getting signatures to meet academic honesty policies. I also sometimes provide templates to help students propose potential projects to do as part of a course in project management or other courses. Since the project teams normally include three to five students, we get more than enough good projects for students to work on.

These two examples illustrate how you can create templates to meet your specific needs. Similar to the example above, if you want to get to know stakeholders on a project, you could design your own survey template to collect information about them. If you want people to propose project ideas, you also could develop a template for that purpose.

As you can see, templates can assist project managers, team members, and instructors in doing their jobs. There are many templates available from this text, and many more available from numerous Web sites. It is also relatively easy to create your own templates based on your unique needs.

Appendix E

Fissure Project Management Simulation

INTRODUCTION

Many people enjoy using simulation software to help them learn. This appendix briefly describes Fissure's project management simulation demo, which is included on the CD-ROM with this text. Fissure (*www.fissure.com*) is a PMI Registered Education Provider (REP) specializing in project management, software management, leadership, and organizational change training. Fissure has used an expanded version of this simulation tool to help thousands of people learn how to apply various project management concepts.

You can access more detailed instructions for using this simulation, including several screen shots and suggestions on how to use it in various learning settings, from the companion Web site (*www.course.com/mis/schwalbe/4e*). This simulation uses many of the concepts, tools, and techniques described throughout this text. You should be familiar with at least Chapters 1 through 7 before running the simulation. Feel free to consult the text to reinforce your understanding of some of the topics included in the simulation.

INTRODUCTION TO FISSURE'S SIMULATION SOFTWARE

Fissure's project management simulation software is based on its SimProject engine, driven by a specific project database. The demo version of the simulation software included with this text includes a fairly simple, 11-week project database consisting of seven tasks and ten potential team members. Fissure estimates that it will take about three to four hours to run the demo simulation. Contact Fissure (*www.fissure.com* or 1-877-877-6333) for more information on other challenging project and leadership simulations, as well as other products and services.

Table E-1 describes the project. A similar description of the project is provided in the project definition document included in the References section of the simulation software.

Table E-1: Simulation Project Information

DESCRIPTION:

The Alliance Prototype project represents a strategic effort to augment the sales and marketing capabilities of Uniworld with an e-commerce equipped Web site. This Web site is needed to help Uniworld regain its global market dominance. To validate this objective and to gain a better understanding of Web site deployment, the company has budgeted $40,000 for the development of a prototype Web site. Marketing has stated that it needs this prototype site to be fully demonstratable within 11 weeks in order to make a determination to proceed with a fully functional site.

DELIVERABLES:

The Alliance Prototype project will provide hardware and software required for establishing an e-commerce Web site that will demonstrate the feasibility of developing a fully functional site capable of taking product orders and generating product fulfillment. The deliverables for this project are the demonstration Web site, site documentation, and a recommended product architecture.

PERFORMANCE:

The Alliance Prototype Web site will be accepted upon successful completion of the customer acceptance test with no more than seven known defects, which will represent an estimated Mean Time To Failure (MTTF) of five days. Furthermore, no known defects can be present that can cause potential system failure or major loss of functionality. The Web site must be capable of demonstrating the feasibility of handling a peak load of 2000 hits per hour without appreciable degradation in system response.

To participate in this simulated project, you will be expected to read about the company, project, and people available to work on this project. You will plan your project and make typical project decisions each week, such as when to assign which staff, when to hold meetings, when to send staff to various training opportunities, and so on. You will run your project a week at a time, analyzing your results each week, referring to your weekly reports, and making your decisions for the next week. As you run each week, you will be presented with communications from people within the company, team members, or other stakeholders related to the project. You will have a choice on how to respond to these communications, and all the decisions you make will impact how your project progresses.

You can close the simulation at any time, but your work will not be saved in this demo version. You can also run the simulation as many times as you like, and the results will vary based on your decisions.

INSTALLING AND LAUNCHING THE SIMPROJECT SIMULATION

Before you begin working with the simulated project, you need to install the simulation software.

To load the simulation software:

1. Insert the Fissure Project Management Simulation Demo CD-ROM into your CD-ROM drive. This CD-ROM is attached to the inside front cover of this text.

2. Click the **Start** button on the taskbar, and then click **Run**.

3. In the Run dialog box, click **Browse**. Click the **Look in list arrow** and select the drive letter associated with your computer's CD-ROM drive. Double-click the **SimProject Demo** icon, and then click **OK**.

4. Follow the installation instructions. The installation takes about 1-2 minutes, and you will need to restart your computer before running the simulation software.

 It is recommended that you use Microsoft Windows 2000 Professional or XP to run the simulation software. If you have any technical problems, contact Course Technology's Support Services at *www.course.com/support*.

Once installation is complete, you can launch SimProject and take a tour.

To launch SimProject and take a tour:

1. Click the **Start** button on the taskbar, select **Programs**, and then select **SimProject**.

2. Move your mouse to the right to click **SimProject**.

3. Click the **green parrot** in the upper-right area of your screen. Your screen should resemble Figure E-1. The green parrot, Peedy, is your project assistant and guide for using the simulation tool. Notice the other main pictures or reference buttons on the screen: References, Planning, Staffing, Team Interaction, Stakeholders, Reports, and Work Week 1. Click **Yes** to explore the software.

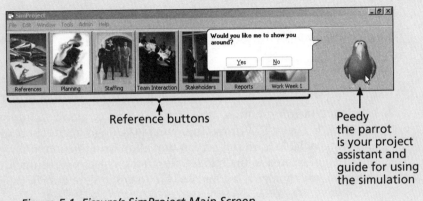

Figure E-1. Fissure's SimProject Main Screen

RUNNING THE SIMPROJECT SIMULATION

To familiarize yourself with the simulation software, Peedy will show you what each of the reference buttons (References, Planning, Staffing, Team Interaction, Stakeholders, Reports, and Work Week 1) contain and explain the basic process for running the simulation. The simulation includes audio, and it also displays the text so you can see what Peedy is saying. Listen closely to the instructions, but you can run them again as needed. Also don't be surprised to hear Peedy sleeping in the background if you open the simulation tool and then neglect it!

One of the last elements Peedy describes are the items available under the Reports button, such as the earned value graph, as shown in Figure E-2. See Chapter 7 for more information on earned value management.

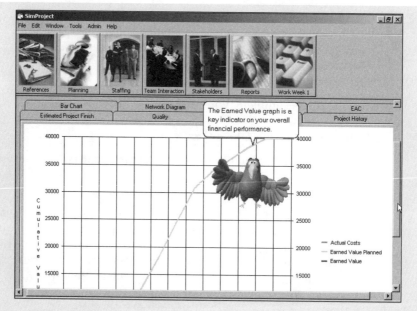

Figure E-2. Earned Value Graph

After you complete Peedy's tour, you can begin the project.

To start running your project:

1. Click the **Reference** button to read important reference information.
2. Click the **Planning** button to enter your resource planning information for Week 1. Figure E-3 shows the initial Planning screen. There are several tabs available from this screen that allow you to enter and view information related to the budget, resources, risks/opportunities, education, and stakeholder relations. Be sure to enter all of the appropriate

planning information, and check to make sure your budget does not exceed the approved budget.

The detailed version of this appendix on the companion Web site provides many more sample screen shots to guide you through the planning process.

Work Breakdown Structure	Task	Approved Budget	Plan Budget	Spent Budget
Alliance Project		**$40000**	**$6660**	$0
1.0 Project Management		**$8500**	**$1500**	$0
1.1 Education		$1500	$0	$0
1.2 Stakeholder Relations		$1000	$0	$0
1.3 Reserves		$4000	$0	$0
1.4 Team / Individual Recognition		$500	$0	$0
1.5 Materials		$1500	$1500	$0
2.0 Design		**$8700**	**$5160**	$0
2.1 Systems Design	1	$4500	$5160	$0
2.2 Network Design / Implementation	2	$4200	$0	$0
3.0 Implementation		**$14500**	**$0**	$0
3.1 Software Implementation	3	$10600	$0	$0
3.2 Systems Integration	5	$3900	$0	$0
4.0 Documentation		**$4100**	**$0**	$0
4.1 Product Documentation	4	$4100	$0	$0
5.0 Systems Test & Review		**$4200**	**$0**	$0
5.1 Customer Acceptance Test	6	$2700	$0	$0
5.2 Documentation Review	7	$1500	$0	$0

Figure E-3. Planning Screen

3. Update your plan, including staffing, training, education, and stakeholder interactions for the week *before* you run the simulation for that week. Select the display of your resource plan and then click the **Staffing** button to see your resource plan in the background as you make your weekly staffing decisions. Decisions you need to make each week before "running" the simulation for the week include the following:

- Staffing: Select individuals you want to add to your team or remove from your team. Note that there is a "quick look" button to your resource plan located in the lower-left corner of the window.
- Task Assignments: Assign individuals to work on specific tasks and request overtime. You can again use the "quick look" button in the lower-left corner of the window to see your resource plan.
- Team Meetings: Select the day, the duration, and topic(s) for meetings.
- Individual Conferences: Select the individual, the day, the duration, and the topic(s).

- Training: Select the individual or team and the specific training.
- Recognition: Select the individual or team and the recognition.
- Stakeholder Relations: Select the stakeholder and the interaction.

4. To run the simulation for the week, click **Work Week**. This will indicate the number of the project week you are running. As you run the week's simulation, you will be presented with communications from various sources. Read the communications carefully and choose your response. The decisions you make prior to running each week's simulation and the decisions you made in earlier weeks will determine the results you see for the project. At the completion of each week, a report of the week's decisions and results will be displayed. This report can be printed, and it is also viewable via the History tab within the Reports screen. Figure E-4 provides an example of the Work Week report. This report was run without entering any new planning information for Week 1 in the simulation. Make sure you enter all planning information *before* running the Work Week report.

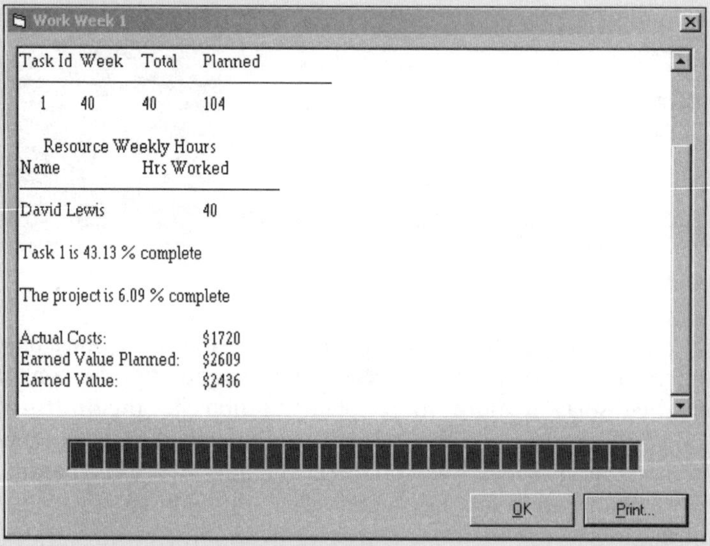

Figure E-4. Sample Work Week Report

5. Click **Reports** to review various project management reports to help you assess your current project status each week. Reports include the following:
- Bar Chart: Displays a Gantt chart showing a project timetable and task status or progress
- Network Diagram: Shows task dependencies/sequences
- Project Performance Index (PPI): Shows the percentage of the budget spent/the percentage of the project complete

- Estimate at Completion (EAC): Provides an estimate for the cost of the project at completion, assuming the project performance continues at the same pace
- Estimated Project Finish: Provides an estimate for the time when the project will be complete, assuming the project performance continues at the same pace
- Quality: Shows the number of open defects reports
- Earned Value: Displays a comparison of planned value (called Earned Value Planned in this simulation), actual cost, and earned value
- Project History: Provides a report of each week's decisions and results

Look at the reports carefully each week and use them to help make your weekly decisions. Don't forget to keep your plans updated, especially your resource plan. Any of this information can be printed at any time. Review the results of your simulation. Did you do a good job managing the project? What did you learn by running the simulation? You might want to run the simulation more than once to practice your project management skills.

Glossary

A

acceptance decisions — decisions that determine if the products or services produced as part of the project will be accepted or rejected

activity — an element of work, normally found on the WBS, that has an expected duration and cost, and expected resource requirements; also called task

activity attributes — information about each activity, such as predecessors, successors, logical relationships, leads and lags, resource requirements, constraints, imposed dates, and assumptions related to the activity

activity definition — identifying the specific activities that the project team members and stakeholders must perform to produce the project deliverables

activity duration estimating — estimating the number of work periods that are needed to complete individual activities

activity list — a tabulation of activities to be included on a project schedule

activity-on-arrow (AOA) or **arrow diagramming method (ADM)** — a network diagramming technique in which activities are represented by arrows and connected at points called nodes to illustrate the sequence of activities

activity sequencing — identifying and documenting the relationships between project activities

actual cost (AC) — the total of direct and indirect costs incurred in accomplishing work on an activity during a given period, formerly called the actual cost of work performed (ACWP)

adaptive software development (ASD) — a software development approach used when requirements cannot be clearly expressed early in the life cycle

agile software development — a method for software development that uses new approaches, focusing on close collaboration between programming teams and business experts

analogous estimates — a cost estimating technique that uses the actual cost of a previous, similar project as the basis for estimating the cost of the current project, also called **top-down estimates**

analogy approach — creating a WBS by using a similar project's WBS as a starting point

appraisal cost — the cost of evaluating processes and their outputs to ensure that a project is error-free or within an acceptable error range

B

backward pass — a project network diagramming technique that determines the late start and late finish dates for each activity in a similar fashion

balanced scorecard — a methodology that converts an organization's value drivers to a series of defined metrics

baseline — the approved project management plan plus approved changes

baseline dates — the planned schedule dates for activities in a Tracking Gantt chart

benchmarking — a technique used to generate ideas for quality improvements by comparing specific project practices or product characteristics to those of other projects or products within or outside the performing organization

bid — also called a tender or quote (short for quotation), a document prepared by sellers providing pricing for standard items that have been clearly defined by the buyer

blogs — easy to use journals on the Web that allow users to write entries, create links, and upload pictures, while readers can post comments to journal entries

bottom-up approach — creating a WBS by having team members identify as many specific tasks related to the project as possible and then grouping them into higher level categories

bottom-up estimates — a cost estimating technique based on estimating individual work items and summing them to get a project total

brainstorming — a technique by which a group attempts to generate ideas or find a solution for a specific problem by amassing ideas spontaneously and without judgment

budget at completion (BAC) — the original total budget for a project

budgetary estimate — a cost estimate used to allocate money into an organization's budget

buffer — additional time to complete a task, added to an estimate to account for various factors

burst — when a single node is followed by two or more activities on a network diagram

C

Capability Maturity Model (CMM) — a five-level model laying out a generic path to process improvement for software development in organizations

Capability Maturity Model Integration (CMMI) — a process improvement model that addresses software engineering, system engineering, and program management and is replacing the older CMM ratings

cash flow — benefits minus costs or income minus expenses

cash flow analysis — a method for determining the estimated annual costs and benefits for a project

champion — a senior manager who acts as a key proponent for a project

change control board (CCB) — a formal group of people responsible for approving or rejecting changes on a project

change control system — a formal, documented process that describes when and how official project documents may be changed

closing processes — formalizing acceptance of the project or project phase and ending it efficiently

COCOMO II — a newer, computerized cost-estimating tool based on Boehm's original model that allows one to estimate the cost, effort, and schedule when planning a new software development activity

coercive power — using punishment, threats, or other negative approaches to get people to do things they do not want to do

communications infrastructure — a set of tools, techniques, and principles that provide a foundation for the effective transfer of information among people

communications management plan — a document that guides project communications

communications planning — determining the information and communications needs of the stakeholders: who needs what information, when will they need it, and how will the information be given to them

compromise mode — using a give-and-take approach to resolving conflicts; bargaining and searching for solutions that bring some degree of satisfaction to all the parties in a dispute

computerized tools — cost-estimating tools that use computer software, such as spreadsheets and project management software

configuration management — a process that ensures that the descriptions of the project's products are correct and complete

conformance — delivering products that meet requirements and fitness for use

conformance to requirements — the project processes and products meet written specifications

confrontation mode — directly facing a conflict using a problem-solving approach that allows affected parties to work through their disagreements

constructive change orders — oral or written acts or omissions by someone with actual or apparent authority that can be construed to have the same effect as a written change order

Constructive Cost Model (COCOMO) — a parametric model developed by Barry Boehm for estimating software development costs

contingency plans — predefined actions that the project team will take if an identified risk event occurs

contingency reserves — dollars included in a cost estimate to allow for future situations that may be partially planned for (sometimes called **known unknowns**) and are included in the project cost baseline

contingency reserves or **contingency allowances** — provisions held by the project sponsor or organization to reduce the risk of cost or schedule overruns to an acceptable level

contract — a mutually binding agreement that obligates the seller to provide the specified products or services, and obligates the buyer to pay for them

control chart — a graphic display of data that illustrates the results of a process over time

cost baseline — a time-phased budget that project managers use to measure and monitor cost performance

cost budgeting — allocating the overall cost estimate to individual work items to establish a baseline for measuring performance

cost control — controlling changes to the project budget

cost estimating — developing an approximation or estimate of the costs of the resources needed to complete the project

cost management plan — a document that describes how cost variances will be managed on the project

cost of capital — the return available by investing the capital elsewhere

cost of nonconformance — taking responsibility for failures or not meeting quality expectations

cost of quality — the cost of conformance plus the cost of nonconformance

cost performance index (CPI) — the ratio of earned value to actual cost; can be used to estimate the projected cost to complete the project

cost plus fixed fee (CPFF) contract — a contract in which the buyer pays the supplier for allowable performance costs plus a fixed fee payment usually based on a percentage of estimated costs

cost plus incentive fee (CPIF) contract — a contract in which the buyer pays the supplier for allowable performance costs along with a predetermined fee and an incentive bonus

cost plus percentage of costs (CPPC) contract — a contract in which the buyer pays the supplier for allowable performance costs along with a predetermined percentage based on total costs

cost-reimbursable contracts — contracts involving payment to the supplier for direct and indirect actual costs

cost variance (CV) — the earned value minus the actual cost

crashing — a technique for making cost and schedule trade-offs to obtain the greatest amount of schedule compression for the least incremental cost

critical chain scheduling — a method of scheduling that takes limited resources into account when creating a project schedule and includes buffers to protect the project completion date

critical path — the series of activities in a network diagram that determines the earliest completion of the project; it is the longest path through the network diagram and has the least amount of slack or float

critical path method (CPM) or **critical path analysis** — a project network analysis technique used to predict total project duration

D

decision tree — a diagramming analysis technique used to help select the best course of action in situations in which future outcomes are uncertain

decomposition — subdividing project deliverables into smaller pieces

defect — any instance where the product or service fails to meet customer requirements

Define, Measure, Analyze, Improve, Control (DMAIC) — a systematic, closed-loop process for continued improvement that is scientific and fact based

definitive estimate — a cost estimate that provides an accurate estimate of project costs

deliverable — a product or service, such as a report, a training session, a piece of hardware, or a segment of software code, produced or provided as part of a project

Delphi technique — an approach used to derive a consensus among a panel of experts, to make predictions about future developments

dependency — the sequencing of project activities or tasks; also called a **relationship**

deputy project managers — people who fill in for project managers in their absence and assist them as needed, similar to the role of a vice president

design of experiments — a quality technique that helps identify which variables have the most influence on the overall outcome of a process

direct costs — costs that can be directly related to producing the products and services of the project

directives — new requirements imposed by management, government, or some external influence

discount factor — a multiplier for each year based on the discount rate and year

discount rate — the rate used in discounting future cash flow; also called the capitalization rate or opportunity cost of capital

discretionary dependencies — sequencing of project activities or tasks defined by the project team and used with care since they may limit later scheduling options

dummy activities — activities with no duration and no resources used to show a logical relationship between two activities in the arrow diagramming method of project network diagrams

duration — the actual amount of time worked on an activity *plus* elapsed time

E

early finish date — the earliest possible time an activity can finish based on the project network logic

early start date — the earliest possible time an activity can start based on the project network logic

earned value (EV) — the percentage of work actually completed multiplied by the planned cost, formerly called the **budgeted cost of work performed (BCWP)**

earned value management (EVM) — a project performance measurement technique that integrates scope, time, and cost data

effort — the number of workdays or work hours required to complete a task

empathic listening — listening with the intent to understand

enterprise project management software — software that integrates information from multiple projects to show the status of active, approved, and future projects across an entire organization

estimate at completion (EAC) — an estimate of what it will cost to complete the project based on performance to date

executing processes — coordinating people and other resources to carry out the project plans and produce the products, services, or results of the project or project phase

expectations management matrix — a tool to help understand unique measures of success for a particular project

executive steering committee — a group of senior executives from various parts of the organization who regularly review important corporate projects and issues

expected monetary value (EMV) — the product of the risk event probability and the risk event's monetary value

expert power — using one's personal knowledge and expertise to get people to change their behavior

external dependencies — sequencing of project activities or tasks that involve relationships between project and non-project activities

external failure cost — a cost related to all errors not detected and corrected before delivery to the customer

extreme programming (XP) — an approach to programming for developing software in rapidly changing environments

extrinsic motivation — causes people to do something for a reward or to avoid a penalty

F

fallback plans — plans developed for risks that have a high impact on meeting project objectives, to be implemented if attempts to reduce the risk are not effective

fast tracking — a schedule compression technique in which you do activities in parallel that you would normally do in sequence

features — the special characteristics that appeal to users

feeding buffers — additional time added before tasks on the critical path that are preceded by non-critical-path tasks

finish-to-finish dependency — a relationship on a project network diagram where the "from" activity must be finished before the "to" activity can be finished

finish-to-start dependency — a relationship on a project network diagram where the "from" activity must be finished before the "to" activity can be started

fishbone diagrams — diagrams that trace complaints about quality problems back to the responsible production operations or root cause; sometimes called **Ishikawa diagrams**

fitness for use — a product can be used as it was intended

fixed-price or lump-sum contracts — contracts with a fixed total price for a well-defined product or service

flowcharts — diagrams that show how various elements of a system relate to each other

forcing mode — using a win-lose approach to conflict resolution to get one's way

forecasts — used to predict future project status and progress based on past information and trends

forward pass — a network diagramming technique that determines the early start and early finish dates for each activity

free slack (free float) — the amount of time an activity can be delayed without delaying the early start of any immediately following activities

functionality — the degree to which a system performs its intended function

functional organizational structure — an organizational structure that groups people by functional areas such as information technology, manufacturing, engineering, and human resources

function points — technology-independent assessments of the functions involved in developing a system

G

Gantt chart — a standard format for displaying project schedule information by listing project activities and their corresponding start and finish dates in a calendar format; sometimes referred to as bar charts

groupthink — conformance to the values or ethical standards of a group

H

hierarchy of needs — a pyramid structure illustrating Maslow's theory that people's behaviors are guided or motivated by a sequence of needs

human resources frame — focuses on producing harmony between the needs of the organization and the needs of people

I

indirect costs — costs that are not directly related to the products or services of the project, but are indirectly related to performing the project

influence diagrams — diagrams that represent decision problems by displaying essential elements, including decisions, uncertainties, and objectives, and how they influence each other

information distribution — making needed information available to project stakeholders in a timely manner

initiating processes — defining and authorizing a project or project phase

intangible costs or **benefits** — costs or benefits that are difficult to measure in monetary terms

integrated change control — identifying, evaluating, and managing changes throughout the project life cycle

integration testing — testing that occurs between unit and system testing to test functionally grouped components to ensure a subset(s) of the entire system works together

interface management — identifying and managing the points of interaction between various elements of a project

internal failure cost — a cost incurred to correct an identified defect before the customer receives the product

internal rate of return (IRR) — the discount rate that results in an NPV of zero for a project

interviewing — a fact-finding technique that is normally done face-to-face, but can also occur through phone calls, e-mail, or instant messaging

intrinsic motivation — causes people to participate in an activity for their own enjoyment

ISO 9000 — a quality system standard developed by the International Organization for Standardization (ISO) that includes a three-part, continuous cycle of planning, controlling, and documenting quality in an organization

ISO 15504 — a framework for the assessment of software processes developed by the International Organization for Standardization (ISO)

issue — a matter under question or dispute that could impede project success

issue log — a tool to document and monitor the resolution of project issues

J

Joint Application Design (JAD) — using highly organized and intensive workshops to bring together project stakeholders—the sponsor, users, business analysts, programmers, and so on—to jointly define and design information systems

K

kickoff meeting — a meeting held at the beginning of a project or project phase where all major project stakeholders discuss project objectives, plans, and so on

L

late finish date — the latest possible time an activity can be completed without delaying the project finish date

late start date — the latest possible time an activity may begin without delaying the project finish date

leader — a person who focuses on long-term goals and big-picture objectives, while inspiring people to reach those goals

learning curve theory — a theory that states that when many items are produced repetitively, the unit cost of those items normally decreases in a regular pattern as more units are produced

legitimate power — getting people to do things based on a position of authority

lessons-learned report — reflective statements written by project managers and their team members to document important things they have learned from working on the project

life cycle costing — considers the total cost of ownership, or development plus support costs, for a project

M

maintainability — the ease of performing maintenance on a product

make-or-buy decision — when an organization decides if it is in its best interests to make certain products or perform certain services inside the organization, or if it is better to buy them from an outside organization

Malcolm Baldrige National Quality Award — an award started in 1987 to recognize companies that have achieved a level of world-class competition through quality management

management reserves — dollars included in a cost estimate to allow for future situations that are unpredictable (sometimes called **unknown unknowns**)

manager — a person who deals with the day-to-day details of meeting specific goals

mandatory dependencies — sequencing of project activities or tasks that are inherent in the nature of the work being done on the project

matrix organizational structure — an organizational structure in which employees are assigned to both functional and project managers

maturity model — a framework for helping organizations improve their processes and systems

mean — the average value of a population

measurement and test equipment costs — the capital cost of equipment used to perform prevention and appraisal activities

merge — when two or more nodes precede a single node on a network diagram

methodology — describes *how* things should be done

milestone — a significant event that normally has no duration on a project; serves as a marker to help in identifying necessary activities, setting schedule goals, and monitoring progress

mind mapping — a technique that can be used to develop WBSs by using branches radiating out from a central core idea to structure thoughts and ideas

mirroring — the matching of certain behaviors of the other person

monitoring and controlling processes — regularly measuring and monitoring progress to ensure that the project team meets the project objectives

Monte Carlo analysis — a risk quantification technique that simulates a model's outcome many times, to provide a statistical distribution of the calculated results

multitasking — when a resource works on more than one task at a time

Murphy's Law — if something can go wrong, it will

Myers-Briggs Type Indicator (MBTI) — a popular tool for determining personality preferences

N

net present value (NPV) analysis — a method of calculating the expected net monetary gain or loss from a project by discounting all expected future cash inflows and outflows to the present point in time

network diagram — a schematic display of the logical relationships or sequencing of project activities

node — the starting and ending point of an activity on an activity-on-arrow diagram

normal distribution — a bell-shaped curve that is symmetrical about the mean of the population

O

opportunities — chances to improve the organization

organizational breakdown structure (OBS) — a specific type of organizational chart that shows which organizational units are responsible for which work items

organizational culture — a set of shared assumptions, values, and behaviors that characterize the functioning of an organization

organizational process assets — formal and informal plans, policies, procedures, guidelines, information systems, financial systems, management systems, lessons learned, and historical information that help people understand, follow, and improve business in a specific organization

overallocation — when more resources than are available are assigned to perform work at a given time

overrun — the additional percentage or dollar amount by which actual costs exceed estimates

P

parametric modeling — a cost-estimating technique that uses project characteristics (parameters) in a mathematical model to estimate project costs

Pareto analysis — identifying the vital few contributors that account for most quality problems in a system

Pareto diagrams — histograms that help identify and prioritize problem areas

Parkinson's Law — work expands to fill the time allowed

payback period — the amount of time it will take to recoup, in the form of net cash inflows, the total dollars invested in a project

performance — how well a product or service performs the customer's intended use

performance reporting — collecting and disseminating performance information, which includes status reports, progress measurement, and forecasting

PERT weighted average =

$$\frac{\text{optimistic time} + 4 \times \text{most likely time} + \text{pessimistic time}}{6}$$

phase exit or kill point — management review that should occur after each project phase to determine if projects should be continued, redirected, or terminated

planned value (PV) — that portion of the approved total cost estimate planned to be spent on an activity during a given period, formerly called the budgeted cost of work scheduled (BCWS)

planning processes — devising and maintaining a workable scheme to ensure that the project addresses the organization's needs

politics — competition between groups or individuals for power and leadership

political frame — addresses organizational and personal politics

power — the potential ability to influence behavior to get people to do things they would not otherwise do

Precedence Diagramming Method (PDM) — a network diagramming technique in which boxes represent activities

predictive life cycle — a software development approach used when the scope of the project can be clearly articulated and the schedule and cost can be accurately predicted

prevention cost — the cost of planning and executing a project so that it is error-free or within an acceptable error range

probability/impact matrix or chart — a matrix or chart that lists the relative probability of a risk occurring on one side of a matrix or axis on a chart and the relative impact of the risk occurring on the other

probabilistic time estimates — duration estimates based on using optimistic, most likely, and pessimistic estimates of activity durations instead of using one specific or discrete estimate

problems — undesirable situations that prevent the organization from achieving its goals

process — a series of actions directed toward a particular result

process adjustments — adjustments made to correct or prevent further quality problems based on quality control measurements

procurement — acquiring goods and/or services from an outside source

profit margin — the ratio between revenues and profits

profits — revenues minus expenses

program — a group of projects managed in a coordinated way to obtain benefits and control not available from managing them individually

Program Evaluation and Review Technique (PERT) — a project network analysis technique used to estimate project duration when there is a high degree of uncertainty with the individual activity duration estimates

progress reports — reports that describe what the project team has accomplished during a certain period of time

project — a temporary endeavor undertaken to create a unique product, service, or result

project archives — a complete set of organized project records that provide an accurate history of the project

project acquisition — the last two phases in a project (implementation and close-out) that focus on delivering the actual work

project buffer — additional time added before the project's due date

project charter — a document that formally recognizes the existence of a project and provides direction on the project's objectives and management

project cost management — the processes required to ensure that the project is completed within the approved budget

project feasibility — the first two phases in a project (concept and development) that focus on planning

project integration management — coordinating all of the other project management knowledge areas throughout a project's life. These processes include developing the project charter, developing the preliminary project scope statement, developing the project management plan, directing and managing the project, monitoring and controlling the project, providing integrated change control, and closing the project

project life cycle — a collection of project phases, such as concept, development, implementation, and close-out

project organizational structure — an organizational structure that groups people by major projects, such as specific aircraft programs

project management — the application of knowledge, skills, tools, and techniques to project activities to meet project requirements

Project Management Institute (PMI) — international professional society for project managers

project management knowledge areas — project integration management, scope, time, cost, quality, human resource, communications, risk, and procurement management

Project Management Office (PMO) — an organizational group responsible for coordinating the project management functions throughout an organization

project management plan — a document used to coordinate all project planning documents and guide project execution and control

project management process groups — the progression of project activities from initiation to planning, executing, monitoring and controlling, and closing

Project Management Professional (PMP) — certification provided by PMI that requires documenting project experience, agreeing to follow the PMI code of ethics, and passing a comprehensive exam

project management tools and techniques — methods available to assist project managers and their teams; some popular tools in the time management knowledge area include Gantt charts, network diagrams, and critical path analysis

project manager — the person responsible for working with the project sponsor, the project team, and the other people involved in a project to meet project goals

project portfolio management — when organizations group and manage projects as a portfolio of investments that contribute to the entire enterprise's success

project procurement management — the processes required to acquire goods and services for a project from outside the performing organization

project quality management — ensuring that a project will satisfy the needs for which it was undertaken

project sponsor — the person who provides the direction and funding for a project

project scope management — the processes involved in defining and controlling what is or is not included in a project

project scope statement — a document that includes, at a minimum, a description of the project, including its overall objectives and justification, detailed descriptions of all project deliverables, and the characteristics and requirements of products and services produced as part of the project

project time management — the processes required to ensure timely completion of a project

proposal — a document prepared by sellers when there are different approaches for meeting buyer needs

prototyping — developing a working replica of the system or some aspect of the system to help define user requirements

Q

qualitative risk analysis — qualitatively analyzing risks and prioritizing their effects on project objectives

quality — the totality of characteristics of an entity that bear on its ability to satisfy stated or implied needs or the degree to which a set of inherent characteristics fulfill requirements

quality assurance — periodically evaluating overall project performance to ensure that the project will satisfy the relevant quality standards

quality audit — structured review of specific quality management activities that helps identify lessons learned and can improve performance on current or future projects

quality circles — groups of nonsupervisors and work leaders in a single company department who volunteer to conduct group studies on how to improve the effectiveness of work in their department

quality control — monitoring specific project results to ensure that they comply with the relevant quality standards and identifying ways to improve overall quality

quality planning — identifying which quality standards are relevant to the project and how to satisfy them

quantitative risk analysis — measuring the probability and consequences of risks and estimating their effects on project objectives

R

RACI charts — charts that show Responsibility, Accountability, Consultation, and Informed roles for project stakeholders

rapport — a relation of harmony, conformity, accord, or affinity

rate of performance (RP) — the ratio of actual work completed to the percentage of work planned to have been completed at any given time during the life of the project or activity

Rational Unified Process (RUP) — an iterative software development process that focuses on team productivity and delivers software best practices to all team members

referent power — getting people to do things based on an individual's personal charisma

reliability — the ability of a product or service to perform as expected under normal conditions

Request for Proposal (RFP) — a document used to solicit proposals from prospective suppliers

Request for Quote (RFQ) — a document used to solicit quotes or bids from prospective suppliers

required rate of return — the minimum acceptable rate of return on an investment

reserves — dollars included in a cost estimate to mitigate cost risk by allowing for future situations that are difficult to predict

residual risks — risks that remain after all of the response strategies have been implemented

resource breakdown structure — a hierarchical structure that identifies the project's resources by category and type

resource histogram — a column chart that shows the number of resources assigned to a project over time

resource leveling — a technique for resolving resource conflicts by delaying tasks

resource loading — the amount of individual resources an existing schedule requires during specific time periods

resources — people, equipment, and materials

responsibility assignment matrix (RAM) — a matrix that maps the work of the project as described in the WBS to the people responsible for performing the work as described in the organizational breakdown structure (OBS)

return on investment (ROI) — (benefits minus costs) divided by costs

reward power — using incentives to induce people to do things

rework — action taken to bring rejected items into compliance with product requirements or specifications or other stakeholder expectations

risk — an uncertainty that can have a negative or positive effect on meeting project objectives

risk acceptance — accepting the consequences should a risk occur

risk-averse — having a low tolerance for risk

risk avoidance — eliminating a specific threat or risk, usually by eliminating its causes

risk breakdown structure — a hierarchy of potential risk categories for a project

risk enhancement — changing the size of an opportunity by identifying and maximizing key drivers of the positive risk

risk events — specific circumstances that may occur to the detriment of the project

risk exploitation — doing whatever you can to make sure the positive risk happens

risk factors — numbers that represent overall risk of specific events, given their probability of occurring and the consequence to the project if they do occur

risk identification — determining which risks are likely to affect a project and documenting the characteristics of each

risk management plan — a plan that documents the procedures for managing risk throughout a project

risk management planning — deciding how to approach and plan the risk management activities for a project, by reviewing the project charter, WBS, roles and responsibilities, stakeholder risk tolerances, and the organization's risk management policies and plan templates

risk mitigation — reducing the impact of a risk event by reducing the probability of its occurrence

risk monitoring and control — monitoring known risks, identifying new risks, reducing risks, and evaluating the effectiveness of risk reduction throughout the life of the project

risk-neutral — a balance between risk and payoff

risk owner — the person who will take responsibility for a risk and its associated response strategies and tasks

risk register — a document that contains results of various risk management processes, often displayed in a table or spreadsheet format

risk response planning — taking steps to enhance opportunities and reduce threats to meeting project objectives

risk-seeking — having a high tolerance for risk

risk sharing — allocating ownership of the risk to another party

risk transference — shifting the consequence of a risk and responsibility for its management to a third party

risk utility or risk tolerance — the amount of satisfaction or pleasure received from a potential payoff

Robust Design methods — methods that focus on eliminating defects by substituting scientific inquiry for trial-and-error methods

rough order of magnitude (ROM) estimate — a cost estimate prepared very early in the life of a project to provide a rough idea of what a project will cost

runaway projects — projects that have significant cost or schedule overruns

S

schedule baseline — the approved planned schedule for the project

schedule control — controlling and managing changes to the project schedule

schedule development — analyzing activity sequences, activity duration estimates, and resource requirements to create the project schedule

schedule performance index (SPI) — the ratio of earned value to planned value; can be used to estimate the projected time to complete a project

schedule variance (SV) — the earned value minus the planned value

scope — all the work involved in creating the products of the project and the processes used to create them

scope baseline — the approved project scope statement and its associated WBS and WBS dictionary

scope control — controlling changes to the project scope

scope creep — the tendency for project scope to keep getting bigger

scope management plan — document that includes descriptions of how the project team will prepare the project scope statement, create the WBS, verify completion of the project deliverables, and control requests for changes to the project scope

scope statement — a document used to develop and confirm a common understanding of the project scope, the first version is often called a preliminary scope statement

scope verification — formalizing acceptance of the project scope, sometimes by customer sign-off

Scrum — an iterative development process for software where the iterations are referred to as sprints, which normally last thirty days

secondary risks — risks that are a direct result of implementing a risk response

sellers — contractors, suppliers, or providers who provide goods and services to other organizations

sensitivity analysis — a technique used to show the effects of changing one or more variables on an outcome

seven run rule — if seven data points in a row on a quality control chart are all below the mean, above the mean, or are all increasing or decreasing, then the process needs to be examined for nonrandom problems

six 9s of quality — a measure of quality control equal to 1 fault in 1 million opportunities

Six Sigma — a comprehensive and flexible system for achieving, sustaining, and maximizing business success that is uniquely driven by close understanding of customer needs, disciplined use of facts, data, statistical analysis, and diligent attention to managing, improving, and reinventing business processes

slack — the amount of time a project activity may be delayed without delaying a succeeding activity or the project finish date; also called **float**

slipped milestone — a milestone activity that is completed later than planned

SMART criteria — guidelines to help define milestones that are specific, measurable, assignable, realistic, and time-framed

smoothing mode — deemphasizing or avoiding areas of differences and emphasizing areas of agreements

software defect — anything that must be changed before delivery of the program

Software Quality Function Deployment (SQFD) model — a maturity model that focuses on defining user requirements and planning software projects

Source Lines of Code (SLOC) — a human written line of code that is not a blank line or comment

staffing management plan — a document that describes when and how people will be added to and taken off a project team

stakeholder analysis — an analysis of information such as key stakeholders' names and organizations, their roles on the project, unique facts about each stakeholder, their level of interest in the project, their influence on the project, and suggestions for managing relationships with each stakeholder

stakeholders — people involved in or affected by project activities

standard deviation — a measure of how much variation exists in a distribution of data

start-to-finish dependency — a relationship on a project network diagram where the "from" activity cannot start before the "to" activity is finished

start-to-start dependency — a relationship on a project network diagram in which the "from" activity cannot start until the "to" activity starts

statement of work (SOW) — a description of the work required for the procurement

statistical sampling — choosing part of a population of interest for inspection

status reports — reports that describe where the project stands at a specific point in time

strategic planning — determining long-term objectives by analyzing the strengths and weaknesses of an organization, studying opportunities and threats in the business environment, predicting future trends, and projecting the need for new products and services

structural frame — deals with how the organization is structured (usually depicted in an organizational chart) and focuses on different groups' roles and responsibilities to meet the goals and policies set by top management

subproject managers — people responsible for managing the subprojects that a large project might be broken into

sunk cost — money that has been spent in the past

symbolic frame — focuses on the symbols, meanings, and culture of an organization

synergy — an approach where the whole is greater than the sum of the parts

system outputs — the screens and reports the system generates

system testing — testing the entire system as one entity to ensure that it is working properly

systems — sets of interacting components working within an environment to fulfill some purpose

systems analysis — a problem-solving approach that requires defining the scope of the system to be studied, and then dividing it into its component parts for identifying and evaluating its problems, opportunities, constraints, and needs

systems approach — a holistic and analytical approach to solving complex problems that includes using a systems philosophy, systems analysis, and systems management

systems development life cycle (SDLC) — a framework for describing the phases involved in developing and maintaining information systems

systems management — addressing the business, technological, and organizational issues associated with creating, maintaining, and making a change to a system

systems philosophy — an overall model for thinking about things as systems

systems thinking — taking a holistic view of an organization to effectively handle complex situations

T

tangible costs or **benefits** — costs or benefits that can be easily measured in dollars

team development — building individual and group skills to enhance project performance

termination clause — a contract clause that allows the buyer or supplier to end the contract

Theory of Constraints (TOC) — a management philosophy that states that any complex system at any point in time often has only one aspect or constraint that is limiting its ability to achieve more of its goal

three-point estimate — an estimate that includes an optimistic, most likely, and pessimistic estimate

time and material contracts — a hybrid of both fixed-price and cost-reimbursable contracts

top-down approach — creating a WBS by starting with the largest items of the project and breaking them into their subordinate items

Top Ten Risk Item Tracking — a qualitative risk analysis tool for identifying risks and maintaining an awareness of risks throughout the life of a project

total slack (total float) — the amount of time an activity may be delayed from its early start without delaying the planned project finish date

Tracking Gantt chart — a Gantt chart that compares planned and actual project schedule information

triggers — indications for actual risk events

triple constraint — balancing scope, time, and cost goals

Tuckman model — describes five stages of team development: forming, storming, norming, performing, and adjourning

U

unit pricing — an approach in which the buyer pays the supplier a predetermined amount per unit of service, and the total value of the contract is a function of the quantities needed to complete the work

unit test — a test of each individual component (often a program) to ensure that it is as defect-free as possible

use case modeling — a process for identifying and modeling business events, who initiated them, and how the system should respond to them

user acceptance testing — an independent test performed by end users prior to accepting the delivered system

V

variance — the difference between planned and actual performance

W

watch list — a list of risks that are low priority, but are still identified as potential risks

WBS dictionary — a document that describes detailed information about each WBS item

weighted scoring model — a technique that provides a systematic process for basing project selection on numerous criteria

withdrawal mode — retreating or withdrawing from an actual or potential disagreement

workarounds — unplanned responses to risk events when there are no contingency plans in place

work breakdown structure (WBS) — a deliverable-oriented grouping of the work involved in a project that defines the total scope of the project

work package — a task at the lowest level of the WBS

Y

yield — represents the number of units handled correctly through the development process

Index